Omics Approaches in Biofilm Research

Siddhardha Busi
Subhaswaraj Pattnaik • Ram Prasad
Editors

Omics Approaches in Biofilm Research

Perspectives and Applications

Springer

Editors
Siddhardha Busi
Department of Microbiology
School of Life Sciences, Pondicherry University
Kalapet, Puducherry, India

Subhaswaraj Pattnaik
PG Department of Zoology
Government Women's College, Sambalpur University
Sambalpur, Odisha, India

Ram Prasad
Department of Botany
Mahatma Gandhi Central University
Motihari, Bihar, India

ISBN 978-3-031-91862-9 ISBN 978-3-031-91863-6 (eBook)
https://doi.org/10.1007/978-3-031-91863-6

This Springer imprint is published by the registered company Springer Nature Switzerland AG
The registered company address is: Gewerbestrasse 11, 6330 Cham, Switzerland

Preface

The world is witnessing the resurgence of resistance to commonly used antibiotics by several nosocomial pathogenic microorganisms leading to alarming health ailments with high rate of mortality and morbidity across the globe. Even the last-resort antibiotics are not able to counteract the chronic infections owing to the emergence of multi-drug resistance (MDR) phenomena. Noticeably, the inherent ability of pathogenic microorganisms to form recalcitrant biofilm matrices critically augments the MDR phenomena by providing an arsenal against environmental stress including antibiotic treatment. As per the recent developments in the increased drug resistance phenomena, it is estimated to mark the death of ten million individuals with an estimated economic loss of 100 trillion dollars by the year 2050. Since the biofilm-mediated drug resistance has catastrophic effect on global health and economy, the early detection of biofilm infections and resistance pathophysiology is of paramount significance which will decide the right course of therapeutic modalities upon such chronic infections. To achieve the same, the differential biofilm pathophysiological responses need to be understood for effective management of biofilm-mediated chronic infections. The lack of understanding on biofilm microenvironment and pathophysiological responses has a critical influence on our diagnostic and therapeutic efforts. In this regard, a holistic approach needs to be spatially developed to provide the right course of actions to understand biofilm pathophysiology and modulate the biofilm management strategies. The advent of advanced molecular tools such as Omics-based approaches has revolutionized the current understanding of pathophysiological role of specific antibiotic-resistant genes (ARGs), their differential expression patterns, and response-based interactions with therapeutic agents upon infection. The inherent ability of several Omics-based approaches has provided a comprehensive platform to understand the biofilm dynamics and resistance profile at various levels of organization such as genes, mRNA levels, proteins, and their regulation. The Omics-based tools such as metagenomics, transcriptomics, proteomics, metabolomics, etc. have provided new horizons to understand biofilm microenvironment and could facilitate the spatial development of strategic therapeutic modalities to regulate the menace of biofilm-mediated antibiotic resistance. Besides, the integrated approach to consider multi-Omics tools (e.g., genomics, transcriptomics, proteomics, lipidomics, metabolomics, etc.) has further improved our current understanding on the several mechanisms associated with biofilm pathophysiological responses and resistance profile. Since the compilation of the comprehensive list of Omics-based approaches for biofilm regulations remains unreported in a single platform, the present book will provide new perspectives of Omics-based tools to recognize several pathophysiological processes, including relative expression profile of ARGs, resistance patterns, virulence mechanisms, and interaction mechanism with therapeutic agents associated with biofilm microenvironment in a single platform. The contents of this book are arranged systematically, ensuring comprehensive insights into the early diagnosis and therapeutic applications of different Omics-based approaches with their relative advantages in the management of biofilms. The book will also emphasize the elaborate elucidation of biofilm microenvironment in resistance severity and their control strategies at different levels of Omics-based tools. The chapters will also critically elucidate the pivotal role of Omics-based approaches in the development of therapeutic modalities to inhibit biofilm formation and control strategies against MDR phenomena. All the chapters in the book emphasize the basic principles of Omics-based molecular tools and how their advanced approaches are recently being employed for several diagnostics and therapeutic applications pertaining to chronic biofilm infections and antimicrobial resistance management. Besides, the chapters in the book also highlight the integration of several advance molecular tools in combination with Omics-based approaches for effective control of bacterial virulence, biofilm mechanics, and drug resistance patterns in pathogenic microorganisms. The book will give an impetus to research enthusiasts, students, and academicians to look for the physiological responses of microbial biofilms and the strategic regulations of biofilm infections.

Kalapet, Puducherry, India — Siddhardha Busi
Sambalpur, Odisha, India — Subhaswaraj Pattnaik
Motihari, Bihar, India — Ram Prasad

Contents

Part I

An Introduction to Biofilms and Omics

Introduction to Microbial Biofilms

1

Z. H. Tawfeeq Ahmed, Prabhakar Singh, A. Saraswathi,
S. Deekshitha, Rejith R. S. Kumar, Aswin Thacharodi,
and Saqib Hassan

1.1 Introduction

Biofilms are communities of microorganisms (such as bacteria and fungi) encased within a matrix of extracellular polysaccharides (EPS) that are rooted in an abiotic or biotic surface and able to reproduce collectively as colonies. Biofilms can be detrimental or useful, depending on their function. Quorum sensing (QS) is the process by which bacteria in a biofilm generate autoinducers (AIs) to communicate. QS is responsible for biofilm formation and the production of EPS. Interfering QS, bacterial adhesion, and disturbance in biofilm design all work to prevent harmful biofilm development. The beneficial biofilms may be built by modifying the ambient parameters, the surface of adhesion, and QS signals (Sonawane et al., 2022). Mature biofilms are highly effective microbial communities on surfaces due to eventual collocations, and their structure is determined by physiological cooperativity (Sportelli et al., 2022). Different bacterial species exhibit various expression patterns and regulatory systems in distinct biofilm phases. Numerous bacterial species coexist in almost every natural environment, which is why microbiologists are interested in studying multispecies biofilms and identifying interspecific interactions within these communities. Additionally, emergent characteristics arising from the social interactions of multiple bacterial species in biofilms have important

Z. H. T. Ahmed · P. Singh · A. Saraswathi · S. Deekshitha · S. Hassan (✉)
Department of Biotechnology, School of Bio and Chemical Engineering, Sathyabama Institute of Science and Technology, Chennai, Tamil Nadu, India

R. R. S. Kumar
Department of Microbiology, Faculty of Medicine, Asian Institute of Medicine, Science and Technology, Bedong, Kedah, Malaysia

A. Thacharodi
Dr. Thacharodi's Laboratories, Department of Research and Development, Puducherry, India

S. Busi et al. (eds.), *Omics Approaches in Biofilm Research*,
https://doi.org/10.1007/978-3-031-91863-6_1

implications for the development of new studies and practical advice for managing and modifying microbial communities (Sadiq et al., 2021).

Microbial biofilms are more concerning from a clinical standpoint because they prevent drugs from entering the body during therapy, which leads to antimicrobial resistance (AMR) in microbial species. With sharp rises in hospital-associated mortality and morbidity, AMR has emerged as a serious healthcare concern burdening the global economy (Ahmed et al., 2024). In 2019, the World Health Organization (WHO) listed AMR as one of the top ten potential health threats that needs utmost attention, as it is predicted to cause 8.22 million deaths globally in 2050 (Naghavi et al., 2024). The escalating threat of AMR necessitates the discovery of novel antimicrobial agents (Fathima et al., 2025). In response to AMR-caused microbial biofilms, several pharmaceutical giants are in a race to find novel compounds that could disrupt microbial biofilms, thereby rendering better treatment strategies (Ferrara et al., 2024). To date, there are several promising approaches to disrupting microbial biofilms that include small molecule inhibitors, phage therapy, nanoparticles, and ultrasound; however, the efficacy of these approaches has to be translated to relevant *in vivo* models to understand their pharmacokinetics and antibiofilm efficacy. The recent development of antimicrobial peptides (AMPs), which have shown higher efficiency, holds the key to eliminating mature biofilms irrespective of whether they are organic or synthetic (Naaz et al., 2023).

Although microbial biofilms are a threat to humans clinically, they do possess a significant benefit to humans when used wisely. Heavy metal pollution from industrial and mining operations pose a significant health risk to humans. As a part of a natural process, certain microorganisms can utilize these toxic pollutants as nutrient sources. Microbial biofilm holds the key, as it can bioremediate toxic pollutants effectively and economically. The biofilm matrix is dynamic, with microbial cells playing a vital role in continually changing and emerging microenvironments. The bioabsorption and bioaccumulation of metal ions in microbial biofilms may help to break complex compounds into simple, unharmful compounds (Syed et al., 2023). Similarly, microbial biofilms hold a significant role in wastewater management. Microbial biofilm-based reactors are used industrially for remediating wastewater as they are both cost-effective and environmentally beneficial. Further, the use of electroactive biofilms (EAB) that have been derived from sewage sludge, activated sludge, or industrial effluents is becoming increasingly popular in the field of bioremediation (Chattopadhyay et al., 2022). Microplastics have detrimental effects on humans that need urgent attention because of their recalcitrant nature (Thacharodi et al., 2024a, b). Plastic-microbial biofilm interactions may alter the polymer's physical and chemical characteristics, resulting in deterioration. As a result, to decrease the anticipated nuisance of microplastic pollution, biofilms are being explored as viable candidates for marine environment restoration (Debroy et al., 2022).

The increasing awareness of the significance of microbial biofilms across a range of industries is opening up new avenues for translational research to develop innovative technologies that could mitigate their drawbacks while maximizing their benefits. Determining the actual financial impact of microbial biofilms has been difficult due to the wide range of regions they affect. However, there are still numerous unresolved scientific and technological obstacles that must be addressed in order to capitalize on these prospects. Moreover, a major barrier to the translation of biofilm science is the discrepancy between industrial processes and scholarly investigations of biofilms. Thus, biofilm research must quickly identify and address industrially relevant challenges in order to produce insights that can be applied elsewhere. Accordingly, we will thoroughly examine the development of bacterial biofilms, the infections they cause, detection techniques, and possible treatment approaches in this chapter.

1.2 Structure and Composition of Biofilms

When bacteria attach themselves to surfaces via the polymer matrices they secrete, microbial biofilms are formed. Extracellular proteins, polysaccharides, and DNA make up the majority of these matrices. By modifying elements like QS during bioremediation, surface properties, and other environmental circumstances, their efficacy can be increased. Organic pollutants are persistently harmful to people, animals, and marine life, and they have become a global problem. These pollutants last in the environment for long periods and are resistant to degradation (Sonawane et al., 2022). Microbial biofilms' structure and makeup are crucial to their operation and provide them with special characteristics like resilience to environmental stresses, host immune systems, and antibiotics. For several disciplines, including environmental research, industrial applications, and medicine, understanding biofilm composition is necessary.

Microbial Cells in Biofilms Microbial cells compose the core of any biofilm. Depending on the conditions in which the biofilm forms, these microorganisms may be bacteria, fungi, algae, or even protozoa. The cells of the most researched type of bacterial biofilms usually stick to a surface in a way that makes it possible for dense clusters, or microcolonies to form. The cells inside a biofilm are not all equally active, as they frequently display a gradient of metabolic rates throughout the biofilm structure. This suggests that because of nutritional constraints, decreased oxygen availability, or waste accumulation, cells deeper within the biofilm may be in a more dormant or slow-growing condition, whereas cells at the biofilm's surface may be more metabolically active (Flemming & Wingender, 2010).

The main component of the biofilm matrix is either solvent or water. Planktonic bacteria are microorganisms that do not stick to surfaces. Planktonic bacterial populations are generally smaller than those seen in biofilms. Biofilm-forming bacteria may endure a variety of adverse environmental factors, such as pH fluctuations, the presence of toxins and free radicals, and restricted nutrition availability (Chattopadhyay et al., 2022). The biofilm's microbial cells are arranged in a very ordered manner. The extracellular matrix material separates the microcolonies that the cells create. In reaction to environmental stressors, these microcolonies may be in varying phases of growth or even experience programmed cell death (Baker et al., 2009).

EPS (Extracellular Polymeric Substances) The microorganisms in the biofilm create EPS, which make up the majority of the matrix. Although EPS contain a range of macromolecules, their main constituents include proteins, polysaccharides, and extracellular DNA; lipids are occasionally seen in minute amounts. The cells create and release these chemicals to help with adherence to surfaces, provide structural scaffolding, and shield the community from environmental stressors like immunological reactions, dehydration, and antimicrobial agents (Mahto et al., 2022). EPS are high-molecular-weight microbiological polymers that include proteins, humic compounds, and hydrocarbons. They play a crucial role in mediating toxic inhibition, adsorbing biotoxins, and protecting organisms from outside dangers. EPS serve as an extracellular physical barrier, with a number of functional groups, including carboxyl, hydroxyl, amino, phosphoryl, and sulfhydryl groups, that improve pollutant adherence as well as reduce exposure and harmful effects. Additionally, EPS are essential for the adsorption of antibiotics, preventing direct interaction with microorganisms and thus promoting antibiotic resistance (Zhao et al., 2023a). The EPS mainly comprise extracellular DNA, polysaccharides, proteins, and lipids.

Extracellular DNA (eDNA) is crucial for the structural composition of biofilms, enhancing their mechanical stability and acting as a repository of genetic information. This DNA is typically released from cells that are lysing, or it is actively secreted by living cells. eDNA contributes to a biofilm's structural integrity by providing additional cohesion among cells, functioning as a "glue" that binds microbial cells together. It is an essential component of the majority of biofilms that grow in suspension, as well as a prevalent extracellular matrix constituent in biofilms that form on the surfaces of biomaterials. In the biofilm community, eDNA can also facilitate horizontal gene transfer, which allows different microbial species to share genetic material, including genes that resist antibiotics. In hospital settings, where biofilms can result in chronic infections that are difficult to cure with conventional antibiotics, this feature presents serious issues (Campoccia et al., 2021).

Polysaccharides Polysaccharides in the EPS are crucial for maintaining the biofilm's structural integrity. The microbes that form biofilms manufacture and release these long chains of sugar molecules, which create a gel-like matrix that stabilizes and envelops the biofilm. Many exopolysaccharides, such as cellulose, fructans, and glucans, are homopolysaccharides. Most, however, are heteropolysaccharides, which are made up of a combination of charged and neutral sugar residues. Alginate, xanthan, and colanic acid are notable examples of exopolysaccharides that are classified as polyanionic. Furthermore, polycationic exopolysaccharides such as pellicle polysaccharide (PEL) and polysaccharide intercellular adhesin (PNAG) are known to exist. By retaining water in the biofilm, these polysaccharides help to improve their hydration, which is essential for microbial growth. Additionally, they shield biofilm cells from environmental stresses such as mechanical shear forces, pH variations, and desiccation. Polysaccharides also aid in the adhesion of microbial cells to solid surfaces, thereby initiating the process of biofilm formation (Balducci et al., 2023).

Proteins are an essential component of EPS, which improve the biofilms' structural stability and functional capabilities. Enzymes within these proteins aid in the breakdown of external substrates, allowing bacteria to draw nutrients from their environment. The initial adhesion of microbial cells to surfaces, which signifies the initial stage of biofilm growth, depends on adhesion proteins, such as pili and fimbriae. Amyloid-like proteins may also be incorporated into some biofilms to increase their stability. Biofilms are important in bioremediation efforts because they may also include enzymes that can break down poisons or pollutants (Orhan-Yanıkan et al., 2020).

Lipids Despite being less prevalent in the biofilm matrix than proteins and polysaccharides, lipids are essential for defining the physical properties of biofilms. For example, Gram-negative bacteria's lipopolysaccharides (LPS) are incorporated into the matrix, increasing the biofilm's hydrophobicity. This hydrophobic characteristic protects the biofilm from contaminants and antibiotics. Lipids can also alter the biofilm's surface characteristics, affecting its permeability and resistance to outside pressures, including shear forces and osmotic stress (Vandana et al., 2023).

1.2.1 Biofilm Architecture and the Role of ECM in Biofilm Integrity

The complex organization of a biofilm's structure is crucial to its operation. Biofilms are usually three-dimensional (3D) structures composed of extracellular matrix and microbial cells. Environmental conditions, nutrition availability, and the particular microbial species present are some of the elements that affect the architecture of these biofilms. Biofilms frequently display a heterogeneous composition with variations in internal structure, density, and thickness. Microbial cells generate microcolonies within a biofilm, which can grow into structures like *mushrooms* or *pillars*. Channels or voids that allow nutrients, gases, and waste products to pass freely are scattered throughout these cellular columns (Kassinger & van Hoek, 2020). The structure of a biofilm is dynamic, capable of evolving in reaction to external factors. For instance, when a biofilm reaches excessive thickness, certain cells may disperse—a phenomenon referred to as *biofilm maturation*—enabling them to establish colonies on new surfaces. This adaptability facilitates the biofilm's response to environmental changes, thereby promoting its sustained existence (Liaqat et al., 2019).

1.2.1.1 Outer Layers of Biofilm

The first areas of a biofilm to interface with the outside world are its outer layers, which provide ideal circumstances for microbial growth and metabolism. Among these outer layers' noteworthy features include the accessibility of nutrients and oxygen, as well as nutrient uptake. Sugars, amino acids, oxygen, and other nutrients found in the surrounding environment are all directly accessible to these layers. Microorganisms are actively metabolizing and growing in this area, and their rapid growth depends on these nutrients. Since oxygen can readily diffuse these cells from the surrounding environment or atmosphere, its content is often higher in the outer layers. Because of this, the cells in these outer layers mostly use aerobic respiration, which uses less energy (Xiao et al., 2021).

The cells in the outer layers are in the best possible position to divide and develop quickly. Because they have easy access to oxygen and nutrients, these cells usually have the highest levels of metabolic activity. They may participate in fermentation or aerobic respiration, depending on the available nutrient sources. Furthermore, the EPS matrix found in these outer layers provides a confined, safe environment that promotes cell proliferation and serves as structural support for cell attachment. To maintain cells affixed to surfaces and shield them from external forces like fluid movement, this matrix is necessary. The outer layers are necessary for the biofilm's early adhesion and colonization phases. They act as a barrier separating the biofilm from its environment, and their metabolic activities are frequently essential in promoting the biofilm's growth and maturity (Rather et al., 2021).

1.2.1.2 Inner Layers of Biofilm

The biofilm's inner layers experience harsher conditions than its outer layers. Cells located farther from the surface are subject to more waste products produced by the cells above them, as well as less oxygen and nutrients. Nutrient transport (including glucose and amino acids) is severely impeded in the lower layers. These cells are not as well-nourished as those on the surface and may be dependent on the biofilm matrix or the diffusion of nutrients from the outer layers. Nutritional gradients cause nutritional deficiencies in certain places, which hinder the growth of microorganisms. Hypoxic conditions (low oxygen levels) are frequently observed in the interior parts of biofilms, particularly those that are dense or thick, due to the limited diffusion of oxygen through the biofilm matrix. Deep within the biofilm, bacteria use different electron acceptors, such as carbon dioxide, nitrate, or sulphate, which results in a change from aerobic to anaerobic metabolism (Stewart & Franklin, 2008). The inner layers' cells can grow more slowly or even enter a latent phase as a result of the limited availability of nutrients and oxygen.

These latent cells exhibit increased resistance to environmental stresses, including antibiotics and immunological reactions, despite having an inactive metabolism. Compared to the quickly dividing cells found in the outer layers, the phenotype of dormant or slow-growing cells is significantly different. Because they are not actively dividing or metabolizing when exposed, these inner layer cells are typically less susceptible to antimicrobial drugs (Stewart, 2003). The deeper layers of biofilms are more resistant to antibiotics because of things like slower cell reproduction and restricted oxygen, along with nutrient availability. Rapidly dividing cells are often the target of many antibiotics, which implies that the cells in the deeper parts of the biofilm often withstand therapy, resulting in chronic infections and diseases. Further shielding the inner layers from therapeutic interventions is the biofilm matrix's ability to prevent antibiotic penetration (Soares et al., 2022).

1.2.1.3 Water Channels in Biofilms

The intricate webs of water-filled channels that make up biofilms act as pathways for the flow of waste materials, nutrients, and signaling molecules from microorganisms. These routes make it easier for nutrients to be efficiently transported to every area of the biofilm, especially into the deeper layers where diffusion is insufficient. These channels promote liquid circulation, which supports the metabolic processes of cells deep within the biofilm. They also facilitate waste removal, which prevents the accumulation of toxins and fosters an atmosphere favorable to microbial activity. The dynamic flow of these channels maintains the biofilm's internal environment and allows waste products, nutrients, and oxygen to be exchanged between the cells and their surroundings. The biofilm's size and layout may change as it develops, matures, or responds to external stimuli like fluid flow and shear stress. These channels also allow signaling molecules, such as quorum-sensing signals, to be sent and received between microbial cells, facilitating communication. This capability promotes coordinated activities, including the production of virulence-promoting compounds and extracellular matrix constituents, along with metabolic process regulation across the biofilm (Quan et al., 2022).

1.2.1.4 Clusters and Pockets of Microcolonies

Microcolonies are a compact collection of bacterial cells embedded in the EPS that are commonly formed by microbial cells within a biofilm. A biofilm's structural and functional organization depends heavily on these microcolonies. These cells clump together to create tiny, compact colonies, each creating a distinct localized ecosystem that can be established by each microcolony, independent of the surrounding matrix (Johnson, 2008). The EPS serve as a shield to protect the bacterial cells from harmful external elements such as immune system reactions, mechanical shear pressures, and antimicrobial chemicals (Mah & O'Toole, 2001). A mutualistic ecosystem where different microbial species or strains work together can be established by microcolonies found within biofilms. To encourage cooperative contact, some bacteria, for example, may release enzymes that aid in the breakdown of complex organic materials, which are then used by other species in the biofilm (Busnelli et al., 2023).

1.2.1.5 Microbial Diversity

The biofilm's phenotypic diversity is mostly a result of the tight cell arrangement in microcolonies. The nutrient supply, oxygen content, and waste accumulation in different parts of a microcolony might vary, which can cause differences in the behavior of the cells (for example, some cells may be metabolically active, while others remain dormant or demonstrate resistance). These behavioral variations improve the overall robustness of the biofilm. To increase the biofilm's ability to respond to environmental challenges, certain cells may be actively dividing while others may be in a dormant or slow-growing phase (Stewart & Franklin, 2008).

1.3 Biofilm Formation Process

1.3.1 Stages of Biofilm Development

When specific biological, chemical, and physical activities take place simultaneously, a complicated, multi-step process known as biofilm development takes place. The reversible initial attachment of planktonic microbes to an appropriate substrate is the first of four steps that serve as a precursor to the formation of biofilms. This is followed by a diversion to irreversible attachment through the production of EPS, the maturation and differentiation of microcolonies, and the release or dispersal of planktonic cells to form new biofilms (Fig. 1.1). The formation of biofilms starts with single-cell adhesion to preconditioned surfaces exposed to an aqueous medium. This preconditioned surface possesses one or more conditioned layers, which are biotic docking stations for initial reversibly attached cells that form a base of growth for biofilms. Biotic surfaces include necrotic tissues and endothelial lesions, whereas abiotic surfaces where biofilms can potentially proliferate include urinary catheters and prostheses in clinical settings (Preda & Săndulescu, 2019). This layer alters the biofilm and makes it more susceptible to attachment by controlling parameters such as surface charge, potential, and tension. Multiple conditioned layers can be formed by the coaggregation of freely floating bacteria. This condition also facilitates properties like the maintenance of an oxygen and nutrient gradient, horizontal gene transfer (HGT), and QS. Factors such as a turbulent flow environment and smooth or rough surfaces are known to enhance bacterial formation (Sharma et al., 2023). Detachment occurs when repulsive forces far surpass attractive forces before conditioning occurs due to weak bonds (Singh et al., 2020).

1.3.1.1 Initial Reversible Attachment

The production of bacterial adhesins promoting attachment is a defining feature of the first phase of the biofilm lifecycle, known as initial attachment. Most frequently, the colonizing bacteria are free-floating, motile planktonic bacteria that adhere to the surface with covalent bonds. In addition to weak physical forces like Van der Waals, hydrophobic, or electrostatic forces, bacteria are attached to the surface via appendages such as bacterial pili, flagella, and fimbriae. These forces occur in unison to facilitate irreversible attachment to form a bacterial monolayer. Adherence to the surface is determined by factors such as the surface material, the bacterial surface, pressure, and temperature. Initial adhesion is also controlled by QS, surface sensing, and nutrient availability. The bacteria in biofilm are sessile and embedded in the EPS matrix. Nonmotile bacteria tend to attach themselves passively to the surface, as seen in *Staphylococci*, while motile bacteria attach actively to the surface, as exhibited by *P. aeruginosa* (Hajiagha & Kafil, 2023).

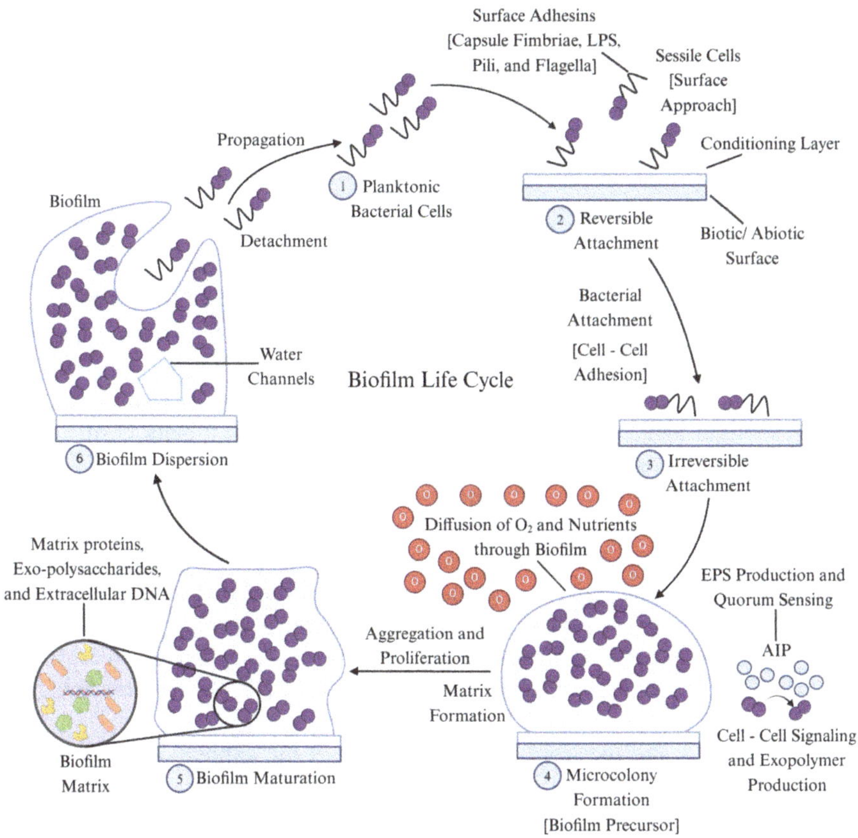

Fig. 1.1 An illustration of the six steps of the biofilm life cycle. The formation of biofilms starts with motile planktonic bacterial cells (1) reversibly attaching (2) to the substrate in a nonspecific manner with weak Van der Waals forces. These bacterial cells irreversibly attach (3) to the substrate with the help of surface adhesins such as fimbriae, capsule, LPS, and pili. The formation of microcolonies, which helps in the production of EPS (4), acts as a precursor to biofilms. The maturation of biofilms (5) can be identified by the formation of mushroom structures due to the production of exopolysaccharides along with intracellular signaling and quorum sensing. The dispersal of biofilms (6), also known as the detachment phase, forms planktonic bacterial cells and rejuvenates the life cycle

1.3.1.2 Irreversible Attachment

The second phase is commonly known as the *latching* or *anchoring* phase, which involves adhesive binding in the outermost layer through the creation of EPS. This binding to the surface is facilitated by interactions with either the surface materials or the receptors present on the bacterial surface in the form of external appendages such as pili. The secretion of EPS forms a strong bond between the bacterial surface and substrate after a certain threshold. Aggregates of planktonic microbes are formed as they stick to each other in clusters and become attached to the surface through adhesion (Alotaibi & Bukhari, 2021).

1.3.1.3 Microcolony Formation

Irreversible attachment is followed by cell division and proliferation, activated by cell-to-cell intercellular signaling within EPS and the microbial community in the biofilm to form three-dimensional microcolonies. These microcolonies form during the exponential growth phase, which takes place after the lag phase, resulting in an increased uptake of nutrition derived from the conditioning film and environment (Alotaibi & Bukhari, 2021).

1.3.1.4 Maturation

During the fourth phase of maturation, the attached bacterial cells develop further with the help of signaling factors and then release EPS, which provides stability to the biofilm structure, limits interaction with antimicrobial agents, and enhances bacterial virulence. Environmental DNA (eDNA) in particular is associated with conferring strength and intercellular signaling within biofilms (Gupta et al., 2016). Maturation occurs through rapid cell division, resulting in the formation of a microcolony. The maturation stage overall encompasses the development of EPS, QS, bacterial aggregation, differentiation of cells, establishment of three-dimensional structures, and maintaining steep oxygen and nutrient gradients, which make bacterial colonies and the immediate environment extremely heterogeneous. The process of maturation involves two steps: First, Gram-negative bacteria produce the AI signaling molecule N-acylated homoserine lactones (AHLs); second, the microcolony grows in size and thickness to about 100 μm, which satisfies the requirements for a microcolony.

Microcolonies and the level of adhesion control the expression of genes and proteins. Postmaturation biofilms possess three distinct layers: a joining layer that binds the biofilm to the surface, a foundational layer of densely packed bacteria, and a surface layer containing freely floating microbes that disseminate in the environment to form biofilms. The joining layer is mostly tertiary and plays a negligible role relative to the other layers (Alotaibi & Bukhari, 2021).

1.3.1.5 Dispersal

The biofilm is separated into two layers—namely, the foundational layer, which is colonized by planktonic bacteria, and a surface layer, which acts as a dispersal zone that can be used by sessile bacteria to disseminate and spread into the immediate environment. In clinical settings, embolic symptoms and persistent infections are also brought on by the metastatic seeding that the dispersal stage causes. Chemotactic, developed, and motile dispersion cells are released simultaneously in enormous quantities during the last step of biofilm formation. The detachment process involves three steps: abrasion, erosion, and sloughing. The loss of suspended biofilm particles, small biofilm particles, and a large biomass biofilm falls within these three steps, respectively. Detachment could occur as a result of nutritional imbalances, such as when a deficiency in carbon increases EPS production (Preda & Săndulescu, 2019).

Dispersal in biofilms is known to be triggered by a deficiency of oxygen and nutrition in aerobic bacteria, whereby autophosphorylation activates molecules like the fatty acid diffusible signal factor (DSF). Cyclic diguanylate guanosine monophosphate (c-di-GMP) is broken down by c-di-GMP phosphodiesterase, which is also activated by autophosphorylation. The degradation of c-di-GMP breaks down bacterial clusters and releases planktonic cells, which in turn breaks down portions of EPS. Aside from the typical gene regulation pathways that are responsible for the breakdown of EPS, enzymes also dissolve the polysaccharide layer and free the top layer of bacterial cells. These bacteria proliferate into new biofilms to renew the biofilm life cycle on other surfaces. They could also disseminate into other organs of the body or simply float in the surface, as they are motile due to the production of certain proteins (Sharma et al., 2023). After dispersal, biofilms have characteristic voids and hollows throughout their structures (Barraud et al., 2015).

1.3.2 Factors Influencing Biofilm Formation

A number of inherent environmental parameters, including pH, temperature, and nutrient availability, influence the formation and makeup of bacterial biofilms. Since the structure of the biofilm is determined by the type of carbon source and iron shortage may result in stunted growth, the type of nutrients and food have a significant impact on regulating the structure of the biofilm (Shrout et al., 2011). Changes in pH are mostly detrimental to bacterial growth, and a gradual increase in acidity could promote the survival rate of cells. The ideal pH required for the production of EPS varies between bacterial species, but a general pH range of 6.5–7.2 is common for the optimal growth of most biofilms (Pal & Lavanya, 2022). The growth of biofilms in optimal temperatures exponentially accelerates growth at the expense of increased nutrient uptake. Temperature affects the growth of biofilms, as it mediates the rate of enzymatic reactions and adhesion behavior. Lower temperatures can promote biofilm formation by synchronizing the characteristics of polysaccharides but can also inhibit growth by altering cell surface hydrophobicity.

Physical properties fluctuate according to changes in environmental temperature. According to the temperature, bacterial cells can exhibit varying numbers of flagella. For instance, bacteria grown at 35°C and 21°C can possess one to three flagella. In contrast, bacteria grown at temperatures less than 10°C can possess several flagella (Singh et al., 2020). Within the temperature range of 20–27°C, microorganisms like *S. aureus* can also exhibit varying degrees of hydrophobicity and adhesion (Chihib, 2017). A decrease in surface polymers in pseudomonads can cause their adherence to diminish when the temperature drops. Temperature variations also affect EPS's viscosity, which causes a gel-like material to progressively dissolve. Lower temperatures can help polysaccharides become more consistent, which can improve adhesion (Garrett et al., 2008).

Hydrodynamic characteristics, including compressive strength, laminar flow, and water velocity, also have an impact on the composition of biofilms and the adherence of bacteria. Shear stress is one property that can affect the generation of EPS, the rate at which biofilms grow, the expression of genes, and the shape of biofilms. For example, it has been demonstrated that shear stress prevents *P. putida* biofilms from maturing (Krsmanovic et al., 2021; Rochex et al., 2008). The surface area of the substrate and bacterial surface is crucial in determining the rate of adhesion, as rougher surfaces with greater surface area tend to get colonized at a higher rate than smooth surfaces with smaller area. Rough surfaces are more hospitable for colonization, as they reduce shear forces, which prevents the detachment of adhering cells. Bacteria in a diverse microcolony may have a competitive advantage due to the characteristics of the cell surface, such as the quantity of exterior appendages, intercellular signaling, and EPS generation (Song et al., 2015).

1.4 Quorum Sensing in Biofilms

Quorum sensing (QS) is a bacterial communication method that enables the bacteria to express specific genes at predetermined population densities by synthesizing and secreting tiny signaling molecules known as autoinducers (Seghal Kiran et al., 2017). Through cell-to-cell intercellular transmission triggered by variations in the bacterial population's density, QS synchronizes gene expression and biofilm formation. The phrase *quorum sensing* originates from the fact that the induction and repression of QS-controlled gene expression necessitate a native cell density or *quorum* of microbes. *P. aeruginosa*, along with several strains of bacteria, can monitor their population densities and appropriately modulate their gene expression (Shrout et al., 2011). AIs are QS-signaling molecules that mediate activities like plasmic conjugation, HGT, and spore formation. They also govern QS-controlled genes that produce virulence factors and AMPs (Subramani & Jayaprakashvel, 2019).

QS systems produce, detect, and respond to AIs that accumulate in the immediate surroundings and stimulate receptor proteins to influence the expression of certain genes after a certain threshold is crossed. QS also gives a competitive edge between competing bacterial communities, as seen in the increased expression of pathogenicity in *P. aeruginosa*, and it immobilizes the densely packed cells inside the EPS (Rather et al., 2022). QS gene activation promotes the maturation and spread of biofilms. The concentration of AI-signaling molecules generated by the bacterial microcolony can be used to quantify the threshold of bacteria that presupposes the QS process. Since quorum quenching (QQ) is essential for biofilm maturation and dispersal, products and substances that promote QQ can be utilized to prevent biofilm infections by preventing the maturation process. Studies on *P. aeruginosa* and *Burkholderia cepacia* complex have also demonstrated that QQ lowers antibiotic resistance (Brackman et al., 2011). Some of the QQ methods used to inhibit QS in biofilms are inhibiting signal molecule synthesis, inactivating signal molecules through enzymatic degradation, blocking signal transduction cascades, such as the formation of AI-receptor complexes to inhibit QQ, and antagonizing signal molecule-receptors with their analogues (D'Aquila et al., 2024; Delago et al., 2016; Lade et al., 2014; Rampioni et al., 2014).

Gram-positive bacteria use the autoinducing peptide (AIP) QS system; Gram-negative bacteria use the AHL QS system; and both of the aforementioned bacteria use the AI-2 system (Fig. 1.2). These are the three main QS systems that bacteria use. Membrane transporters create AIPs, which raise the concentration of AIPs locally. When the concentration hits a certain point, AIPs attach to the histidine kinase receptors, causing phosphorylation and the subsequent expression of genes (Preda & Săndulescu, 2019). In Gram-negative bacteria, the AI AHL produces the *LuxI*-type enzyme, which breaks through the cell membrane and serves as the foundation for determining the population density of bacteria based on the quantity of proliferating cells. Cell functions like conjugation, biofilm formation, and virulence factor production are mediated by AHLs (Carlier et al., 2015). The activation of the *LuxR* receptor proteins, which is shown in bacteria like *Vibrio fischeri* and *P. aeruginosa*, triggers the transcription of target genes after a particular cell threshold is reached (Abisado et al., 2018; Whiteley et al., 2017). In contrast, Gram-positive

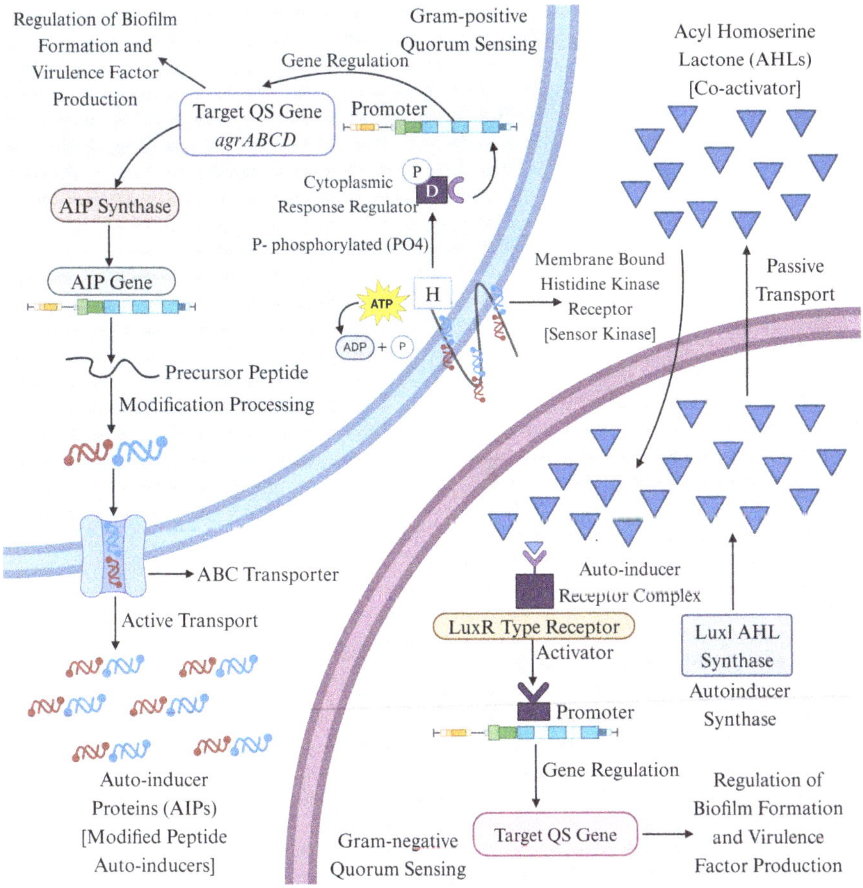

Fig. 1.2 An illustration of the quorum sensing systems in both Gram-positive and Gram-negative bacteria that aid in controlling the production of virulence factors and biofilm development. The transcription of QS genes is activated by signaling molecules (AIs), such as AIPs and AHPs, to bind to their corresponding receptors in bacterial cells

bacteria use short oligopeptide signals as part of their AI molecules, whereas the QS system mediates gene expression using a two-component system that includes cytoplasmic transcription factors and a membrane-bound histidine kinase receptor. The QS *agr* system is used by gram-positive bacteria like *S. aureus* and *S. mutans* to secrete exotoxins, which mediate the creation of biofilms and the generation of virulence factors (Li & Tian, 2012; Paluch et al., 2020).

1.5 Characteristics and Properties of Biofilm

1.5.1 Introduction to Biofilm Characteristics

Microbial communities known as biofilms are extremely well-organized, stick to surfaces, and create an EPS matrix. These structures are not random aggregates but sophisticated systems with specialized adaptive functions such as nutrient acquisition, waste management, and enhanced protection against environmental stressors. Such complexity allows biofilms to thrive in diverse settings, from natural environments to the built environment, impacting ecosystems and human activity alike.

In extreme environments, biofilms are an adaptation strategy, enabling microorganisms to maintain homeostasis under harsh conditions like extreme pH, ultraviolet radiation, and temperature fluctuations. These biofilms were likely essential during the early evolution of life on Earth, where they facilitated complex cell-to-cell interactions and survival under unstable conditions.

The durability and functionality of biofilms are significantly influenced by the EPS matrix, which is very hydrated (often more than 90% water by mass). The characteristics of the EPS matrix depend on biological factors, such as the microbial species present, chemical factors like the fluid composition in contact with the biofilm, and physical factors, including hydrodynamic and thermal conditions during formation. This intricate interplay of factors underscores the biofilm's adaptability and resilience across varying environments (Bamford et al., 2023; Parrilli et al., 2022).

1.5.2 Physical and Structural Properties

1.5.2.1 Matrix Composition

The self-produced EPS matrix that makes up biofilms is made up of proteins, extracellular DNA, and polysaccharides. Depending on the species, a microcolony's composition might vary, but generally speaking, it contains 10–25% cells and 75–90% EPS matrix. The cells that are implanted in this matrix are protected and given structural integrity. Biofilms are made up of microbial populations residing in a three-dimensional extracellular matrix. A complex array of EPS makes up the matrix, which adds to the distinctive traits and pathogenicity of the biofilm lifestyle.

The term *matrixome* refers to this group of biomolecules with a variety of chemical and functional characteristics (Singh et al., 2020; Karygianni et al., 2020). We now know that eDNA, lipids, exopolysaccharides, and extracellular proteins—many of which have amyloid-like characteristics—commonly make up the biofilm matrix that most organisms manufacture (Hobley et al., 2015).

The three-dimensional architecture of the matrix (the dense areas, pores, and channels), as well as the concentration, cohesion, charge, sorption capacity, specificity, and nature of the individual EPS components, dictates the mode of life in a particular biofilm. The final biofilm can have a variety of morphologies, including filamentous, fluffy, rough, smooth, and flat. It can also have different levels of porosity, with water-filled spaces encircling mushroom-like macrocolonies (Flemming & Wingender, 2010).

More than 70% of biofilms are made of water, which can be categorized as either free or bound water. Bound water exhibits unique nuclear magnetic resonance and infrared signals and gets adsorbed to bacterial surfaces or the biofilm matrix. This kind of water is not the same as the water restricted inside of bacterial cells. Bacteria use lytic self-sacrifice, dispersion, and swimming to assist in the formation of temporary water-filled structures called *channels* and *pores*.

As long as the cargo does not adhere to the channel walls and the channels have a high length-to-width ratio, they can carry antimicrobials, waste products, nutrients, and signaling chemicals. A length-to-width ratio of nearly one is characteristic of pores, which are mostly used for storing nutrients and diluted waste products (Quan et al., 2022). Some of the environmental variables that affect the composition of biofilms are the availability of nutrients, pH, and temperature. They mostly consist of EPS, water, and bacterial cells, though particular compositions can vary. Since water can make up as much as 97% of a biofilm's mass, it forms an absorbent, porous structure with pathways that disperse microbes, nutrients, and oxygen. About 1–2% of the matrix is made up of EPS, whereas 2–5% is made up of microbial cells (Alotaibi & Bukhari, 2021). The arrangement of bacteria in a dense biofilm, which has more than 100 layers, is determined by their oxygen tolerance and metabolic activity.

Anaerobic bacteria, for example, prefer deeper layers to avoid oxygen. These bacteria communicate and engage in specific activities, producing biofilm scaffolds like proteins, DNA, and polysaccharides as the biofilm grows. Once fully developed, the biofilm enters the dissemination phase, dispersing due to factors like nutrient scarcity, competition, and overpopulation. Dispersal can occur throughout the biofilm or in specific areas, releasing planktonic bacteria that can form new biofilms elsewhere (Saharan et al., 2024). EPS in biofilms make up about 50–90% of their composition and serve as the main component of the biofilm matrix. Despite being primarily polysaccharides, EPS exhibit varied chemical and physical properties. In Gram-negative bacteria, these polysaccharides can be either polyanionic or neutral, depending on the presence of pyruvates or uronic acids (Sonawane et al., 2022).

1.5.3 Biochemical Properties

In clonal populations, all cells have the same genes, which govern the spatial organization of enzymes based on local gene expression regulated by gene networks. Cells adapt their gene expression to their microenvironment and alter it through the uptake and release of chemicals. In multispecies communities, enzyme arrangement is constrained by differing gene sets among species. Biofilms generally range from micrometres to millimetres in size. While diffusion can quickly equalize chemical concentrations without active sources, significant variations typically occur when cells actively uptake and release chemicals, leading to spatial concentration differences driven by cell activity (Ackermann & Van Vliet, 2023).

The glycocalyx, a key biofilm component, enhances adhesion and maturation while protecting bacteria from antibiotics and environmental stress. Population heterogeneity within biofilms increases adaptability, with nutrient availability shaping bacterial growth and metabolism. Persister cells, highly resistant to antimicrobial agents, ensure biofilm regrowth after treatment. Genetic adaptations contribute to multidrug resistance, as seen in *E. coli*, while *Mycobacterium tuberculosis* evades detection by entering dormancy. Bacteria also produce enzymes that counter oxidative stress, reinforcing resilience and facilitating survival, making biofilms a significant factor in antimicrobial resistance (AMR). This complex interplay underscores biofilms' critical role in persistent infections and treatment challenges (Singh et al., 2017).

As indicated by the minimum bactericidal concentration (the lowest dose required to kill 99.9% of cells), antibiotic tolerance enables bacteria to withstand brief exposure to high antibiotic doses. Tolerance, in contrast to resistance, is transient; extended exposure will result in bacterial death. Bacteria undergo metabolic changes as a result of this adaptive response, which turns them from active to inactive. Antibiotics that become stuck in the extracellular matrix of biofilms make tolerance more difficult. A subgroup of bacteria known as persisters can survive without raising the minimum inhibitory concentration (MIC) thanks to persistence, a type of tolerance. Persisters come in two varieties: type I, which is impacted by cues from the environment, and type II, which develops on its own. Since persisters might eventually be destroyed with prolonged exposure, persistence can be referred to as heterotolerance, which is different from heteroresistance (Uruén et al., 2020).

In clinical settings, ESKAPE pathogens—*Enterococcus faecium*, *Staphylococcus aureus*, *Klebsiella pneumoniae*, *Acinetobacter baumannii*, *Pseudomonas aeruginosa*, and *Escherichia coli*— present a serious risk in clinical settings. These pathogens use various strategies to survive, including producing enzymes that inactivate antibiotics and altering their internal target sites to prevent drug binding. They can restrict antibiotic entry by modifying cell permeability or actively expelling drugs through efflux pumps (Fig. 1.3). Additionally, many can form biofilms, which protect them even in the presence of antibiotics (Omwenga & Awuor, 2024).

Research indicates that EPS contributes to bacterial tolerance to aminoglycosides. EPS can inhibit the action of antibiotics in biofilms through mechanisms such as chelation or enzymatic degradation. Bacterial biofilm communities survive

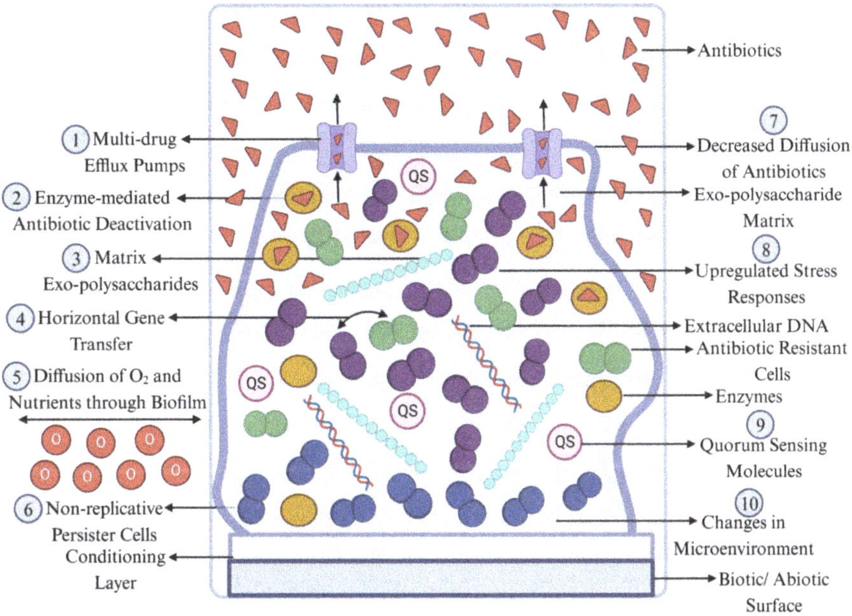

Fig. 1.3 An illustration of the most common antibiotic resistance mechanisms employed by microbial biofilms. Intracellular mechanisms such as enzyme-mediated antibiotic deactivation, horizontal gene transfer, and quorum sensing help in enhancing the antimicrobial efficacy of microbial biofilms. Multidrug efflux pumps help in the expulsion of antibiotics out of the exopolysaccharide matrix, limiting antibiotic diffusion and interaction

antibiotic stress by entering the stationary phase (a gradual or zero-growth process) or the viable-but-nonculturable state (VBNC). Cells in the stationary phase exhibit reduced antibiotic susceptibility, with at least 1% developing tolerance. Older biofilms are more likely to be in the stationary phase, demonstrating increased tolerance to antibiotics (Sharma et al., 2019).

Bacteria can coordinate group behaviors like bioluminescence, sporulation, antibiotic synthesis, and virulence factor secretion through the process known as QS. AIs are extracellular signal molecules that facilitate this connection. AIs attach to QS receptors and alter gene expression when their concentration reaches a critical point (Yi et al., 2021). AIs are undetected at low cell density (LCD), but when they accumulate at high cell density (HCD), a reaction is triggered. AIs create a feedback loop by attaching to receptors in the membrane or cytoplasm, which activates genes for cooperative behaviors and further AI creation. High quantities of AIPs are used by Gram-positive bacteria to activate signaling pathways.

Gram-negative bacteria, on the other hand, employ tiny molecules that are easily able to diffuse across their membranes, such as AHLs. AIPs are big precursor peptides that are broken down into smaller cyclic peptides and moved across the membrane in two different ways—two-component signaling and transcription factor signaling that binds to AIPs. Histidine kinases on nearby cells are bound by AIPs in two-component signaling, which causes kinase phosphorylation. According to Grutsch et al. (2018), this triggers a response regulator that affects target gene transcription. The buildup of AIs at HCD produces a concentration that enables response and detection. When AIs are detected by cytoplasmic or membrane receptors, genes that promote cooperative behavior are activated, which results in the creation of further AIs (Rutherford & Bassler, 2012). By serving as mirrors that reflect the inoculum density, these molecules control the expression of the attached genes once bacterial cultures have reached a particular stage of growth. Gram-negative bacteria employ fatty acids for signaling, whereas Gram-positive bacteria use peptide derivatives. Both can be used by many bacteria to control the expression of certain genes. Quorum-sensing inhibitors (QSIs), both natural and synthetic, have been designed to decrease microbial pathogenicity and are vital for water treatment, agriculture, fisheries, and human health (Naga et al., 2023).

1.5.4 Mechanical Properties

Bacterial adhesion is vital for biofilm development, occurring in three stages: adsorption, attachment, and colonization. The conditioning layer enhances substrate properties for easier access. Initial adhesion can be reversible, influenced by environmental factors and physical forces. Some bacteria use appendages to achieve irreversible adhesion. Surface hydrophobicity or hydrophilicity affects adhesion strength. Following adhesion, bacteria divide and form clusters, creating structures that improve nutrient access within the biofilm (Garrett et al., 2008). Because EPS promotes cohesion and strong surface adhesion, biofilms are stable.

Different strategies are used by different species to establish biofilm adhesion. Certain matrix proteins found in *V. cholerae* combine with the EPS to induce the entire biofilm to attach. But for surface adherence, *P. aeruginosa* biofilms depend on pili and their main exopolysaccharide.

Staphylococci use a variety of biofilm-forming techniques. While some tactics rely on the production of amyloid fibres, others rely on extracellular matrix (ECM)-binding adhesins and EPS (Hasan & Aggarwal, 2025). Bacteria release PIA polymers during biofilm formation, which strengthen cell-to-cell bonds and support the biofilm's structural integrity. By reinforcing the structure of the biofilm and promoting the development of complex communities, these aggregative polysaccharides function as molecular glue. To keep cells attached in nutrient-rich settings, polysaccharides aid bacteria in resisting physical forces such as shear stress from fluid movement by encouraging adhesion (Limoli et al., 2015).

Bacteria manufacture EPS primarily to maintain proper hydration and nutrition availability, to adhere bacterial cells to solid surfaces, and to hold the bacterial community together. However, biofilms are thought to be one of the main reasons indwelling medical devices (IMDs), like surgical implants and catheters, can develop chronic infections. Additionally, they have a connection to infections like rhinosinusitis, osteomyelitis, wound infections, and cystic fibrosis. Both the development of biofilms and their resistance to host defenses and antimicrobial treatments depend on EPS (Singh et al., 2021). Changes in gene expression accompany the transition from planktonic to sessile states; sessile bacteria produce fewer motility-related appendages and more surface proteins and excretion products. In comparison to planktonic microbiota, sessile bacteria are more active but also more vulnerable to preservation stresses. Due to butyrate producers' redundancy, the loss of important taxa during preservation may not affect functionality but can upset the organization of the community (Bircher et al., 2020). The differential expression of genes linked to transport, allowance for movement (including flagellar and convulsing movements), and hampering of exopolysaccharide formation in planktonic cells suggests that genes related to nitrogen assimilation and nodulation play a role in the formation of biofilms (Reyes-Pérez et al., 2016). In natural settings, bacteria frequently exist in the sessile stage and are not restricted to any particular group. Every sessile bacterium starts as a planktonic organism. To revert to planktonic life, these creatures have evolved several strategies, such as releasing daughter cells, changing their hydrophobicity, reversible adhesion, and perhaps dissolving sticky polymers.

On the other hand, planktonic bacteria employ a variety of techniques to stick to surfaces, such as creating sticky polymers on their own or using premade ones (Marshall, 2006). Cell dispersion, the last phase of biofilm formation, occurs when sessile cells switch to a planktonic growth mode to colonize new locations. This dynamic process lowers c-di-GMP levels and is brought on by environmental factors like oxygen deprivation and nutrition availability, as well as self-synthesized signaling molecules. A dynamic cycle of attachment and detachment is demonstrated by the bacteria's production of enzymes that break down the polysaccharide matrix during the dying phase of biofilm growth. This allows surface bacteria to be released for the colonization of other substrates (Singh et al., 2021). Genes involved in cell motility, including those for chemotaxis and flagella formation, as well as enzymes that break down the EPS matrix, are upregulated when c-di-GMP levels are low. In contrast, genes pertinent to attachment and EPS synthesis are downregulated. Consequently, phenotypes linked to low c-di-GMP include greater dispersion, decreased matrix production, decreased adhesiveness, and increased motility (Amankwah et al., 2021).

Like other organic fluids such as skin or blood, biofilms have properties of both viscoelastic solids and liquids. The composition and structure of biofilms, especially the EPS matrix and bacterial interactions, affect the stress relaxation seen in these environments. Biofilms combine different elastic and viscous reactions to external stress in a time-dependent fashion. A gradual viscous relaxation follows the instantaneous elastic response (Peterson et al., 2015). The characteristics of the cell surface (such as hydrophobicity, flagellation, and motility), the substrate (such as roughness and hydrophobicity), and the environment (such as temperature, pH, and nutrient availability) all affect bacterial adhesion. Hydrophobic bacteria tend to adhere to surfaces more easily, particularly rough surfaces with conditioning coatings applied. Environmental influences can change the substratum's surface characteristics. As long as these variables stay within critical bounds, higher flow velocity, water temperature, and nutrient concentration can improve microbial adhesion (Alotaibi & Bukhari, 2021).

1.5.5 Ecological Properties

Biofilms play a crucial role in nutrient cycling by housing diverse microbial populations that perform complementary metabolic functions. In soil, biofilms are essential, as they protect microorganisms from harsh conditions and improve soil quality by promoting the formation of cohesive macroaggregates (Wu et al., 2019). They are also effective in removing various pollutants, such as organic contaminants and heavy metals, while enhancing plants' stress tolerance and resistance to pathogens. Furthermore, biofilms contribute to the degradation of pollutants through the action of extracellular enzymes and the process of horizontal gene transfer, aiding in soil detoxification and fertility (Zhang et al., 2024). Biofilms are homes for microorganisms that engage in symbiotic, commensal, and competitive interactions. These interactions significantly impact the biofilms' function, biomass, diversity, and pathogenesis. Each human body has a unique microbiome that is vital for health. Most indigenous species in microbial communities maintain stable and harmless relationships with their hosts, even amid environmental changes and host defenses. This stability, or homeostasis, is achieved through a balance of microbial–microbial and microbial–host interactions (Li & Tian, 2016).

1.5.6 Functional Properties

Biofilms protect microorganisms from environmental stresses like desiccation, UV radiation, and antimicrobial agents. They provide resilience in extreme conditions, acting as protective clothing against factors such as extreme temperatures, pH levels, high salinity, pressure, nutrient scarcity, and antibiotics. In harsh environments, microorganisms regulate biofilm-forming genes through mechanisms like QS, enhancing their resistance to these extreme conditions. Biofilm formation increases bacterial resistance to the body's defenses and antimicrobial treatments, leading to chronic infections. These biofilms can accumulate various bacterial species in specific locations, potentially harming host cells through their combined activities (Vestby et al., 2020). In wastewater treatment, biofilms are important for processes like QS and QQ, which contribute to biofouling, granulation, aggregation, colonization, the metabolism of organic matter, and nutrient removal (Saini et al., 2023).

1.6 Biofilms in Medical Context

Because biofilm development is a major issue in the medical field, a great deal of study has been done to better comprehend the processes involved (Fathima et al., 2023).

1.6.1 Introduction to Biofilms in Medicine

Due to their association with chronic infections and role in antibiotic resistance, biofilms are significant in the medical field. They pose serious difficulties in clinical settings and make treatment plans more difficult. Biofilms can potentially cause chronic inflammation as a result of ongoing infections. Following a positive diagnosis of a medical issue linked to biofilms, surgical removal or replacement and medication should be taken into consideration. As the prevalence of biofilm-associated infections rises due to increasing rates of antibiotic resistance, these bacteria can grow on both living and dead tissues, as well as on medical devices like pacemakers, breast implants, contact lenses, prosthetic heart valves, catheters, and cerebrospinal fluid shunts (Gondil & Subhadra, 2023).

1.6.2 Biofilm-Associated Infections

Biofilm stimulates a low-grade inflammatory response that impedes wound healing and is linked to compromised epithelialization and granulation tissue development. It has been demonstrated that polymicrobial biofilms, which are always present in chronic wounds, cause more healing delays than single-species biofilms. Wound biofilm most likely develops as a mysterious ecosystem that eventually becomes well-established enough to obstruct wound healing and, if left unchecked, can lead to infection. Biofilms frequently develop on heart valves, prosthetic joints, and catheters, which can result in systemic infections and device failure.

Cutaneous wound healing is a complex process involving various cell types that secrete growth factors, cytokines, and chemokines. Disruptions can hinder skin barrier restoration, leading to chronic wounds characterized by increased keratinocyte proliferation, impaired migration, and elevated fibrosis levels. Beneficial interactions between commensal microorganisms and skin cells may aid in modulating the innate immune response during healing, while pathogenic microorganisms are believed to contribute to delayed wound healing.

In biofilms, bacteria often express proteins that mediate immune evasion more abundantly than in planktonic states. These can include various toxins and immune-modulatory factors, which help to subvert immune responses during infection. Some bacteria in biofilms exhibit *agr* dysfunction, which alters their ability to regulate gene expression, impacting biofilm density and protecting against immune penetration even in the absence of some toxins. This complexity allows for a mixed population within the biofilm that can balance protective structures with active immune evasion strategies.

Biofilms support polymicrobial colonies where microorganisms adapt to adverse conditions through phenotypic and genetic changes. These bacteria inhibit cell proliferation, prevent migration, and can cause cell death by producing various bioactive factors. This leads to persistent low-grade inflammation in the wound bed, primarily due to bacterial activity. Their ability to survive in low-oxygen environments allows them to thrive in hypoxic conditions and with increased bioburden.

Bacteria in chronic infections mainly grow in biofilms, complicating treatment for wounds, ear and eye infections, and prosthetic device-related infections.

Lung infections with cystic fibrosis are difficult to treat because of the production of biofilms, which contain a variety of bacterial populations and provide antibiotic resistance. Active planktonic microbes infiltrate tissue and cause an immediate inflammatory response, which results in acute wound infections. In contrast, chronic wound infections feature persistent biofilm, which is difficult to eradicate and triggers a prolonged inflammatory response. This biofilm attracts neutrophils that damage surrounding tissue without effectively eliminating the bacteria. Thus, acute infections have a direct inflammatory response, while chronic infections involve an indirect response due to biofilm presence (Hurlow & Bowler, 2022).

1.6.3 Mechanisms of Pathogenicity

Biofilms are composed of EPS, which provide mechanical protection against immune cells such as neutrophils and macrophages. The dense structure of biofilms can hinder the ability of phagocytes to effectively engulf and dismantle bacterial aggregates, especially when they exceed a certain size (Zhao et al., 2023b). Neutrophils can struggle to phagocytose larger bacterial aggregates, resulting in frustrated phagocytosis. The presence of certain EPS components, like PNAG, has been shown to protect bacteria from phagocytosis by neutrophils and macrophages. Additionally, neutrophils lack enzymes that would degrade these protective substances, further enhancing the biofilm's resistance to immune cell activity.

PNAG and other EPS components help shield bacteria from cationic and anionic AMPs. Charge repulsion mediated by PNAG protects against positively charged AMPs, while its ability to sequester anionic AMPs also contributes to a reduced antimicrobial effect, preventing the effective action of these immune defense molecules. Biofilm EPS often incorporates various molecules that can mask pathogen-associated molecular patterns (PAMPs) on the bacterial surface, diminishing the functioning of the immune apparatus to recognize and target the bacteria effectively. For example, *S. aureus* can incorporate fibrin into its biofilm, which helps to obscure bacteria from immune detection. The ability of PNAG to function as an adhesin enables bacteria to stick to other cells or abiotic surfaces, making it a crucial component of bacterial intercellular adhesion as well as microbial adhesion on the surfaces of polymeric scaffolds and biomaterials. Components like PNAG can act as decoys in immune opsonization. While PNAG can activate the complement system, this activation may not effectively lead to opsonization close to the bacterial surface, thereby shielding the bacteria themselves from opsonization and subsequent phagocytosis. The presence of PNAG can also capture the antibiotics in the biofilm and reduce their availability to bind to bacterial surfaces, which diminishes the likelihood of opsonization and antibody-mediated killing.

Bacterial biofilms resist antibiotics primarily by reducing drug penetration through their polysaccharide layer. This thick biofilm acts as a barrier, slowing down the diffusion of antibiotics into the biofilm. The effectiveness of antibiotic penetration depends on factors such as biofilm thickness, the diffusion capacity of the agents, their interactions with the biofilm, and the concentration and duration of exposure (Singh et al., 2021).

1.6.4 Clinical Diagnosis of Biofilm-Associated Infections

Because of their nonspecific symptoms and resistance to conventional culture methods, biofilm-associated illnesses are challenging to detect. Because biofilms can elude the host immune system and traditional treatment methods, they are especially significant in long-term infections such as those linked to intramedical devices and long-term wounds. Biofilms have been connected to medical device infections, which lead to worse patient outcomes, more prolonged hospital stays, and higher medical expenses. They can grow on orthopedic devices, urine catheters, endotracheal tubes, and central venous catheters. Treatment becomes challenging because of the biofilm's self-produced EPS matrix, which develops antimicrobial resistance. These cells have strong antibiotic resistance and are immersed in EPS, a matrix that is often made up of proteins, polysaccharides, and eDNA. Through indwelling devices, it is one of the main reasons why infections persist, particularly in nosocomial settings (Sharma et al., 2019).

By inducing a protracted inflammation reaction, biofilms can result in tissue damage and enduring symptoms, presumably hindering the host's capacity to eradicate the infection. Due to their frequent overlap with symptoms of various disorders, the diagnosis might be challenging (Vestby et al., 2020). Chronic lung infections in patients with cystic fibrosis, chronic osteomyelitis, chronic prostatitis, chronic rhinosinusitis, chronic otitis media, chronic wounds, recurrent urinary tract infections, endocarditis, periodontitis, and dental caries are among the many chronic host tissue-related biofilm infections (Sharma et al., 2019). Biofilms have a big influence on the economy, industry, and public health, among other facets of human life. They may result in illnesses, contaminated products, and equipment deterioration. Scientists now know that biofilms are complex cell populations encased in an EPS matrix, with permeable water channels and slime clusters, thanks to advanced technology like scanning and confocal imaging (Sharma et al., 2023). To see and distinguish between distinct biofilms, advanced microscopy methods like fluorescence microscopy, confocal laser scanning microscopy (CLSM), and scanning electron microscopy (SEM) are frequently employed.

However, many laboratories are unable to use these techniques due to their high cost and complexity. Conversely, less complex methods such as Congo red and crystal violet staining are insufficient because they are unable to differentiate bacterial cells from the biofilm matrix.

1.7 Treatment and Management of Biofilms

Biofilms' resistance mechanisms typically prevent antibiotics from completely eliminating them. Poor antimicrobial penetration through a biofilm to its depth is a common cause of resistance. On a self-produced protective matrix of EPS, poor penetration is caused by decreased antimicrobial diffusion and adsorption.

1.7.1 Biofilm Degrading Enzymes

Numerous enzymes have demonstrated efficacy in preventing the production of biofilms and breaking down preexisting biofilms of harmful and foodborne bacteria. By focusing on EPS, proteinase K, a serine protease, prevents the production of new biofilms and breaks down existing ones of *Staphylococcus aureus* and *Listeria monocytogenes*, improving antimicrobial penetration (Al-Madboly et al., 2024).

Pseudomonas fluorescens and *Macrococcus caseolyticus* biofilms are efficiently eliminated by subtilisin, another serine protease, particularly in multi-species biofilms. A combination of endo- and exopeptidases called flavorzyme prevents the production of biofilms and lowers the expression of virulence genes in bacteria such as *Salmonella typhimurium* and *Escherichia coli*. Dual-species biofilms can be effectively broken down by enzymes such as Pronase and Dispase II. Dispase II is a peptidase that hydrolyzes the N-terminal peptide bonds of nonpolar amino acids, while Pronase, a combination of protease enzymes, was efficient against the dual-species biofilm of *E. coli* and *L. monocytogenes* isolated from seafood processing factories. It dramatically hindered the production of *B. cereus* biofilm and drastically decreased the preformed biofilm of *Salmonella*.

Enzymes that break down polysaccharides, such as cellulase and amylase, impair the integrity of biofilms and prevent the growth of bacteria like *Salmonella* and *Pseudomonas aeruginosa*. Although cellulase works well against *Salmonella*, each strain has a different level of effectiveness. Glycosyl hydrolases block *E. coli* and *Listeria monocytogenes* biofilms, while amyloglucosidase stops *Vibrio parahaemolyticus* from forming biofilms. Biofilms are also broken down by endo-1,4-β-d-glucanase and β-1,6-hexosaminidase, the latter of which is very efficient against *E. coli*. Deoxyribonuclease-I (DNase-I) is an essential part of biofilm control tactics because it targets extracellular DNA in biofilms, destroying the structure (Brown et al., 2015).

1.7.2 Quorum Sensing Inhibition

Numerous cutting-edge tactics and variables are being investigated to counteract QS in bacteria. Developing QSIs for quorum quenching (QQ) is a crucial strategy. These QSIs, especially those made from natural sources like traditional medicinal plants, can interfere with bacterial communication without affecting the growth of the bacteria, reducing the possibility of resistance development. C-di-GMP, which is generated by diguanylate cyclases (DGCs) with GGDEF domains and broken down by phosphodiesterases with either EAL or HD-GYP domains, adversely regulates motility in a variety of bacterial species. Changing these environmental cues may affect the overall regulation of behaviors mediated by QS and c-di-GMP, as c-di-GMP levels can override QS inputs depending on additional environmental signals. The creation of QS signals may be inhibited by strategies that directly disrupt the production or reception of AIs, which could impact downstream signaling pathways associated with c-di-GMP (Isenberg & Montero, 2024). Additionally, enzymatic QQ is essential, and enzymes like acylase and AHL lactonases may efficiently break down QS signaling molecules, interfering with communication that is required for virulence and biofilm development (Paluch et al., 2020).

1.8 Biofilms and Public Health

As much as 70% of nosocomial illnesses in patients in the intensive care unit (ICU) are caused by pathogenic bacteria, either Gram-positive or Gram-negative. One of the primary virulence mechanisms and characteristics that set pathogenic bacteria apart is the formation of biofilms. In solid and air–liquid mediums, the majority of bacterial diseases form biofilms. The existence of a base film is one requirement for biofilm development. This base film creates an environment that is conducive to the production of biofilms and offers the first substrate for bacteria to cling to. By creating microenvironments or sending chemical cues, the conditioning film enhances microbial adhesion. This film can provide nutrition for microorganisms. Microbes can obtain nutrients from this film. The film either produces these components internally or collects both organic and inorganic materials from its surroundings. These nutritional materials support early proliferation of adhesive bacteria and help form biofilms due to their enriching properties. This film serves as a source of nutrients for microorganisms. Both organic and inorganic materials are collected by the film from its environment, or bacteria within the film produce these materials. By serving as a food supply, these nutrients promote the adherent bacteria's initial growth and aid in the creation of biofilms.

Biofilms make it easier for bacteria to stick to IMDs such as pacemakers, contact lenses, catheters, and prosthetic joints, putting patients at risk for drug-resistant infections. Additionally, they can develop on high-touch surfaces such as sterilizing bottles, clipboards, and patient rooms in medical settings. Moreover, complex chemicals found in hospital wastewater pose a threat since they can harbor genes and microbes resistant to drugs that can infect the population.

Risks can be reduced by putting in place stringent hygiene procedures, creating materials that are resistant to biofilm, and using preventive treatments.

1.9 Environmental and Industrial Applications of Biofilms

1.9.1 Role of Biofilms in Natural Environments

Water Large amounts of micropollutants (MPs) can enter natural aquatic environments all over the world through a variety of pathways, including runoff from agricultural streams and wastewater treatment plant (WWTP) effluents. As a result, MPs like medications, cosmetics, artificial sweeteners, and pesticides wind up in streams and rivers as complex mixes. These MPs have the potential to endanger both human and environmental health. They may, among other things, have a variety of unfavorable effects on organisms such as fish, invertebrates, and microorganisms that are not their intended targets. Ecosystems' makeup and functioning may eventually alter as a result of this. In natural streams, microbial stream biofilms are the most common lifeform. Stream biofilms, often referred to as periphyton, are made up of a variety of bacterial, fungal, algal, and other microorganism populations that may stick to any surface in a stream because they are embedded in an extracellular polymeric matrix (EPM). Because they are primary producers vital to biogeochemical cycles and ecosystem respiration, stream biofilms form the basis of the food web. It has also been suggested that stream biofilms can be used to monitor the effects of MP on river ecosystems. The concept of pollution-induced community tolerance (PICT) in particular may be useful in figuring out whether MPs have an impact on how communities respond to chemical stress. Numerous studies have demonstrated that stream biofilms can remove MPs from the aquatic environment by acting as a natural sink (Desiante et al., 2021).

Soil A universal characteristic of bacteria adhering to organic or manmade materials forms the basis of the lifecycle of biofilms. Because they are essential to many different industries, including agriculture, the environment, industry, and health, microbial biofilms are an intriguing subject. Biofilms have garnered attention in the agricultural industry due to their immense potential for the cultivation of crops, safeguarding, and improvement through their role in colonizing surface soils, plant roots, and shoots, allowing proliferation in the designated niche and boosting soil fertility. Gram-positive and Gram-negative bacteria, cyanobacteria, archaea, fungi, and microalgae are among the microorganisms that have been shown to create biofilms worldwide. *Actinobacteria*, *Ascomycota*, *Bacteroidetes*, *Basidiomycota*, *Chloroflexi*, *Crenarchaeota*, and *Cyanobacteria* have all been reported to produce microbial biofilms in their respective phyla. The most important genera involved in biofilm formation include *Agrobacterium*, *Anabaena*, *Azospirillum*, *Azotobacter*, *Bacillus*, *Bradyrhizobium*, *Burkholderia*, *Gluconacetobacter*, *Paenibacillus*, *Pseudomonas*, *Rhizobium*, *Trichoderma*, *Xanthomonas*, and *Xylella*. Agriculturally significant microbial communities and their interactions can have a variety of effects on climate change, plant nutrition, plant protection, biofertilization, and bioremediation for sustainable agriculture and habitats (Rana et al., 2020).

Human Body Numerous microbial illnesses in the body contain biofilms. The National Institutes of Health (NIH) estimates that 60–80% of all microbial diseases are caused by biofilm formation. In addition to medical equipment such as contacts, catheters, prosthetic devices, vascular valves, and heart rate monitors, biofilms can form on a range of physiological substrates, such as the epidermis and mucosal membranes of the gastrointestinal and respiratory systems. Furthermore, environmental biofilms serve as both an intermediary for the emergence of novel infections and a stage whereby facultative pathogens that are not part of the host system are likely to endure. Compared to their planktonic counterparts, bacteria in biofilms are more resistant to antimicrobial treatments, as many investigations have shown. A significant rise in antibiotics has been attributed to metabolic alterations in the microbes within the biofilm, decreased penetration of antibiotics by the extracellular matrix, antibiotic suppression by extracellular matrix compounds, inoculum effects related to the high number of bacteria in the biofilm relative to the available antibiotic molecules, and increased exchange of bacterial resistance mechanisms because of the close proximity of bacteria. Additionally, by avoiding phagocytosis or immune cell modulation and malfunction through the production of bacterial byproducts or toxins, bacterial biofilms aid in immune evasion (Schulze et al., 2021).

1.9.2 Impact of Biofilms in Medical Settings

Globally, chronic wounds represent a significant public health concern. Skin disorders are a serious human issue in addition to being expensive. The prevalence of chronic wounds is most likely underreported. Numerous reasons, such as an aging population and the early onset of chronic diseases like diabetes mellitus, are contributing to the growth in these ailments. According to recent estimates, four million people in Europe and seven million in the US are currently impacted, and that number is predicted to increase by 2% over the course of the next 10 years. Pressure ulcers, vascular ulcers (including venous and arterial ulcers), diabetic foot ulcers (DFUs), and vascular ulcers are the four conditions that cause chronic wounds. Infection is a common side effect of these ulcers that impedes healing. They are associated with morbidities that require daily wound care, frequent medical visits, antibiotic medication, and surgery—all of which have high costs.

Treating wounds that do not heal is made more difficult by the polymicrobial nature of the skin bed, where bacteria are grouped to form harmful biofilms. To better understand biofilms and create innovative antibiofilm techniques, it is essential to look into the social traits of bacteria and how they interact inside the EPM. Since *Pseudomonas aeruginosa* and *Staphylococcus aureus* are both quite common in chronic wounds, their interaction is the most commonly seen and studied interaction in these wounds (Pouget et al., 2022).

1.10 Advantages and Disadvantages of Using Biofilms in Biotechnology

Advantages Biofilm-based fermentation holds great potential due to its intrinsic characteristics, which include long-term activity, strong resistance to reactants, and self-immobilization. The goals of this conversation are to develop biofilm engineering research in order to optimize the benefits of biofilms and successfully utilize them in biofilm-mediated fermentation (Xu et al., 2024).

The impact of biofilms on healthcare, which has been extensively studied, is one area where a substantial amount of knowledge is available. Over 80% of microbial infections in the body are caused by biofilms, according to the US National Institutes of Health. Surgery may be required to treat persistent infections because many of these biofilms are resistant to standard antibiotic therapies. The presence of persistent cells in biofilms displaying antibiotic tolerance is a significant hurdle in the development of innovative chemical therapies, in addition to the issues caused by antibiotic resistance (Cámara et al., 2022).

Certain costly pretreatment, saccharification, and product recovery processes that convert lignocellulose resources into biofuel can benefit from the use of biofilms. A layered microbial structure that enables multiple species to sequentially convert complex substrates, the concentration of hydrolytic enzymes linked to cells that increase reaction rates at the interface, the conferment of hexose and pentose as hydrolysates diffuse outward between biofilm and substrate, and the possibility of fungal-bacterial symbioses that permit simultaneous delignification and saccharification are some advantages of using biofilms in the production of biofuel (Kazemi et al., 2022).

Disadvantages Biofilms often develop on the surface of solid fermented foods or at the liquid-air interface for fermented sauces and drinks. The thickness of the biofilms increases as the bacteria develop and mature, resulting in turbidity or "spots," further restricting consumer acceptability (Gong et al., 2021).

Eighty percent of persistent microbial human infections are caused by biofilms, which raises hospitalization rates, medical costs, and rates of death and morbidity. Upper and lower respiratory tract infections, native valve endocarditis, chronic otitis media, eye infections, chronic wounds, DFUs, urinary tract infections (UTIs), and periodontal disease are among the conditions linked to biofilms. Additionally, biofilms can develop on abiotic surfaces such as prosthetic heart valves, coronary stents, intravascular and urinary catheters, dentures, neurosurgical, cochlear, and breast implants, ventricular support, and ophthalmic devices (Magana et al., 2018).

In many different industries, such as shipping and maritime (ship damage), chemical processing, and water treatment (water pipelines, heat exchangers, and stainless-steel tanks), biofilm-associated corrosion is a significant problem. These bacteria can withstand pH values between 4 and 9 and temperatures between 10°C and 50°C. One of the main causes of marine industrial corrosion, for instance, is sulphate-reducing bacteria, which thrive in anoxic environments and affect physicochemical parameters like the metal's redox potential and the pH of the surrounding environment. Metal corrosion results from the reduction of sulfate to metal sulfide and the subsequent hydrogen sulfide gas. Every year, microbial-induced corrosion causes billions of dollars' worth of losses to man-made infrastructure (Theerthankar & Brandon, 2022).

1.11 Detection and Analysis of Biofilms

1.11.1 Methods for Studying Biofilms in the Laboratory

Microscopy Contemporary microbiological techniques in confocal microscopy, including the numeration of colony-forming units (CFUs), often require human counting, making it challenging to quantify biofilm formation on surfaces. These are laborious, resource-intensive methods that are more likely to be made by mistake. A high-resolution method for visualizing biofilm architecture in three dimensions is CLSM. It can be applied to both transparent and opaque surfaces to assess the viability of biofilms in combination with a live/dead stain. Nonetheless, there is no consensus regarding the optimal approach for processing confocal micrographs (Mountcastle et al., 2021).

Electron Microscopy Encased in extracellular polysaccharides (EPS), bacterial biofilms are organized sessile populations of bacteria. The complex structure of biofilms must be maintained in order to use scanning electron microscopy (SEM) to examine the organization of bacteria and EPS at high resolution and magnification. In order to determine which fixative would best preserve the intricate structure of bacterial biofilms, tests were conducted on the fixing qualities of formalin, glutaraldehyde, and methacarn (methanol/chloroform/acetic acid-6:3:1; Dassanayake et al., 2020). Comprehensive imaging techniques are essential for studying the formation and removal of biofilms. The merits and disadvantages of conventional SEM and variable pressure SEM are examined in this article, which contrasts traditional and state-of-the-art microscopy techniques for examining biofilm structure, ultrastructure, and 3D architecture. When ultrastructural evaluation of the EPS and its rooted bacterial cells is necessary, as in research studies on the consequences of therapeutic interventions on biofilm, SEM with tailored procedures such as osmium tetroxide (OsO4), ruthenium red (RR), tannic acid (TA), and ionic liquid (IL) need to be selected across different techniques due to its unparalleled superior image, magnification and resolution, negligible sample loss, and actual sample arrangement (Relucenti et al., 2021).

Biochemical Assays In addition to the industrial sector, biofilms have caused a number of problems in our daily lives. We must therefore continue to exercise control over them by carrying out appropriate assessments, such as crystal violet staining. The formed biofilm can be qualitatively evaluated using a variety of experimental methods of analysis. These techniques include Raman spectroscopy and various microscopes. However, there are a number of biological processes, including gene analysis and staining. Staining techniques seem to be the most effective from a practical perspective for several reasons. Therefore, a staining method based on crystal violet solution is recommended (Kamimura et al., 2022).

Metabolic Activity Assays One of the main causes of delayed wound healing is bacterial biofilms. Research on wound biofilms has therefore gained importance, especially in host-relevant contexts. The majority of *in vitro* studies analyze biofilms, which reflect conditions unrelated to the state of infection, using sophisticated laboratory media. *In vitro* biofilm research commonly uses clot or matrix-based biofilm models, as well as serum or plasma in growth settings, to mimic the wound environment. A straightforward method is to use blood or plasma alone, but more complex *in vitro* wound models are both technically difficult and incompatible with standard biofilm assays. To replicate wound fluid, an *in vitro* wound milieu (IVWM) comprising matrix materials, biochemical components, and serum was developed based on previous results regarding the composition of clinical wound fluid. Planktonic growth, biofilm properties, and interspecies interactions of frequent wound infections by *Staphylococcus aureus* and *Pseudomonas aeruginosa* were assessed using the IVWM, with controls being fetal bovine serum (FBS) and Luria-Bertani broth (Kadam et al., 2021).

1.12 Strategies to Control and Inhibit Biofilms

1.12.1 Inhibition of Biofilm Maturation

1.12.1.1 Prevention of Bacterial Adhesion

Compared to established biofilms, bacteria are less resistant in the early phases of biofilm formation. Preventing the initial bacterial adhesion is, therefore, a crucial biofilm management tactic. Surface coating modification is a commonly utilized strategy to lower implant-related infections. A universal technique for producing natural nanodaggers on a variety of bacterial surfaces that have remarkable sterilizing efficacy was developed based on the properties of bactericidal and repellent nanotopological formations found on cicada and dragonfly wings (Yuan & Zhang, 2017). Additionally, Gu et al. created a repellent surface based on active landscape, which was driven by the contractile motion found in human lung cell epithelium. The emergence of multidrug resistant *S. aureus*, *E. coli*, and *P. aeruginosa* biological films inside the catheter may be stopped by the micrometer-sized column on the surface of the material traveling horizontally in a predictable way when exposed to an electromagnetic field (Gu et al., 2020). A zwitterionic hydrogel coating comprising [2-(methacryloyloxy) ethyl] dimethyl-(3-sulfopropyl) (SBMA) as a main component, gentamicin sulfate (GS) grafted to hyaluronic acid (HA) via Schiff base linkages, and HA serving as the crosslinker. A very small proportion of bacteria that have been able to survive the coating defenses and adhere to the surface will create a slightly acidic structure that breaks the Schiff base bonds and releases GS, killing the bacteria. These bacteria produce a substance called hyaluronidase (HAase), which can rapidly break down surfaces and create new stain-resistant surfaces (Zhang et al., 2022).

1.12.1.2 Inhibition of Extracellular Matrix (ECM) Formation

The extracellular matrix is one of the most robust metabolic and physical barriers found in biofilms. A theoretical foundation for employing ECM inhibition to address the problem of antimicrobial resistance was provided by Dieltjens et al. (2020), who demonstrated that blocking ECM could reduce bacterial adhesion and make bacteria more vulnerable to antibiotics (Dieltjens et al., 2020). One crucial element present in biofilms is eDNA. Exogenous injections of DNase-I or restriction enzymes rapidly degrade eDNA, destroy biofilms, and stop biofilm reconstitution (Brown et al., 2015).

Amyloid is also necessary for early adhesion, the formation of microbial communities, and the intermediary for the preservation of biofilm integrity. The synthesis of amyloid curli depends on two proteins, Csg A and Csg B. Remarkably, D-enantiomeric peptides prevented the fibrillation of the Csg A spine and hindered the formation of *S. typhimurium* biofilms (Perov et al., 2019).

Furthermore, the production of biofilms depends on lectin, a type of protein that attaches to carbohydrates and has several glycan-binding sites. Of them, lectins Lec B and Lec A are crucial components of *P. aeruginosa* biofilms and virulence factors (Zhou & Shi, 2012). Inhibition of the function of these proteins may hinder the creation of biofilms and the development of drug-resistant bacteria. High specificity and selectivity Lec B inhibitors were effective in stopping the formation of *P. aeruginosa* biofilms in inorganic environments.

1.12.1.3 Modulating the Signaling Pathways

During the QS process, many bacteria can create and release messenger cells called AIs. AI density rises as bacterial density increases (Yadav et al., 2022). Blocking QS is a secure and efficient way to prevent biofilm formation. As of recently, numerous compounds display excellent QS system-deterrent effects (Kang et al., 2023).

Low quantities of c-di-GMP, which promotes bacterial motility and toxin expression, are found in many bacteria. Nonetheless, the high concentration of c di-GMP encourages the synthesis of adhesive proteins and biofilm development. Furthermore, the bacterium's c-di-GMP content is a key factor in deciding whether or not cells can form biofilms or stay in a planktonic state as individual cells. This makes it a possible target for defense against diseases caused by bacteria and biofilms (Liu et al., 2022). The proline-rich peptide Gup-Gup-Nap-Arg, which binds to c-di-GMP selectively and prevents *P. aeruginosa* biofilms from growing, was recently found (Foletti et al., 2018). However, compared to their Gram-negative equivalents, Gram-positive bacteria have far smaller levels of GGDEF and EAL domains, suggesting that c-di-GMP suppression has less bearing on them.

1.12.1.4 Modulating the Metabolic Pathway

Upon transitioning from the planktonic state to an actualized biofilm, microbes edit numerous intracellular metabolic systems to better suit a non-nomadic existence.

Both *S. aureus* and *E. faecalis* need a range of purines to form biofilms, as evidenced by Gélinas's research, which shows that the *de novo* purine biosynthesis pathway increases during biofilm development (Gélinas et al., 2021). Sinha et al. (2023) states that the ceramidase produced by host lipids is necessary for *P. aeruginosa* to form biofilms, which stops the host's wounds from healing (Sinha et al., 2023). In the future, biofilms may be inhibited by targeting these modified metabolic pathways.

The proliferation of biofilms corrodes the optimal equilibrium between O_2 availability and consumption of bacteria, creating an anoxic microenvironment. This increases the resilience exhibited by bacteria coated in biofilms. Therefore, lowering hypoxia may be a helpful strategy in biofilm therapy's battle against antibiotic resistance.

In contrast to mature mushroom biofilms, *P. aeruginosa* strains with siderophore mutants can only form flat biofilms, indicating that iron ions are also important for biofilm formation (Oliveira et al., 2021; Zhou & Shi, 2012). Biofilms have been demonstrated to provide a degree of defense by vying for the same iron ions with neutrophils and other innate immune cells. Therefore, one effective strategy to treat biofilms and activate the immune response is to regulate iron metabolism. The development of biofilms may be hindered or disturbed by high iron ion levels. The technique of *iron-actuated Janus ion therapy (IJIT)* was proposed by Zhu and colleagues to stop the formation of biofilms. It effectively regulates iron metabolism in both immune cells and biofilms (Zhu et al., 2022).

The aforementioned events suggest that disrupting bacterial metabolism can prevent the production of bacterial biofilms, offering a new strategy to fight biofilms.

1.12.2 Disruption of Mature Biofilm

1.12.2.1 Reactive Oxygen Species (ROS) Treatment

ROS can be produced in many ways and have bactericidal and antibiofilm properties. When a photosensitizer is stimulated by laser irradiation at a certain wavelength during photodynamic therapy (PDT), energy is transferred to neighboring oxygen molecules, emitting ROS. The development of the bimetallic metal-organic frameworks (MOFs) PCN-224 (Zr/Ti) has been shown to drastically increase ROS generation under naked light and demonstrated potent bactericidal properties (Chen et al., 2020). The active nanoenzymes known as peroxidase, which catalyze the oxygen-independent conversion of zymolyte into ROS, are commonly used in antibiofilm techniques. An ideal antibacterial agent, nitrous oxide (NO) can disintegrate bacterial biofilms and prevent antibiotic resistance. NO can accelerate wound healing during skin restoration by promoting the synthesis of collagen and myofibroblasts (Cai et al., 2021; Shi et al., 2022; Wang et al., 2023). Consequently, it eventually plays a big part in treating biofilm infections. The subsequent activation of ROS cascades minimizes damage to healthy cells by precisely localizing ROS emission at the infection site. Additionally, applying NO inhibits the formation of biofilms and encourages wound healing.

1.12.2.2 Photothermal Therapy (PTT)

The PTT approach uses high conversion efficiency photothermal agents and nonintrusive radiation from light to create regional hyperthermia. As a result, disinfection works better. Among its many advantages are its tissue permeability, minimal intrusion, and spatial autonomy. Materials such as metals, biological π-conjugated compounds, graphene, carbon quantum dots, and polydopamine (PDA) all display unique PTT properties whenever bombarded with near-infrared (NIR) light (Dash et al., 2021; Lagos et al., 2021; Lu et al., 2021; Lv et al., 2021). However, PTT requires temperatures above 70°C on its own, and high temperatures might damage healthy tissue. Using gold (Au) nanorods and antibiotics, Zhao and colleagues developed a cationic heat-sensitive liposome to treat MRSA-induced osteomyelitis. The permeability of the biofilms can be increased by positively charged nanoparticles, making it easier for bacteria to access drugs. Under 808 nm laser irradiation at a regional temperature of 50°C, over 80% of the antibiotics were quickly lost, destroying the overwhelming majority of biofilms (Zhao, You, et al., 2023a). Additionally, PDT and PTT are frequently used in conjunction to research the treatment of infections caused by biofilms.

1.12.2.3 Nanomotors

In recent years, bacterial biofilms have been broken up by nanomotors because of their substantial sensitivity and dynamic movement. Among these, Fe_3O_4-based magnetic nanomotors have recently been extensively employed to synthetically break down the membrane, greatly increasing the medicines' antibacterial properties and capacity to penetrate biofilms. Biological hybrids have been developed that could be introduced into established *E. coli* biofilms using drug-loaded mesoporous silica microtubules and benign magnetotactic microbes (MSR-1). The antibiotic Cip can then be released in response to the biofilm's corrosive surroundings, guaranteeing effective medication administration (Stanton et al., 2017).

Using metallic nanoparticles (MNPs) and mesoporous silica, Ji and colleagues developed a magnetically driven multifunctional micromotor system that uses MnO_2 as a catalyst and H_2O_2 as fuel. By combining precise magnetic guiding, ROS, and significant mechanical damage, this technique produces an antibiofilm and rapid sterilizing effect (Ji et al., 2022). When paired with Au's PTT action, SiO_2/Au nanomotors exhibit enhanced biofilm dispersion and penetration into the extracellular matrix. After 30 s and 3 min of laser irradiation, the biofilm eradication rate rose by 20% and 71%, respectively (Maric et al., 2023).

1.12.2.4 Phage Invasion

For the treatment of biofilms, phages are seen as a viable secondary to the administration of antibiotics, as they are targeted, inexpensive, and remain external to normal microbiota (Chang et al., 2022). Furthermore, some phages disintegrate carbohydrates and contribute to the fragile nature of the membrane (Kaszowska et al., 2021; Zurabov & Zhilenkov, 2021). Using intravenous phage therapy, the effectiveness of phages in treating recurrent Klebsiella pneumoniae infections on artificial knees was demonstrated. The regional symptoms were considerably lessened after therapy, and the artificial prosthetic's functionality steadily enhanced with no related side effects (Cano et al., 2021). Phages are working hard to go from the testing phase to clinical trials. These bioengineered bacteriophages could be a practical way to decrease dangerous pathogenic biofilms in the future (Sun et al., 2022; Tang et al., 2023).

1.12.2.5 Probiotics

The antibacterial and antibiofilm properties of *E. faecium* and *P. pentosaceus* extracted from aged fish and chicken may be effective against pathogenic microbes, including *B. cereus*, *S. enterica*, and *E. coli* (Tatsaporn & Kornkanok, 2020). Furthermore, *T. halophilus* works against *S. aureus* and *S. typhimurium* by preventing biofilm formation (Yao et al., 2022). These technologies use beneficial bacteria to combat harmful germs and are finding wide application in the food industry. These methods are affordable, unlikely to develop resistance to, safe for both humans and animals, and have long-lasting effects with minimal environmental impact.

1.13 Conclusion

A highly structured and dynamic form of microbial life, biofilms have significant implications for a range of habitats, including industrial settings, natural ecosystems, and healthcare settings. They pose a recurring problem in industrial and healthcare operations due to their structural complexity, resistance to environmental stresses, and ability to elude antimicrobial treatments. Developing effective control and eradication strategies requires an understanding of biofilm formation, growth, and resistance mechanisms. Interdisciplinary approaches combining microbiology, biotechnology, and materials science will be crucial as biofilm research advances and explores their potential beneficial applications.

References

Abisado, R. G., Benomar, S., Klaus, J. R., Dandekar, A. A., & Chandler, J. R. (2018). Bacterial quorum sensing and microbial community interactions. *MBio, 9*(3). https://doi.org/10.1128/mbio.02331-02317

Ackermann, M., & van Vliet, S. (2023). Spatial self-organization of metabolism in microbial systems: A matter of enzymes and chemicals. *Cell Systems, 14*(2), 98–108.

Ahmed, S. K., Hussein, S., Qurbani, K., Ibrahim, R. H., Fareeq, A., Mahmood, K. A., & Mohamed, M. G. (2024). Antimicrobial resistance: Impacts, challenges, and future prospects. *Journal of Medicine, Surgery, and Public Health, 2*, 100081.

Al-Madboly, L. A., Aboulmagd, A., El-Salam, M. A., et al. (2024). Microbial enzymes as powerful natural anti-biofilm candidates. *Microbial Cell Factories, 23*, 343.

Alotaibi, G. F., & Bukhari, M. A. (2021). Factors influencing bacterial biofilm formation and development. *American Journal of Biomedical Science and Research, 12*(6), 617–626.

Amankwah, S., Abdella, K., & Kassa, T. (2021). Bacterial biofilm destruction: A focused review on the recent use of phage-based strategies with other antibiofilm agents. In *Nanotechnology, science and applications* (pp. 161–177).

Baker, K. L., Langenheder, S., Nicol, G. W., Ricketts, D., Killham, K., Campbell, C. D., & Prosser, J. I. (2009). Environmental and spatial characterisation of bacterial community composition in soil to inform sampling strategies. *Soil Biology and Biochemistry, 41*(11), 2292–2298. https://doi.org/10.1016/j.soilbio.2009.08.010

Balducci, E., Papi, F., Capialbi, D. E., & Del Bino, L. (2023). Polysaccharides' structures and functions in biofilm architecture of antimicrobial-resistant (AMR) pathogens. *International Journal of Molecular Sciences, 24*(4), 4030.

Bamford, N. C., MacPhee, C. E., & Stanley-Wall, N. R. (2023). Microbial primer: An introduction to biofilms–what they are, why they form, and their impact on built and natural environments. *Microbiology, 169*(8), 001338.

Barraud, N., Kjelleberg, S., & Rice, S. A. (2015). Dispersal from microbial biofilms. In *Microbial biofilms* (pp. 343–362).

Bircher, L., Schwab, C., Geirnaert, A., Greppi, A., & Lacroix, C. (2020). Planktonic and sessile artificial colonic microbiota harbor distinct composition and reestablish differently upon frozen and freeze-dried long-term storage. *Msystems, 5*(1), 10–1128.

Brackman, G., Cos, P., Maes, L., Nelis, H. J., & Coenye, T. (2011). Quorum sensing inhibitors increase the susceptibility of bacterial biofilms to antibiotics in vitro and in vivo. *Antimicrobial Agents and Chemotherapy, 55*(6), 2655–2661.

Brown, H. L., Hanman, K., Reuter, M., Betts, R. P., & van Vliet, A. H. (2015). Campylobacter jejuni biofilms contain extracellular DNA and are sensitive to DNase I treatment. *Frontiers in Microbiology, 6*, 699.

Busnelli, M. P., Lazzarini Behrmann, I. C., Monroy, A. M., Daniel, M. A., & Vullo, D. L. (2023). Chapter 38: Biofilms as sustainable tools for environmental biotechnologies: An interdisciplinary approach. In S. Das & N. A. Kungwani (Eds.), *Understanding microbial biofilms* (pp. 683–698). Academic. https://doi.org/10.1016/B978-0-323-99977-9.00028-4

Cai, Y.-M., Zhang, Y.-D., & Yang, L. (2021). NO donors and NO delivery methods for controlling biofilms in chronic lung infections. *Applied Microbiology and Biotechnology, 105*, 3931–3954.

Cámara, M., Green, W., MacPhee, C. E., Rakowska, P. D., Raval, R., Richardson, M. C., Slater-Jefferies, J., Steventon, K., & Webb, J. S. (2022). Economic significance of biofilms: A multidisciplinary and cross-sectoral challenge. *npj Biofilms and Microbiomes, 8*(1), 42.

Campoccia, D., Montanaro, L., & Arciola, C. R. (2021). Extracellular DNA (eDNA). A major ubiquitous element of the bacterial biofilm architecture. *International Journal of Molecular Sciences, 22*(16), 9100.

Cano, E. J., Caflisch, K. M., Bollyky, P. L., Van Belleghem, J. D., Patel, R., Fackler, J., Brownstein, M. J., Horne, B. A., Biswas, B., & Henry, M. (2021). Phage therapy for limb-threatening prosthetic knee Klebsiella pneumoniae infection: Case report and in vitro characterization of anti-biofilm activity. *Clinical Infectious Diseases, 73*(1), e144–e151.

Carlier, A., Pessi, G., & Eberl, L. (2015). Microbial biofilms and quorum sensing. In *Principles of plant-microbe interactions: Microbes for sustainable agriculture* (pp. 45–52).

Chang, C., Yu, X., Guo, W., Guo, C., Guo, X., Li, Q., & Zhu, Y. (2022). Bacteriophage-mediated control of biofilm: A promising new dawn for the future. *Frontiers in Microbiology, 13*, 825828.

Chattopadhyay, I., Rajesh Banu, J., Usman, T. M. M., & Varjani, S. (2022). Exploring the role of microbial biofilm for industrial effluents treatment. *Bioengineered, 13*(3), 6420–6440. https://doi.org/10.1080/21655979.2022.2044250

Chen, M., Long, Z., Dong, R., Wang, L., Zhang, J., Li, S., Zhao, X., Hou, X., Shao, H., & Jiang, X. (2020). Titanium incorporation into Zr-porphyrinic metal–organic frameworks with enhanced antibacterial activity against multidrug-resistant pathogens. *Small, 16*(7), 1906240.

Chihib, N. E. (2017). Effect of incubation duration, growth temperature, and abiotic surface type on cell surface properties, adhesion and pathogenicity of biofilm-detached Staphylococcus aureus cells. *AMB Express, 7*, 191. https://doi.org/10.1186/s13568-017-0492-0

D'Aquila, P., De Rose, E., Sena, G., Scorza, A., Cretella, B., Passarino, G., & Bellizzi, D. (2024). Quorum quenching approaches against bacterial-biofilm-induced antibiotic resistance. *Antibiotics, 13*(7), 619.

Dash, B. S., Das, S., & Chen, J.-P. (2021). Photosensitizer-functionalized nanocomposites for light-activated cancer theranostics. *International Journal of Molecular Sciences, 22*(13), 6658.

Dassanayake, R. P., Falkenberg, S. M., Stasko, J. A., Shircliff, A. L., Lippolis, J. D., & Briggs, R. E. (2020). Identification of a reliable fixative solution to preserve the complex architecture of bacterial biofilms for scanning electron microscopy evaluation. *PLoS One, 15*(5), e0233973.

Debroy, A., George, N., & Mukherjee, G. (2022). Role of biofilms in the degradation of microplastics in aquatic environments. *Journal of Chemical Technology & Biotechnology, 97*(12), 3271–3282.

Delago, A., Mandabi, A., & Meijler, M. M. (2016). Natural quorum sensing inhibitors–small molecules, big messages. *Israel Journal of Chemistry, 56*(5), 310–320.

Desiante, W. L., Minas, N. S., & Fenner, K. (2021). Micropollutant biotransformation and bioaccumulation in natural stream biofilms. *Water Research, 193*, 116846. https://doi.org/10.1016/j.watres.2021.116846

Dieltjens, L., Appermans, K., Lissens, M., Lories, B., Kim, W., Van der Eycken, E. V., Foster, K. R., & Steenackers, H. P. (2020). Inhibiting bacterial cooperation is an evolutionarily robust anti-biofilm strategy. *Nature Communications, 11*(1), 107.

Fathima, A., Arafath, Y., Hassan, S., Prathiviraj, R., Kiran, G. S., & Selvin, J. (2023). Microbial biofilms: A persisting public health challenge. In *Understanding microbial biofilms* (pp. 291–314). Academic.

Fathima, S. A., Arafath, A. Y., Prathiviraj, R., Hassan, S., Kiran, G. S., & Selvin, J. (2025). Bioactive fraction of Streptomyces thinghirensis MSA1 effectively inhibits biofilm forming clinically significant AMR pathogens. *Current Microbiology, 82*(6), 234.

Ferrara, F., Castagna, T., Pantolini, B., Campanardi, M. C., Roperti, M., Grotto, A., Fattori, M., Dal Maso, L., Carrara, F., & Zambarbieri, G. (2024). The challenge of antimicrobial resistance (AMR): Current status and future prospects. *Naunyn-Schmiedeberg's Archives of Pharmacology, 397*, 1–13.

Flemming, H. C., & Wingender, J. (2010). The biofilm matrix. *Nature Reviews. Microbiology, 8*(9), 623–633. https://doi.org/10.1038/nrmicro2415

Foletti, C., Kramer, R. A., Mauser, H., Jenal, U., Bleicher, K. H., & Wennemers, H. (2018). Functionalized proline-rich peptides bind the bacterial second messenger c-di-GMP. *Angewandte Chemie International Edition, 57*(26), 7729–7733.

Garrett, T. R., Bhakoo, M., & Zhang, Z. (2008). Bacterial adhesion and biofilms on surfaces. *Progress in Natural Science, 18*(9), 1049–1056.

Gélinas, M., Museau, L., Milot, A., & Beauregard, P. B. (2021). The de novo purine biosynthesis pathway is the only commonly regulated cellular pathway during biofilm formation in TSB-based medium in Staphylococcus aureus and Enterococcus faecalis. *Microbiology Spectrum, 9*(3), e00804–e00821.

Gondil, V. S., & Subhadra, B. (2023). Biofilms and their role on diseases abstract BMC. *Microbiology, 23*(1). https://doi.org/10.1186/s12866-023-02954-2

Gong, C., He, Y., Tang, Y., Hu, R., Lv, Y., Zhang, Q., Tardy, B. L., Richardson, J. J., He, Q., & Guo, J. (2021). Biofilms in plant-based fermented foods: Formation mechanisms, benefits and drawbacks on quality and safety, and functionalization strategies. *Trends in Food Science & Technology, 116*, 940–953.

Grutsch, A. A., Nimmer, P. S., Pittsley, R. H., & McKillip, J. L. (2018). Bacillus spp. as pathogens in the dairy industry. In *Foodborne diseases* (pp. 193–211). Academic.

Gu, H., Lee, S. W., Carnicelli, J., Zhang, T., & Ren, D. (2020). Magnetically driven active topography for long-term biofilm control. *Nature Communications, 11*(1), 2211.

Gupta, P., Sarkar, S., Das, B., Bhattacharjee, S., & Tribedi, P. (2016). Biofilm, pathogenesis and prevention—A journey to break the wall: A review. *Archives of Microbiology, 198*, 1–15.

Hajiagha, M. N., & Kafil, H. S. (2023). Efflux pumps and microbial biofilm formation. *Infection, Genetics and Evolution, 112*, 105459.

Hasan, M. I., & Aggarwal, S. (2025). Matrix matters: How extracellular substances shape biofilm structure and mechanical properties. *Colloids and Surfaces B: Biointerfaces, 246*, 114341.

Hobley, L., Harkins, C., MacPhee, C. E., & Stanley-Wall, N. R. (2015). Giving structure to the biofilm matrix: An overview of individual strategies and emerging common themes. *FEMS Microbiology Reviews, 39*(5), 649–669.

Hurlow, J., & Bowler, P. G. (2022). Acute and chronic wound infections: Microbiological, immunological, clinical and therapeutic distinctions. *Journal of Wound Care, 31*(5), 436–445. https://doi.org/10.12968/jowc.2022.31.5.436

Isenberg, J. S., & Montero, E. (2024). Tolerating CD47. *Clinical and Translational Medicine, 14*(2), e1584.

Ji, H., Hu, H., Tang, Q., Kang, X., Liu, X., Zhao, L., Jing, R., Wu, M., Li, G., & Zhou, X. (2022). Precisely controlled and deeply penetrated micro-nano hybrid multifunctional motors with enhanced antibacterial activity against refractory biofilm infections. *Journal of Hazardous Materials, 436*, 129210.

Johnson, L. R. (2008). Microcolony and biofilm formation as a survival strategy for bacteria. *Journal of Theoretical Biology, 251*(1), 24–34. https://doi.org/10.1016/j.jtbi.2007.10.039

Kadam, S., Madhusoodhanan, V., Dhekane, R., Bhide, D., Ugale, R., Tikhole, U., & Kaushik, K. S. (2021). Milieu matters: An in vitro wound milieu to recapitulate key features of, and probe new insights into, mixed-species bacterial biofilms. *Biofilms, 3*, 100047.

Kamimura, R., Kanematsu, H., Ogawa, A., Kogo, T., Miura, H., Kawai, R., Hirai, N., Kato, T., Yoshitake, M., & Barry, D. M. (2022). Quantitative analyses of biofilm by using crystal violet staining and optical reflection. *Materials, 15*(19), 6727.

Kang, X., Mao, T., Hu, W., Gu, X., Tang, Q., Zhao, X., Wu, L., & Qin, Y. (2023). Strategies and materials for the prevention and treatment of biofilms. *Materials Today Bio, 23*, 100827.

Karygianni, L., Ren, Z., Koo, H., & Thurnheer, T. (2020). Biofilm matrixome: Extracellular components in structured microbial communities. *Trends in Microbiology, 28*(8), 668–681.

Kassinger, S. J., & van Hoek, M. L. (2020). Biofilm architecture: An emerging synthetic biology target. *Synthetic and Systems Biotechnology, 5*(1), 1–10. https://doi.org/10.1016/j.synbio.2020.01.001

Kaszowska, M., Majkowska-Skrobek, G., Markwitz, P., Lood, C., Jachymek, W., Maciejewska, A., Lukasiewicz, J., & Drulis-Kawa, Z. (2021). The mutation in wbaP cps gene cluster selected by phage-borne depolymerase abolishes capsule production and diminishes the virulence of Klebsiella pneumoniae. *International Journal of Molecular Sciences, 22*(21), 11562.

Kazemi, K., Ghahramani, Y., & Kalashgrani, M. Y. (2022). Nano biofilms: An emerging biotechnology applications. *Benefits, 9*, 11.

Krsmanovic, M., Biswas, D., Ali, H., Kumar, A., Ghosh, R., & Dickerson, A. K. (2021). Hydrodynamics and surface properties influence biofilm proliferation. *Advances in Colloid and Interface Science, 288*, 102336.

Lade, H., Paul, D., & Kweon, J. H. (2014). Quorum quenching mediated approaches for control of membrane biofouling. *International Journal of Biological Sciences, 10*(5), 550.

Lagos, K. J., Buzzá, H. H., Bagnato, V. S., & Romero, M. P. (2021). Carbon-based materials in photodynamic and photothermal therapies applied to tumor destruction. *International Journal of Molecular Sciences, 23*(1), 22.

Li, Y.-H., & Tian, X. (2012). Quorum sensing and bacterial social interactions in biofilms. *Sensors, 12*(3), 2519–2538.

Li, Y. H., & Tian, X. L. (2016). *Microbial interactions in biofilms: Impacts on homeostasis and pathogenesis*. IntechOpen.

Liaqat, I., Liaqat, M., Ali, S., Ali, N., Haneef, U., Mirza, S., & Tahir, H. (2019). Biofilm formation, maturation and prevention: A review. *Journal of Bacteriology and Mycology, 6*(1), 1092.

Limoli, D. H., Jones, C. J., & Wozniak, D. J. (2015). Bacterial extracellular polysaccharides in biofilm formation and function. *Microbial Biofilms*, 223–247.

Liu, X., Cao, B., Yang, L., & Gu, J.-D. (2022). Biofilm control by interfering with c-di-GMP metabolism and signaling. *Biotechnology Advances, 56*, 107915.

Lu, J., Cai, L., Dai, Y., Liu, Y., Zuo, F., Ni, C., Shi, M., & Li, J. (2021). Polydopamine-based nanoparticles for photothermal therapy/chemotherapy and their synergistic therapy with autophagy inhibitor to promote antitumor treatment. *The Chemical Record, 21*(4), 781–796.

Lv, Z., He, S., Wang, Y., & Zhu, X. (2021). Noble metal nanomaterials for NIR-triggered photothermal therapy in cancer. *Advanced Healthcare Materials, 10*(6), 2001806.

Magana, M., Sereti, C., Ioannidis, A., Mitchell, C. A., Ball, A. R., Magiorkinis, E., Chatzipanagiotou, S., Hamblin, M. R., Hadjifrangiskou, M., & Tegos, G. P. (2018). Options and limitations in clinical investigation of bacterial biofilms. *Clinical Microbiology Reviews, 31*(3). https://doi.org/10.1128/cmr.00084-16

Mah, T.-F. C., & O'Toole, G. A. (2001). Mechanisms of biofilm resistance to antimicrobial agents. *Trends in Microbiology, 9*(1), 34–39. https://doi.org/10.1016/S0966-842X(00)01913-2

Mahto, K. U., Vandana, Priyadarshanee, M., Samantaray, D. P., & Das, S. (2022). Bacterial biofilm and extracellular polymeric substances in the treatment of environmental pollutants: Beyond the protective role in survivability. *Journal of Cleaner Production, 379*, 134759. https://doi.org/10.1016/j.jclepro.2022.134759

Maric, T., Løvind, A., Zhang, Z., Geng, J., & Boisen, A. (2023). Near-infrared light-driven mesoporous SiO2/Au Nanomotors for eradication of Pseudomonas aeruginosa biofilm. *Advanced Healthcare Materials, 12*(13), 2203018.

Marshall, K. C. (2006). Planktonic versus sessile life of prokaryotes. *The Prokaryotes, 2*, 3–15.

Mountcastle, S. E., Vyas, N., Villapun, V. M., Cox, S. C., Jabbari, S., Sammons, R. L., Shelton, R. M., Walmsley, A. D., & Kuehne, S. A. (2021). Biofilm viability checker: An open-source tool for automated biofilm viability analysis from confocal microscopy images. *npj Biofilms and Microbiomes, 7*(1), 44.

Naaz, T., Lahiri, D., Pandit, S., Nag, M., Gupta, P. K., Al-Dayan, N., Rai, N., Chaubey, K. K., & Gupta, A. K. (2023). Antimicrobial peptides against microbial biofilms: Efficacy, challenges, and future prospect. *International Journal of Peptide Research and Therapeutics, 29*(3), 48.

Naga, N. G., El-Badan, D. E., Ghanem, K. M., & Shaaban, M. I. (2023). It is the time for quorum sensing inhibition as alternative strategy of antimicrobial therapy. *Cell Communication and Signaling, 21*(1), 1–14.

Naghavi, M., Vollset, S. E., Ikuta, K. S., Swetschinski, L. R., Gray, A. P., Wool, E. E., Aguilar, G. R., Mestrovic, T., Smith, G., & Han, C. (2024). Global burden of bacterial antimicrobial resistance 1990–2021: A systematic analysis with forecasts to 2050. *The Lancet, 404*(10459), 1199–1226.

Oliveira, F., Rohde, H., Vilanova, M., & Cerca, N. (2021). The emerging role of iron acquisition in biofilm-associated infections. *Trends in Microbiology, 29*(9), 772–775.

Omwenga, E. O., & Awuor, S. O. (2024). The bacterial biofilms: Formation, impacts, and possible management targets in the healthcare system. *Canadian Journal of Infectious Diseases and Medical Microbiology, 2024*(1), 1542576.

Orhan-Yanıkan, E., Gülseren, G., & Ayhan, K. (2020). Protein profile of bacterial extracellular polymeric substance by Fourier transform infrared spectroscopy. *Microchemical Journal, 156*, 104831. https://doi.org/10.1016/j.microc.2020.104831

Pal, M. K., & Lavanya, M. (2022). Microbial influenced corrosion: Understanding bioadhesion and biofilm formation. *Journal of Bio-and Tribo-Corrosion, 8*(3), 76.

Paluch, E., Rewak-Soroczyńska, J., Jędrusik, I., Mazurkiewicz, E., & Jermakow, K. (2020). Prevention of biofilm formation by quorum quenching. *Applied Microbiology and Biotechnology, 104*, 1871–1881.

Parrilli, E., Tutino, M. L., & Marino, G. (2022). Biofilm as an adaptation strategy to extreme conditions. *Rendiconti Lincei. Scienze Fisiche e Naturali, 33*(3), 527–536.

Perov, S., Lidor, O., Salinas, N., Golan, N., Tayeb-Fligelman, E., Deshmukh, M., Willbold, D., & Landau, M. (2019). Structural insights into curli CsgA cross-β fibril architecture inspire repurposing of anti-amyloid compounds as anti-biofilm agents. *PLoS Pathogens, 15*(8), e1007978.

Peterson, B. W., He, Y., Ren, Y., Zerdoum, A., Libera, M. R., Sharma, P. K., et al. (2015). Viscoelasticity of biofilms and their recalcitrance to mechanical and chemical challenges. *FEMS Microbiology Reviews, 39*(2), 234–245.

Pouget, C., Dunyach-Remy, C., Bernardi, T., Provot, C., Tasse, J., Sotto, A., & Lavigne, J.-P. (2022). A relevant wound-like in vitro media to study bacterial cooperation and biofilm in chronic wounds. *Frontiers in Microbiology, 13*, 705479.

Preda, V. G., & Săndulescu, O. (2019). Communication is the key: Biofilms, quorum sensing, formation and prevention. *Discover, 7*(3), e10.
Quan, K., Hou, J., Zhang, Z., Ren, Y., Peterson, B. W., Flemming, H.-C., Mayer, C., Busscher, H. J., & van der Mei, H. C. (2022). Water in bacterial biofilms: Pores and channels, storage and transport functions. *Critical Reviews in Microbiology, 48*(3), 283–302.
Rampioni, G., Leoni, L., & Williams, P. (2014). The art of antibacterial warfare: Deception through interference with quorum sensing–mediated communication. *Bioorganic Chemistry, 55*, 60–68.
Rana, K. L., Kour, D., Yadav, A. N., Yadav, N., & Saxena, A. K. (2020). Chapter 16: Agriculturally important microbial biofilms: Biodiversity, ecological significances, and biotechnological applications. In M. K. Yadav & B. P. Singh (Eds.), *New and future developments in microbial biotechnology and bioengineering: Microbial biofilms* (pp. 221–265). Elsevier. https://doi.org/10.1016/B978-0-444-64279-0.00016-5
Rather, M. A., Gupta, K., & Mandal, M. (2021). Microbial biofilm: Formation, architecture, antibiotic resistance, and control strategies. *Brazilian Journal of Microbiology, 52*, 1–18.
Rather, M. A., Saha, D., Bhuyan, S., Jha, A. N., & Mandal, M. (2022). Quorum quenching: A drug discovery approach against Pseudomonas aeruginosa. *Microbiological Research, 264*, 127173.
Relucenti, M., Familiari, G., Donfrancesco, O., Taurino, M., Li, X., Chen, R., Artini, M., Papa, R., & Selan, L. (2021). Microscopy methods for biofilm imaging: Focus on SEM and VP-SEM pros and cons. *Biology, 10*(1), 51.
Reyes-Pérez, A., Vargas, M. D. C., Hernández, M., Aguirre-von-Wobeser, E., Pérez-Rueda, E., & Encarnacion, S. (2016). Transcriptomic analysis of the process of biofilm formation in Rhizobium etli CFN42. *Archives of Microbiology, 198*, 847–860.
Rochex, A., Godon, J.-J., Bernet, N., & Escudié, R. (2008). Role of shear stress on composition, diversity and dynamics of biofilm bacterial communities. *Water Research, 42*(20), 4915–4922.
Rutherford, S. T., & Bassler, B. L. (2012). Bacterial quorum sensing: Its role in virulence and possibilities for its control. *Cold Spring Harbor Perspectives in Medicine, 2*(11), a012427.
Sadiq, F. A., Burmølle, M., Heyndrickx, M., Flint, S., Lu, W., Chen, W., Zhao, J., & Zhang, H. (2021). Community-wide changes reflecting bacterial interspecific interactions in multispecies biofilms. *Critical Reviews in Microbiology, 47*(3), 338–358.
Saharan, B. S., Beniwal, N., & Duhan, J. S. (2024). From formulation to function: A detailed review of microbial biofilms and their polymer-based extracellular substances. *The Microbe, 5*, 100194.
Saini, S., Tewari, S., Dwivedi, J., & Sharma, V. (2023). Biofilm-mediated wastewater treatment: A comprehensive review. *Materials Advances, 4*(6), 1415–1443.
Schulze, A., Mitterer, F., Pombo, J. P., & Schild, S. (2021). Biofilms by bacterial human pathogens: Clinical relevance – Development, composition and regulation – Therapeutical strategies. *Microbial Cell, 8*(2), 28–56. https://doi.org/10.15698/mic2021.02.741
Seghal Kiran, G., Hassan, S., Sajayan, A., & Selvin, J. (2017). Quorum quenching compounds from natural sources. In *Bioresources and bioprocess in biotechnology: Volume 2: Exploring potential biomolecules* (pp. 351–364).
Sharma, D., Misba, L., & Khan, A. U. (2019). Antibiotics versus biofilm: An emerging battleground in microbial communities. *Antimicrobial Resistance & Infection Control, 8*, 1–10.
Sharma, S., Mohler, J., Mahajan, S. D., Schwartz, S. A., Bruggemann, L., & Aalinkeel, R. (2023). Microbial biofilm: A review on formation, infection, antibiotic resistance, control measures, and innovative treatment. *Microorganisms, 11*(6), 1614.
Shi, Y., Cao, Y., Cheng, J., Yu, W., Liu, M., Yin, J., Huang, C., Liang, X., Zhou, H., & Liu, H. (2022). Construction of self-activated nanoreactors for cascade catalytic anti-biofilm therapy based on H2O2 self-generation and switch-on NO release. *Advanced Functional Materials, 32*(20), 2111148.
Shrout, J. D., Tolker-Nielsen, T., Givskov, M., & Parsek, M. R. (2011). The contribution of cell-cell signaling and motility to bacterial biofilm formation. *MRS Bulletin, 36*(5), 367–373.
Singh, S., Singh, S. K., Chowdhury, I., & Singh, R. (2017). Understanding the mechanism of bacterial biofilm resistance to antimicrobial agents. *The Open Microbiology Journal, 11*, 53.

Singh, M. P., Singh, P., Li, H.-B., Song, Q.-Q., & Singh, R. K. (2020). Microbial biofilms: Development, structure, and their social assemblage for beneficial applications. In *New and future developments in microbial biotechnology and bioengineering: Microbial biofilms* (pp. 125–138). Elsevier.

Singh, S., Datta, S., Narayanan, K. B., & Rajnish, K. N. (2021). Bacterial exo-polysaccharides in biofilms: Role in antimicrobial resistance and treatments. *Journal of Genetic Engineering and Biotechnology, 19*, 1–19.

Sinha, M., Ghosh, N., Wijesinghe, D. S., Mathew-Steiner, S. S., Das, A., Singh, K., El Masry, M., Khanna, S., Inoue, H., & Yamazaki, K. (2023). Pseudomonas aeruginosa theft biofilm require host lipids of cutaneous wound. *Annals of Surgery, 277*(3), e634–e647.

Soares, A., Gomes, L. C., Monteiro, G. A., & Mergulhão, F. J. (2022). Hydrodynamic effects on biofilm development and recombinant protein expression. *Microorganisms, 10*(5). https://doi.org/10.3390/microorganisms10050931

Sonawane, J. M., Rai, A. K., Sharma, M., Tripathi, M., & Prasad, R. (2022). Microbial biofilms: Recent advances and progress in environmental bioremediation. *Science of the Total Environment, 824*, 153843.

Song, F., Koo, H., & Ren, D. (2015). Effects of material properties on bacterial adhesion and biofilm formation. *Journal of Dental Research, 94*(8), 1027–1034.

Sportelli, M. C., Kranz, C., Mizaikoff, B., & Cioffi, N. (2022). Recent advances on the spectroscopic characterization of microbial biofilms: A critical review. *Analytica Chimica Acta, 1195*, 339433.

Stanton, M. M., Park, B.-W., Vilela, D., Bente, K., Faivre, D., Sitti, M., & Sánchez, S. (2017). Magnetotactic bacteria powered biohybrids target E. coli biofilms. *ACS Nano, 11*(10), 9968–9978.

Stewart, P. S. (2003). Diffusion in biofilms. *Journal of Bacteriology, 185*(5), 1485–1491.

Stewart, P. S., & Franklin, M. J. (2008). Physiological heterogeneity in biofilms. *Nature Reviews Microbiology, 6*(3), 199–210.

Subramani, R., & Jayaprakashvel, M. (2019). Bacterial quorum sensing: Biofilm formation, survival behaviour and antibiotic resistance. In *Implication of quorum sensing and biofilm formation in medicine, agriculture and food industry* (pp. 21–37).

Sun, R., Yu, P., Zuo, P., Villagrán, D., Mathieu, J., & Alvarez, P. J. (2022). Biofilm control in flow-through systems using polyvalent phages delivered by peptide-modified M13 Coliphages with enhanced polysaccharide affinity. *Environmental Science & Technology, 56*(23), 17177–17187.

Syed, Z., Sogani, M., Rajvanshi, J., & Sonu, K. (2023). Microbial biofilms for environmental bioremediation of heavy metals: A review. *Applied Biochemistry and Biotechnology, 195*(9), 5693–5711.

Tang, M., Huang, Z., Zhang, X., Kong, J., Zhou, B., Han, Y., Zhang, Y., Chen, L., & Zhou, T. (2023). Phage resistance formation and fitness costs of hypervirulent Klebsiella pneumoniae mediated by K2 capsule-specific phage and the corresponding mechanisms. *Frontiers in Microbiology, 14*, 1156292.

Tatsaporn, T., & Kornkanok, K. (2020). Using potential lactic acid bacteria biofilms and their compounds to control biofilms of foodborne pathogens. *Biotechnology Reports, 26*, e00477.

Thacharodi, A., Hassan, S., Meenatchi, R., Bhat, M. A., Hussain, N., Arockiaraj, J., Ngo, H. H., Sharma, A., Nguyen, H., & Pugazhendhi, A. (2024a). Mitigating microplastic pollution: A critical review on the effects, remediation, and utilization strategies of microplastics. *Journal of Environmental Management, 351*, 119988.

Thacharodi, A., Meenatchi, R., Hassan, S., Hussain, N., Bhat, M. A., Arockiaraj, J., Ngo, H. H., Le, Q. H., & Pugazhendhi, A. (2024b). Microplastics in the environment: A critical overview on its fate, toxicity, implications, management, and bioremediation strategies. *Journal of Environmental Management, 349*, 119433.

Theerthankar, D., & Brandon, C. Y. (2022). Introductory chapter: Highlighting pros and cons of bacterial biofilms. In D. Theerthankar (Ed.), *Focus on bacterial biofilms* (p. Ch. 1). IntechOpen. https://doi.org/10.5772/intechopen.106775

Uruén, C., Chopo-Escuin, G., Tommassen, J., Mainar-Jaime, R. C., & Arenas, J. (2020). Biofilms as promoters of bacterial antibiotic resistance and tolerance. *Antibiotics, 10*(1), 3.

Vandana, Priyadarshanee, M., & Das, S. (2023). Bacterial extracellular polymeric substances: Biosynthesis and interaction with environmental pollutants. *Chemosphere, 332*, 138876. https://doi.org/10.1016/j.chemosphere.2023.138876

Vestby, L. K., Grønseth, T., Simm, R., & Nesse, L. L. (2020). Bacterial biofilm and its role in the pathogenesis of disease. *Antibiotics, 9*(2), 59.

Wang, T.-Y., Zhu, X.-Y., & Wu, F.-G. (2023). Antibacterial gas therapy: Strategies, advances, and prospects. *Bioactive Materials, 23*, 129–155.

Whiteley, M., Diggle, S. P., & Greenberg, E. P. (2017). Bacterial quorum sensing: The progress and promise of an emerging research area. *Nature, 551*(7680), 313.

Wu, Y., Cai, P., Jing, X., Niu, X., Ji, D., Ashry, N. M., et al. (2019). Soil biofilm formation enhances microbial community diversity and metabolic activity. *Environment International, 132*, 105116.

Xiao, P., Zhou, J., Luo, X., Kang, B., Guo, L., Yuan, G., Zhang, L., & Zhao, T. (2021). Enhanced nitrogen removal from high-strength ammonium wastewater by improving heterotrophic nitrification-aerobic denitrification process: Insight into the influence of dissolved oxygen in the outer layer of the biofilm. *Journal of Cleaner Production, 297*, 126658.

Xu, A., Xin, F., Dong, W., & Jiang, M. (2024). Maximizing the benefits of biofilms in fermentation processes. *Trends in Biotechnology, 42*(6), 677–679. https://doi.org/10.1016/j.tibtech.2023.11.012

Yadav, J., Das, S., Singh, S., Jyoti, A., Srivastava, V. K., Sharma, V., Kumar, S., Kumar, S., & Kaushik, S. (2022). Deciphering the role of S-adenosyl homocysteine nucleosidase in quorum sensing mediated biofilm formation. *Current Protein and Peptide Science, 23*(4), 211–225.

Yao, S., Hao, L., Zhou, R., Jin, Y., Huang, J., & Wu, C. (2022). Formation of biofilm by Tetragenococcus halophilus benefited stress tolerance and anti-biofilm activity against S. aureus and S. typhimurium. *Frontiers in Microbiology, 13*, 819302.

Yi, L., Dong, X., Grenier, D., Wang, K., & Wang, Y. (2021). Research progress of bacterial quorum sensing receptors: Classification, structure, function, and characteristics. *Science of the Total Environment, 763*, 143031.

Yuan, Y., & Zhang, Y. (2017). Enhanced biomimic bactericidal surfaces by coating with positively-charged ZIF nano-dagger arrays. *Nanomedicine: Nanotechnology, Biology and Medicine, 13*(7), 2199–2207.

Zhang, J., Wu, M., Peng, P., Liu, J., Lu, J., Qian, S., & Feng, J. (2022). "Self-defensive" antifouling zwitterionic hydrogel coatings on polymeric substrates. *ACS Applied Materials & Interfaces, 14*(50), 56097–56109.

Zhang, B., Hu, X., Zhao, D., Wang, Y., Qu, J., Tao, Y., et al. (2024). Harnessing microbial biofilms in soil ecosystems: Enhancing nutrient cycling, stress resilience, and sustainable agriculture. *Journal of Environmental Management, 370*, 122973.

Zhao, W., You, J., Yin, S., Yang, H., He, S., Feng, L., Li, J., Zhao, Q., & Wei, L. (2023a). Extracellular polymeric substances—Antibiotics interaction in activated sludge: A review. *Environmental Science and Ecotechnology, 13*, 100212. https://doi.org/10.1016/j.ese.2022.100212

Zhao, Z., Zhang, Y., Cheng, Y., Li, J., Wang, W., Huang, S., Ma, X., & Zhang, X. (2023b). Thermosensitive nanotherapeutics for localized photothermal ablation of MRSA-infected osteomyelitis combined with chemotherapy. *ACS Applied Materials & Interfaces, 15*(10), 12842–12854.

Zhou, X., & Shi, W. (2012). *Microbial biofilms and infection*. People's Medica l Publishing House, China.

Zhu, W., Mei, J., Zhang, X., Zhou, J., Xu, D., Su, Z., Fang, S., Wang, J., Zhang, X., & Zhu, C. (2022). Photothermal nanozyme-based microneedle patch against refractory bacterial biofilm infection via iron-actuated Janus ion therapy. *Advanced Materials, 34*(51), 2207961.

Zurabov, F., & Zhilenkov, E. (2021). Characterization of four virulent Klebsiella pneumoniae bacteriophages, and evaluation of their potential use in complex phage preparation. *Virology Journal, 18*, 1–20.

2 Microbial Biofilms and Resistome: Concept and Applications

Simi Asma Salim, Siddhardha Busi, Asad Syed, and Rajaneesh Anupam

2.1 Introduction

Biofilms are a community of microbial cells, enclosed in a self-produced matrix of exopolymeric substances (EPS). A biofilm matrix consists of various proteins, polysaccharides, extracellular DNA, and other substances. Biofilm cells exhibit a different lifestyle compared to planktonic cells and show varied expression of genes and proteins in both lifestyles. Biofilm mode is the preferred lifestyle of microbes, owing to their high resistance to different chemical and physical stresses, and this can be seen in a variety of environments. The formation of biofilms is a complicated process that involves sequential stages of attachment followed by microcolony formation and then biofilm maturation and dispersion (Liu et al., 2024). Multiple factors associated with biofilm contribute to higher antibiotic resistance than that of planktonic cells, causing difficulty in management of biofilm-associated infections. These factors include less penetration of antibiotics, specific microenvironment conditions (e.g., for hypoxia), persister cells, and close association between cells (Fig. 2.1). The close proximity of cells within a biofilm acts as the major contributor to antimicrobial resistome transfer through horizontal gene transfer (Shree et al., 2023).

S. A. Salim · S. Busi (✉)
Department of Microbiology, School of Life Sciences, Pondicherry University, Kalapet, Puducherry, India

A. Syed
Department of Botany and Microbiology, College of Science, King Saud University, Riyadh, Saudi Arabia

R. Anupam
Department of Biochemistry and Molecular Biology, School of Life Sciences, Pondicherry University, Puducherry, India
e-mail: ranupam@pondiuni.ac.in

S. Busi et al. (eds.), *Omics Approaches in Biofilm Research*,
https://doi.org/10.1007/978-3-031-91863-6_2

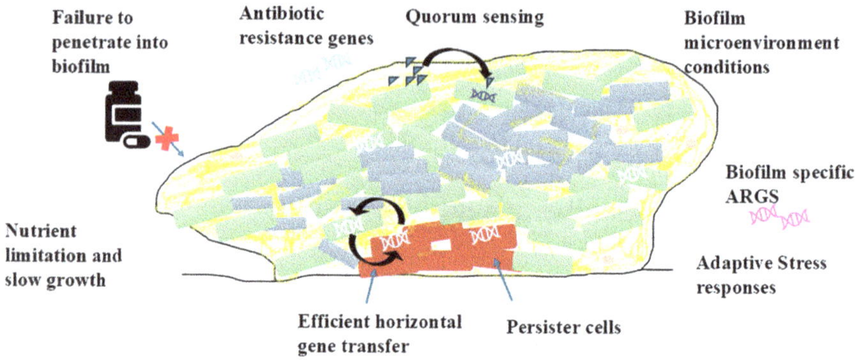

Fig. 2.1 Mechanisms of antimicrobial resistance in the biofilm

The antibiotic resistome comprises all the genes that exhibit resistance towards antibiotics. ARGs may either exist in chromosomal DNA or in mobile genetic elements (MGEs) including plasmids and transposons. The genes responsible for ARGs can either be fully functional or have the potential to become an active resistance gene (protoresistance genes or silent genes) upon mutation or alteration in expression. The transfer of these resistance genes can be either vertical (mostly in case of inherent resistance) or through horizontal gene transmission (Perry et al., 2014).

In this chapter, we focus specifically on the resistome in biofilms. For ease of understanding, the antibiotic resistomes of biofilms under various environmental and geographical conditions (including aquatic environments, drinking water distribution systems, wastewater treatment plants, plastic surfaces, and clinical settings such as hospitals) are categorized and discussed.

2.2 Horizontal Gene Transfer of Antibiotic Resistomes in Biofilms

Due to the proximity of cells within biofilms and their prolonged retention within the encasing EPS, biofilms promote genetic exchange and nucleic acid rearrangements. Here, genetic exchange can occur through either vertical or horizontal gene transfer (HGT), increasing the likelihood that harmless bacteria become multidrug-resistant pathogens. Various mechanisms of HGT include conjugation, transformation, and transduction, which happen with the help of various mobile genetic elements (MGEs) including plasmids, integrons, and transposons (Michaelis & Grohmann, 2023). A higher transfer efficiency of ARG-harboring MGEs in biofilm cells compared to planktonic cells was reported by various researchers. For instance, there was a study of the conjugation frequency of a $bla_{\text{NDM-1}}$-harbouring plasmid among *Klebsiella pneumoniae* strains. They found a significantly higher transfer frequency in biofilm than planktonic cells (Element et al., 2023). Likewise, biofilm formation enhanced the transfer of integrative and conjugative element ICE*Bs1* in *Bacillus subtilis* (Lécuyer et al., 2018). The natural transformation of MGE associated with β-lactam resistance—SCC*mec*—was demonstrated in *Staphylococcus* *aureus* biofilm. In biofilms, the two-component system up-regulates natural competence genes, which enhances natural transformation efficiency (Maree et al., 2022).

Apart from conjugation, transduction, and transformation, a newly identified mechanism of HGT is gene exchange by outer-membrane vesicles (OMVs; Michaelis & Grohmann, 2023). Gram-negative bacteria produce these membrane-bound vesicles, which help in packaging various cargo materials including DNA and therefore can be transferred to other bacteria or host cells. Both planktonic and biofilm cells can form these OMVs; however, studies on these cargo packages in *Pseudomonas aeruginosa* showed that, regardless of the small size of biofilm-derived OMVs, the plasmid DNA content was higher than planktonic-derived larger OMVs (Johnston et al., 2023).

Persister cells, subsets of cells which are extremely resistant to antimicrobial substances, in biofilms, also contribute to the transmission of ARGs through HGT. Recently, research by Nasu and Maeda (2024) revealed the transformation of plasmids containing ARGs in *E. coli*. Another contributing factor for enhanced HGT and ARG transmission is the microbial communication system quorum sensing (QS). Even though QS and biofilm formation is interdependent, their direct role in facilitating the HGT is underexplored. Zhu et al. (2020) demonstrated how the QS works in multidrug plasmid RP4 transfer in biofilms. Autoinducer acyl-homoserine-lactone molecules enhance the expression of conjugation-related genes (Mpf and Dtr system related). Similarly, the enhancement of HGT by autoinducer molecules CAI-I and AI-2 of multispecies biofilm has been reported (Antonova & Hammer, 2011).

2.3 Antibiotic Resistomes in Biofilm Infections

Direct studies and resistome analyses of the biofilm from infections are very limited. In most of the studies, isolation of specific microbes and their antimicrobial resistance is evaluated. Polymicrobial communities in biofilms of diabetic foot infections complicate infection and chronic wound formation with synergistic action (Afonso et al., 2021). Different biofilm-forming *S. aureus* isolates from diabetic foot ulcers of Jordanian patients showed high antibiotic resistance. The highest resistance was observed for the antibiotic Oxacillin, followed by Clindamycin, Erythromycin, and Cefazolin (Owais et al., 2024).

Shotgun metagenomics analysis on oral biofilm microbiome of healthy, caries-active, and periodontally diseased individuals revealed that it can serve as an ARG reservoir. They identified more than 60 ARGs conferring resistance to β-lactams, tetracycline, macrolide-lincosamide-streptograminin etc. Interestingly, higher occurrence was observed in healthy and caries-associated individuals than individuals with periodontal disease (Anderson et al., 2023). Conversely, the qualitative and quantitative PCR study by Kim et al. (2011) showed a higher occurrence of resistance genes (ARGs) *tet(Q)*, *tet(M)*, *cfxA*, and *bla* $_{TEM}$ in both the periodontitis patient and the healthy subject group. A slightly higher occurrence of ARGs was observed in the patient group, for all except *erm(F)*. The *erm(F)* predominated in the healthy group. Sedlacek and Walker developed an *in vitro* subgingival biofilm model to assess the antibiotic resistance. Biofilms showed higher resistance than planktonic forms, and inhibitory concentrations of up to 250 times greater were found (Sedlacek & Walker, 2007). Bessa et al. conducted a metagenomic analysis of subgingival peri-implant biofilms. However, the results were contradictory to the works discussed above. No significant difference in the resistome profile of healthy and diseased individuals was observed. Above all, they observed higher ARG occurrence in salivary samples than in subgingival biofilm samples (Bessa et al., 2025).

Biofilm-forming *P. aeruginosa* isolated from patients with keratitis and endophthalmitis was observed for *ndvB and tssC1*, which specifically enhance biofilm resistance towards antibiotics. This resistance was observed against colistin, gentamicin, imipenem, aztreonam, piperacillin-tazobactam, ceftazidime, and ciprofloxacin. Out of the total isolates, more than 90% harbored the genes *ndvB* and *tssC1* (Saffari et al., 2017). Other than direct observation of biofilm from infections, plentiful studies have been conducted on *in vitro* biofilm formed by clinically relevant microbes.

Biofilm formation in medical indwelling devices such as urinary stents, catheters, and endotracheal tubes is a major concern. *S. aureus* biofilms are the main reason for spinal cord stimulation device–linked infections. Strains isolated from infected patients were analyzed for biofilm formation ability, biofilm matrix composition, and resistome profile. All isolates were observed to have the *ica*A/D gene and *in vitro* biofilm-forming ability. A high biofilm resistance against vancomycin, trimethoprim/sulfamethoxazole, and levofloxacin was also observed (Sivori et al., 2024). Singhai et al. studied antibiotic resistance in biofilms of different indwelling devices. They found that a significant percentage of biofilm-forming bacterial isolates were multidrug resistant. Several biofilm-producing Gram-negative isolates were identified as ESβL producers, with many harboring the $bla_{\mathrm{CTX\text{-}M}}$ gene among the various ESβL genotypes (Singhai et al., 2012). Biofilm-forming *Proteus mirabilis* isolates from urinary catheters were found to be predominantly multidrug resistant. Significant occurrence of *int1* and *int2* genes were also observed in these isolates. Moreover, bla_{TEM} is counted as the most frequently detected EsBL gene (Mirzaei et al., 2021).

2.4 Biofilm-Specific Antibiotic Resistance Gene

In most cases, the biofilm-specific genes that confer antibiotic resistance have been studied in the pathogen *P. aeruginosa*. The expression of certain genes can significantly impact the biofilm state compared to the planktonic state. For example, the genes coding for the efflux pump, identified as PA1874 to PA1877 in *P. aeruginosa*, have been found to be crucial for biofilm-specific resistance to a group of antibiotics. Complete deletion of these genes resulted in increased sensitivity to tobramycin, gentamicin, and ciprofloxacin, especially when the mutant strain was growing in a biofilm. Higher expression of this efflux pump was observed in biofilm cells compared to planktonic cells, which helps explain why these genes are significant for antibiotic resistance in a biofilm context (Zhang & Mah, 2008).

The most studied biofilm-specific ARG is *ndvB*, which codes for the transferase enzyme involved in the periplasmic glucan biosynthesis in *P. aeruginosa*. Suppression of this gene causes glucan production to be blocked but will not affect growth, biofilm formation kinetics, or the structure of the biofilms. Interestingly, in biofilms, mutation of *ndvB* caused higher sensitivity to the same aminoglycosides and fluoroquinolone mentioned above. The mechanism of this biofilm-specific resistance is supposed to be due to sequestration of antibiotic molecules away from their cell targets by glucans. Apart from this mechanism, another proposed mechanism of the *ndvB* gene in antibiotic resistance is by the activation of ethanol oxidation genes. From DNA microarray analysis, a higher expression of different ethanol oxidation genes was observed in wild-type *ndvB* biofilm cells, whereas this was not observed in the case of *ndvB*-mutated biofilms. Inactivation of these genes also increased biofilm tobramycin sensitivity (Beaudoin et al., 2012). Another gene *brlR*, which encodes the *MerR* family (of multidrug transport activators) transcriptional activator BrlR, was shown to have a role in the elevated drug resistance of *P. aeruginosa* biofilms. This gene's expression was found to be specific to biofilms, and its inactivation did not affect biofilm formation, motility, or *pslA* expression. However, *blrR* inactivation caused higher expression of the gene *ndvB* (Liao & Sauer, 2012).

Another gene, *tssC1*, in *P. aeruginosa* was demonstrated to be important in biofilm-specific resistance. This gene is associated with the type VI secretion T6S. Similar to that of *ndvB*, mutation of *tssC1* did not disrupt biofilm formation. They demonstrated that *tssC1* expression is induced in biofilms and confirmed it is necessary for T6S. The minimum bactericidal concentration (MBC) for biofilm cells of the drugs tobramycin, gentamicin, and ciprofloxacin was reduced by two- to fourfold following deletion of the *tssC1* gene, whereas, for planktonic cells, only a minor effect was observed. They proposed that an uncharacterized effector of the T6S contributes to a biofilm-specific resistance mechanism (Zhang et al., 2011). Another operon, PA0756–0757, which codes for a putative two-component regulatory system, and the genes PA2070 and PA5033, which are predicted to encode hypothetical proteins of unknown function, demonstrated higher expression in biofilm cells and also showcased biofilm-specific antibiotic resistance (Zhang et al., 2013).

The gene *rapA* is also predicted to have a role in biofilm-specific resistance in *E. coli.* Mutation in this gene showed no phenotypic difference in biofilm formation but was shown to have less resistance to penicillin G, whereas no resistance change was observed in planktonic cells. Down-regulation of 22 genes, including a gene called *yhcQ*, was observed in mutant biofilm by transcriptomic analysis, which codes for a putative multidrug resistance pump. Gene mutants of *yhcQ* also formed biofilms with lowered resistance; however, the effect was less significant than that of the *rapA* mutation (Lynch et al., 2007).

Differential gene expression in the *S. aureus* biofilm state was assessed by Resch et al. The *ica* genes code for proteins involved in the production of glucosamine polymer PIA. This polymer has been previously reported to have a role in biofilm formation and virulence. Therefore, changes in expression of *icaADBC* (SA2459 to SA2462) were studied along its growth cycle. The results showed that these genes were largely expressed under biofilm conditions—more than five-fold higher expression under biofilm than in planktonic conditions (Resch et al., 2005). A higher expression of succinate dehydrogenase genes (*sdhCABC*) was reported in *S. aureus* biofilms. Deletion of these genes showed little increase in resistance towards aminoglycosides (Gaupp et al., 2010). Methicillin resistance genes were also found upregulated in biofilms formed on different surfaces, including stainless steel (Crawford et al., 2016). Comprehensive information on various antibiotic resistance–related genes encoding different transporters, outer membrane proteins, and multidrug-resistant proteins, which were found up-regulated in *Salmonella* biofilms, was given in the review by Aleksandrowicz et al. (Aleksandrowicz et al., 2023).

2.5 Antimicrobial Resistomes in Drinking Water Distribution System Biofilms

Over time in water distribution systems, biofilm formation occurs and causes various functional, environmental, and health challenges. Although pipe corrosion and diminished water quality are the primary factors (Gholipour et al., 2023), our concern is the role of biofilms in protecting and transferring various antibiotic-resistant pathogenic microbes, ARGs, and MGEs to the environment, humans, and other animals. Chen et al. (2020) compared the biofilm and water (planktonic) microbiome with their resistome compositions from drinking water distribution systems. Unsurprisingly, a higher prevalence of ARGs and opportunistic pathogens was observed in biofilms than in water samples. They also identified a significant correlation between the relative abundance of the integrase gene *int*I1 and that of ARGs in opportunistic pathogens.

Studies reported that a biofilm's microbiome and resistome in water distribution systems for drinking water showed seasonal variations. During summer, a higher prevalence of multidrug resistance and bacitracin resistance genes were identified. In spring, Fosmidomycin-resistance-conferring genes predominated, whereas in autumn, polymycin and β-lactam resistance genes predominated. A higher abundance of ARG and associated MGE-possessing pathogenic bacteria (*Stenotrophomonas maltophilia*) was observed in summer (Ke et al., 2023). Resistome variations in biofilms found on water distribution systems were influenced by factors besides seasonal variations, including the consistency of the pipe material and natural organic matter on it. ARGs conferring resistance to bacitracin, fosmidomycin, and multidrugs were prevalent in both polyethylene and cast-iron pipe biofilms. Among mobile elements, plasmid was the predominant one coexisting with ARGs. *Achromobacter xylosoxidans* and *Ralstonia pickettii* were the most significant pathogens that harbored various ARGs such as *qacEΔ1, OXA-22*, and *aadA*, as well as MGEs such as integrase, plasmid, and transposase (Ke et al., 2024).

Contrasting to all other studies which analyzes total ARG content, Li and coworkers analyzed intracellular and extracellular ARG independently. Relative abundance of both the ARGs in biofilms of drinking water supply systems were found to be more than two-fold higher than that of the water samples (Li et al., 2023).

2.6 Antibiotic Resistome in Wastewater Biofilms and Impact on Surface Water Biofilms

Biofilms forms on pipes at wastewater treatment plants, implicating them as a major source of ARGs. The study conducted by Buelow et al. (2023) evaluated how hospital and urban wastewater biofilms contribute to the transmission of ARGs and MGEs, both *in vitro* and *in situ*. The research found that both wastewater biofilms exhibited different potential for disseminating ARGs through MGEs. Notably, the expression of the resistome and plasmid-borne resistance genes in hospital wastewater biofilms was significantly higher, underscoring their role in the spread of ARGs.

Mostly, rivers serve as a temporary sink of treated or untreated wastewater and leaching water. To no surprise, the microbial and biofilm communities in these rivers are shaped according to the input they receive. Many studies, mostly metagenomically based, have been conducted by researchers to evaluate the effect of wastewater inputs and compare the water samples from upstream and downstream discharge points. Most often in these studies, water served as samples, resulting in concentration mainly on the planktonic communities. However, very few studies have been conducted on biofilms and resistomes within biofilm. For instance, wastewater has been shown to influence and favor the ARG transmission in small river biofilms. To be precise, low ARG concentrations (consisting of principally *ermB* and *sul 1* genes) were observed in biofilm samples collected upstream from the wastewater treatment plant, whereas higher concentrations were observed in downstream samples. Noticeable concentrations of ARGs were observed at a distance of up to 1 km downstream, indicating the biofilm's capability as an ideal setting for ARG acquisition and transmission (Proia et al., 2016). Epilithic biofilms collected downstream from different wastewater plants had the influence of their discharge studied by evaluating the differences in biofilm composition and comparing with corresponding upstream biofilm. They specifically analyzed the concentrations of pharmaceuticals, and all downstream biofilm samples were found to have significant levels of these substances. Pertinent to our topic of concern, the amount of exposure to the discharge point increased the occurrence of resistance integrons, with increases ranging from 3 to 31 times the amount for Class 1 (Aubertheau et al., 2017).

Another advanced study based on metagenomics uncovered evidence indicating significant influence of wastewater effluents on both microbiome and resistomes in river biofilm samples. Here they employed a sophisticated model comparison technique to describe ARG and host relationships from unassembled metagenomic DNA. This method effectively mitigated the risks of false associations commonly occurring in correlation-based ARG-host analyses. They found specific bacterial orders to be strong candidates for hosting several families of ARGs, especially those that confer resistance to beta-lactams. Notably, strong associations were found of the genes *bla*FOX and *cph*A with *Aeromonadales*, as well as *bla*PER with *Chromatiales*. Moreover, for other ARG families such as *bla*IMP and *tet*, they found clusters of bacterial orders that potentially harbor a substantial portion of host species. However, for certain ARGs like *ant* and *erm*, no specific specific host candidates were identified, suggesting their broader dissemination across various taxonomic groups (Kneis et al., 2022). Estuary biofilms were also evaluated for the antibiotic resistome to analyze the effect of ARGs in the presence of antibiotics, metals, and total organic carbon. Higher abundances of ARGs were detected in biofilm samples than sediment and water samples. Furthermore, they found that ARG abundance was predominantly from the extracellular DNA (eDNA) in biofilm and sediment samples. A significant correlation between eDNA ARGS such as *sul*1, *sul*2, *tet*A, and *tet*W was found with total organic carbon in both samples (Guo et al., 2018).

Apart from disturbed freshwater sources, the biofilm resistome profile of a pristine environment was also studied. It is important to study the resistance profile of a pristine environment (areas with minimal human influence), because antimicrobial resistance arises from natural ecological interactions, such as competition among organisms within various ecological niches. One group of researchers evaluated the resistance determinants on epilithic biofilms from a glacier-fed stream and detected 29 distinct AMR-categories encoded by both bacteria and eukaryotes. Among these, higher enrichment was found for beta-lactam, aminoglycoside, and multidrug resistance. Taxonomical distribution of these genes was found in most of the bacterial and eukaryotic phyla (Busi et al., 2023). Similarly, the microbiome, resistome, and MGEs of river epilithic biofilms were analyzed using MinION long-read sequencing methods. They identified enrichment of pathogenic bacteria from *Enterobacteriaceae* and *Shigella flexneri* in locations upstream and downstream of the large city. Although different classes of ARGs in the biofilm resistome were detected, they stated that it is of less clinical relevance. *Escherichia*, *E. coli*, and *Muvirus* were found to be mostly associated with ARGs and MGEs. Moreover, they found *Pseudomonas* and *Cyanobacteria* as frequent hosts for MGEs, but not ARGs, indicating the potential role of these taxonomical groups in HGT within and around biofilms (Mills et al., 2024).

2.7 Plastisphere Biofilms and the 'Plastiome'

Plastics in various types and sizes (nano, micro, and macro form) are polluting the environment. These materials, therefore, serve as a substratum for microbes and algae to colonize and habituate, thereby creating an ecological habitat called the *plastisphere*. As stated, plastic surfaces can act as reservoirs for various microbes, including potential pathogens, promoting the development of a mobile resistome carrying diverse antibiotic and biocide/metal resistance genes (BMRGs/MGEs). This collective genetic repertoire, called the *plastiome*, can potentially disseminate antimicrobial resistance across various environments. The colonizing microbes form biofilm, or nonbiofilm formers become entrapped in the resulting biofilm, so the discussion of the plastisphere in this chapter focuses on biofilm communities on plastic surfaces.

Mostly, aquatic microplastics have been studied to evaluate the microbiome composition and resistome composition of plastisphere biofilm. For instance, recently Guruge et al. studied the plastisphere of rivers. The plastiome analysis revealed the role of microplastics as carriers of various drug-resistant pathogenic bacteria. Moreover, selective enrichment of ARGs, including beta-lactam and glycopeptide antibiotics, biocides/metals, and plasmid-associated MGEs, were observed (Guruge et al., 2024). Similarly, Zadjelovic et al. (2023) found that the plastisphere community showed similarity with a naturally co-occurring community, wood biofilm, highlighting the potential risk of resistance transfer. The potential opportunistic pathogens of biofilms included *P. aeruginosa*, *Acinetobacter*, and *Aeromonas*, and ARGs conferring resistance to sulfonamides, macrolides/lincosamides, and disinfectants predominated, whereas the planktonic samples were predominated with pathogens *Escherichia*, *Streptococcus*, *Klebsiella*, and *Salmonella*, as well as ARGs for aminoglycosides, tetracycline, fluoroquinolones, nitroimidazole, aminocoumarin, oxazolidinone, and fosfomycin.

Limited studies exist on the soil plastisphere. An enrichment of a variety of potential pathogens and ARGs was observed in the plastisphere of the corresponding soil sample by Zhu et al., 2022. The composition showed variations based on microplastic and soil types. They also found enhancement in plastisphere ARG from manure addition, elevated soil temperature, and moisture, whereas the enrichment of potential pathogens was observed with increased soil moisture. Enrichment of actinobacteria and proteobacteria was reported in the soil plastisphere. Enrichment of total ARG in the plastisphere was also detected in varying amounts across plastic types, with low-density polyethylene plastisphere having one with highest total ARGs in abundance. Enrichment of the pathogens *Acinetobacter johnsonii* and *E. coli* was shown by polypropylene plastisphere, highlighting higher pathogenic risk, and remarkably, the coexistence of multidrug resistance genes such as *ceoB* and *muxB* occurred in more than four microbial taxa, indicating the possible risk of ARG spread among pathogenic microbes (Yang et al., 2022).

Other than aquatic soil plastispheres, other plastisphere community environments were studied by researchers to evaluate ARG dissemination. For example, Li and coworkers evaluated the microbiome composition, pathogenic bacteria, and ARG composition in the activated sludge plastisphere. They observed enrichment of various potential ARG hosts and a stronger correlation with ARGs in plastisphere than activated sludge. Moreover, they studied the effect of tetracyclines and/or Cu (II) pressures on ARG abundance. There was an observed increase in ARG abundance, particularly in tetracycline resistance genes. The results further indicated that these resistance genes, which have different mechanisms of resistance, were separately enriched in both the plastisphere and activated sludge (Li et al., 2022). Biofilm formed on plastic wastes in an estuarine environment was evaluated for the presence of ARGs. In agreement with most of the studies discussed above, water and sediment samples in this one showed a lower abundance of ARGs than the biofilm samples. Among the biofilm samples from different plastic types, polyethylene biofilms showed a higher relative abundance of ARGs. The probable association of *Firmicutes* with ARG hosts was predicted as a reason for higher ARG abundance in polyethylene. Moreover, they found that metal content, phosphorus content, and salinity can act as regulating factors of ARGs in biofilms formed on plastics (Guo et al., 2020).

2.8 Metagenomic Approaches for Resistome Detection and Monitoring

For an efficient battle against antimicrobial resistance and its monitoring, different genes that confer resistance to antibiotics have to be understood, and their location should be identified (whether in chromosomal DNA or MGE). For this, entire microbiota should be investigated. However, for the identification of ARG in uncultivable organisms, metagenomic approaches need to be utilized. The rise of the next generation of sequencing technology has revolutionized metagenomic studies, making them an essential tool for exploring the complexities of microbial communities. It not only enhanced knowledge on microbial communities, but also helped in quantifying, tracking, and assessing antimicrobial resistance. Three metagenomic-based approaches to investigating the resistome include (1) targeted metagenomics, which is based on PCR and/or microarray, (2) sequence-based metagenomics, and (3) functional metagenomics (Fig. 2.2).

PCR primers are designed to amplify the resistance gene of interest in targeted (PCR-based) metagenomics and can be used for tracking its presence within and between different ecosystems. A number of studies have been conducted using the real-time polymerase chain reaction (qPCR) method for the detection and prevalence of ARGs in various environments. However, the limitation is that it is only possible to track known resistance genes. Although microarrays help in the

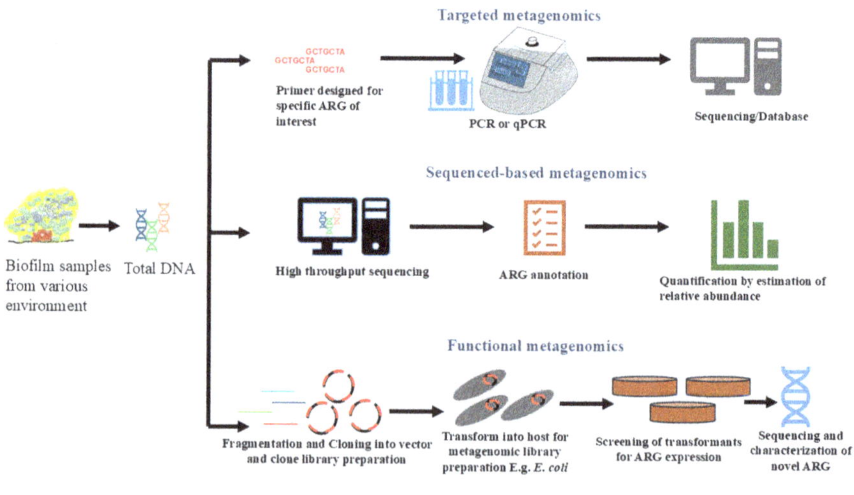

Fig. 2.2 Diagrammatic representation of several types of metagenomic techniques for the detection and monitoring of ARGs from biofilms of various environmental samples. Major steps involved in each technique are depicted in the figure

identification of all known resistance genes, PCR-based methods cannot account for all that exist. Other limitations include higher variation of DNA templates and complication of samples from environment. Also, due to possible nonspecific amplification, the target genes may be overestimated or could result in false-positive findings. Moreover, both array and PCR-based approaches are not useful for obtaining information regarding the microbial host of the ARGs in metagenomes (Mullany, 2014; Li & Yan, 2021).

As mentioned earlier, high throughput sequencing has made sequenced-based metagenomics practical. Here, complete sequences are obtained from DNA extracted from various environments. Different public databases can be utilized to analyze and detect ARGs from the sequences obtained. This method will help to overcome the limitation of tracking unknown resistance genes in a metagenome. Still, it is possible to detect resistance genes that have been identified by either functional metagenomics or experimental results on cultivable antibiotic-resistant bacteria. Moreover, it will not give any information on gene expression. So, continuous reassessment of the sequenced based metagenomic studies should be done for updating functional information.

Functional metagenomics does not necessitate any previous information on resistance genes. With this technique, genes from metagenomic clones are expressed in surrogate hosts. Briefly, the DNA sample extracted from the specific environment is prepared and shotgun cloned into cloning vectors to create metagenomic libraries. Usually, *E. coli* is used as host for cloning. After library preparation, the clones are screened for resistance by plating onto antibiotic-containing agar. The antibiotic dose will be high enough to kill the host unless it carries a resistance gene. After that, a nucleotide sequence determination of the inserted gene will be performed. Currently, functional metagenomics is the efficient method capable of isolating completely novel ARG. However, the limitation is that expression of the ARG is needed. Apart from ARGs, the functional metagenomics can also be exploited to isolate MGEs including simple transposons, conjugative transposons, plasmids, and bacteriophages (Mullany, 2014). Various metagenomics-based studies for the detection and occurrence of ARGs in biofilms sampled from various sources are explained in Table 2.1.

2.9 Databases for Detection of ARGs

In order to analyze a large set of data from metagenomic analysis, a suitable analysis method and reference database is required. Specifically, for ARG detection and analysis, researchers have created various databases and analysis pipelines. These include the Comprehensive Antibiotic Resistance Database (CARD), ResFinder, and ARG-ANNOT (Table 2.2). CARD is a widely used bioinformatic database for resistance genes, their products, and associated phenotypes. Its Resistance Gene Identifier software predicts the resistome based on homology and SNP models (Alcock et al., 2023). ResFinder is another online database for ARG identification from the input of next-generation sequencing data. The database is improved

Table 2.1 Recent studies of the antibiotic resistome in biofilms sampled from various environments

Sample Source	Objective of Study	References
River biofilm	Shotgun metagenomics for tracking ARG and taxonomic markers along a gradient of fecal pollution represented by crAssphage signatures	Kneis et al. (2022)
Drinking water supply system biofilm	For the characterization of biofilm activity, microbial community, ARGs, MGEs, and pathogenic ARG hosts	Ke et al. (2024)
Oral biofilm	shotgun metagenomics combined with culture technique for examining the resistome and phenotypic resistance	Anderson et al. (2023)
River biofilm	Influence of sub-MIC antibiotic concentrations on microbiome and resistome by microscopy and metagenomics	Flores-Vargas et al. (2023)
Sewer biofilm	ARG abundance, 16S rRNA gene quantification with qPCR	Medina et al. (2020)
Drinking water distribution system biofilm	Evaluation of the seasonal variation in microbiomes, ARGs, coexistence of MGEs with ARGs, pathogenic ARG hosts, and the driving elements of ARG in biofilms	Ke et al. (2023)
Hospital water distribution system	Analysis of ARGs and integrase 1 gene (*int*1) gene presence in biofilm and water samples using PCR and qPCR techniques	Gholipour et al. (2023)
Riverine plastisphere biofilm	Metagenomic analysis for taxonomic characterization, examination of the ARGs and virulence factors of the microbial community	Zadjelovic et al. (2023)
Esturay biofilm	ARG occurrence abundance ARG in intracellular or extracellular DNA	Guo et al. (2018)
Plastisphere biofilm in river	Metagenomic and metatranscriptomic analysis for Presence and relative abundance of ARG, Transcriptional level analysis ARGs, plasmid associated ARGs, microbial composition and bacterial hosts of ARG	Wu et al. (2022)
Microplastic pastisphere in sludge anerobic digestion	Metagenomic analysis of ARGs and MGEs	Li and Yuan (2024)

Most of the studies followed sequenced-based metagenomics to characterize the microbiome composition, ARG composition, and associated MGEs

continuously and now includes point mutations and phenotype prediction for some selected bacterial species. ARG-ANNOT (Antibiotic Resistance Gene-ANNOTation) is a tool that helps in the detection of both existing and putative antibiotic resistance genes in bacteria. They gather resistance determinants from published studies and online resources, while nucleotide and protein sequences are obtained from the NCBI GenBank database (Gupta et al., 2014).

Apart from ARG detection software or databases, a distinctive but needed database (LCPDb-ARG) was developed by Gorecki et al. This is a literature-based database of PCR primers for the detection of ARGs. The database clarifies the question, "Which is the best primer for detection of a particular ARG?" Based on specificity, efficacy, and taxonomic efficacy values generated, we can select the optimal primer pair for the detection of target ARGs by PCR (Gorecki et al., 2019).

Table 2.2 Different bioinformatic tools developed by various researchers for the identification, annotation, classification, and mobility prediction of antimicrobial resistance genes

Name of the Database/Tool/ Software	Features	Input	References
DeepARG	Identification of ARG Annotation of ARG	Nucleotide and Protein sequences	Arango-Argoty et al. (2018)
HMD-ARG	Identification of ARG Antibiotic family the ARG resistant Resistant mechanism Prediction of intrinsic and acquired nature of ARG	Protein sequences	Li and Yan (2021)
ARGNet	Identification of ARG Annotation of ARG	Protein and nucleotide sequences (short and long)	Pei et al. (2024)
PLM-ARG	Identification of ARG Classification of ARG	Protein sequences	Wu et al. (2023)
ProtAlign-ARG	Identification of ARG Classification of ARG	DNA sequencing data	Ahmed et al. (2024)
TGC-ARG	Prediction of ARGs	Protein sequences	Dong et al. (2024)
NanoARG	Identification of ARGs Co-occurrence of ARGs, MGEs and metal resistance genes Pathogenicity	Long reads from nanopore sequencing data	Arango-Argoty et al. (2019)
GNN-ARG	Prediction of ARGs	Protein sequences	Abbas et al. (2024)
AMR-Meta	Classification of ARG	Short reads from high throughput sequencing	Marini et al. (2022)
ARG-SHINE	Classification of ARG	Protein sequences	Wang et al. (2021)

Machine learning can be applied to predict and develop novel ARG detection databases, tools, and software. Deep learning, owing to its intrinsic superiority in feature extraction from raw data, can be exploited to study the statistical patterns of ARGs (Li et al., 2021). For instance, a group of researchers developed two deep learning models to predict ARGs from short and full-length sequences. This database, called DeepARG, is meticulously curated and filled with ARGs that have been identified with a high level of confidence, making it a valuable resource for further research and applications (Arango-Argoty et al., 2019). Its limitations, including dependency on alignment-based methods (which will not provide any information on putative resistance genes) and inability to predict the mobility nature of ARG, are taken into account for developing new tools by Li et al. Their tool, HMD–ARG, based on deep learning, can identify multiple properties, including whether the input sequence encodes for an ARG. It can also identify the antibiotic family it is resistant to, the resistance mechanism of that ARG, and the nature of ARG mobility—whether it is acquired or intrinsic. Beyond this, it's able to predict the subclassification of beta-lactam resistance (Li et al., 2021). Recently, another deep learning-based too, named ARGNet, was developed. This method also helps to detect known and new ARGs. Both amino acid and nucleotide sequences of different lengths can serve as input here. This enables detection of ARGs from target sequencing and metagenomic sequencing. They claimed greater efficiency for their tool than for the previous deep learning–based tools DeepARG and HMD-ARG (Pei et al., 2024).

2.10 Conclusion

Biofilms, or structured microbial communities, account for enhanced antimicrobial resistance. This resistance is not due to single factors. Multiple factors contribute to it, including less penetration of antibiotics, enhanced horizontal gene transfer, antibiotic resistance genes, biofilm-specific antibiotic resistance genes, quorum sensing, and slow growth rate. Antibiotic resistome detection, the monitoring and tracking of planktonic communities in various environments, is well studied, while exploration of antimicrobial resistomes in biofilm communities is in its beginning stage. Developed sequencing technologies and metagenomic approaches have helped researchers in this field to study microbiome composition in biofilm, ARG composition, and mobile genetic elements. Researchers have also developed various bioinformatics tools to analyze large data from metagenomic studies.

References

Abbas, M. M., Ranjan, A., Mukhopadhyay, S., & Hou, A. (2024). GNN-ARG: A graph neural network-based framework for predicting antibiotic resistance genes. In *2024 IEEE international conference on big data (BigData)* (pp. 4283–4291). IEEE.

Afonso, A. C., Oliveira, D., Saavedra, M. J., Borges, A., & Simões, M. (2021). Biofilms in diabetic foot ulcers: Impact, risk factors and control strategies. *International Journal of Molecular Sciences, 22*(15), 8278.

Ahmed, S., Emon, M. I., Moumi, N. A., Huang, L., Zhou, D., Vikesland, P., Pruden, A., & Zhang, L. (2024). ProtAlign-ARG: Antibiotic resistance gene characterization integrating protein language models and alignment-based scoring. *bioRxiv*, 2024-03.

Alcock, B. P., Huynh, W., Chalil, R., Smith, K. W., Raphenya, A. R., Wlodarski, M. A., Edalatmand, A., Petkau, A., Syed, S. A., Tsang, K. K., & Baker, S. J. (2023). CARD 2023: Expanded curation, support for machine learning, and resistome prediction at the comprehensive antibiotic resistance database. *Nucleic Acids Research, 51*(D1), D690–D699.

Aleksandrowicz, A., Carolak, E., Dutkiewicz, A., Błachut, A., Waszczuk, W., & Grzymajlo, K. (2023). Better together–salmonella biofilm-associated antibiotic resistance. *Gut Microbes, 15*(1), 2229937.

Anderson, A. C., von Ohle, C., Frese, C., Boutin, S., Bridson, C., Schoilew, K., Peikert, S. A., Hellwig, E., Pelz, K., Wittmer, A., & Wolff, D. (2023). The oral microbiota is a reservoir for antimicrobial resistance: Resistome and phenotypic resistance characteristics of oral biofilm in health, caries, and periodontitis. *Annals of Clinical Microbiology and Antimicrobials, 22*(1), 37.

Antonova, E. S., & Hammer, B. K. (2011). Quorum-sensing autoinducer molecules produced by members of a multispecies biofilm promote horizontal gene transfer to vibrio cholerae. *FEMS Microbiology Letters, 322*(1), 68–76.

Arango-Argoty, G., Garner, E., Pruden, A., Heath, L. S., Vikesland, P., & Zhang, L. (2018). DeepARG: A deep learning approach for predicting antibiotic resistance genes from metagenomic data. *Microbiome, 6*, 1–15.

Arango-Argoty, G. A., Dai, D., Pruden, A., Vikesland, P., Heath, L. S., & Zhang, L. (2019). NanoARG: A web service for detecting and contextualizing antimicrobial resistance genes from nanopore-derived metagenomes. *Microbiome, 7*, 1–18.

Aubertheau, E., Stalder, T., Mondamert, L., Ploy, M. C., Dagot, C., & Labanowski, J. (2017). Impact of wastewater treatment plant discharge on the contamination of river biofilms by pharmaceuticals and antibiotic resistance. *Science of the Total Environment, 579*, 1387–1398.

Beaudoin, T., Zhang, L., Hinz, A. J., Parr, C. J., & Mah, T. F. (2012). The biofilm-specific antibiotic resistance gene ndvB is important for expression of ethanol oxidation genes in Pseudomonas aeruginosa biofilms. *Journal of Bacteriology, 194*(12), 3128–3136.

Bessa, L. J., Egas, C., Botelho, J., Machado, V., Alcoforado, G., Mendes, J. J., & Alves, R. (2025). Unveiling the Resistome landscape in Peri-implant health and disease. *Journal of Clinical Medicine, 14*(3), 931.

Buelow, E., Dauga, C., Carrion, C., Mathé-Hubert, H., Achaibou, S., Gaschet, M., Jové, T., Chesneau, O., Kennedy, S. P., Ploy, M. C., & Da Re, S. (2023). Hospital and urban wastewaters shape the matrix and active resistome of environmental biofilms. *Water Research, 244*, 120408.

Busi, S. B., de Nies, L., Pramateftaki, P., Bourquin, M., Kohler, T. J., Ezzat, L., Fodelianakis, S., Michoud, G., Peter, H., Styllas, M., & Tolosano, M. (2023). Glacier-fed stream biofilms harbor diverse resistomes and biosynthetic gene clusters. *Microbiology Spectrum, 11*(1), e04069–e04022.

Chen, J., Li, W., Zhang, J., Qi, W., Li, Y., Chen, S., & Zhou, W. (2020). Prevalence of antibiotic resistance genes in drinking water and biofilms: The correlation with the microbial community and opportunistic pathogens. *Chemosphere, 259*, 127483.

Crawford, E. C., Singh, A., Gibson, T. W., & Weese, J. S. (2016). Biofilm-associated gene expression in Staphylococcus pseudintermedius on a variety of implant materials. *Veterinary Surgery, 45*(4), 499–506.

Dong, Y., Quan, H., Ma, C., Shan, L., & Deng, L. (2024). TGC-ARG: Anticipating antibiotic resistance via transformer-based modeling and contrastive learning. *International Journal of Molecular Sciences, 25*(13), 7228.

Element, S. J., Moran, R. A., Beattie, E., Hall, R. J., van Schaik, W., & Buckner, M. M. (2023). Growth in a biofilm promotes conjugation of a bla NDM-1-bearing plasmid between Klebsiella pneumoniae strains. *Msphere, 8*(4), e00170–e00123.

Flores-Vargas, G., Korber, D. R., & Bergsveinson, J. (2023). Sub-MIC antibiotics influence the microbiome, resistome and structure of riverine biofilm communities. *Frontiers in Microbiology, 14*, 1194952.

Gaupp, R., Schlag, S., Liebeke, M., Lalk, M., & Götz, F. (2010). Advantage of upregulation of succinate dehydrogenase in Staphylococcus aureus biofilms. *Journal of Bacteriology, 192*(9), 2385–2394.

Gholipour, S., Shamsizadeh, Z., Gwenzi, W., & Nikaeen, M. (2023). The bacterial biofilm resistome in drinking water distribution systems: A systematic review. *Chemosphere, 329*, 138642.

Gorecki, A., Decewicz, P., Dziurzynski, M., Janeczko, A., Drewniak, L., & Dziewit, L. (2019). Literature-based, manually-curated database of PCR primers for the detection of antibiotic resistance genes in various environments. *Water Research, 161*, 211–221.

Guo, X. P., Yang, Y., Lu, D. P., Niu, Z. S., Feng, J. N., Chen, Y. R., Tou, F. Y., Garner, E., Xu, J., Liu, M., & Hochella, M. F., Jr. (2018). Biofilms as a sink for antibiotic resistance genes (ARGs) in the Yangtze Estuary. *Water Research, 129*, 277–286.

Guo, X. P., Sun, X. L., Chen, Y. R., Hou, L., Liu, M., & Yang, Y. (2020). Antibiotic resistance genes in biofilms on plastic wastes in an estuarine environment. *Science of the Total Environment, 745*, 140916.

Gupta, S. K., Padmanabhan, B. R., Diene, S. M., Lopez-Rojas, R., Kempf, M., Landraud, L., & Rolain, J. M. (2014). ARGANNOT, a new bioinformatic tool to discover antibiotic resistance genes in bacterial genomes. *Antimicrobial agents and chemotherapy, 58*(1), 212–220.

Guruge, K. S., Goswami, P., Kanda, K., Abeynayaka, A., Kumagai, M., Watanabe, M., & Tamamura-Andoh, Y. (2024). Plastiome: Plastisphere-enriched mobile resistome in aquatic environments. *Journal of Hazardous Materials, 471*, 134353.

Johnston, E. L., Zavan, L., Bitto, N. J., Petrovski, S., Hill, A. F., & Kaparakis-Liaskos, M. (2023). Planktonic and biofilm-derived Pseudomonas aeruginosa outer membrane vesicles facilitate horizontal gene transfer of plasmid DNA. *Microbiology Spectrum, 11*(2), e05179–e05122.

Ke, Y., Sun, W., Chen, X., Zhu, Y., Guo, X., Yan, W., & Xie, S. (2023). Seasonality determines the variations of biofilm microbiome and antibiotic resistome in a pilot-scale chlorinated drinking water distribution system deciphered by metagenome assembly. *Environmental Science & Technology, 57*(31), 11430–11441.

Ke, Y., Sun, W., Xue, Y., Yuan, Z., Zhu, Y., Chen, X., Yan, S., Li, Y., & Xie, S. (2024). Pipe material and natural organic matter impact drinking water biofilm microbial community, pathogen profiles and antibiotic resistome deciphered by metagenomics assembly. *Environmental Research, 262*, 119964.

Kim, S. M., Kim, H. C. & Lee, S. W. S. (2011). Characterization of antibiotic resistance determinants in oral biofilms. *The Journal of Microbiology, 49*, 595–602.

Kneis, D., Berendonk, T. U., Forslund, S. K., & Hess, S. (2022). Antibiotic resistance genes in river biofilms: A metagenomic approach toward the identification of sources and candidate hosts. *Environmental Science & Technology, 56*(21), 14913–14922.

Lécuyer, F., Bourassa, J. S., Gélinas, M., Charron-Lamoureux, V., Burrus, V., & Beauregard, P. B. (2018). Biofilm formation drives transfer of the conjugative element ICE Bs1 in Bacillus subtilis. *Msphere, 3*(5), 10–1128.

Li, B., & Yan, T. (2021). Next generation sequencing reveals limitation of qPCR methods in quantifying emerging antibiotic resistance genes (ARGs) in the environment. *Applied Microbiology and Biotechnology, 105*, 2925–2936.

Li, Z., & Yuan, D. (2024). Metagenomic analysis reveals the effects of microplastics on antibiotic resistance genes in sludge anaerobic digestion. *Toxics, 12*(12), 920.

Li, Y., Xu, Z., Han, W., Cao, H., Umarov, R., Yan, A., Fan, M., Chen, H., Duarte, C. M., Li, L., & Ho, P. L. (2021). HMD-ARG: Hierarchical multi-task deep learning for annotating antibiotic resistance genes. *Microbiome, 9*, 1–12.

Li, Q., Tian, L., Cai, X., Wang, Y., & Mao, Y. (2022). Plastisphere showing unique microbiome and resistome different from activated sludge. *Science of the Total Environment, 851*, 158330.

Li, S., Niu, Z., Wang, M., & Zhang, Y. (2023). The occurrence and variations of extracellular antibiotic resistance genes in drinking water supply system: A potential risk to our health. *Journal of Cleaner Production, 402*, 136714.

Liao, J., & Sauer, K. (2012). The MerR-like transcriptional regulator BrlR contributes to Pseudomonas aeruginosa biofilm tolerance. *Journal of Bacteriology, 194*(18), 4823–4836.

Liu, H. Y., Prentice, E. L., & Webber, M. A. (2024). Mechanisms of antimicrobial resistance in biofilms. *npj Antimicrobials and Resistance, 2*(1), 27.

Lynch, S. V., Dixon, L., Benoit, M. R., Brodie, E. L., Keyhan, M., Hu, P., Ackerley, D. F., Andersen, G. L., & Matin, A. (2007). Role of the rapA gene in controlling antibiotic resistance of Escherichia coli biofilms. *Antimicrobial Agents and Chemotherapy, 51*(10), 3650–3658.

Maree, M., Thi Nguyen, L. T., Ohniwa, R. L., Higashide, M., Msadek, T., & Morikawa, K. (2022). Natural transformation allows transfer of SCC mec-mediated methicillin resistance in Staphylococcus aureus biofilms. *Nature Communications, 13*(1), 2477.

Marini, S., Oliva, M., Slizovskiy, I. B., Das, R. A., Noyes, N. R., Kahveci, T., Boucher, C., & Prosperi, M. (2022). AMR-meta: a k-mer and metafeature approach to classify antimicrobial resistance from high-throughput short-read metagenomics data. *GigaScience, 11*, giac029.

Medina, W. R. M., Eramo, A., Tu, M., & Fahrenfeld, N. L. (2020). Sewer biofilm microbiome and antibiotic resistance genes as function of pipe material, source of microbes, and disinfection: Field and laboratory studies. *Environmental Science: Water Research & Technology, 6*(8), 2122–2137.

Michaelis, C., & Grohmann, E. (2023). Horizontal gene transfer of antibiotic resistance genes in biofilms. *Antibiotics, 12*(2), 328.

Mills, M., Wittum, T., & Lee, J. (2024). Dynamic microbiome and mobile resistome are revealed in river biofilms from a multi-use watershed through long-read sequencing. *Frontiers in Environmental Science, 12*, 1440635.

Mirzaei, A., Nasr Esfahani, B., Raz, A., Ghanadian, M., & Moghim, S. (2021). From the urinary catheter to the prevalence of three classes of integrons, β-lactamase genes, and differences in antimicrobial susceptibility of Proteus mirabilis and clonal relatedness with Rep-PCR. *BioMed Research International, 2021*(1), 9952769.

Mullany, P. (2014). Functional metagenomics for the investigation of antibiotic resistance. *Virulence, 5*(3), 443–447.

Nasu, T., & Maeda, S. (2024). Escherichia coli persisters in biofilm can perform horizontal gene transfer by transformation. *Biochemical and Biophysical Research Communications, 738*, 150549.

Owais, D., Al-Groom, R. M., AlRamadneh, T. N., Alsawalha, L., Khan, M. S. A., Yousef, O. H., & Burjaq, S. Z. (2024). Antibiotic susceptibility and biofilm forming ability of Staphylococcus aureus isolated from Jordanian patients with diabetic foot ulcer. *Iranian Journal of Microbiology, 16*(4), 450.

Pei, Y., Shum, M. H. H., Liao, Y., Leung, V. W., Gong, Y. N., Smith, D. K., Yin, X., Guan, Y., Luo, R., Zhang, T., & Lam, T. T. Y. (2024). ARGNet: Using deep neural networks for robust identification and classification of antibiotic resistance genes from sequences. *Microbiome, 12*(1), 84.

Perry, J. A., Westman, E. L., & Wright, G. D. (2014). The antibiotic resistome: What's new? *Current Opinion in Microbiology, 21*, 45–50.

Proia, L., von Schiller, D., Sànchez-Melsió, A., Sabater, S., Borrego, C. M., Rodríguez-Mozaz, S., & Balcázar, J. L. (2016). Occurrence and persistence of antibiotic resistance genes in river biofilms after wastewater inputs in small rivers. *Environmental Pollution, 210*, 121–128.

Resch, A., Rosenstein, R., Nerz, C., & Götz, F. (2005). Differential gene expression profiling of Staphylococcus aureus cultivated under biofilm and planktonic conditions. *Applied and Environmental Microbiology, 71*(5), 2663–2676.

Saffari, M., Karami, S., Firoozeh, F., & Sehat, M. (2017). Evaluation of biofilm-specific antimicrobial resistance genes in Pseudomonas aeruginosa isolates in Farabi Hospital. *Journal of Medical Microbiology, 66*(7), 905–909.

Sedlacek, M. J., & Walker, C. (2007). Antibiotic resistance in an in vitro subgingival biofilm model. *Oral Microbiology and Immunology, 22*(5), 333–339.

Shree, P., Singh, C. K., Sodhi, K. K., Surya, J. N., & Singh, D. K. (2023). Biofilms: Understanding the structure and contribution towards bacterial resistance in antibiotics. *Medicine in Microecology, 16*, 100084.

Singhai, M., Malik, A., Shahid, M., Malik, M. A., & Goyal, R. (2012). A study on device-related infections with special reference to biofilm production and antibiotic resistance. *Journal of Global Infectious Diseases, 4*(4), 193–198.

Sivori, F., Cavallo, I., Truglio, M., Pelagalli, L., Mariani, V., Fabrizio, G., Abril, E., Santino, I., Fradiani, P. A., Solmone, M., & Pimpinelli, F. (2024). Biofilm-mediated antibiotic tolerance in Staphylococcus aureus from spinal cord stimulation device-related infections. *Microbiology Spectrum, 12*(12), e01683–e01624.

Wang, Z., Li, S., You, R., Zhu, S., Zhou, X. J., & Sun, F. (2021). ARG-SHINE: Improve antibiotic resistance class prediction by integrating sequence homology, functional information and deep convolutional neural network. *NAR Genomics and Bioinformatics, 3*(3), lqab066.

Wu, X., Liu, Z., Li, M., Bartlam, M., & Wang, Y. (2022). Integrated metagenomic and metatranscriptomic analysis reveals actively expressed antibiotic resistomes in the plastisphere. *Journal of Hazardous Materials, 430*, 128418.

Wu, J., Ouyang, J., Qin, H., Zhou, J., Roberts, R., Siam, R., Wang, L., Tong, W., Liu, Z., & Shi, T. (2023). PLM-ARG: Antibiotic resistance gene identification using a pretrained protein language model. *Bioinformatics, 39*(11), btad690.

Yang, Y., Li, T., Liu, P., Li, H., & Hu, F. (2022). The formation of specific bacterial communities contributes to the enrichment of antibiotic resistance genes in the soil plastisphere. *Journal of Hazardous Materials, 436*, 129247.

Zadjelovic, V., Wright, R. J., Borsetto, C., Quartey, J., Cairns, T. N., Langille, M. G., Wellington, E. M., & Christie-Oleza, J. A. (2023). Microbial hitchhikers harbouring antimicrobial-resistance genes in the riverine plastisphere. *Microbiome, 11*(1), 225.

Zhang, L., & Mah, T. F. (2008). Involvement of a novel efflux system in biofilm-specific resistance to antibiotics. *Journal of Bacteriology, 190*(13), 4447–4452.

Zhang, L., Hinz, A. J., Nadeau, J. P., & Mah, T. F. (2011). Pseudomonas aeruginosa tssC1 links type VI secretion and biofilm-specific antibiotic resistance. *Journal of Bacteriology, 193*(19), 5510–5513.

Zhang, L., Fritsch, M., Hammond, L., Landreville, R., Slatculescu, C., Colavita, A., & Mah, T. F. (2013). Identification of genes involved in Pseudomonas aeruginosa biofilm-specific resistance to antibiotics. *PLoS One, 8*(4), e61625.

Zhu, L., Chen, T., Xu, L., Zhou, Z., Feng, W., Liu, Y., & Chen, H. (2020). Effect and mechanism of quorum sensing on horizontal transfer of multidrug plasmid RP4 in BAC biofilm. *Science of the Total Environment, 698*, 134236.

Zhu, D., Ma, J., Li, G., Rillig, M. C., & Zhu, Y. G. (2022). Soil plastispheres as hotspots of antibiotic resistance genes and potential pathogens. *The ISME Journal, 16*(2), 521–532.

3 Fundamentals and Applications of Omics in Microbiology

Zarin Taj, Sochanphy Keishing, and Indranil Chattopadhyay

3.1 Introduction

"Omics" refers to analysing and exploring extensive datasets that capture the structure and function of an entire biological system at a specific level. This approach has revolutionised how we study and understand biological systems (Dai & Shen, 2022). The suffix "omics" comes from the word "ome", which is a Greek word meaning "all" or "every". From a biological standpoint, omics refers to molecular systems' comprehensive or universal description (Darshna & Tiwari, 2023).

The development of omics technologies has made conducting in-depth quantitative and qualitative analyses of cellular systems feasible. Omics studies generate accurate data that promote understanding and can fully decode an entire organism, including insights into metabolites and genes (Priyadarsini, 2023). Following the central dogma, researchers employ omics technologies to record post-translational modifications (PTMs), RNA splicing mechanisms, temporal and spatial protein regulation, temporal transcriptomic perturbations, and stable genetic changes (Dai & Shen, 2022). Additionally, omics technologies have advanced the approaches and tools used to manage complex, large-scale datasets (Kinaret et al., 2020). Furthermore, scientists have advanced omics technologies to study multiple epi-level molecular landscapes, which account for all modifications of these omics beyond what is captured inside a solitary cell, multilayered biological networks, and disease-related features such as the metabolic profile and immune system profile (Dai & Shen, 2022).

Z. Taj · S. Keishing · I. Chattopadhyay (✉)
Department of Biotechnology, School of Integrative Biology, Central University of Tamil Nadu, Neelakudi, Thiruvarur, Tamil Nadu, India
e-mail: indranil@cutn.ac.in

S. Busi et al. (eds.), *Omics Approaches in Biofilm Research*,
https://doi.org/10.1007/978-3-031-91863-6_3

3.1.1 Genomics

The study of genomics focuses on the DNA code found in the cells of viruses, bacteria, and fungi as well as in the cytoplasm of plant and animal cells. It can provide insights into an organism's disease resistance or susceptibility, its evolutionary history, and its relative relatedness to others. Additionally, it can reveal details about an organism's genome size and complexity and how two organisms are both similar and different (Darshna & Tiwari, 2023). Genomic techniques explore variations between individuals at germline and somatic levels by target genome sequencing. Advances such as DNA microarray, Sanger sequencing, next-generation sequencing (NGS), and Nanopore sequencing have enabled comprehensive genome or exome sequencing. These developments provide precise insights into the mutational environment of the sample (Dai & Shen, 2022.). The increased advancements in sequencing speed, data quality, and affordability have made genome sequencing a significant branch (Ghosh et al., 2022). The procedures used in genomic research are (1) DNA isolation, (2) sequencing, and (3) data analysis (Darshna & Tiwari, 2023).

3.1.2 Transcriptomics

Unlike DNA's high stability, microorganisms' RNA is dynamically changing. Therefore, transcriptomics aims to identify genes differentially expressed under specific conditions at any given time (Yuan et al., 2023a, b). It studies an organism's transcriptome, encompassing its RNA transcripts. The information stored in an organism's genome is expressed through transcription, following the principles of the central dogma (Darshna & Tiwari, 2023). The transcriptome, in contrast to the genome, is ever-changing and consists of diverse components. It undergoes changes influenced by the changes in cell development, any stimuli, and signal measurement timing. The traditional transcriptome primarily focuses on mRNA transcripts but also includes transcripts such as regulatory small RNAs (Dai & Shen, 2022), functional non-coding RNAs, and circularised RNAs (Chen et al., 2023). Transcriptomic techniques aim to characterise and quantify the actively transcribed RNA pool at a specific moment (Dai & Shen, 2022.) From this viewpoint, mRNA functions as a transient intermediary, while non-coding RNAs play a variety of other roles (Darshna & Tiwari, 2023). RNA sequencing (RNA-Seq) is a broadly adopted method in transcriptome research, enabling the identification and quantification of transcribed RNA sequences even from limited sample sizes. With technological advancements, single-cell transcriptomics is gaining prominence (Chen et al., 2023). The procedures used in transcriptomic research are (1) RNA isolation, (2) transcriptome reconstruction, (3) expression quantification, and (4) differential expression analysis (Darshna & Tiwari, 2023).

3.1.3 Proteomics

Proteomics explores all the proteins expressed by a genome, and proteomes refer to all the proteins an organism expresses. Unlike the genome, the proteome is dynamic, not static. A single gene can produce multiple proteins through alternative splicing, contributing to a variable and versatile proteome (Darshna & Tiwari, 2023). Proteomics examines the levels of protein expression, the effects of co-translational modifications and PTMs on protein functionality and phenotype, the regulation of enzymatic activity, and the impact of protein–protein interactions on an organism's health or disease state (Dai & Shen, 2022). This field enables comprehensive characterisation and measurement of proteins within cells (Chen et al., 2023). It explores the functions of every protein expressed in a cell, tissue, or organism by analysing information flow through protein signalling pathways (Dai & Shen, 2022). Techniques such as mass spectrometry (MS), affinity proteomics, protein chips, and reverse phase protein microarrays facilitate large-scale protein studies. In addition, researchers can identify proteins and their interactions by combining immunoprecipitation with mass spectrometry (Chen et al., 2023). The procedures used in proteomic research are (1) protein isolation, (2) differential separation, (3) protein detection, and (4) characterisation (Darshna & Tiwari, 2023).

3.1.4 Metabolomics

Metabolomics examines the small molecules produced by cellular biochemical processes. A meaningful number of metabolites having low molecular weight (<1500 Da) play important roles in the internal metabolism of cells. These molecules serve as energy sources, signalling molecules, and metabolic intermediates while simultaneously modulating numerous proteins within detailed biological systems (Dai & Shen, 2022). Several of these metabolites comprise small molecule substrates, intermediate molecules, and a large number of metabolic end products, for example, carbohydrates, fatty acids, and amino acids (Chen et al., 2023). Furthermore, a significant number of metabolites exert their influence on several cellular functions by binding to a large proportion of macromolecules, thereby easing signalling pathways, regulatory mechanisms, and other cellular processes (Wang et al., 2021). Several metabolites function as biological modulators across multiple layers of omics data, thereby supporting the maintenance of cellular homeostasis. The metabolome comprises the entire collection of small molecule metabolites within a single cell. This includes all biomolecules, except genome components, the transcriptome, the proteome, and metals (Dai & Shen, 2022).

Several methods exist for metabolomic analysis. These are Fourier transform infrared (FT-IR) spectroscopy, Raman spectroscopy, and nuclear magnetic resonance (NMR) spectroscopy as well as mass spectrometry (MS)-based approaches such as tandem mass spectrometry (MS/MS), liquid chromatography–mass spectrometry (LC/MS), and gas chromatography–mass spectrometry (GC/MS), amongst others (Dai & Shen, 2022). Metabolite profiling can rapidly indicate dynamic alterations in cellular physiology (Chen et al., 2023). The procedures used in metabolomic research are (1) metabolite isolation, (2) separation, and (3) detection (Darshna & Tiwari, 2023).

3.1.5 Lipidomics

Lipids are fundamental building blocks of microbial life and are extensively distributed across various environmental niches (Bale et al., 2021). They are widely recognised as fat molecules, commonly found in cell membranes, and are linked to proteins such as lipoproteins. Lipids are hydrophobic molecules composed of fatty acids with long carbon chains. The entire set of lipids present in a biological entity is referred to as its lipidome. Scientists also consider it a subset of the metabolome (Darshna & Tiwari, 2023). Lipidomics is a rapidly advancing "omics" field that offers valuable insights into lipid pools related to structure, energy source/storage, and signal transduction (Wood & Erol, 2023). It involves the study of lipids within biological systems, including cells, tissues, and biochemical pathways (Darshna & Tiwari, 2023). Bacteria possess several unique lipids that are absent in their mammalian hosts, enabling lipid mapping to generate valuable insights into extra-mammalian lipid profiles. These data can be used to (i) identify host-associated microbial infections, (ii) track the course of infections, (iii) evaluate the therapeutic efficacy, and (iv) identify potential candidates for antimicrobial drug development (Wood & Erol, 2023). As the primary structural components of cell membranes, lipids influence membrane properties such as fluidity and curvature as well as various membrane interactions. Additionally, lipids play essential roles in energy transport, storage, and signal transmission within cells (Walczak-Skierska et al., 2023). Lipidomics research primarily focuses on understanding the structure, functions, interactions, and dynamics of lipids and other molecules (Darshna & Tiwari, 2023). Early biomarker lipid studies in the environment used gas chromatography (GC), frequently in conjunction with mass spectrometry (MS), to detect lipid biomarkers. In recent decades, researchers have extended biomarker lipids to directly analyse more complex inositol phospholipids (IPLs) using liquid chromatography (LC) integrated with MS (Bale et al., 2021). Microbial lipidomics has emerged as an innovative method for lipid analysis, utilising matrix-assisted laser desorption/ionisation time-of-flight mass spectrometry (MALDI-TOF MS) to overcome some challenges associated with proteomic approaches. This technique serves as a complementary tool for identifying and classifying bacteria. Some lipids show species-specific characteristics, allowing researchers to use them as bacterial chemical barcodes during MS analysis to enhance conventional protein-based methods for microbial identification (Walczak-Skierska et al., 2023). The procedures used in lipidomic research are (1) lipid isolation, (2) separation, (3) detection, and (4) profiling (Darshna & Tiwari, 2023).

Microbial entities are often enhanced by integrating several different characteristics into their genomes, facilitating the expression of the intended phenotype. Integrated omics platforms make it possible to develop tailored cells to produce a range of economically significant products and create distinctive, enticing industrial processes. The omics technique aids in making precise changes to the industrial process environment or selecting among the best genetic trait combinations for desired phenotypic features from various possibilities. With their expertise, scientists have produced strains that perform exceptionally well relatively quickly, enabling the synthesis of both conventional and novel compounds. Researchers have studied omics technology in many medical and health science areas because of its enormous potential (Darshna & Tiwari, 2023). With single-cell profiling providing enhanced resolution for individual cell investigations, multi-omics integration has gained traction as a powerful approach for establishing comprehensive correlations between molecular profiles and disease-associated phenotypes. Furthermore, with omics, it is now possible to unravel the intricate molecular mechanism underlying diverse phenotypic expressions of aberrant characteristics overwhelmingly and systematically at high accuracy (Dai & Shen, 2022).

3.2 Biofilms

Microorganisms predominantly thrive in two distinct forms in nature: the unattached planktonic phase, where they can remain dispersed in liquid as masses or granule-shaped clusters (González-Plaza et al., 2022; Seneviratne et al., 2020), and the adhered biofilm structure (Seneviratne et al., 2020). Composed of diverse microorganisms, biofilms form structured, surface-attached communities encased in a complex matrix of extracellular polymeric substances (EPSs) that support their survival and growth. These EPSs are composed of secreted glycoproteins, exopolysaccharides, extracellular DNA (eDNA), extracellular RNA (eRNA) (Karygianni et al., 2020), and glycolipids, which restrict the penetration of antibiotics and chemical stresses into the bacteria (Anantharaman et al., 2023). The "matrixome" is a network of varying chemicals and functions comprising of those constituents (Karygianni et al., 2020). This structure increases bacteria's antibiotic resistance (ABR) by allowing them to withstand challenging environmental circumstances such as low oxygen levels, extreme pH, ultraviolet (UV) radiation, and temperature variations (Anantharaman et al., 2023). Because the biofilm matrix creates a protective shield and causes physiological changes, this stationary bacterial state has long been used in the form of a highly coordinated survival mechanism for microbes to combat external stress conditions that bacteria typically encounter in both natural and artificial environments (Ghosh et al., 2022). They can colonise both living organisms and inanimate surfaces and macroaggregates (Seneviratne et al., 2020). The development of biofilms is governed by interactions between cells and surfaces and cell-to-cell interactions, which are integral to a complex regulatory network.

Changes in bacterial gene expression and remodelling of metabolic processes often accompany the formation of biofilms. Numerous transcriptomic and metabolomic analyses have highlighted the importance of stress response, quorum sensing (QS), and exopolysaccharide (EPS) synthesis in the formation of biofilms. At the same time, metabolites of amino acids may act as signalling molecules to promote their formation (Wang et al., 2024). At least five developmental phases are believed to be involved in the production of bacterial biofilms. Initially, planktonic bacteria use fluid flow or motility to attach to a solid surface reversibly. This is followed by a transition to irreversible attachment, where bacteria produce extracellular polymeric substances (EPSs) or use specific adhesins on their surface structures to establish a stronger connection with the surface. In the third stage, the early architecture of the biofilm begins to take shape. Microcolonies evolve into mature biofilms as the process goes on, exhibiting consistent EPS production, which acts as an adhesive matrix. The mature biofilm displays intricate features such as elevated formations, aqueous channels, and porous networks. Finally, as the biofilm actively degrades, cells detach and disperse, allowing them to return to their free-floating, planktonic form (Lianou et al., 2020).

Bacteria initiate the reversible attachment stage upon successfully reaching surfaces. Weakly adhering bacteria return to the surrounding liquid, whereas the strongly adhering ones move on to the irreversible attachment stage. In this stage, bacteria activate a large number of adhesion-related genes. In addition, they produce EPSs (Zhang et al., 2023). The EPS matrix plays a crucial role in the structurality and functionality of biofilms. Its composition is exceptionally dynamic and varied but generally consists of a variety of water-bound macromolecules (Lianou et al., 2020). It comprises water-soluble polysaccharides, proteins, eDNA, and other compounds. The evolution of EPSs within a biofilm occurs in distinct stages, influenced by resource availability and the surrounding environmental factors (Zhang et al., 2023). Besides, EPSs increase surface adherence and cellular cohesion, which enables bacteria to establish biofilms on inorganic materials, including ventilators, catheters, and other medical devices (Anantharaman et al., 2023). Once bacteria firmly adhere, biofilm formation progresses to the maturation stage (Zhang et al., 2023). The microbial species and strains that constitute the components of a biofilm, along with its metabolic variations, external environmental factors, and coordinated cell signalling referred to as "quorum sensing" (QS), all seem to have an impact on the course of biofilm development (Lianou et al., 2020). QS is a cell density-dependent process through which microorganisms respond coherently to diverse physiological and environmental stresses (Shanmugam et al., 2020). It can get triggered when the cell density hits a threshold, facilitating the coordinated transition to a biofilm environment (Zhang et al., 2023). It works through small signalling molecules produced by bacteria. These molecules help bacteria regulate their activity by triggering a positive feedback loop called autoinducers. Bacteria generate autoinducers, which are absorbed and interact with their specific receptors or transcription factors at a sufficient extracellular concentration. This interaction leads to the activation of functions, including the upregulation of genes involved in virulence, drug resistance, and biofilm formation (An et al., 2021). The final phase of biofilm formation is dispersion, during which bacteria exit the matrix and return to their free-floating planktonic state. This stage is closely linked to the breakdown and decrease of matrix components (Zhang et al., 2023). Previous research on anti-biofilm interventions was largely centred around preventing biofilm formation, but recent investigations have delved into the mechanisms behind bacterial dispersion. Consequently, focusing on bacterial dispersion has emerged as a viable strategy for creating novel anti-biofilm medications (Wang et al., 2023).

Researchers regard biofilm formation as an environmental adaptation strategy (Lianou et al., 2020). It provides microorganisms various survival benefits, especially in adverse conditions, nutrient-deficient environments, and during antibiotic therapy. Microbial biofilms exhibit distinct physiological, growth, and behavioural traits compared to their planktonic forms, allowing them to endure and persist in conditions unfavourable for planktonic cells (Seneviratne et al., 2020). Moreover, the metabolic activity of biofilm-state bacteria is lower than that of planktonic-state bacteria. According to metabolomics data, biofilm-state bacteria downregulate numerous crucial metabolic pathways. Additionally, there is a decreasing trend in the transcription levels of the enzymes that catalyse the metabolic processes in these pathways (Wang et al., 2024).

3.2.1 Characteristics of Biofilms

Biofilm formation is a widespread characteristic among bacteria. Biofilms consist of microbial cells that frequently attach to a wide range of surfaces. A biofilm can exist as a single layer directly attached to a surface, form floating clusters called flocs, remain mobile, or develop into intricate microbial communities (Dutt et al., 2022). The extracellular matrix (ECM) of a typical biofilm protects against drying out, antibiotics, biocides, ultraviolet radiation, and host immune defences. Because of their mechanical stability, biofilms are incredibly resilient and challenging to remove (Seneviratne et al., 2020). The matrix of a biofilm is primarily composed of water (97%). Its architecture and functionality are composed of soluble components like gel-like polysaccharides, polypeptides, and extracellular DNA (eDNA), along with insoluble elements such as amyloids, cellulosic fibres, adhesive appendages, pili, and flagella (Dutt et al., 2022). Through its ability to link cells, proteins, carriers, and extracellular polysaccharides, eDNA plays a vital role. It shapes the biofilm structure and helps maintain its extracellular matrix. Additionally, it operates like a "kite-line", limiting bacterial displacement along the biofilm surface. The progression of a biofilm, from establishment to disintegration, is dependent on eDNA (Ghosh et al., 2022).

Biofilms generally serve two purposes: they create a barrier that protects against environmental stresses and also contribute to disease when they become stagnant and dysbiotic (Shanmugam et al., 2020). Numerous bacteria use different processes to form biofilms, which are often influenced by environmental factors and the specific characteristics of the bacterial strain (Dutt et al., 2022). Some of these mechanisms are comparable with those driving resistance in planktonic cells but are amplified during the biofilm development phase. Nonetheless, specialised genetic regulators and effectors may influence the process (An et al., 2021). When compared to their free-floating planktonic form, microbial biofilms exhibit remarkable resistance to adverse environmental stressors (Seneviratne et al., 2020) such as dehydration, host defence responses, UV radiation, biocides (Vohra et al., 2024), metal toxicity, pH and osmotic fluctuations, protozoan predation, and antimicrobial agents by temporarily adopting a distinct phenotype (Lianou et al., 2020). These increased adaptive resistance characteristics have been attributed to several possible occurrences that may take place within biofilm communities. The underlying phenomena include mutational changes or gene upregulation, reduced cellular metabolism, and extracellular protective structures, including cell confinement and space limitations. Additionally, the extracellular matrix also acts as a diffusion barrier for molecules like nutrients and vitamins, leading to slower cellular proliferation (Lianou et al., 2020). The existence of persisters, a latent subpopulation, is a potential additional driver of the adaptive resistance against pharmaceuticals and environmental pressures. Originally found in the *Staphylococcus* species, persister cells have more recently been found in the *Candida* species, *Pseudomonas aeruginosa*, *Escherichia coli*, and *Lactobacillus acidophilus* (Seneviratne et al., 2020). Despite having the same genetic make-up as the rest of the biofilm population, these cells do not divide and, by expressing toxin–antitoxin systems, show better resilience to environmental challenges (Lianou et al., 2020). Efflux pump systems represent another mechanism that enhances bacterial survival in extreme environments (Seneviratne et al., 2020).

All higher species, including humans, are naturally colonised by microorganisms and often form biofilms. These biofilms are commonly found in ecosystems, industrial systems, and healthcare environments and can develop on both living hosts and artificial materials, such as medical prosthetics and dental implants. Because of their resistance to antibiotics, biofilms contribute to persistent and chronic infections across humans and animals (Dutt et al., 2022). Biofilm-related infections pose a considerable threat to human well-being (Long et al., 2022). According to the National Institutes of Health (NIH) data, biofilms are responsible for 80–90% of all chronic illnesses and 65–80% of all microbial infections (An et al., 2021). These infections persist due to microorganisms' capacity to bypass the host's immune defences and resist most antimicrobial treatments (Silva et al., 2021) Antimicrobial resistance (AMR) includes antibiotic resistance (ABR), in which bacteria develop particular resistance to the antibiotics intended to eradicate them. The mechanisms behind AMR include restricted or reduced antimicrobial entry, mutational alterations or modifications in antibiotic targets, and breaking down antibiotics through enzymatic hydrolysis or chemical transformation (Dutt et al., 2022). Several nosocomial pathogenic microbes, including *Acinetobacter baumannii*, *Bacillus subtilis*, and *Staphylococcus aureus*, show tolerance to antibiotics by enhancing the activity of their efflux pump systems, which frequently leads to a multidrug-resistant phenotype (Seneviratne et al., 2020).

The resistance of biofilms arises from a number of benefits associated with their transformation from dispersed planktonic cells to an organised biofilm community. Increased horizontal gene transfer and improved stress tolerance are two examples. Biofilms can withstand up to 1000 times as many antibiotics as planktonic cells, making them significantly more complicated to treat (Vohra et al., 2024). This distinctiveness highlights the immense challenge of developing efficient therapies to combat the production of biofilms (Wang et al., 2023). Further contributing to biofilms' increased drug resistance are genetic polymorphisms and enzymes that modify cell walls. Planktonic cells may also dislodge from the biofilm, enter systemic circulation, where they may cause bacteraemia, colonise multiple organs, or initiate a septic shock (An et al., 2021). However, there are currently few methods for controlling biofilms, and the standard treatment for biofilm infections involves high-dose combinational therapies administered over an extended period (Long et al., 2022).

Many disease-related biofilms are now known to be polymicrobial—comprising several different microbial species. Because the various species cooperate to strengthen their overall tolerance through a variety of resistance mechanisms, this diversity increases their resistance (Seneviratne et al., 2020). Within any biofilm, numerous microcolonies coexist and interact with one another using metabolites and signalling molecules. A biofilm's development, survival, and mutualistic cooperation all depend on this communication (Shanmugam et al., 2020). Interactions between bacteria, usually cooperative, neutral, or competitive, could affect the structure and functions of mixed-species biofilms (Yuan et al., 2023a, b), which influence species' geographical distribution, behaviour, and virulence. Examples of interspecies interactions that have been shown to affect the composition and structure of biofilms include competition for resources, metabolic cooperation, interspecies adherence, communication through signalling molecules, and the synthesis of antimicrobial compounds to eliminate other members of the community selectively (Shokeen et al., 2021). Interestingly, through interspecies communication networks, pathogenic species in microbial communities have a greater chance of surviving or committing altruistic suicide to maintain the ecological balance as a response to stress (Shanmugam et al., 2020). Another notable characteristic of biofilm communities is the use of secondary compounds called siderophores to efficiently capture and utilise available iron for their growth and survival (González-Plaza et al., 2022). Crucially, biofilm formation represents an adaptive growth stage, and the bacteria within these clusters are extremely resistant to antibiotics (An et al., 2021). Biofilm formation not only provides a protective environment for bacteria, enabling their survival under harsh conditions, but also presents a significant challenge when it comes to treatment and therapy (Wang et al., 2024).

While researchers have categorised biofilm formation into five stages, comprehensive investigations are required to uncover the specific genes, metabolites, and processes facilitating biofilm development across various bacterial species (Zhang et al., 2023). Numerous approaches, such as genetic analysis and phenotypic characterisation, have been employed to comprehend the crucial elements regulating the formation of biofilms (Seneviratne et al., 2020). A deeper exploration of biofilms is necessary to understand their contributions to disease, environmental interactions, and evolutionary processes (Vohra et al., 2024).

3.3 Integration of Omics

A single omics approach can only reveal a limited amount of information about the whole biofilm system, considering biofilm formation is a highly dynamic and multifaceted process (Wang et al., 2024). Understanding the intricate alterations that regulate the biofilm phenotype necessitates comprehensive, systems-based methodologies to determine microbial functional shifts. Emerging advancements in "omics" technologies—spanning transcriptomics, proteomics, metabolomics, and multi-omics data fusion—offer transformative potential for studying biofilm regulation at a systems-wide scale (Seneviratne et al., 2020). These omics analyses can analyse complex systems, offering a deeper understanding of biofilm adaptation than isolated, reductionist analyses focused on single targets. Nevertheless, by integrating data from multiple omics techniques, scientists can reveal novel connections and interactions that might be missed using individual methods alone (An et al., 2021). Integrating omics data across three or more layers has become feasible with significant advancements in software development (Yuan et al., 2023a, b). A thorough grasp of biofilm formation is obtained by multi-omics techniques, which combine information (Vohra et al., 2024) and cross-validate data analysis (Wang et al., 2024) from various omics disciplines (Vohra et al., 2024). These methods also minimise false positives associated with single-omics analyses, thereby contributing to uncovering a more detailed and accurate picture of the regulatory networks driving biofilm formation (Wang et al., 2024). The integration of transcriptomics and proteomics has also provided additional information that could explain the cross-kingdom interactions in dual-species biofilms. Researchers can comprehensively analyse the increased stress tolerance of biofilms using multi-omics tools (Yuan et al., 2023a, b) to gather additional information on the biofilm control mechanism based on novel strategies (Yuan et al., 2023a, b). Further understanding of the mechanisms underlying biofilm growth will pave the way for creating more potent strategies to manage biofilms in both clinical and environmental settings (Vohra et al., 2024). Nonetheless, the fusion of multi-omics datasets remains challenging because of their fundamental differences in data structures. For instance, data from genomics are qualitative, precise, and highly reproducible. On the other hand, although proteomics and metabolomics deliver comprehensive qualitative and quantitative data, they are frequently less reproducible and more susceptible to noise. Multi-omics data typically undergo pre-treatment using deconvolution, normalisation, scaling, and transformation, often supported by specialised software before integration. IT support and specialists in the corresponding omics domains are also necessary for multi-omics studies to validate the data. While this approach enhances the accuracy of data interpretation, it also adds complexity to data acquisition and analysis (Amer & Baidoo, 2021). Despite these challenges, the utilisation of a multi-omics approach offers a significant synergistic benefit of improving the predictive accuracy of biofilm studies, ultimately supporting the development of more effective biofilm control strategies (Yuan et al., 2023a, b). Regardless of its specific functions, omics provides a vast understanding of the entire microbiome, which is enormous when combined with statistical methods, mathematical modelling, and artificial intelligence (AI) techniques. AI-assisted virtual and predictive technologies can significantly enhance meta-analyses and microbiome-wide studies, providing deeper insights and more precise predictions (Shanmugam et al., 2020). Using metabolomics, proteomics, and transcriptomics, AI technologies will help study patterns and repetitive behaviours, easing the prediction and analysis of non-phenotypic traits and features. Researchers can personalise therapeutic processes based on the specific functional aetiology to achieve effective and efficient containment, culmination, and cure (Shanmugam et al., 2020). Compiling and integration are key learning elements, with advanced knowledge achieved daily (Shanmugam et al., 2020). Therefore, developing adequate statistical tools for integrating and analysing multi-omics data and increasing the size and availability of normalised databases could significantly accelerate biofilm studies using multi-omics tools (Yuan et al., 2023a, b).

A multitude of integration techniques is available for application today. One approach is using a web-based tool like NetworkAnalyst to create a network of protein–protein interactions. Metabolomic data are integrated into this network via protein–metabolite interactors characterised by MetaBridge. Another sophisticated approach available is tools such as the mixOmic package (An et al., 2021).

3.3.1 Different Omics Techniques in Analysing Biofilm Formation

Studies on microbial communities, particularly those on biofilms, frequently employ omics techniques (Seneviratne et al., 2020), for instance, transcriptomics and proteomics, to identify new genetic and protein targets that can aid in developing innovative anti-biofilm agents (An et al., 2021). Multi-omics analysis, instead of single-omics approaches, yield a more thorough view of biological systems by integrating information across multiple levels, including genomic sequences, mRNA transcription rate, protein production and regulation, and cellular metabolic processes (Zhang et al., 2023). Omics techniques have provided valuable insights into how distinct species respond in the presence of relevant microbial partners despite the current technology's ineffectiveness to fully capture the transcriptional, translational, or metabolic responses of individual species during their interactions with others in the complex biofilm environment (Shokeen et al., 2021). Determining the key genetic targets, molecular pathways, and processes necessary for biofilm formation is greatly aided by multi-omics techniques (An et al., 2021).

3.3.1.1 Metatranscriptomics

Metatranscriptomics studies gene expression across an entire population or microbial community. This approach aids in a deeper comprehension of how a community adapts to environmental changes by examining gene transcription, regulatory processes, and metabolic pathways (Priyadarsini et al., 2023). Transcriptome sequencing has also become an important tool for studying interspecies and interkingdom relationships (Shokeen et al., 2021).

The characterisation of the transcriptome, which is made up of all the mRNA transcripts that a genome produces, is known as transcriptomics (Seneviratne et al., 2020). The transcriptome encompasses the detection of various RNAs (Chen et al., 2023) that dictate the composition of the proteome, making transcriptome studies a valuable tool for providing a first-level biochemical picture of gene regulation in a particular environment, like the transformation from planktonic existence to biofilm formation. Transcriptome analysis utilises conventional techniques, like quantitative polymerase chain reaction (qPCR), or comprehensive screening methods, including microarray assays and RNA sequencing (Seneviratne et al., 2020), to extensively uncover genes that are differentially expressed for distinct bacterial species living in planktonic and biofilm environments (An et al., 2021). Moreover, even metatranscriptomics could decipher potential molecular mechanisms for biofilm formation and the response of biofilms to stress conditions (Yuan et al., 2023a, b). The development of advanced technologies has enhanced the capability of scRNA-seq to accurately identify transcripts in specific cell populations during disease conditions (Chen et al., 2023).

qPCR is an essential assay for analysing gene expression, but it has limitations (Kinaret et al., 2020). Although efficient and economical, this technique focuses solely on known sequences and is limited in the number of genes it can study (Kinaret et al., 2020; Seneviratne et al., 2020). The qPCR method monitors gene expression in real time and detects changes in gene expression during the biofilm state or after compound exposure. It also confirms or validates microarray or RNA-Seq results (Seneviratne et al., 2020).

After qPCR, DNA microarrays are regarded as one of the most frequently applied tools in transcriptomic research to simultaneously analyse the expression levels of many genes, achieved by placing collections of targeted DNA sequences in wells on a chip (Seneviratne et al., 2020). This method can independently identify significant transcriptomic changes, producing consistent biological signals across multiple platforms and research settings. The relative maturity of microarray technology, which has several well-known commercial and freely available data analysis software, is another advantage (Kinaret et al., 2020). Using DNA microarrays, a recent transcriptomic study has compared the immunological responses of cells of the oral epithelium in planktonic and biofilm states brought on by multi-species oral pathogenic bacteria. Microarrays offer a comparatively inexpensive solution, but their scanners operate within a limited dynamic range (Seneviratne et al., 2020).

While DNA microarrays have enabled the production of extensive gene expression data, cutting-edge sequencing advancements have significantly broadened this capability (Kinaret et al., 2020). Nowadays, RNA-Seq with next-generation sequencing (NGS) is a highly sought-after option for detecting and quantifying known and unknown transcripts (Seneviratne et al., 2020). It can detect transcriptomic signatures that remarkably differ from those of biofilm-facilitated microbial migration. This approach can also help evaluate the impact of antimicrobial agents and possible biofilm inhibitors on biofilm growth (Dutt et al., 2022). The detailed and comprehensive data provided by RNA-Seq make it an excellent choice for gene expression studies despite the fact that NGS platforms are comparatively expensive and laborious (Seneviratne et al., 2020). With the declining costs of RNA-Seq, scientists are enhancing technical techniques and creating improved in vitro and in vivo systems for biofilm research. This progress enables dual RNA-Seq to simultaneously study the transcriptomes of both the host and pathogen, offering a more thorough comprehension of host–pathogen interactions (An et al., 2021). High-throughput transcriptomic tools have provided more profound insights into the drug resistance of microbial biofilms (Seneviratne et al., 2020). These advancements significantly boost the efficiency of transcriptomic data output. Notably, these diverse methods make it possible to profile dose–response relationships at a minimal cost (Kinaret et al., 2020).

Metatranscriptomics is a tool that provides knowledge on the whole gene expression profiles of mixed-species biofilm communities due to changes in environmental conditions. This technique has unravelled the complex assembly and bacterial functions in biofilms on teeth, fixed-bed biofilm reactors, and sediments (Yuan et al., 2023a, b). Most current transcriptomic studies assume that biofilm populations are homogeneous. However, biofilms consist of a heterogeneous community of cells exhibiting varying degrees of metabolic activity, which can significantly impact their resistance to antimicrobials. To better understand the mechanisms underlying biofilm-mediated antimicrobial resistance, analysing the transcriptomic profile at the single-cell level is crucial. Enhancements in NGS methodologies now make it possible to examine these transcriptomes in remarkable detail (Seneviratne et al., 2020). The first challenge of the metatranscriptome technique is that most RNA, typically over 95%, consists of ribosomal RNA (rRNA). Therefore, rRNA must be removed before sequencing to concentrate on mRNA (Procópio, 2020). Transcriptomic approaches offer a comprehensive understanding of the critical biological processes underlying biofilm formation, drug resistance, and more, rather than focusing on a small set of predetermined genes. Nevertheless, the choice of model utilised for transcriptomic profiling can impact the observed patterns of the transcription process (Seneviratne et al., 2020). Comprehensive RNA sequencing has recently revealed the community's functional potential during health and disease (Shokeen et al., 2021). Transposon insertion sequencing (Tn-Seq) is a modern technique offering a high-throughput strategy for determining multiple genes essential for survival under particular circumstances, such as within a biofilm. It has also been applied to study the development of persister cells, which contribute to the difficulty of eliminating biofilms (An et al., 2021). Furthermore, Tn-Seq has incorporated novel techniques, such as segregation of mutant cell types and the functional analysis of essential gene with regulatable promoters (Dutt et al., 2022) (Fig. 3.1).

3.3.1.2 Metaproteomics

Metaproteomics is the study of the large-scale characterisation of all proteins that are expressed in a specific environment at a given time to ascertain the structure, metabolism, and physiology of community components (Amer & Baidoo, 2021; Seneviratne et al., 2020). It plays a key role in systems biology and is essential for understanding the functional efficiency of microorganisms (Seneviratne et al.,

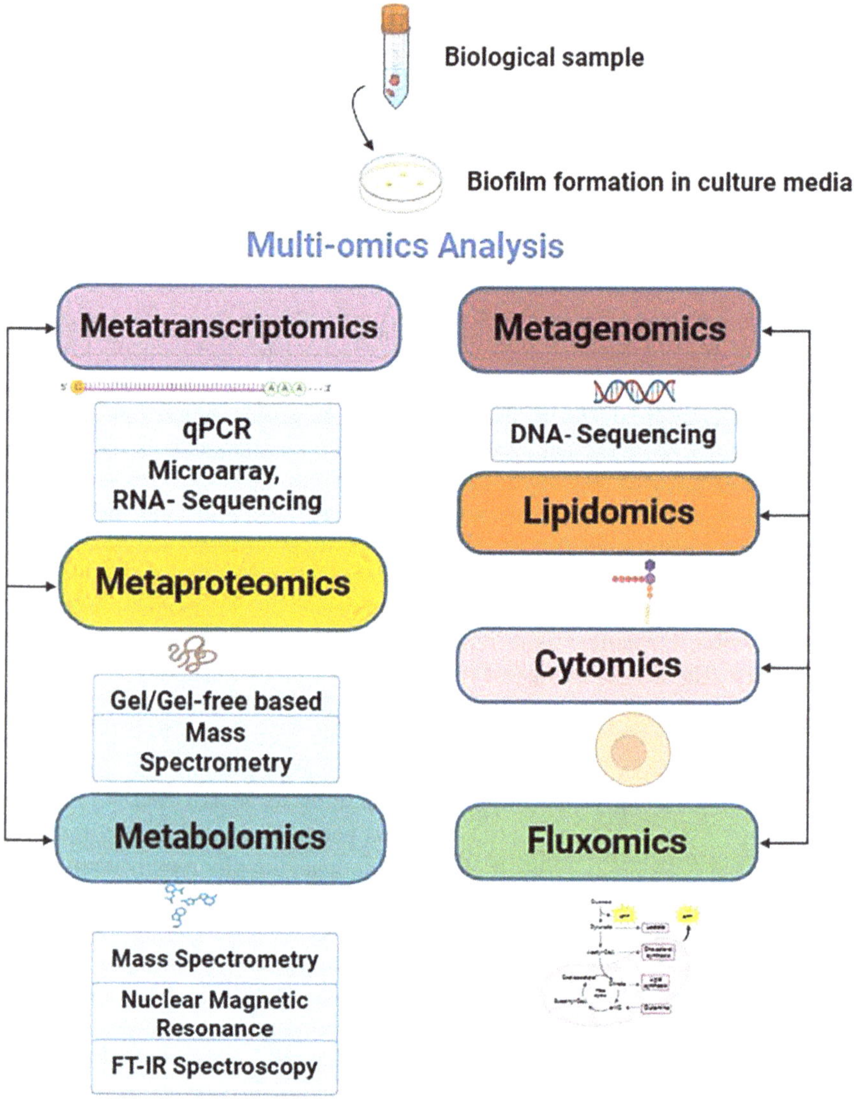

Fig. 3.1 This schematic illustrates different omics approaches used for comprehensive molecular analysis, integrating metatranscriptomics (RNA-sequencing, qPCR), metagenomics (DNA-sequencing), metaproteomics (mass spectrometry), metabolomics (NMR, FT-IR spectrometry), lipidomics, cytomics, and fluxomics. These integrative approaches provide insights into microbial communities, their functional activities, and their interactions within the biofilm environment

2020). It represents an exciting field in proteomics research, where the abundance of proteins at a specific moment, often in response to interactions with other species, is measured. This offers insights into the proteins essential for the functioning of multi-species populations (An et al., 2021; Shokeen et al., 2021). There are numerous variables that can affect protein levels, such as protein lifespan or mRNA turnover (Shokeen et al., 2021). Although gel electrophoresis has historically been employed for protein identification, its limitations concerning sensitivity and quantification have improved the demand for methods like liquid chromatography/mass spectrometry (LC/MS). More than 80% of the proteome can be analysed using LC/MS techniques, providing increased sensitivity and coverage (An et al., 2021). Recent advancements in LC and high-resolution MS have made it possible to identify and quantify large-scale peptide and protein profiling exceeding 10,000 per sample using metaproteomic analysis. This approach also allows for measuring interactions between community components and evaluating substrate consumption (Amer & Baidoo, 2021). Additionally, even in single-species communities, it can determine the abundance of proteins that each species employs in critical metabolic and energy-related pathways, including amino acid metabolism and microbial fermentation (Dutt et al., 2022).

The term "proteome" refers to the complete collection of expressed proteins under defined conditions within a given period. The transcriptome primarily showcases gene-level alterations, whereas the proteome offers a more comprehensive and reliable depiction of biological transformations in an organism. Proteomics involves qualitative and quantitative analysis of the proteome, allowing for precise measurement of changes during growth, development, and response to various environmental factors (Seneviratne et al., 2020). Proteomics has successfully identified precise changes in proteins and provided key insights into the behaviour and pathways of microorganisms during biofilm formation. At this stage, the two proteomics methods currently employed to clarify potential biofilm-related components are two-dimensional polyacrylamide gel electrophoresis (2D-PAGE) paired with MS and label-based proteomics (Yuan et al., 2023a, b). Historically, proteomics research was largely based on gel-based methods; however, gel-free techniques have gained popularity recently because of their more straightforward sample preparation procedures and improvements in MS techniques (Seneviratne et al., 2020). Shotgun quantification-based MS techniques can measure the quantity of proteins in a sample. By analysing the mass and abundance of peptides, fragment patterns can be used to identify full-length proteins through sequence homology stored in database entries. For instance, metaproteomic approaches have revealed the effects of a temperature gradient on biofilm physiology at an environmental site (Procópio, 2020). MS-based techniques enable identifying and quantifying more proteins relatively quickly, covering proteins present in low quantities and complex protein groups such as membrane-bound proteins. Label-free or isotopic labelling MS methods can quantify differential protein expression.

For the relative quantification of proteins, recently, a number of microbial biofilm studies have been employing isotopic labelling strategies like isotope-coded affinity tags (ICAT), stable isotope labelling by amino acids in cell culture (SILAC), isobaric tags for relative and absolute quantitation (iTRAQ), and tandem mass tag (TMT) in a gel-free proteomics workflow.(Seneviratne et al., 2020). Additionally, studies on biofilms—particularly those involving bacteria and fungi—have extensively used label-free techniques, which count distinct spectra or peak intensities (Seneviratne et al., 2020).

Proteins are the key factors in supporting and maintaining a biofilm's structure (Wang et al., 2024). They often physically mediate cell–surface and cell–cell interactions in microorganisms, playing a key role in their ability to colonise biotic and abiotic surfaces advantageously. Studying these interactions yields important information about microbial behaviour and adaptation. Proteomics provides the added advantage of revealing multiple layers of the components and players involved in these interactions. Various methods extract proteins from the extracellular space (secretome) and identify those located on the cell membrane (surfaceome) and analyse intracellular proteins globally or within specific organelles (González-Plaza et al., 2022). Only a few studies have investigated host–biofilm interactions using proteomics; however, the intriguing mechanisms underlying the enhanced resistance observed in microbial biofilms have been better understood through advancements in proteomics studies. The enhanced resilience of microbial biofilms stems from reduced growth dynamics, protein expression, and metabolic processes within biofilm structures (Seneviratne et al., 2020). The induction of cellular antioxidant defence mechanisms is one hypothesis explaining the rise in resistance to drugs in microbial biofilms. Research supports the role of antioxidant defence systems in enhancing biofilms' resistance to antibiotics (Seneviratne et al., 2020). In order to identify markers influencing biofilm formation, proteomics studies on biofilms have analysed the differences in protein expression between robust and ineffective strains that form biofilms. Additionally, an exoproteome study on exopolysaccharides indicates that the proteins of ECM may contribute to stress tolerance, nutrition uptake, and pathogenesis (Seneviratne et al., 2020).

Findings from a proteomics study on biofilms suggest a correlation between heightened lipid biosynthesis and antibiotic resistance. The proteome of the persister's population has been the focus of specific proteomics studies to understand the mechanism underlying drug resistance better. These studies imply that quorum sensing helps form new subpopulations of antibiotic-tolerant cells, and type IV pili facilitate these cells to shift towards the exterior of the biofilm (Seneviratne et al., 2020). Paleoproteomics, the study of ancient proteins, represents a new frontier in whole proteome analysis. Owing to the enhanced stability of proteins relative to DNA, paleoproteomics can explore much older samples and is less susceptible to the presence of recently introduced contaminants. With respect to bacterial colonisation of non-biological materials, proteomics provides valuable insights into biofilm formation, revealing how sessile microbial cells organise and the molecular changes that occur in bacterial physiology within biofilms. Additionally, proteomics can illuminate how microbes attach and form biofilms on various biomaterials, including metals, alloys, ceramics, and polymers. Researchers continuously develop new coatings to restrict bacterial colonisation and biofilm formation on medical instruments, so studying how bacteria adhere to biomaterials is essential (González-Plaza et al., 2022) (Fig. 3.1).

3.3.1.3 Metabolomics

Metabolites are typically molecules with a molecular weight of less than 1.5 kDa that are frequently the result of numerous biological processes and interactions (González-Plaza et al., 2022) in microorganisms, which better reflect bacterial activities at the phenotypic level (Yuan et al., 2023a, b). Metabolite analysis is a valuable method for identifying key molecules and indirectly revealing the kinds of bacteria that are probably present. It involves both untargeted and targeted chemical approaches (Afshari et al., 2020) to examine the variation in the synthesis of particular metabolites or intermediate molecules such as carbohydrates, nucleotides, amino acids (Dutt et al., 2022), and organic acids (Afshari et al., 2020). By studying these smaller molecules, metabolomics provides insights into their production and a quick overview of the functional alterations resulting from transcriptomic and proteomic alterations. Additionally, analysing metabolites can help predict biofilm activity (Dutt et al., 2022) and characterise the metabolic precursors and products. It is a practical approach to evaluating environmental phenotypic traits that involves examining the complex network of relationships among microbial community constituents (Seneviratne et al., 2020).

The thorough recognition, classification, and measurement of an organism's metabolome is known as metabolomics. Profiling metabolites can reveal information on a cell's physiological, pathological, and biochemical states (Seneviratne et al., 2020). This technique enables accurate measurement of metabolite levels at different stages of growth. It makes production optimisation easier, and it facilitates the identification of strains with advantageous genetic modifications (González-Plaza et al., 2022). This technique employs chromatographic techniques to detect and quantify small molecules, including metabolites. To achieve this, high-performance liquid chromatography (HPLC) resolves the sample into its individual constituents while MS measures and detects them (Procópio, 2020). Since metabolite profiles are more directly related to the traits observed than genes or proteins, metabolomics serves as a conclusive assessment technique (Seneviratne et al., 2020). Furthermore, metabolomics exhibits a measurable correlation with other omics approaches (Chen et al., 2023).

The formation of biofilms relies on the functional roles of primary and secondary metabolites. Metabolome profiling, combined with biological system modelling, helps identify pathways involved in this process. Additionally, metabolomics also helps uncover small molecules that regulate biofilm growth at both intra- and intercellular levels, potentially uncovering new drug targets (Seneviratne et al., 2020).

Owing to the intricate interconnections of metabolic processes and their susceptibility to even slight alterations in the system, a single method may not always yield comprehensive results. It may require complementary analytical platforms (Seneviratne et al., 2020). Data from various omics experiments may not correlate with one another, and single-omics analysis is limited to characterising data at the nucleic acid, protein, or metabolite level. Therefore, single-omics data are insufficient to profile detailed and exact mechanisms to describe biofilm formation, the complex interactions in mixed-species biofilms, or their behaviour when exposed to stressors (Yuan et al., 2023a, b). The associated databases and related technologies for metabolomics include nuclear magnetic resonance (NMR), gas chromatography–mass spectrometry (GC/MS), liquid chromatography–tandem mass spectrometry (LC/MS), and capillary electrophoresis–mass spectrometry (CE/MS) (Chen et al., 2023). NMR has established itself as a valuable tool in metabolomics (Wang et al., 2021). Its non-invasive nature allows samples to remain intact and unaltered, making it possible to analyse metabolite fluxes in real time without pretreatment. NMR is particularly suited for analysing extremely polar substances and is adaptable enough to handle fluids and solids (González-Plaza et al., 2022). While NMR spectroscopy has lower sensitivity compared to techniques like LC/MS or GC/MS, it offers straightforward sample processing and consistent molecular analysis. It serves as a robust method for identifying metabolic biomarkers (Eom et al., 2021). Numerous NMR methods study particular interactions between tiny and big molecules. These methods frequently use the differences in physical characteristics to differentiate between tiny molecules and complex ones (Wang et al., 2021). MS primarily integrates with gas chromatography (GC/MS) or liquid chromatography, utilising single- or tandem-stage detection (LC/MS or MS/MS). GC/MS requires fewer samples than other methods due to its high sensitivity (González-Plaza et al., 2022). Metabolite profiling employs untargeted or targeted approaches determined by the study's goals and the number of metabolites analysed (Seneviratne et al., 2020). Non-targeted metabolomics aids in developing biological theories by profiling a wider variety of metabolites without predetermined targets, whereas targeted metabolomics concentrates on distinct biological theories by offering quantitative data for metabolites involved in particular metabolic pathways.

The untargeted or global metabolomics approach provides an in-depth snapshot of metabolic composition under defined conditions. It is particularly well suited for drug discovery and other exploratory research. Untargeted metabolomics provides an unbiased analysis of samples, identifying numerous metabolites and typically generating hypotheses (Seneviratne et al., 2020). However, LC/MS-based untargeted metabolomics encounters limitations of unresolved metabolite identities by the variability of ionisation and fragmentation patterns, complicating the use or creation of specialised spectral databases for identifying metabolites (Afshari et al., 2020).

A specific hypothesis or biological question drives targeted metabolomics approaches. For instance, a targeted approach is ideal for studies focused on measuring particular polysaccharides in the EPS matrix of a biofilm. Focusing on specific pathways directly addresses the biological question being investigated. Targeted metabolomics quantifies metabolites of interest directly with exceptional sensitivity and has extremely low detection limits (Seneviratne et al., 2020).

Integrative analysis investigates the interrelationships between the microbiota and metabolomes by combining 16S rRNA-derived microbial assessment with unbiased metabolomic analysis using LC/tandem MS and GC/MS. Integrated data processing from these studies is carried out using multiple co-inertia analysis

(MCIA), which is a multidimensional reduction method designed to merge more than two omics datasets systematically. MCIA facilitates improved biological translation, understanding, and pathway analysis by finding correlations between several large-scale datasets and examining the variations within each. Beyond this, the approach holds promise for applications in examining the origin and integrity of samples while detecting biomarkers. This method could be implemented in extensive genomic and metagenomic research to connect the genetic pathways that produce metabolites for their synthesis (Afshari et al., 2020) (Fig. 3.1).

3.4 Other Omics Techniques Applied in Microbial Research

3.4.1 Metagenomics

Genomics involves studying whole genomes to identify, characterise, and measure an organism's genes, understand how they interact, and determine their effects on the organism. It typically includes processes such as NGS, genome assembly, functional annotation, and genome-wide association studies (GWASs). Key technologies in genomics include genomic profiling arrays, long-read whole genome sequencing (WGS), and targeted exome sequencing (TES) (Chen et al., 2023).

Genome-encoded functional analysis provides a precise understanding of microbial functions and their interactions with their surroundings. This study approach is exceptionally well-optimised for research where sample sensitivity is a key consideration (González-Plaza et al., 2022). The term "metagenomics" was first used to describe a method for analysing the total genetic composition of a microbial community. This approach allows researchers to explore genetic diversity, microbiological composition, and bacterial metabolic processes within specific environments. Microbial genomics builds on this by examining functional gene properties, metabolic pathways, and interactions between regulatory factors (Ou et al., 2020). While traditional microbiological methods require culturing, metagenomics directly captures and studies diverse genetic material from environmental samples and employs DNA sequencing to gather target information. This makes it a more effective tool for identifying microbial diversity and dynamics within a defined habitat. Methodologies like sequence- and function-based metagenomics obtain metagenomes from environmental data (Seneviratne et al., 2020). The reads produced by current sequencing platforms are too short to cover the whole microbial genome in a single read. Over a long span, massively parallel sequencing platforms, such as Illumina (Illumina, Inc.), have dominated genomic and transcriptomic analysis by delivering millions of accurate short sequencing reads within the 150–500 bp range (González-Plaza et al., 2022). NGS technologies, or single-molecule real-time (SMRT) sequencing, eliminate the PCR bias associated with clonal amplification in traditional approaches (Procópio, 2020). NGS facilitates the parallel sequencing of specific genomic regions (TS, WES, WGS) (Pruneri et al., 2021). Unlike traditional methods, NGS sequences single DNA strands directly rather than relying on clonally amplified templates. This approach reduces the need for extensive biochemical steps and reagents, streamlining the process and enabling sequencing at the nanoscale level (Procópio, 2020). Cutting-edge sequencing technologies such as those from PacBio and Oxford Nanopore further expand the capabilities of DNA analysis (González-Plaza et al., 2022). SMRT sequencing is an advanced technology that enables real-time analysis at the single-molecule level (Xiao & Zhou, 2020), delivering up to 10-kb reads. Two main approaches have been developed for obtaining whole genomes using these sequencing technologies: de novo genome assembly and reference-dependent sequencing, often known as resequencing (González-Plaza et al., 2022). In recent years, genomic tools, including metagenomics and whole genome sequencing, have been widely used to analyse the microbial communities and related genes implicated in biofilms due to their increasing throughput and decreasing sequencing cost (Yuan et al., 2023a, b). TGS technologies enable real-time long-read sequencing with minimal alignment and mapping mistakes during library building (Xiao & Zhou, 2020). Advances in DNA sequencing have significantly expanded our comprehension of taxonomic distribution, genomic functional potential, and dynamic shifts of health and disease-associated biofilms, providing valuable information about their role in transitions between health and disease (Shokeen et al., 2021) (Fig. 3.1).

3.4.2 Cytomics

Cytomics focuses on analysing and modifying individual cells, assessing their spatial organisation, function, and physiological responses in biological systems. Unlike other omics technologies, cytomics quantifies changes in community structure. Microbial cytomics offers a novel perspective on understanding community interactions and the ecological principles guiding natural or engineered microbial communities (Darshna & Tiwari, 2023). Over the past decade, advancements in single-cell cytometric analysis through flow cytometry have enabled the development of "cytometric fingerprints", which provide phenotypic insights on a single-cell scale (De Vrieze et al., 2021). Even with its distinct capabilities and high-throughput efficiency, this optics-based approach remains underexplored in optimised microbial systems (De Vrieze et al., 2021) (Fig. 3.1).

3.4.3 Fluxomics

The term "fluxomics" encompasses various methods to study metabolic flux analysis (MFA). This approach mathematically represents changes in the concentrations of intracellular metabolite pools influenced by the reaction stoichiometry and directional movement of metabolites moving in and out of the pool (Darshna & Tiwari, 2023). MFA identifies active pathways and distributes carbon fluxes across various biological systems. Adapting MFA techniques to other biological systems is generally simple as long as the metabolic pathways remain unchanged in the target species (Volke et al., 2023). Microorganisms have played a crucial role in human societies for centuries, contributing to the production of food, beverages, and other goods. Harnessing omics-based techniques in microbial communities facilitates the advancement of novel and optimised production methods (Darshna & Tiwari, 2023) (Fig. 3.1).

3.5 Applications of Omics

The integration of omics technologies into microbial engineering increases the yield of biofuels as well as the synthesis of therapeutic molecules, proteins, and enzymes while also promoting agricultural advancements (Priyadarsini et al., 2023). Successful biological engineering relies on accurately predicting phenotype from genotype (Amer & Baidoo, 2021).

3.5.1 Applications in Food Processing

The field dedicated to applying omics technologies in food and nutritional studies is called "foodomics". Cutting-edge omics approaches are applied to evaluate the nutritional content of food, screen for foodborne infections connected to disease outbreaks, and analyse how food metabolites affect critical biochemical pathways to prevent disease and improve the overall value of food products. Genomics and transcriptomics aim to study the molecular regulation of pigment production in fruits and vegetables (Priyadarsini et al., 2023). Applying omics technologies to starch bioprocessing has deepened our comprehension of the key enzymes involved in starch formation. This gained knowledge has facilitated more accurate prediction and modification of starch characteristics, advancing research in rice starch biotechnology.

Additionally, omics technologies address food quality and traceability challenges, verify food origins, and identify biomarkers for potential food safety issues (Amer & Baidoo, 2021). With its rapid progress, NGS is revolutionising food microbiology by enhancing quality and reducing food production costs. NGS is used in food microbiology in two ways: metagenomics and whole genome sequencing (WGS). Metagenomics is in its early stages of application for improving food quality and safety but offers promising opportunities. It enables the engineering of microbial biofilms, forecasting of alterations in microbial populations, and characterisation of previously unknown microbiota (Ghosh et al., 2022). Combining various omics technologies monitors the preparation and quality assurance of seafood and meat products. Proteomics and immunoassays are practical tools for identifying biomarkers of foodborne bacterial and fungal pathogens. Metabolomics aids in the detection and quantification of toxic metabolites, assesses their concentrations, and monitors the prevalence of foodborne illnesses. Transcriptomics-based methods, such as NGS, RNA-Seq, and microarrays, are used to investigate the formation of biofilms and the mechanisms behind food spoilage (Priyadarsini et al., 2023) (Fig. 3.2).

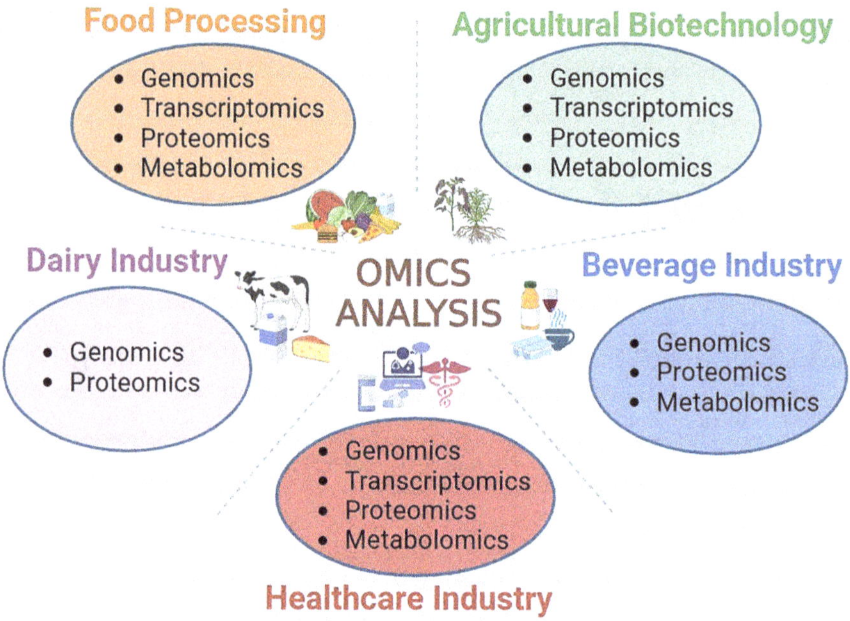

Fig. 3.2 Omics technologies (genomics, transcriptomics, proteomics, and metabolomics) enhance quality, safety, and efficiency across industries. In food processing and agriculture, they improve crops, disease resistance, and food safety. Healthcare applies omics for disease diagnosis, drug development, and personalised medicine. The dairy and beverage industries use them for microbial quality control, fermentation, and product optimisation, driving biotechnological advancements

3.5.2 Applications in the Dairy Industry

The biofilm lifestyle is the predominant mode of microorganisms in various food sectors, including the dairy industry (Yuan et al., 2023a, b). Microbial biofilms' most distinguishing behaviour is their extraordinary capacity to resist control measures, like the cleaning-in-place (CIP) procedures, which are typically used in the dairy sector. This results in a negative impact on the quality assurance and safety standards of dairy products by making it harder to eradicate microbial infestation. Considering the complex nature of biofilms in the dairy industry, the use of omics techniques and even their integrations, significantly benefits the biofilm studies industry (Yuan et al., 2023a, b).

Additionally, omics methodologies have been fundamental in discovering substitute sources of proteinases from plants and microorganisms. The quality assessment of milk and dairy products extensively relies on these methods, which focus on the analysis of major proteins like whey and casein. Using omics technologies, researchers have identified more than 151 proteins in milk, providing valuable insights into its composition and functionality. Scientists extensively use omics technologies in lactation research on dairy cows (Priyadarsini et al., 2023). Metagenomics has shown high-throughput and in-depth identification of microbial composition, including unculturable cells in mixed-species biofilms in dairy environments (Yuan et al., 2023a, b). Genomic mapping of livestock animals has enabled selective breeding for desirable traits. By analysing somatic cell counts, genomic studies have also been essential in detecting mastitis in dairy cows, which has helped lower the financial losses related to milk production (Priyadarsini et al., 2023). Metagenomics also helps confirm the critical control points during milk powder processing by tracing the remarkable changes in the bacterial community, which is significant for the quality guarantee of dairy products (Yuan et al., 2023a, b). Proteomics techniques analyse alterations in protein expression patterns, availability, and post-translational modifications (Priyadarsini et al., 2023). The combination of metagenomics and proteomics provides comprehensive insights into microbial communities and their metabolic profiles. For instance, this integration has been effectively implemented in industrial biofilters to enhance nitrate removal prior to treatment through denitrification process. Moreover, applying multi-omics tools would unravel a comprehensive microbial interaction between lactic acid bacteria (LAB) biofilms and gut microbiome and further provide the specific biological processes by which the gut microbiota supports health (Yuan et al., 2023a, b) (Fig. 3.2).

3.5.3 Applications in the Beverage Industry

Omics technologies offer valuable insights into the metabolic processes of microbes and how they affect beverage quality, including flavour, aroma, and nutritional value. They support the beverage sector in detecting and controlling pathogenic contaminants. To satisfy the growing demand worldwide, omics techniques have created genetically modified tea and coffee varieties. Additionally, high-throughput omics methods have improved our knowledge of the wine microbiome, which has helped refine its flavour and fragrance (Priyadarsini et al., 2023). In the wine industry, the fermentation-associated microbiome plays a crucial role in transforming grapes into wine by influencing key factors such as flavour and aroma. Omics technologies aid in describing the intricate relationships that shape wine production between microbes, substrates, and the environment. Combining omics technologies with genome editing makes it possible to create improved probiotic strains, develop innovative biotherapeutics, and modify the microbial community structure in food matrices (Amer & Baidoo, 2021). Multiple omics disciplines are integrated to evaluate the nutritional value of probiotic beverages and how they affect human health (Priyadarsini et al., 2023) (Fig. 3.2).

3.5.4 Applications in the Healthcare Industry

With the advancements in medical technology, researchers must thoroughly examine diseases' pathophysiology. Omics technologies—including genomics and transcriptomics, incorporating cellular and spatial transcriptomics—offer high-throughput screening methods to effectively and rapidly reveal detailed biological insights into human diseases. Each omics approach identifies specific molecular changes linked to disease, which act as key biomarkers for tracking disease advancement and help reveal variations in biological pathways between diseased and healthy states. However, the full potential of single-omics approaches in capturing disease complexity cannot be realised. Through the integration of multi-omics data, researchers can identify important signalling pathways, create more accurate disease biomarkers, and discover new relationships between biomolecules and disease phenotypes. This multi-omics integration is crucial for linking molecular mechanisms to disease phenotypes and environmental factors, ultimately improving disease screening, prognosis, and therapeutic strategies (Chen et al., 2023).

The pharmaceutical industry significantly relies on omics technologies to enhance drug discovery, optimise development, refine drug delivery, personalise treatments, and ensure drug safety. Understanding how inherited genes affect the body's reaction to drugs, especially in diseases like cancer and cardiotoxicity, has been made easier by pharmacogenomics. The use of transcriptomics to analyse gene regulation patterns in neurofibrillary tangles has led to the identification of biomarkers associated with Alzheimer's disease. Proteomics plays a crucial role in identifying significant molecular markers for cancer, such as cathepsin B, pfetin, and HSP27, as well as in discovering novel therapeutic agents targeting cancer (Priyadarsini et al., 2023). The fast-paced growth of single-cell sequencing platforms is driving significant innovations in clinical research. This technique provides deep insights into cellular heterogeneity by analysing variations in transcriptomics, genomics, and epigenomics at the single-cell level, enabling researchers to track dynamic molecular changes and improve disease understanding and treatment strategies. Other omics technologies widely known for medical research are spatial transcriptomics, epigenomics, microbiomics, and radiomics (Chen et al., 2023) (Fig. 3.2).

3.5.5 Applications in Agricultural Biotechnology

Recent advancements in agricultural biotechnology have enabled the development of new plant varieties engineered through recombinant DNA technology to meet market demands and address environmental challenges. Nowadays, omics technologies are used in agricultural biotechnology to enhance desired characteristics like pesticide resistance, colour, taste, and drought tolerance (Amer & Baidoo, 2021). Omics approaches play a pivotal role in developing biofertilizers containing vital microbes that improve plants' uptake of nutrients from the soil. Leveraging different omics approaches has been instrumental in uncovering regulatory genes that enable the development of crop varieties with enhanced nutrient absorption and reduced water requirements, contributing to more sustainable agriculture (Priyadarsini et al., 2023). Recent findings from a comparative multi-omics study have revealed that certain rhizosphere bacteria selectively metabolise particular aromatic acid metabolites secreted by plants. This suggests a dynamic interaction between plant secretion patterns and microbial metabolic uptake strategies, influencing microbial populations' composition and structure and ultimately shaping microbial community assembly patterns (Amer & Baidoo, 2021). Omics technologies have allowed scientists to identify key genes and metabolic pathways in crop improvement. Through selective breeding, these insights have led to enhanced food crops with more health advantages, improved nutritional content, and higher levels of essential minerals and vitamins. Omics technologies uncover interactions between plants and pathogens, providing valuable data to prevent plant diseases (Priyadarsini et al., 2023).

Metagenomic analyses are divided into sequence- and function-based screening, enabling the identification of novel genes and natural compounds from environmental samples. In agriculture, metagenomics significantly contributes to the study of soil microbial populations, identifying microbes that facilitate macro- and micronutrient cycling, and detecting essential enzymes, both of which enhance crop productivity (Amer & Baidoo, 2021). While proteomics aids in examining the effects of pathogenic bacteria and fungi on plant protein expression patterns, transcriptomics has proven crucial in analysing shifts in the genetic adaptations of staple crops to environmental and biological stressors. Through extensive applications of omics technologies, it has been made possible to create crop varieties that are pesticide- and herbicide-resistant (Priyadarsini et al., 2023) (Fig. 3.2).

3.6 Conclusions

Omics technologies have revolutionised our ability to investigate biological responses to surface interactions at an unmatched level of resolution, offering a comprehensive understanding of cellular responses in diverse environments. By integrating all the omics disciplines, the amalgamated approach provides a systems-level perspective, linking genotype to phenotype and unveiling the intricate mechanisms governing biofilm formation. These technologies augment our fundamental knowledge of microbial communities and hold significant potential for industrial and clinical applications. In industrial biotechnology, systems metabolic engineering can leverage omics insights to optimise microbial production processes, while, in clinical diagnostics, omics-based analyses offer rapid and precise pathogen identification, improving disease management. As research advances, the continued integration of omics data will drive innovation, refine diagnostic methodologies, and expand our knowledge of complex biological systems, paving the way for new scientific and technological breakthroughs.

References

Afshari, R., Pillidge, C. J., Read, E., Rochfort, S., Dias, D. A., Osborn, A. M., & Gill, H. (2020). New insights into cheddar cheese microbiota-metabolome relationships revealed by integrative analysis of multi-omics data. *Scientific Reports, 10*(1), 3164. https://doi.

org/10.1038/s41598-020-59617-9. Erratum in: Sci Rep. 2021 25;11(1):2680. doi:10.1038/s41598-021-82097-4. PMID: 32081987; PMCID: PMC7035325.

Amer, B., & Baidoo, E. E. K. (2021). Omics-driven biotechnology for industrial applications. *Frontiers in Bioengineering and Biotechnology, 9*, 613307. https://doi.org/10.3389/fbioe.2021.613307. PMID: 33708762; PMCID: PMC7940536.

An, A. Y., Choi, K. G., Baghela, A. S., & Hancock, R. E. W. (2021). An overview of biological and computational methods for designing mechanism-informed anti-biofilm agents. *Frontiers in Microbiology, 12*, 640787. https://doi.org/10.3389/fmicb.2021.640787. PMID: 33927701; PMCID: PMC8076610.

Anantharaman, S., Guercio, D., Mendoza, A. G., Withorn, J. M., & Boon, E. M. (2023). Negative regulation of biofilm formation by nitric oxide sensing proteins. *Biochemical Society Transactions, 51*(4), 1447–1458. https://doi.org/10.1042/BST20220845. PMID: 37610010; PMCID: PMC10625800.

Bale, N. J., Ding, S., Hopmans, E. C., Arts, M. G. I., Villanueva, L., Boschman, C., Haas, A. F., Schouten, S., & Sinninghe Damsté, J. S. (2021). Lipidomics of environmental microbial communities. I: Visualization of component distributions using untargeted analysis of high-resolution mass spectrometry data. *Frontiers in Microbiology, 12*, 659302. https://doi.org/10.3389/fmicb.2021.659302. PMID: 34367080; PMCID: PMC8343106.

Chen, C., Wang, J., Pan, D., Wang, X., Xu, Y., Yan, J., Wang, L., Yang, X., Yang, M., & Liu, G. P. (2023). Applications of multi-omics analysis in human diseases. *MedComm (2020), 4*(4), e315. https://doi.org/10.1002/mco2.315. PMID: 37533767; PMCID: PMC10390758.

Dai, X., & Shen, L. (2022). Advances and trends in omics technology development. *Frontiers in Medicine (Lausanne), 9*, 911861. https://doi.org/10.3389/fmed.2022.911861. PMID: 35860739; PMCID: PMC9289742.

Darshna, D., & Tiwari, S. S. (2023). Basic of omics and its applications. In *Industrial microbiology and biotechnology: Emerging concepts in microbial technology* (pp. 1–22). Springer Nature Singapore.

De Vrieze, J., Heyer, R., Props, R., Van Meulebroek, L., Gille, K., Vanhaecke, L., Benndorf, D., & Boon, N. (2021). Triangulation of microbial fingerprinting in anaerobic digestion reveals consistent fingerprinting profiles. *Water Research, 202*, 117422. https://doi.org/10.1016/j.watres.2021.117422. Epub 2021 Jul 10.

Dutt, Y., Dhiman, R., Singh, T., Vibhuti, A., Gupta, A., Pandey, R. P., Raj, V. S., Chang, C. M., & Priyadarshini, A. (2022). The association between biofilm formation and antimicrobial resistance with possible ingenious bio-remedial approaches. *Antibiotics (Basel), 11*(7), 930. https://doi.org/10.3390/antibiotics11070930. PMID: 35884186; PMCID: PMC9312340.

Eom, J. S., Kim, E. T., Kim, H. S., Choi, Y. Y., Lee, S. J., Lee, S. S., Kim, S. H., & Lee, S. S. (2021). Metabolomics comparison of rumen fluid and milk in dairy cattle using proton nuclear magnetic resonance spectroscopy. *Animal Bioscience, 34*(2), 213–222. https://doi.org/10.5713/ajas.20.0197. Epub 2020 Jun 24. PMID: 32819075; PMCID: PMC7876715.

Futo, M., Široki, T., Koska, S., Čorak, N., Tušar, A., Domazet-Lošo, M., & Domazet-Lošo, T. (2022). A novel time-lapse imaging method for studying developing bacterial biofilms. *Scientific Reports, 12*(1), 21120. https://doi.org/10.1038/s41598-022-24431-y. PMID: 36476631; PMCID: PMC9729682.

Ghosh, S., Nag, M., Lahiri, D., Sarkar, T., Pati, S., Kari, Z. A., Nirmal, N. P., Edinur, H. A., & Ray, R. R. (2022). Engineered biofilm: Innovative Nextgen strategy for quality enhancement of fermented foods. *Frontiers in Nutrition, 9*, 808630. https://doi.org/10.3389/fnut.2022.808630. PMID: 35479755; PMCID: PMC9036442.

González-Plaza, J. J., Furlan, C., Rijavec, T., Lapanje, A., Barros, R., Tamayo-Ramos, J. A., & Suarez-Diez, M. (2022). Advances in experimental and computational methodologies for the study of microbial-surface interactions at different omics levels. *Frontiers in Microbiology, 13*, 1006946. https://doi.org/10.3389/fmicb.2022.1006946. PMID: 36519168; PMCID: PMC9744117.

Karygianni, L., Ren, Z., Koo, H., & Thurnheer, T. (2020). Biofilm Matrixome: Extracellular components in structured microbial communities. *Trends in Microbiology, 28*(8), 668–681. https://doi.org/10.1016/j.tim.2020.03.016. Epub 2020 Apr 21. PMID: 32663461.

Kinaret, P. A. S., Serra, A., Federico, A., Kohonen, P., Nymark, P., Liampa, I., Ha, M. K., Choi, J. S., Jagiello, K., Sanabria, N., Melagraki, G., Cattelani, L., Fratello, M., Sarimveis, H., Afantitis, A., Yoon, T. H., Gulumian, M., Grafström, R., Puzyn, T., & Greco, D. (2020). Transcriptomics in toxicogenomics, part I: Experimental design, technologies, publicly available data, and regulatory aspects. *Nanomaterials (Basel), 10*(4), 750. https://doi.org/10.3390/nano10040750. PMID: 32326418; PMCID: PMC7221878.

Lianou, A., Nychas, G. E., & Koutsoumanis, K. P. (2020). Strain variability in biofilm formation: A food safety and quality perspective. *Food Research International, 137*, 109424. https://doi.org/10.1016/j.foodres.2020.109424. Epub 2020 Jun 8. PMID: 33233106.

Long, L., Sulaiman, J. E., Xiao, Y., Cheng, A., Wang, R., Malit, J. J., Wong, W. C., Liu, W., Li, Y. X., Chen, F., Lam, H., & Qian, P. Y. (2022). Mode of action of elasnin as biofilm formation eradicator of methicillin-resistant Staphylococcus aureus. *Frontiers in Microbiology, 13*, 967845. https://doi.org/10.3389/fmicb.2022.967845. PMID: 36003935; PMCID: PMC9393526.

Ou, Y. J., Ren, Q. Q., Fang, S. T., Wu, J. G., Jiang, Y. X., Chen, Y. R., Zhong, Y., Wang, D. D., & Zhang, G. X. (2020). Complete genome insights into Lactococcus petauri CF11 isolated from a healthy human gut using second- and third-generation sequencing. *Frontiers in Genetics, 11*, 119. https://doi.org/10.3389/fgene.2020.00119. PMID: 32174973; PMCID: PMC7054480.

Priyadarsini, M., Rani, J., Kushwaha, J., Pandey, K. P., Singh, Y., & Dhoble, A. S. (2023). An introduction to omics in relevance to industrial microbiology. In *Industrial microbiology and biotechnology: Emerging concepts in microbial technology* (pp. 23–39). Springer Nature Singapore.

Procópio, L. (2020). The era of 'omics' technologies in the study of microbiologically influenced corrosion. *Biotechnology Letters, 42*(3), 341–356. https://doi.org/10.1007/s10529-019-02789-w. Epub 2020 Jan 3. PMID: 31897850.

Pruneri, G., De Braud, F., Sapino, A., Aglietta, M., Vecchione, A., Giusti, R., Marchiò, C., Scarpino, S., Baggi, A., Bonetti, G., Franzini, J. M., Volpe, M., & Jommi, C. (2021). Next-generation sequencing in clinical practice: Is it a cost-saving alternative to a single-gene testing approach? *PharmacoEconomics - Open, 5*(2), 285–298. https://doi.org/10.1007/s41669-020-00249-0. Epub 2021 Mar 4. PMID: 33660227; PMCID: PMC8160052.

Seneviratne, C. J., Suriyanarayanan, T., Widyarman, A. S., Lee, L. S., Lau, M., Ching, J., Delaney, C., & Ramage, G. (2020). Multi-omics tools for studying microbial biofilms: Current perspectives and future directions. *Critical Reviews in Microbiology, 46*(6), 759–778. https://doi.org/10.1080/1040841X.2020.1828817. Epub 2020 Oct 8. PMID: 33030973.

Shanmugam, K., Sarveswari, H. B., Udayashankar, A., Swamy, S. S., Pudipeddi, A., Shanmugam, T., Solomon, A. P., & Neelakantan, P. (2020). Guardian genes ensuring subsistence of oral Streptococcus mutans. *Critical Reviews in Microbiology, 46*(4), 475–491. https://doi.org/10.1080/1040841X.2020.1796579. Epub 2020 Jul 28. PMID: 32720594.

Shokeen, B., Dinis, M. D. B., Haghighi, F., Tran, N. C., & Lux, R. (2021). Omics and interspecies interaction. *Periodontology 2000, 85*(1), 101–111. https://doi.org/10.1111/prd.12354. Epub 2020 Nov 23. PMID: 33226675.

Silva, N. B. S., Marques, L. A., & Röder, D. D. B. (2021). Diagnosis of biofilm infections: Current methods used, challenges and perspectives for the future. *Journal of Applied Microbiology, 131*(5), 2148–2160. https://doi.org/10.1111/jam.15049. Epub 2021 Mar 8. PMID: 33629487.

Vohra, M., Kour, A., Kalia, N. P., Kumar, M., Sharma, S., Jaglan, S., Kamath, N., & Sharma, S. (2024). A comprehensive review of genomics, transcriptomics, proteomics, and metabolomic insights into the differentiation of Pseudomonas aeruginosa from the planktonic to biofilm state: A multi-omics approach. *International Journal of Biological Macromolecules, 257*(Pt 1), 128563. https://doi.org/10.1016/j.ijbiomac.2023.128563. Epub 2023 Dec 7. PMID: 38070800.

Volke, D. C., Gurdo, N., Milanesi, R., & Nikel, P. I. (2023). Time-resolved, deuterium-based fluxomics uncovers the hierarchy and dynamics of sugar processing by pseudomonas putida.

Metabolic Engineering, 79, 159–172. https://doi.org/10.1016/j.ymben.2023.07.004. Epub 2023 Jul 16. PMID: 37454792.

Walczak-Skierska, J., Monedeiro, F., Maślak, E., & Złoch, M. (2023). Lipidomics characterization of the microbiome in people with diabetic foot infection using MALDI-TOF MS. *Analytical Chemistry, 95*(44), 16251–16262. https://doi.org/10.1021/acs.analchem.3c03071. Epub 2023 Oct 25. PMID: 37877781; PMCID: PMC10633811.

Wang, Z., Pisano, S., Ghini, V., Kadeřávek, P., Zachrdla, M., Pelupessy, P., Kazmierczak, M., Marquardsen, T., Tyburn, J. M., Bouvignies, G., Parigi, G., Luchinat, C., & Ferrage, F. (2021). Detection of metabolite-protein interactions in complex biological samples by high-resolution Relaxometry: Toward Interactomics by NMR. *Journal of the American Chemical Society*. https://doi.org/10.1021/jacs.1c01388. Epub ahead of print. PMID: 34133154.

Wang, S., Ma, C., Long, J., Cheng, P., Zhang, Y., Peng, L., Fu, L., Yu, Y., Xu, D., Zhang, S., Qiu, J., He, Y., Yang, H., & Chen, H. (2023). Impact of CRAMP-34 on Pseudomonas aeruginosa biofilms and extracellular metabolites. *Frontiers in Cellular and Infection Microbiology, 13*, 1295311. https://doi.org/10.3389/fcimb.2023.1295311. PMID: 38162583; PMCID: PMC10757720.

Wang, H., Fan, Q., Wang, Y., Yi, L., & Wang, Y. (2024). Multi-omics analysis reveals genes and metabolites involved in Streptococcus suis biofilm formation. *BMC Microbiology, 24*(1), 297. https://doi.org/10.1186/s12866-024-03448-5. PMID: 39127666; PMCID: PMC11316374.

Wood, P. L., & Erol, E. (2023). Construction of a bacterial Lipidomics analytical platform: Pilot validation with bovine Paratuberculosis serum. *Metabolites, 13*(7), 809. https://doi.org/10.3390/metabo13070809. PMID: 37512516; PMCID: PMC10383236.

Xiao, T., & Zhou, W. (2020). The third generation sequencing: The advanced approach to genetic diseases. *Translational Pediatrics, 9*(2), 163–173. https://doi.org/10.21037/tp.2020.03.06. PMID: 32477917; PMCID: PMC7237973.

Yuan, L., Dai, H., He, G., Yang, Z., & Jiao, X. (2023a). Invited review: Current perspectives for analyzing the dairy biofilms by integrated multiomics. *Journal of Dairy Science, 106*(12), 8181–8192. https://doi.org/10.3168/jds.2023-23306. Epub 2023 Aug 23. PMID: 37641326.

Yuan, L., Fan, L., Dai, H., He, G., Zheng, X., Rao, S., Yang, Z., & Jiao, X. A. (2023b). Multi-omics reveals the increased biofilm formation of Salmonella Typhimurium M3 by the induction of tetracycline at sub-inhibitory concentrations. *Science of the Total Environment, 899*, 165695. https://doi.org/10.1016/j.scitotenv.2023.165695. Epub 2023 Jul 22. PMID: 37487904.

Zhang, T., Liu, Z., Wang, H., Zhang, H., Li, H., Lu, W., & Zhu, J. (2023). Multi-omics analysis reveals genes and metabolites involved in Bifidobacterium pseudocatenulatum biofilm formation. *Frontiers in Microbiology, 14*, 1287680. https://doi.org/10.3389/fmicb.2023.1287680. PMID: 38029154; PMCID: PMC10666050.

Microbiomes and Biofilms: Fundamentals and Translational Applications

4

Naji Naseef Pathoor, Pitchaipillai Sankar Ganesh, Rajesh Kanna Gopal, Arun Prabhu Govindasamy, Sam Jabaraj Ramesh, Gulshan Wadhwa, Esaki M. Ponmalar, Vanitha Mariappan, and Esaki M. Shankar

4.1 Introduction

The study of microbiomes and biofilms has become a cornerstone of modern microbiology, fundamentally reshaping our understanding of how microorganisms exist and function in natural environments (Shree et al., 2023). This scientific journey began with Antoni van Leeuwenhoek's pioneering observations of microscopic life and has evolved dramatically over centuries, driven by research and technological advances. While early microbiologists focused on studying isolated species in controlled lab settings, we now know that most microorganisms thrive within complex communities that profoundly influence their behavior and capabilities. The concept of the microbiome gained significant momentum in the early twenty-first century,

N. N. Pathoor · P. S. Ganesh (✉) · R. K. Gopal · A. P. Govindasamy · S. J. Ramesh
Quorum Sensing and Biofilm Research Lab, Department of Microbiology, Centre for Infectious Diseases, Saveetha Dental College and Hospitals, Saveetha Institute of Medical and Technical Sciences (SIMATS), Saveetha University (Deemed to be University), Chennai, Tamil Nadu, India

G. Wadhwa
Department of Biotechnology, Ministry of Science and Technology, New Delhi, India

E. M. Ponmalar
Sri Sairam Siddha Medical College and Research Centre, West Tambaram, Chennai, Tamil Nadu, India

V. Mariappan
Centre for Toxicology and Health Risk Studies (CORE), Faculty of Health Sciences, University Kebangsaan Malaysia, Kuala Lumpur, Malaysia

E. M. Shankar
Infection and Inflammation, Department of Biotechnology, School of Integrative Biology, Central University of Tamil Nadu, Thiruvarur, Tamil Nadu, India

S. Busi et al. (eds.), *Omics Approaches in Biofilm Research*,
https://doi.org/10.1007/978-3-031-91863-6_4

particularly with the launch of the Human Microbiome Project in 2007 (Joseph & Niddk, 2013). This initiative marked a pivotal moment, shedding light on the critical roles microbial communities play in human health. Advances in sequencing technologies and bioinformatics have revealed that the human body hosts approximately 100 trillion microorganisms, which collectively shape immunity, digestion, and even mental health. The impact of microbiome research extends beyond human health to environmental and industrial settings, where these microbial ecosystems are crucial for nutrient cycling, waste decomposition, and biotechnological innovations (Malard et al., 2021).

Alongside the rise of microbiome studies, our understanding of biofilms has undergone a similarly transformative evolution. Since Claude ZoBell's 1940s observations of bacterial attachment, biofilms are now recognized as highly organized, three-dimensional microbial communities encased in self-produced extracellular matrices (Donlan, 2002). Within these microbial communities, bacteria communicate through complex signaling mechanisms like quorum sensing (QS), share resources, and exhibit coordinated, collective behaviors. Biofilms are remarkably resilient, often resisting antimicrobial agents and immune responses, and are implicated in a wide range of infections and industrial challenges (Samrot et al., 2021).

Microbiomes and biofilms have a profound impact on human health. Microbial communities influence everything from nutrient absorption and immune development to drug metabolism and mental well-being. The gut microbiome, for example, plays a central role in overall health, with disruptions linked to conditions such as inflammatory bowel disease, obesity, and neurological disorders like autism and depression (Hou et al., 2022). Consequently, the microbiome has become a target for therapeutic interventions, including probiotics, prebiotics, and fecal microbiota transplants, which aim to restore a healthy balance of microbes.

In industrial settings, biofilms present both challenges and opportunities. They can cause problems such as biofouling in water systems and contamination of medical devices, resulting in economic losses and health risks. Yet, their unique properties can also be harnessed for beneficial uses, including wastewater treatment, bioremediation, and sustainable energy production. In the food industry, while biofilms must be carefully managed to prevent spoilage and foodborne illnesses, they can also be advantageous in specific fermentation processes (Shineh et al., 2023).

Research has shown that microbiomes and biofilms are not just random collections of microbes but highly coordinated, multicellular-like communities (Sharma et al., 2023). These collectives communicate via sophisticated systems like QS, organize spatially to optimize resource use, and adapt to environmental stressors through collective metabolic and genetic shifts. Such discoveries have inspired novel strategies for medical treatment and industrial applications, though significant challenges remain in fully harnessing this knowledge for practical solutions.

One of the significant challenges in microbiome and biofilm research lies in developing methods to manipulate these microbial communities effectively (Coenye et al., 2024). Therapeutic strategies must carefully adjust the microbiome without causing unintended consequences, while approaches to biofilm control must strike a balance between eradicating harmful communities and promoting beneficial ones.

Additionally, understanding the causality between microbiomes and diseases remains complex, as these microbial ecosystems are highly dynamic and influenced by multiple factors, such as diet, medications, and environmental conditions. The convergence of microbiome and biofilm research with other scientific fields has unlocked exciting opportunities for innovation (Ma et al., 2024). As the field progresses, it stands at a pivotal intersection where fundamental science merges with practical applications. Continued technological advancements, paired with a deeper understanding of microbial ecology, promise to address some of society's most critical challenges, from medical breakthroughs to sustainable industry and environmental stewardship.

This chapter lays the foundation for comprehending the significance of microbiomes and biofilms in modern microbiology. It explores key concepts, terminology, types, and methods of studying microbiomes while also discussing microbial diversity, community interactions, and the factors shaping their development. We then delve into biofilm formation, outlining the stages from initial attachment to dispersal, and examine structural components such as extracellular polymeric substances (EPSs) and communication systems as well as regulatory mechanisms like QS and stress responses. This chapter also explores the interactions between microbiomes and biofilms, focusing on their ecological relationships, impact on host systems, and health benefits. Practical applications in medicine, and industry, will also be discussed, from therapeutic interventions to bioremediation. We will then cover advanced analysis techniques, including molecular methods and imaging technologies, shaping our understanding of these microbial systems. Finally, we will explore emerging technologies and future directions, highlighting their potential in transforming medicine, industry, and environmental sustainability.

4.2 Fundamental Concepts of Microbiomes

The entire community of microorganisms that live in a particular environment, together with their genes and the surrounding environmental conditions, constitute the components of the microbiome, one of the most intriguing and intricate biological systems in nature. Although its current usage has broadened to encompass the organisms themselves and their complex interactions with their host or environment, the term "microbiome" was initially used to refer to the collective genome of all microorganisms in a specific habitat (Berg et al., 2020).

When exploring microbiomes, it is important to grasp several foundational terms that underpin the field of microbiome science. The term "microbiota" refers to the living microscopic organisms present within a specific environment, including bacteria, archaea, fungi, protists, and viruses (Matijašić et al., 2020). The concept of the metagenome encompasses all the genetic material found within a microbiome sample, offering researchers valuable insights into the composition and potential functions of the microbial community (Kim et al., 2024). Another key concept, metabolomics, involves studying the chemical signatures left behind by cellular processes, providing scientists with a deeper understanding of how microbiomes interact with and influence their environments (Puig-Castellví et al., 2023).

Microbiomes exist in various environments, each with unique compositions and characteristics. Among these, the human microbiome is the most extensively studied, encompassing specialized communities in the gut, skin, oral cavity, and respiratory tract (Deo & Deshmukh, 2019). In soil, microbiomes are essential for maintaining ecosystem balance, supporting plant health, driving nutrient cycling, and shaping soil structure (Wang et al., 2024). Marine microbiomes, found in habitats ranging from sunlit surface waters to deep-sea hydrothermal vents, play a vital role in regulating global nutrient cycles and sustaining marine food webs (Orcutt et al., 2020). Meanwhile, microbiomes in built environments, such as homes and offices, are gaining interest for their influence on human health and building maintenance (Young et al., 2023).

The study of microbiomes has advanced significantly with the development of modern technologies. While traditional culture-based methods remain useful, they are limited in scope, as many microorganisms cannot be cultivated under laboratory conditions. The advent of advanced sequencing techniques has transformed microbiome research. One key approach is 16S rRNA gene sequencing, which identifies and classifies bacteria and archaea by analyzing this highly conserved genetic marker (Johnson et al., 2019). For a broader perspective, shotgun metagenomic sequencing examines all genetic material in a sample, providing insights into both the microbial community's composition and its functional potential. These methods have greatly expanded our understanding of microbiomes and their complexities.

In addition to sequencing, researchers use a range of techniques to gain a deeper understanding of microbiomes. Meta-transcriptomics focuses on the active gene expression within microbial communities, shedding light on their functional responses under specific conditions (Zampieri et al., 2023). Metabolomics analyzes the chemical compounds produced by these communities, revealing how they interact with and impact their environments (Bedia, 2022). Furthermore, advanced imaging methods such as fluorescence microscopy and electron microscopy enable scientists to observe microbial interactions and community structures with remarkable detail, providing visual insights into their organization and behavior at different scales. Together, these methods offer a comprehensive view of microbiome dynamics (Relucenti et al., 2021).

Bioinformatics plays a crucial role in microbiome research, as the analysis of large-scale sequencing data requires sophisticated computational tools and approaches. Researchers use various software packages and pipelines to process raw sequencing data, identify microbial species, analyze community diversity, and understand functional potential. Machine learning algorithms are increasingly being applied to predict microbiome functions and identify patterns in complex datasets (Ghannam & Techtmann, 2021).

The study of microbiomes has revealed their essential roles in human health, environmental processes, and agricultural systems. Understanding these complex communities requires an interdisciplinary approach, combining molecular biology, ecology, computer science, and other fields. As technology continues to advance, our ability to study and understand microbiomes grows, leading to new insights into how these microscopic communities shape our world and potential applications in medicine, agriculture, and environmental conservation.

4.3 Microbiome Composition and Diversity

Intricate interactions between various microbial communities that coexist and compete within particular environments shape microbiomes, which are extremely complex and dynamic ecosystems. The most prevalent and well-researched of these are bacterial communities, which are the main forces behind the functioning of microbial ecosystems and exhibit remarkable diversity in a range of environments.

Through genetic mutations and horizontal gene transfer mechanisms, individual microbial populations can quickly adapt to shifting environmental conditions, demonstrating the remarkable genetic variability and metabolic versatility of bacterial communities (Groussin et al., 2021). Different bacterial species occupy distinct ecological niches and carry out specialized tasks within their respective environments in these communities, which display complex stratification (Baquero et al., 2021). For example, bacterial communities in human microbiomes differ greatly between body sites, with gut microbiomes supporting remarkably diverse and dense populations in contrast to other anatomical regions (Dekaboruah et al., 2020).

Numerous factors, such as host genetics, environmental conditions, diet, lifestyle, location, and prior exposure to different ecological pressures, affect the structure of bacterial communities (Cahana & Iraqi, 2020). While some bacteria are permanent inhabitants of microbiomes, establishing enduring relationships with their host or surroundings, others are part of transient populations that can quickly colonize and vanish in response to shifting environmental factors (Ma et al., 2024). Both cooperative and competitive relationships are present in the intricate web of bacterial interactions, where various species may cooperate or fight for resources.

Despite being less numerous than bacterial populations, fungal communities are just as important to the dynamics of the microbiome and the operation of ecosystems. With their participation in nutrient cycling, decomposition processes, and intricate symbiotic relationships with other microorganisms, these eukaryotic microorganisms exhibit remarkable metabolic capabilities (Wu et al., 2024). Fungal communities in soil microbiomes play a crucial role in decomposing organic matter and promoting the exchange of nutrients among various ecological components. Numerous fungal populations can be found in human microbiomes; common fungal inhabitants that have a substantial impact on host health and metabolism include *Candida* and *Saccharomyces* (Yiallouris et al., 2024).

The viral components of microbiomes, often underrepresented in traditional studies, are equally dynamic and impactful. Bacteriophages, viruses that specifically target bacteria, regulate bacterial populations through predation and genetic exchange. These processes not only shape microbial community structures by selectively reducing specific bacterial populations but also promote genetic diversity through horizontal gene transfer. Far from being passive participants, viruses actively contribute to the intricate ecological interactions that define and sustain microbial ecosystems (Naureen et al., 2020).

The interactions and networking within microbial communities are some of the most complex and intriguing features of microbiome composition. These interactions involve advanced communication and exchange processes between different microbial populations (Kennedy & Chang, 2020). Broadly, microbial interactions can be classified into several types: mutualistic relationships, where species benefit each other; competitive interactions, where organisms compete for limited resources; and commensalistic associations, where one species benefit while the other remains unaffected (Weiland-Bräuer, 2021). Through QS, microbial communities can respond collectively to environmental changes, acting almost like multicellular organisms. This coordinated behavior allows them to synchronize gene expression and adapt to conditions based on factors like population density and external stimuli (Prescott & Decho, 2020).

Microbial interactions form a highly intricate and interconnected network, going beyond simple binary relationships to create complex systems where multiple species engage through various molecular mechanisms. These interactions occur through metabolic exchanges, production of antimicrobial compounds, competition for resources, and transfer of genetic information, all of which contribute to the functionality and stability of microbial communities (Chiquito-Contreras et al., 2024).

4.4 Factors Influencing Microbiome Development

Microbiome development represents an intricate and dynamic process shaped by a complex interplay of environmental, host, dietary, and age-related factors that collectively determine the composition, diversity, and functionality of microbial communities (Sarkar et al., 2024). The formation and evolution of microbiomes begin even before birth, with initial microbial colonization influenced by multiple interconnected variables that continuously reshape microbial ecosystems throughout an organism's lifetime (Suárez-Martínez et al., 2023).

Environment-related factors have a significant impact on the architecture of microbial communities in various habitats, making them a crucial component of microbiome formation (Sessitsch et al., 2023). Significant differences in microbiome composition are introduced by geographic location, with microbial diversity and population dynamics directly influenced by environmental factors such as climate, soil composition, air quality, and local biodiversity (Matthews et al., 2024). Urban and rural areas often exhibit distinct microbiome features, with anthropogenic activity, pollution, and population density all having a substantial impact on the topologies of microbial ecosystems (Vinogradova et al., 2024). Furthermore, there are different ways to being exposed to various environmental bacteria, such as by coming into direct touch with soil, water, air, and other living things. While highly sterilized or artificially regulated environments can limit microbial exposure and possibly jeopardize microbiome development, natural environments with high biodiversity are likely to support more resilient and diverse microbial populations. By adding chemical contaminants and upsetting natural microbial equilibriums, industrial operations, agricultural methods, and human interventions can significantly change environmental microbiomes (Ahn & Hayes, 2021).

Host factors represent another crucial dimension in microbiome development, with genetic background, immune system characteristics, and physiological conditions significantly influencing microbial colonization and community establishment (Zheng et al., 2020). Genetic variations can predispose individuals to specific microbiome compositions, with certain genetic traits potentially promoting or inhibiting the growth of particular microbial populations (Bubier et al., 2021). The host's immune system plays a sophisticated regulatory role, simultaneously protecting against potential pathogens while allowing beneficial microorganisms to colonize and establish symbiotic relationships (Zheng et al., 2020).

Hormonal changes and metabolic processes within the host organism continuously interact with microbial communities, creating dynamic feedback loops that modify microbiome structures (Zhang et al., 2023). Stress responses, endocrine function, and immune system modulation can all significantly impact microbial population dynamics, demonstrating the intricate interconnectedness between host physiology and microbial ecosystems (Beurel, 2024).

Dietary influences represent a powerful and immediate mechanism for shaping microbiome development, with nutritional intake directly impacting microbial community composition and functionality (Zhang, 2022). Different dietary patterns introduce specific nutrients, prebiotics, and substrates that selectively promote or inhibit particular microbial populations. Plant-based diets typically encourage greater microbial diversity, while processed and high-fat diets can lead to reduced microbial complexity and potentially disrupt community structures (Sidhu et al., 2023). Additionally, specific dietary components like fibers, complex carbohydrates, and polyphenols serve as crucial substrates for microbial metabolism, supporting the growth of beneficial bacterial populations (Ademosun et al., 2025). Probiotics and fermented foods introduce live microorganisms that can temporarily or permanently modify microbiome compositions, while dietary diversity ensures a more robust and adaptable microbial ecosystem. Nutritional interventions have emerged as potential strategies for modulating microbiome health and addressing various physiological conditions (Hitch et al., 2022). Furthermore, age-related changes fundamentally transform microbiome dynamics throughout an organism's developmental trajectory. Initial microbiome establishment occurs during birth, with the delivery method significantly influencing early microbial colonization. Vaginal births expose infants to maternal vaginal and intestinal microbes, while caesarean sections introduce different initial microbial populations (Zhang et al., 2021). Early childhood represents a critical period of microbiome development, with rapid colonization and establishment of foundational microbial communities. As individuals progress through different life stages, microbiome compositions continuously evolve, reflecting changing physiological conditions, environmental exposures, and metabolic requirements (Ronan et al., 2021). Adolescence, reproductive years, and elderly stages each present unique microbiome characteristics, with gradual shifts in microbial diversity and functional capabilities. Immune system maturation, hormonal changes, and accumulated environmental exposures collectively contribute to these age-related microbiome transformations (Carson et al., 2023).

Understanding the multifaceted nature of microbiome development requires an integrative approach that recognizes the dynamic and interconnected relationships between environmental, host, dietary, and age-related factors, highlighting the extraordinary complexity of microbial ecosystem formation.

4.5 Biofilm Formation and Development

Biofilm formation is a remarkable microbial survival strategy, enabling individual bacterial cells to transition into highly organized and resilient communities through a carefully orchestrated developmental process (Sharma et al., 2023). This transformation begins with the initial attachment to a surface, a pivotal phase driven by intricate molecular interactions between bacterial surface adhesins and environmental substrates. Key components such as flagella, pili, and EPSs facilitate this process, overcoming repulsive forces and enabling bacteria to establish a stable connection with the surface, laying the foundation for biofilm development (Rather et al., 2021).

During the initial attachment stage, bacterial cells undergo profound physiological changes, activating specific gene expression patterns to enhance adhesion. This process is further influenced by surface conditioning layers composed of organic and inorganic molecules that form a preliminary matrix, promoting bacterial attachment (Kreve & Reis, 2021). Key molecular interactions, including hydrophobic interactions, electrostatic forces, and Van der Waals forces, work in concert to facilitate firm adhesion to diverse surfaces, ranging from living tissues to industrial materials (Chen et al., 2023). These mechanisms establish the foundation for biofilm maturation and persistence.

Rapid cell proliferation and the first production of extracellular matrix components are characteristics of the crucial subsequent phase known as microcolony formation (Shree et al., 2023). At this point, QS mechanisms play a crucial role in enabling bacteria to coordinate gene expression according to population density (Juszczuk-Kubiak, 2024). Exopolysaccharides, proteins, and extracellular DNA are secreted by bacterial cells, forming a three-dimensional scaffolding that offers structural integrity and defense against external stresses (Kaur & Dey, 2023).

The maturation stage represents the most complex phase of biofilm development, where the community develops a sophisticated architecture with distinct functional zones. Sophisticated channel systems emerge within the biofilm, facilitating nutrient distribution and waste removal through a process resembling primitive circulatory networks (Sauer et al., 2022). Different bacterial populations establish spatial segregation, creating microenvironments with distinct chemical gradients, pH levels, and oxygen concentrations that support diverse metabolic activities (Sharma et al., 2023).

Dispersal, the final stage, demonstrates the biofilm's most dynamic adaptation strategy (Wille & Coenye, 2020). Triggered by environmental cues such as nutrient depletion, overcrowding, or external stress, a subset of bacterial cells detaches from the primary biofilm. These mobile cells, often exhibiting altered phenotypic characteristics, possess enhanced survival capabilities that enable colonization of new environments (Sharma et al., 2023). Molecular mechanisms like cyclic-di-GMP signaling play a crucial role in regulating the transition between sessile and motile states (Park & Sauer, 2022).

The complexity of biofilm formation extends beyond mere bacterial aggregation, representing a sophisticated multicellular behavior that challenges traditional perspectives on microbial existence. Each stage involves intricate molecular communication, genetic regulation, and collective adaptive responses that transform individual bacterial cells into a resilient, functional community (Almatroudi, 2024).

4.6 Structural Components of Biofilms

Biofilms represent far more than simple bacterial aggregations, emerging as complex architectural systems defined by intricate structural components that enable remarkable collective survival strategies. The fundamental scaffolding of these microbial communities is the EPS, a sophisticated matrix that transforms individual bacterial cells into a cohesive, protective environment (Karygianni et al., 2020).

EPSs represent the architectural cornerstone of biofilm structure, comprising a complex molecular matrix that transcends simple chemical composition to become a dynamic, multifunctional biological system (Pinto et al., 2020). At the molecular level, EPSs typically consist of heteropolysaccharides, proteins, nucleic acids, and lipids, with polysaccharides constituting 50–90% of the total organic content, creating a sophisticated three-dimensional scaffold that provides mechanical stability, protection, and communication infrastructure (Li et al., 2021).

The molecular complexity of EPSs reveals intricate structural variations across different bacterial species (Netrusov et al., 2023). For instance, *Pseudomonas aeruginosa* produces alginate-rich EPSs with unique guluronic and mannuronic acid compositions that provide exceptional mechanical strength and environmental resistance (Guillaume et al., 2022). In *Staphylococcus epidermidis*, poly-*N*-acetyl-β-(1-6)-glucosamine (PNAG) serves as a primary EPS component, facilitating cellular aggregation and providing critical protective functions against host immune responses (Tuon et al., 2023).

The architectural organization within biofilms demonstrates remarkable spatial complexity, with microcolonies creating distinct functional zones characterized by precise metabolic gradients (Sauer et al., 2022). Microscopic channel networks enable nutrient distribution and waste removal, resembling primitive circulatory systems. These channels are not randomly distributed but strategically organized to optimize resource utilization, with oxygen-rich outer regions supporting aerobic metabolic processes and deeper zones supporting anaerobic bacterial populations.

Communication systems within biofilms utilize sophisticated molecular signaling mechanisms, primarily through QS systems. Specific molecular signals like *N*-acyl homoserine lactones (AHLs) in Gram-negative bacteria and autoinducing peptides (AIPs) in Gram-positive organisms enable population-wide gene expression coordination. These communication networks allow bacteria to collectively respond to environmental challenges, regulate virulence factors, and make strategic decisions about biofilm development and dispersal (Boo et al., 2021).

From an evolutionary perspective, biofilm structural components represent a profound adaptation strategy that emerged approximately 3.5 billion years ago (Parrilli et al., 2022). The ability to form structured communities provided significant survival advantages, allowing microorganisms to withstand environmental stressors, share genetic information, and access resources more efficiently than individual planktonic cells (Kang et al., 2023). This evolutionary innovation enabled bacterial populations to colonize extreme environments and develop complex symbiotic relationships across diverse ecological niches.

The molecular mechanisms underlying biofilm formation demonstrate intricate genetic regulation, involving multiple gene clusters that control adhesion, EPS production, and communication (Wang et al., 2023). For example, the *ica* operon in *Staphylococcus aureus* directly regulates biofilm formation by controlling PNAG production, while the *pel* and *psl* operons in *Pseudomonas aeruginosa* manage polysaccharide synthesis critical for biofilm architectural development (Peng et al., 2022).

Specific molecular examples highlight the sophisticated nature of these structures. The protein TasA in *Bacillus subtilis* serves as a primary amyloid fiber component, providing mechanical strength to the biofilm matrix (Böhning et al., 2022). Similarly, extracellular DNA (eDNA) released during cell lysis acts as a critical structural component, facilitating cellular adhesion and providing additional matrix stability through electrostatic interactions (Panlilio & Rice, 2021).

The multifunctionality of biofilm structural components extends beyond mere physical protection. These molecular systems enable genetic exchange through horizontal gene transfer, provide resistance against antimicrobial agents, and create microenvironments that support diverse metabolic activities (Mishra et al., 2023). This complexity represents a sophisticated biological strategy that transforms individual bacterial cells into highly adaptive, collective organisms capable of surviving in challenging environments.

4.7 Regulatory Mechanisms

Biofilms represent sophisticated microbial communities regulated by intricate molecular mechanisms that enable collective adaptation and survival. QS emerges as the primary communication system through which bacterial populations coordinate their behaviors, transforming individual cellular responses into synchronized community-wide strategies (Rattray et al., 2022).

QS emerges as the most sophisticated regulatory mechanism in bacterial communities, representing a molecular communication system that enables collective behavior regulation through population density-dependent gene expression (Spacapan et al., 2023). At the molecular level, QS operates through specialized signaling molecules called autoinducers, which act as linguistic molecules allowing bacteria to "count" their population and coordinate complex collective responses (Vadakkan et al., 2018).

Different bacterial species utilize distinct QS molecular architectures. In Gram-negative bacteria, AHLs serve as primary signaling molecules, with specific synthetase enzymes producing these chemical messengers (Chan et al., 2015). Gram-positive bacteria employ more complex autoinducing peptides (AIPs) that interact with two-component regulatory systems, enabling intricate genetic response mechanisms. The LuxI/LuxR system in *Vibrio fischeri* represents a classic example, where signal production and receptor proteins work synergistically to modulate gene expression based on population density (Rutherford & Bassler, 2012).

Molecular-level details reveal extraordinary complexity in these regulatory networks. The cyclic-di-GMP signaling pathway plays a crucial role in controlling the transition between motile and sessile bacterial states (Park & Sauer, 2022). This second messenger molecule acts as a molecular switch, regulating biofilm formation through precise interactions with various regulatory proteins. When cyclic-di-GMP concentrations increase, bacteria reduce motility and enhance EPS production, facilitating biofilm establishment (Poulin & Kuperman, 2021).

From an evolutionary perspective, QS represents a profound adaptation that emerged approximately 2.5 billion years ago, transforming bacterial populations from isolated entities to cooperative communities (Szydlowski et al., 2024). This regulatory mechanism provided significant survival advantages by enabling collective resource management, enhanced defense strategies, and more efficient environmental adaptation (Weiskopf et al., 2020). The ability to coordinate behaviors allowed bacterial populations to survive in challenging environments, develop complex symbiotic relationships, and ultimately contribute to fundamental ecological processes (Bang et al., 2018).

The practical implications of these regulatory mechanisms extend across multiple domains. In medical contexts, understanding QS enables the development of novel therapeutic strategies targeting bacterial communication systems (Su & Ding, 2023). Antimicrobial approaches like QS inhibition can potentially disrupt pathogenic biofilm formation without directly killing bacteria, reducing the risk of developing antimicrobial resistance (Zhao et al., 2020).

Industrial applications leverage these regulatory mechanisms in biotechnological processes. Biofilm-based systems for wastewater treatment, bioremediation, and sustainable energy production utilize the remarkable adaptive capabilities of these microbial communities (Kuppan et al., 2024). By manipulating regulatory mechanisms, researchers can engineer more efficient microbial systems for environmental and industrial applications. Gene regulation within biofilms involves complex multilayered control systems. Regulatory genes like *luxS*, *rpoS*, and various two-component regulatory systems coordinate sophisticated genetic responses. These systems enable bacteria to rapidly modify gene expression in response to environmental changes, activating stress response mechanisms, metabolic adaptations, and virulence factor production (Hajiagha & Kafil, 2023).

Stress response mechanisms represent another critical regulatory strategy, involving molecular defense systems that provide enhanced survival capabilities. Alternative sigma factors like RpoS enable rapid physiological adaptations, activating DNA repair mechanisms, efflux pumps, and protective metabolic pathways. These systems allow bacterial communities to withstand antimicrobial agents, oxidative stress, and extreme environmental conditions (Dawan & Ahn, 2022).

The metabolic adaptation capabilities of biofilms demonstrate remarkable flexibility, creating microenvironments with specialized metabolic zones. Different bacterial populations establish cooperative nutrient utilization strategies, implementing complex resource-sharing mechanisms that enable survival in nutrient-limited environments. Metabolic flexibility allows biofilms to switch between different metabolic pathways, ensuring community survival under diverse conditions (Zhao et al., 2023).

Evolutionary evidence suggests these regulatory mechanisms represent a fundamental survival strategy that transformed microbial life. By enabling collective behaviors, bacteria developed capabilities far beyond individual cellular functions, creating adaptive systems capable of responding to environmental challenges with unprecedented sophistication.

4.8 Microbiome–Biofilm Interactions

Microbiome–biofilm interactions represent a complex and dynamic ecological phenomenon that transcends simple microbial associations, revealing intricate networks of communication, competition, and cooperation within microbial communities (Weiland-Bräuer, 2021). These interactions form the foundation of ecological relationships that shape the function and survival of diverse microbial ecosystems across multiple environments.

Symbiotic interactions emerge as a fundamental mechanism of microbiome and biofilm relationships, where different microbial species establish mutually beneficial associations that enhance survival and function. These relationships range from direct nutritional exchanges to complex metabolic collaborations that enable survival in challenging environmental conditions (Afridi et al., 2022). For instance, certain bacterial species within a microbiome can provide essential nutrients to other community members, creating intricate interdependence networks that support overall community resilience (Daisley et al., 2021).

Competition represents another critical dimension of microbiome–biofilm interactions, where different microbial populations vie for limited resources and environmental niches (Foster & Bell, 2012). Competitive mechanisms include the production of antimicrobial compounds, rapid nutrient consumption, and strategic spatial positioning within the microbial community. These competitive interactions drive evolutionary adaptations and help maintain microbial diversity by preventing any single species from completely dominating the ecosystem (Granato et al., 2019).

Cooperation among microbial populations demonstrates sophisticated collective behaviors that extend beyond simple resource sharing (Cremer et al., 2019). Through complex communication mechanisms like QS, different bacterial species coordinate their activities, share genetic information, and collectively respond to environmental challenges (Preda & Săndulescu, 2019). These cooperative strategies enable microbiomes to develop adaptive responses that individual species could not achieve independently.

Resource sharing emerges as a fundamental strategy in microbiome–biofilm interactions, where different microbial populations exchange metabolites, nutrients, and genetic information (Valiei et al., 2024). Metabolic cross-feeding allows bacteria to utilize diverse substrates, creating sophisticated ecological networks where waste products from one species become nutritional resources for another. This intricate resource management enables microbial communities to survive and thrive in nutrient-limited environments (Culp & Goodman, 2023).

Spatial organization within microbiome–biofilm systems reveals remarkable architectural complexity, where different microbial populations establish strategic positioning to optimize resource utilization and community survival (Dal Co et al., 2023). Microcolonies develop distinct functional zones with specialized metabolic capabilities, creating microenvironments that support diverse bacterial populations and enable sophisticated ecological interactions (Dey et al., 2025).

The interactions between microbiomes and biofilms represent a dynamic, adaptive system that challenges the traditional understanding of microbial life. These relationships demonstrate how microbial communities function as complex, coordinated ecosystems capable of remarkable collective behaviors that ensure survival and functionality across diverse environmental conditions (Li et al., 2024).

4.9 Impact on Host Systems: A Comprehensive Exploration

The interplay between microbiomes and host systems is a cornerstone of biological science, offering a window into the symbiotic and sometimes antagonistic relationships that define life. Far beyond their roles as resident microbes, these communities wield immense influence over host physiology, demonstrating their ability to modulate immunity, transform metabolism, and contribute to both health and disease states. This exploration delves into the key aspects of microbiome–host interactions, unravelling their multifaceted impact on host systems.

4.9.1 Immune Modulation: Orchestrating Host Immune Responses

The microbiome plays a pivotal role in shaping immune system development and functionality through intricate and multifaceted mechanisms. From early life, microbial interactions are critical in programming immune responses, helping distinguish between beneficial and harmful microorganisms. Commensal microbes stimulate the maturation of immune cells, particularly T-regulatory cells, which are essential for maintaining immune homeostasis and preventing excessive inflammatory responses (Zhao et al., 2020).

The interactions between host immune systems and microbiomes involve complex signaling pathways that enable bidirectional communication. Immune cells continuously sample and interact with microbial populations, developing tolerance to beneficial microbes while maintaining robust defensive capabilities against potential pathogens. This delicate balance is mediated through pattern recognition receptors that can detect microbial molecular signatures, triggering appropriate immune responses (Kogut et al., 2020).

4.9.2 Metabolic Influences: Transforming Nutrient Processing

Microbiomes serve as intricate metabolic engines, dramatically expanding the host's metabolic capabilities beyond its inherent physiological limitations. These microbial communities facilitate the breakdown of complex nutrients that human digestive enzymes cannot process independently, particularly in carbohydrate metabolism and short-chain fatty acid production (Rowland et al., 2018). Gut microbiota is particularly adept at fermenting dietary fibers, generating metabolites that provide energy sources for intestinal epithelial cells and contribute to overall metabolic health (Myhrstad et al., 2020).

The metabolic interactions extend beyond nutrient processing, influencing hormone regulation, neurotransmitter production, and systemic metabolic homeostasis (Reed et al., 2024). Microbial metabolites can directly modulate host metabolism, affecting insulin sensitivity, lipid metabolism, and energy expenditure. Some bacterial strains produce vitamins and essential amino acids, effectively acting as supplementary metabolic organs within the host system (Zhang et al., 2023).

4.9.3 Disease Associations: Mapping Microbial Contributions to Pathological Conditions

Emerging research reveals profound connections between microbiome dysbiosis and various pathological conditions, suggesting that microbial imbalances can significantly contribute to disease development (Madhogaria et al., 2022). Chronic inflammatory conditions like inflammatory bowel disease, rheumatoid arthritis, and certain autoimmune disorders demonstrate strong correlations with alterations in microbial composition and function. Neurological conditions, including depression, anxiety, and neurodegenerative diseases, are increasingly linked to microbiome-related inflammatory and metabolic processes (Kim et al., 2023).

The concept of a "microbial signature" for different diseases is gaining traction, with researchers identifying specific microbial patterns associated with cancer, metabolic syndrome, and cardiovascular diseases (Chakaroun et al., 2020). Biofilms can potentially harbor persistent infection reservoirs, contributing to chronic disease states by creating protective environments for potentially pathogenic microorganisms. These complex microbial communities can modulate host immune responses, potentially triggering or exacerbating inflammatory cascades (Perry & Tan, 2023).

4.9.4 Beneficial Effects: Protective and Supportive Host Interactions

Microbiomes and biofilms are not merely passive inhabitants but active contributors to host health and resilience. Beneficial microbes provide critical protective functions, competing with potentially harmful pathogens and preventing their colonization through mechanisms like competitive exclusion and production of antimicrobial compounds. The skin and mucosal microbiomes, for instance, form dynamic protective barriers against external microbial threats (Aggarwal et al., 2023).

Probiotic microorganisms demonstrate remarkable capabilities in supporting host physiological functions, including enhancing mucosal barrier integrity, modulating inflammatory responses, and supporting digestive processes (Mazziotta et al., 2023). Some microbial strains produce compounds that directly support cellular repair mechanisms, boost immune function, and contribute to overall systemic homeostasis. The emerging field of microbiome therapeutics explores leveraging these beneficial interactions for targeted interventions in various health conditions (Gulliver et al., 2022).

4.9.5 Integrative Perspectives: Holistic Understanding of Host–Microbiome Interactions

The impact of microbiomes on host systems transcends traditional disciplinary boundaries, representing a complex, dynamic ecosystem of interactions. Each host can be conceptualized as a holobiont a superorganism where host and microbial components are intricately interconnected and coevolved. This perspective emphasizes the need for comprehensive, interdisciplinary approaches to understanding health and disease (Simon et al., 2019).

As research methodologies advance, particularly in multi-omics technologies and computational modeling, our understanding of these intricate host–microbiome interactions continues to deepen. The emerging paradigm suggests that health should be viewed not as a binary state but as a dynamic equilibrium maintained through constant, sophisticated interactions between the host and microbial systems (Abavisani et al., 2024).

This comprehensive exploration underscores the profound significance of microbiomes and biofilms in shaping host physiological landscapes, promising revolutionary insights into health, disease, and potential therapeutic interventions.

4.10 Translational Applications: Bridging Scientific Discovery and Clinical Innovation

The field of microbiome research represents a revolutionary frontier in medical science, transforming our understanding of health and disease through intricate microbial interactions. Translational applications bridge the gap between fundamental scientific discoveries and practical clinical interventions, offering unprecedented insights into human biology. Microbiome research provides a holistic approach to understanding complex biological systems, moving beyond traditional reductionist medical models. The convergence of advanced technologies and a deep microbial understanding enables researchers to develop innovative diagnostic, therapeutic, and preventive strategies. Figure 4.1 illustrates the dynamic interactions between the human body and microbiota, highlighting their roles in digestion, immunity, pathogen competition, and disease prevention. These approaches promise to reshape our comprehension of human health by viewing the body as an interconnected ecosystem of microbial and human components.

4.10.1 Medical Applications: Innovative Microbiome-Based Interventions

The convergence of microbiome research and medical science represents a groundbreaking frontier in healthcare innovation. Microbiome-based interventions offer unprecedented opportunities to understand, diagnose, and treat complex medical conditions through intricate microbial ecosystem manipulation (Sasso et al., 2023).

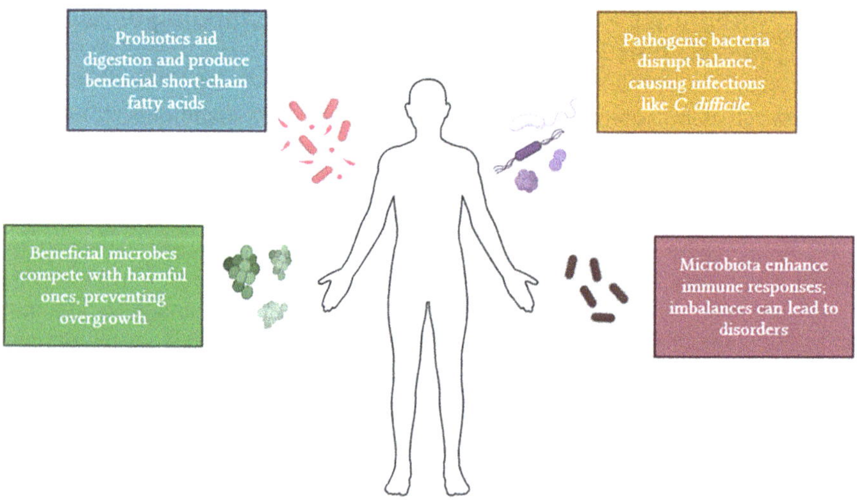

Fig. 4.1 The intricate interactions between the human body and microbiota, emphasizing their contributions to digestion, immune function, pathogen competition, and disease prevention

By leveraging the dynamic interactions between microbial communities and human physiological systems, researchers are developing targeted, personalized therapeutic strategies that move beyond traditional medical paradigms. These innovative approaches promise to revolutionize our understanding of health, disease prevention, and precision medicine by treating the human body as a complex, interconnected biological network (Shukla et al., 2024).

4.10.2 Therapeutic Interventions

Microbiome-based therapeutic strategies represent a paradigm shift in medical treatment methodologies. Scientists are developing targeted interventions that can precisely modulate microbial communities to restore physiological balance and prevent disease progression. These approaches move beyond traditional pharmacological treatments by understanding and manipulating the complex microbial ecosystems within human bodies (Kamel et al., 2024). Personalized microbiome therapies offer the potential to address individual variations in microbial composition and its impact on health. The ability to engineer microbial consortia with specific therapeutic capabilities opens new frontiers in precision medicine (Huang et al., 2023).

4.10.3 Diagnostic Tools

Advanced diagnostic technologies are emerging that utilize microbiome signatures as comprehensive health indicators. Metagenomic sequencing enables unprecedented detailed mapping of microbial populations within human systems. These diagnostic approaches can potentially predict disease risks, track health trajectories, and provide early warning signs of physiological disruptions (Abavisani et al., 2024). Molecular techniques allow for high-resolution characterization of microbial communities, offering insights far beyond traditional diagnostic methods. The integration of multi-omics data creates a more holistic understanding of an individual's health status through microbial profiling (Babu & Snyder, 2023).

4.10.4 Drug Delivery Systems

Microbiome research is revolutionizing drug delivery methodologies by developing innovative transportation and release mechanisms. Researchers are engineering probiotic strains that can serve as targeted drug delivery agents, enhancing treatment precision and effectiveness (Abouelela & Helmy, 2024). Biofilm-inspired encapsulation technologies provide advanced methods for protecting and releasing therapeutic compounds. These approaches enable site-specific drug targeting, minimizing systemic side effects and improving treatment outcomes. Microbial metabolism can be leveraged to create adaptive and responsive drug delivery systems (Sousa et al., 2023).

4.10.5 Antimicrobial Strategies

Contemporary antimicrobial approaches transcend traditional antibiotic treatments by developing more sophisticated intervention strategies. Researchers are exploring selective targeting of pathogenic microorganisms while preserving beneficial microbial populations (Muteeb et al., 2023). Innovative techniques like bacteriophage therapies and quorum sensing inhibition offer alternatives to broad-spectrum antibiotics. These methods aim to disrupt harmful microbial communities without causing widespread ecological disruption in the microbiome. Competitive exclusion mechanisms provide novel ways of managing microbial populations and preventing infectious processes (Jo et al., 2023).

4.10.6 Integrative Perspectives

Translational microbiome research represents a fundamental reimagining of medical science, emphasizing the interconnectedness of biological systems. The approach recognizes that health is not a static condition but a dynamic equilibrium maintained through complex microbial interactions. Interdisciplinary collaboration becomes crucial in advancing our understanding and application of microbiome-based interventions (Tomasulo et al., 2024). Future medical treatments will likely be personalized, precise, and guided by comprehensive microbial ecosystem analysis. Additionally, the future of medical science lies in understanding and harnessing the power of microbial communities (Shukla et al., 2024). Microbiome research offers a holistic, nuanced approach to health management that goes beyond treating symptoms to addressing fundamental biological interactions (Tomasulo et al., 2024). As technologies advance, we can anticipate increasingly sophisticated, personalized medical interventions derived from microbiome insights. The potential for improving human health through these innovative approaches is profound and largely unexplored (Table 4.1).

4.11 Industrial Applications: Microbiome and Biofilm Innovations

Microbiome research represents a transformative technological frontier across multiple industrial sectors. Microbial communities offer unprecedented opportunities for sustainable and efficient industrial processes. These innovative strategies leverage sophisticated ecosystem interactions to address critical global challenges (Fig. 4.2). The unique metabolic and adaptive capabilities of microorganisms provide revolutionary solutions to complex technological problems.

Table 4.1 An overview of the pivotal aspects of translational microbiome research, encompassing medical applications, therapeutic interventions, diagnostic tools, drug delivery systems, and antimicrobial strategies

Applications	Highlights	References
Medical applications	Microbiome-based interventions for personalized therapeutic strategies	Sasso et al. (2023)
	Innovations in understanding, diagnosing, and treating complex conditions by integrating microbial insights	Shukla et al. (2024)
Therapeutic interventions	Targeted modulation of microbial communities to restore balance and prevent disease	Kamel et al. (2024)
	Engineering microbial consortia for precision medicine	Huang et al. (2023)
Diagnostic tools	Metagenomic sequencing for detailed mapping of microbial populations	Abavisani et al. (2024)
	Multi-omics integration for holistic health profiling	Babu and Snyder (2023)
Drug delivery systems	Probiotic strains and biofilm-inspired technologies for site-specific drug targeting	Abouelela and Helmy (2024)
	Leveraging microbial metabolism for adaptive drug delivery systems	Sousa et al. (2023)
Antimicrobial strategies	Selective pathogen targeting while preserving beneficial microbes	Muteeb et al. (2023)
	Techniques like bacteriophage therapies and QS inhibition	Jo et al. (2023)
Integrative perspectives	Health viewed as dynamic equilibrium maintained through microbial interactions	Tomasulo et al. (2024)
	Interdisciplinary collaboration for personalized, precise treatments	Shukla et al. (2024)

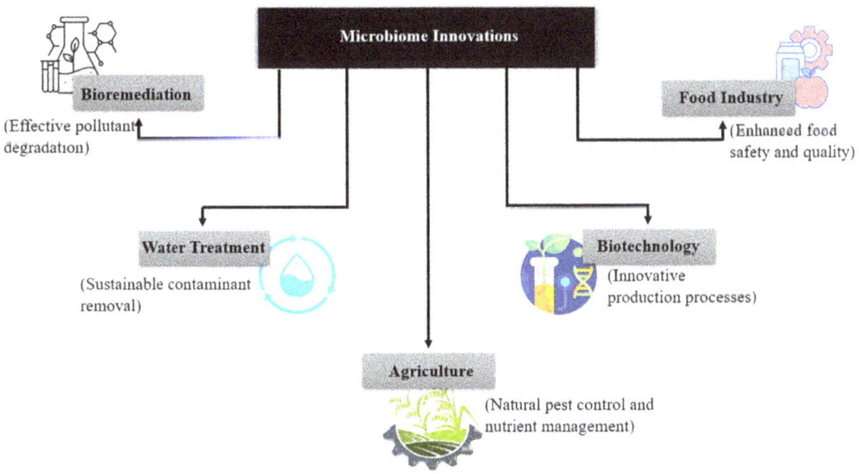

Fig. 4.2 The transformative role of microbiomes and biofilms in bioremediation, food production, water treatment, biotechnology, and agriculture, offering sustainable solutions to global challenges

4.11.1 Industrial Applications: Bioremediation

Bioremediation technologies leverage the extraordinary metabolic capabilities of microbial communities to effectively target and degrade environmental pollutants (Kondakindi et al., 2024). Specialized microbial consortia are designed to break down specific contaminants, while engineered microbial communities enable comprehensive strategies for environmental cleanup. These approaches are particularly effective in addressing complex contamination challenges in soil, water, and industrial sites (Kuppan et al., 2024). Among the various uses of bioremediation are the degradation of petroleum hydrocarbons, the removal of heavy metals, and the neutralization of toxic chemicals through specialized microbial interactions. Furthermore, pollution-affected ecosystems are restored through the use of sophisticated microbial engineering techniques (Das & Chandran, 2011). These creative techniques highlight the critical role bioremediation plays in addressing important environmental issues.

4.11.2 Food Industry: Microbial Innovation Narrative

Microbiome research is revolutionizing food production through advanced microbial management strategies. Fermentation processes benefit from precisely engineered microbial populations. Probiotic development enables enhanced food safety and quality management. Microbial technologies create more sustainable and nutritionally optimized food production methodologies (Abbaspour, 2024). This work enables the development of sophisticated starter cultures via microbial engineering, which improves fermentation results in the food business. It also promotes the creation of functional meals, which use intricate microbial metabolism to offer extra health advantages. Research on the microbiome is advancing sustainable production methods, and focused microbial control improves preservation tactics, which eventually results in safer and better food items (Sun et al., 2022).

4.11.3 Water Treatment: Microbial Solution Approach

Microbiome technologies are revolutionizing water treatment by employing ecological management strategies to enhance sustainability and efficiency. Biofilm-based filtration systems offer advanced methods for contaminant removal, while microbial communities contribute to the development of resilient water treatment systems. These innovative approaches enable sustainable and ecologically sensitive water management (Lago et al., 2024). Through microbial engineering, this discovery makes it easier to create sophisticated starter cultures, which improves fermentation results in the food business. Additionally, it promotes the creation of functional foods that take advantage of intricate microbial metabolism to offer more health advantages. Additionally, research on the microbiome is promoting sustainable production methods, and focused microbial management improves preservation methods, which eventually leads to safer and better food items (Iqbal et al., 2023).

4.11.4 Biotechnology: Microbial Innovation Frontiers

Microbiome research is driving transformative biotechnological advancements with significant industrial applications. Bioprocess engineering has enabled the development of advanced microbial production systems, while innovative microbial manipulation techniques have paved the way for the creation of novel biomaterials. These breakthroughs promise more sustainable and technologically advanced manufacturing processes (Mcmeeking et al., 2024). Notable new uses include the development of innovative enzymes based on in-depth microbiome research and biofuel production systems powered by sophisticated microbial metabolic engineering. The complex relationships found in microbial ecosystems are also improving the production of sustainable chemicals. Moreover, precision microbial genetic manipulation is enabling the synthesis of novel biomaterials (Sartaj et al., 2022).

4.11.5 Transformative Potential of Microbiome Technologies

Microbiome and biofilm technologies represent a revolutionary industrial innovation approach. Continuous research promises increasingly sophisticated microbial-based solutions. The integration of microbiome research transforms technological and environmental problem-solving strategies. Understanding microbial ecosystem capabilities becomes crucial for future industrial development (Berg et al., 2020). The future of industrial innovation lies in harnessing the adaptive capabilities of microbial communities. Interdisciplinary research will continue to unlock unprecedented technological possibilities. Microbial ecosystems offer a sustainable, efficient, and innovative approach to addressing global industrial challenges (Rame et al., 2024).

4.11.6 Agricultural Applications: Microbiome and Biofilm Perspectives in Agriculture

Research on microbiomes and biofilms has transformed agricultural science, creating new possibilities for enhancing crop production, sustainability, and ecosystem health. Microbiomes consist of intricate communities of microorganisms that interact with plants, soil, and their environment. These interactions significantly affect agricultural productivity, plant vitality, and the resilience of ecosystems. By investigating microbiomes and biofilms, scientists gain valuable insights into the roles of microbes in agricultural systems, enabling the creation of innovative strategies to address global agricultural challenges (Kabir et al., 2024). One of the most promising areas of research is the role of microbiomes in crop protection. Beneficial microorganisms within these communities serve as natural defenders against plant pathogens, providing sustainable alternatives to chemical pesticides. These microbes generate antimicrobial compounds, compete with harmful pathogens for resources, and stimulate plant defense mechanisms, forming a protective network around plants (Suman et al., 2022). Biofilms—structured communities of microorganisms embedded in self-produced extracellular matrices—further enhance plant protection by creating biofilm layers on plant roots and surfaces. These biofilms not only protect plants from pathogenic microbes but also aid in nutrient absorption and promote growth, establishing an effective natural defense system (Pandit et al., 2020).

Soil health is a crucial factor influenced by the microbiome. Soil microbiomes play a significant role in nutrient cycling, enhancing soil structure, and fostering ecological resilience. The microbial communities present in the soil are essential for converting organic matter into vital nutrients, thereby improving soil fertility and maintaining ecosystem balance (Etesami, 2024). Additionally, biofilms in the soil contribute to stabilizing aggregates, facilitating nutrient cycling, and protecting soil microbes from environmental stressors. These interactions among microbes support soil health by promoting a diverse and resilient community, which is vital for sustaining soil quality and agricultural productivity (Bhattacharyya et al., 2023). Sustainable farming practices are increasingly adopting microbiome-based strategies to minimize chemical inputs and enhance resource efficiency. By utilizing the natural capabilities of microbiomes—such as nitrogen fixation, organic matter degradation, and plant growth regulation—farmers can lessen their reliance on synthetic fertilizers and pesticides (Mahmud et al., 2021). Biofilms also play a role in sustainable agriculture by establishing natural nutrient management systems and boosting soil microbial diversity. These practices not only reduce environmental impact but also enhance the long-term sustainability of farming systems by nurturing balanced ecosystems that support both crop health and biodiversity (Samantaray et al., 2024). Translational applications of microbiome and biofilm research are effectively connecting scientific discoveries to practical agricultural practices. The rapid development of microbial inoculants, precision farming technologies, and eco-friendly agricultural methods is underway (Trivedi et al., 2021). Advanced molecular techniques and bioinformatics tools enable the creation of tailored microbial interventions that address specific agricultural issues such as disease management, nutrient deficiencies, and soil degradation (Abdul Aziz & Masmoudi, 2024).

Altogether, research on microbiomes and biofilms holds transformative potential for agriculture by providing solutions for sustainable farming, crop protection, and soil health. By leveraging the capabilities of microbial ecosystems, agriculture can reduce its environmental impact while enhancing productivity and resilience. As research progresses, these strategies will increasingly contribute to meeting the rising demands of global food production while maintaining ecological balance.

4.12 Modern Analysis Methods: Molecular Techniques in Microbiome Research

Molecular techniques have revolutionized microbiome research by providing unprecedented insights into microbial ecosystems. Advanced technological approaches enable comprehensive exploration of microbial communities at molecular levels. These sophisticated methodologies transcend traditional observational techniques, offering a deep, nuanced understanding of microbial interactions and compositions (Fig. 4.3). Scientific researchers can now decode complex microbial networks with remarkable precision and detail.

4.12.1 Molecular Techniques: Comprehensive Analytical Approaches

Molecular techniques provide powerful tools to explore the complexities of biological systems at the genetic, cellular, and molecular levels. With advancements like next-generation sequencing (NGS), meta-omics approaches, and single-cell analysis, researchers can unravel intricate biological processes and interactions (Satam et al., 2023). These techniques, supported by advanced bioinformatics tools, enable precise analysis and integration of large datasets, driving innovations in research, diagnostics, and therapeutic development (Rahimian & Panahi, 2024).

4.12.1.1 Next-Generation Sequencing (NGS)

NGS has revolutionized microbiome analysis through its ability to rapidly and comprehensively examine genetic material with exceptional depth and accuracy. This transformative technology enables the simultaneous analysis of multiple microbial genetic samples, offering unprecedented insights into microbial diversity and genetic composition (Wensel et al., 2022). The key features of NGS include its capacity for massive parallel sequencing, allowing the processing of vast amounts

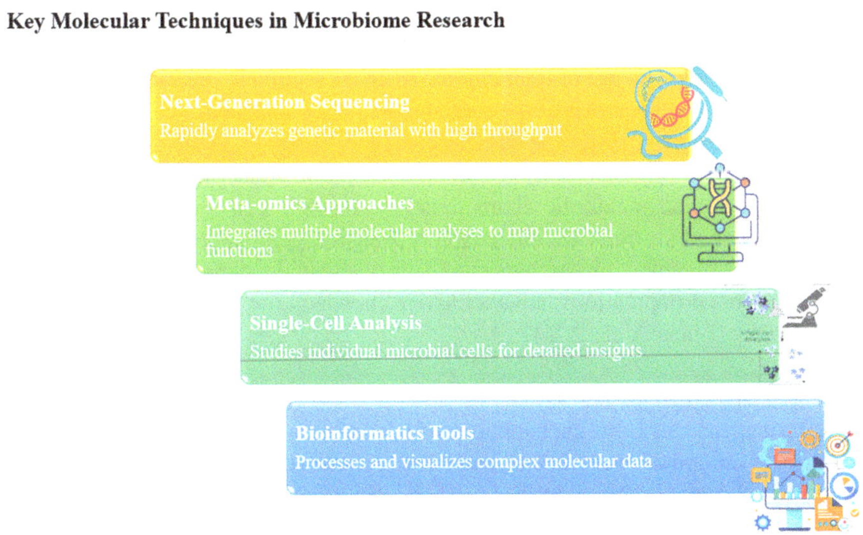

Fig. 4.3 Advanced molecular techniques driving microbiome research, including next-generation sequencing (NGS), meta-omics approaches, single-cell analysis, and bioinformatics tools. These methods enable deep exploration of microbial communities, providing insights into their genetic, functional, and metabolic interactions at unprecedented resolution and precision

of genetic information in a single run. High-throughput platforms can sequence millions of DNA fragments efficiently, while advanced algorithms ensure rapid data generation and precise interpretation. With minimal sample requirements, NGS enables comprehensive exploration of genomic landscapes and unparalleled mapping of genetic diversity at extraordinary levels of detail and precision (Buermans & den Dunnen, 2014).

4.12.1.2 Meta-omics Approaches

Meta-omics methodologies offer a comprehensive framework for exploring the intricate complexities of microbial ecosystems. By integrating multiple molecular-level analyses including genomics, transcriptomics, proteomics, and metabolomics these approaches enable researchers to map the functional capabilities and interactions within microbial communities with exceptional depth (Aplakidou et al., 2024).

Genomics provides a detailed analysis of the genetic material from entire microbial communities, while transcriptomics reveals active gene expression patterns, shedding light on functional dynamics. Proteomics focuses on the composition and interactions of proteins, and metabolomics examines metabolic products and biochemical transformation processes (Jain et al., 2024). Collectively, these integrated meta-omics techniques deliver multidimensional insights, advancing our understanding of microbial ecosystem dynamics and their broader implications.

4.12.1.3 Single-Cell Analysis

Single-cell analysis techniques provide an unparalleled opportunity to study individual microbial cells with exceptional detail. These methodologies offer high-resolution insights into microbial heterogeneity, enabling the exploration of genetic, metabolic, and functional variations within microbial populations (Brehm-Stecher & Johnson, 2004). The use of laser capture microdissection to precisely isolate individual microbial cells is one of the fundamental skills that single-cell analysis provides. Furthermore, cellular architecture may be seen in great detail owing to advanced microscopy techniques (Hu et al., 2016). Molecular profiling uncovers the unique genetic and metabolic characteristics of single cells, while sophisticated computational algorithms facilitate comprehensive data interpretation. Through heterogeneity mapping, researchers can identify and understand complex variations within microbial populations, advancing our knowledge of microbial diversity and functionality (Liang et al., 2024).

4.12.1.4 Bioinformatics Tools

Bioinformatics tools form the backbone of advanced microbiome research, providing critical computational infrastructure for processing, interpreting, and visualizing molecular data. These tools leverage sophisticated algorithms to integrate diverse datasets, enabling the creation of comprehensive models of microbial ecosystems. Using complex computational algorithms to process big molecular datasets efficiently is one of the essential functions offered by bioinformatics tools. Predictive modeling is enhanced by machine learning techniques, and difficult data interpretation is facilitated by strong statistical approaches. Additionally, sophisticated visualization technologies convert complex molecular data into intelligible and straightforward visuals (Auslander et al., 2021). Integrated platforms further streamline cross-platform data analysis and comparison, fostering a holistic understanding of microbial ecosystems.

4.12.2 Technological Integration and Future Perspectives

Modern molecular techniques take a synergistic approach to microbiome research, combining multiple analytical strategies to achieve a comprehensive understanding of microbial ecosystems (Mohseni & Ghorbani, 2024). As technological advancements continue to expand research capabilities, they enable increasingly detailed and nuanced ecological investigations. Molecular techniques have fundamentally transformed microbiome research, offering advanced technological approaches that provide unprecedented insights into microbial ecosystems. These comprehensive strategies allow researchers to decode complex microbial interactions with remarkable precision (Ejaz et al., 2024). The integration of cutting-edge technologies ensures ongoing scientific discovery and innovation, further advancing our understanding of the microbial world.

4.13 Imaging Technologies in Microbiome and Biofilm Research

Imaging technologies have revolutionized our understanding of microbial ecosystems by providing unprecedented visual insights into complex microbial interactions. Advanced visualization techniques enable researchers to observe microscopic biological processes with extraordinary detail and precision. These technologies bridge the gap between theoretical understanding and direct observation of microbial communities, transforming our ability to comprehend the intricate world of microorganisms.

4.13.1 Imaging Technologies: Comprehensive Visualization Strategies

Microscopy serves as the foundational technology for exploring microbial ecosystems at cellular and molecular levels. Advanced microscopic techniques offer diverse approaches to visualizing microbial structures and interactions. Cutting-edge microscopy enables researchers to observe intricate details of microbial communities with unprecedented resolution, while different microscopy methods provide unique perspectives on the complexities of microbial ecosystems (Brehm-Stecher & Johnson, 2004). Confocal laser scanning microscopy allows three-dimensional visualization of microbial structures (Mhade & Kaushik, 2023). Electron microscopy provides ultra-high-resolution imaging of cellular components, while fluorescence microscopy enables real-time tracking of specific molecular interactions (Jeong & Kim, 2022). Super-resolution microscopy breaks traditional resolution limitations, and atomic force microscopy offers detailed surface structural investigations (Dankovich & Rizzoli, 2021).

Biofilm visualization technologies provide comprehensive insights into the complex architectures of microbial communities. Advanced imaging techniques reveal the structural and functional characteristics of biofilm formations, allowing researchers to observe biofilm development, communication, and structural dynamics with remarkable precision. These visualization technologies have transformed our understanding of biofilm ecosystem complexities (Agarwal et al., 2025). Confocal microscopy reveals three-dimensional biofilm architectural structures, while fluorescent labeling techniques enable tracking of specific microbial populations. Live-cell imaging demonstrates dynamic biofilm formation processes, molecular imaging techniques map complex interactions within biofilm communities, and advanced computational reconstruction provides comprehensive biofilm ecosystem models (Barbosa et al., 2023).

In situ monitoring technologies enable direct observation of microbial communities within their native environments. These advanced techniques offer real-world insights into microbial ecosystem dynamics and interactions. Researchers can now study microbial behaviors under natural conditions with unprecedented detail, bridging the gap between laboratory observations and environmental realities (Gallardo-Navarro et al., 2024). Environmental scanning electron microscopy allows observations in native conditions, and microfluidic devices enable controlled in situ experimental investigations. Probe-based monitoring techniques track microbial activities in real-time, noninvasive imaging technologies minimize experimental disturbances, and integrated sensor systems provide continuous ecosystem monitoring (Tihlaříková et al., 2013).

Real-time analysis technologies have revolutionized the study of dynamic microbial processes. Advanced imaging techniques now enable instantaneous observation of microbial interactions and transformations (Arnold et al., 2016). Researchers can track molecular-level changes with extraordinary temporal resolution, providing unprecedented insights into microbial ecosystem dynamics. Time-lapse microscopy captures sequential microbial behavioral changes, while rapid molecular imaging tracks instantaneous cellular transformations (Balomenos et al., 2021). Live-cell tracking demonstrates complex microbial interaction mechanisms, high-speed computational processing enables immediate data interpretation, and integrated sensor technologies provide continuous ecosystem monitoring (Dutta et al., 2024).

Imaging technologies continue to evolve, offering increasingly sophisticated observation capabilities. Integrated approaches that combine multiple visualization techniques are enabling comprehensive understanding of microbial ecosystems. Technological advancements promise more nuanced and detailed insights, fostering continued scientific discovery and innovation (Ahuja & Zaheer, 2025).Imaging technologies have fundamentally revolutionized microbiome and biofilm research methodologies. Advanced visualization approaches provide unprecedented insights into the complexities of microbial ecosystems. Comprehensive imaging strategies empower researchers to observe and understand intricate microbial interactions with remarkable precision. The convergence of cutting-edge technologies promises continued innovation and transformative scientific discovery (Mohseni & Ghorbani, 2024).

4.14 Future Perspectives

The future of microbiome and biofilm research lies in leveraging innovative technologies to address critical challenges in health, industry, and the environment. Emerging technologies such as novel therapeutic approaches, advanced diagnostic methods, engineered microbiomes, and smart materials are poised to transform our understanding and application of microbial ecosystems (Kwa et al., 2023). The development of targeted microbiome therapies, including phage therapy and quorum sensing inhibitors (QSIs), offers precise interventions against pathogenic biofilms while preserving beneficial microbial communities. These therapies aim to mitigate antimicrobial resistance and enhance patient outcomes by manipulating microbial ecosystems (Rizvi et al., 2025). Additionally, high-throughput sequencing and real-time imaging technologies are paving the way for more accurate and rapid diagnostic tools. These advancements allow for detailed profiling of microbial communities, enabling early detection of infections and personalized therapeutic strategies based on microbiome composition and dynamics (Hemmati et al., 2024). Even though synthetic biology and genetic engineering are enabling the design of custom microbiomes tailored for specific applications, challenges remain in ensuring their stability and safety in complex environments. Engineered microbial communities can be utilized for bioremediation, agricultural enhancement, and even in therapeutic contexts to restore or modify host-associated microbiomes. Furthermore, the integration of biofilm research with materials science has led to the development of smart materials capable of interacting with microbial communities. These include antimicrobial coatings for medical devices, responsive materials for controlled drug delivery, and biofilm-resistant surfaces for industrial applications (Rafeeq et al., 2023).

These emerging technologies hold the potential to revolutionize microbiome and biofilm research, fostering innovations that address global challenges and improve human health, environmental sustainability, and industrial efficiency.

4.15 Conclusions

The study of microbiomes and biofilms has transformed our understanding of microbial ecosystems, revealing their significant effects on health, industry, and the environment. Breakthroughs in sequencing technologies, imaging techniques, and bioinformatics have provided exceptional insights into the composition, functionality, and dynamics of these microbial communities. Microbiomes play a crucial role in human health by influencing immune responses, metabolic processes, and disease mechanisms. Meanwhile, biofilms exhibit remarkable resilience and adaptability due to their complex structural and regulatory systems.

Despite significant advancements in microbiome and biofilm research, challenges persist in translating scientific findings into practical applications. The intricate nature and diversity of microbial ecosystems require interdisciplinary approaches that combine molecular biology, computational modeling, and materials science to fully leverage their potential. Emerging technologies such as engineered microbiomes, innovative diagnostic tools, and smart materials present transformative opportunities across medicine, industrial, and agriculture management.

As research continues to evolve, it is essential to maintain a focus on balancing the manipulation of microbial communities to tackle global challenges while safeguarding their natural functions in maintaining ecosystem stability. The future of microbiome and biofilm research hinges on utilizing these innovations to foster sustainable solutions, enhance human health, and create new avenues for industrial and environmental progress. This intersection of science and technology heralds a new era of microbial innovation, redefining our relationship with these microscopic yet profoundly influential systems.

Author Contributions N.N.P., P.S.G., R.KG., A.P.G., S.J.R., G.W., E.M.P., V.M., and E.M.S. wrote this chapter and N.N.P., P.S.G., E.M.S. and R.K.G. reviewed it.

Declarations **Competing Interests**: The authors declare no competing interests.

Conflict of Interest: The authors have no relevant financial or nonfinancial interest to disclose.
Data AvailabilityNo datasets were generated or analyzed during the current study.

References

Abavisani, M., Khoshrou, A., Foroushan, S. K., et al. (2024). Deciphering the gut microbiome: The revolution of artificial intelligence in microbiota analysis and intervention. *Current Research in Biotechnology, 7*, 100211. https://doi.org/10.1016/j.crbiot.2024.100211

Abbaspour, N. (2024). Fermentation's pivotal role in shaping the future of plant-based foods: An integrative review of fermentation processes and their impact on sensory and health benefits. *Applied Food Research, 4*, 100468. https://doi.org/10.1016/j.afres.2024.100468

Abdul Aziz, M., & Masmoudi, K. (2024). Molecular breakthroughs in modern plant breeding techniques. *Horticultural Plant Journal, 11*, 15. https://doi.org/10.1016/j.hpj.2024.01.004

Abouelela, M. E., & Helmy, Y. A. (2024). Next-generation probiotics as novel therapeutics for improving human health: Current trends and future perspectives. *Microorganisms, 12*, 430. https://doi.org/10.3390/microorganisms12030430

Ademosun, A. O., Ajeigbe, O. F., Ademosun, M. T., et al. (2025). Improving gut microbiome through diet rich in dietary fibre and polyphenols: The case for orange peels. *Human Nutrition & Metabolism, 39*, 200295. https://doi.org/10.1016/j.hnm.2024.200295

Afridi, M. S., Fakhar, A., Kumar, A., et al. (2022). Harnessing microbial multitrophic interactions for rhizosphere microbiome engineering. *Microbiological Research, 265*, 127199. https://doi.org/10.1016/j.micres.2022.127199

Agarwal, H., Gurnani, B., Pippal, B., & Jain, N. (2025). Capturing the micro-communities: Insights into biogenesis and architecture of bacterial biofilms. *BBA Advances, 7*, 100133. https://doi.org/10.1016/j.bbadva.2024.100133

Aggarwal, N., Kitano, S., Puah, G. R. Y., et al. (2023). Microbiome and human health: Current understanding, engineering, and enabling technologies. *Chemical Reviews, 123*, 31–72. https://doi.org/10.1021/acs.chemrev.2c00431

Ahn, J., & Hayes, R. B. (2021). Environmental influences on the human microbiome and implications for noncommunicable disease. *Annual Review of Public Health, 42*, 277–292. https://doi.org/10.1146/annurev-publhealth-012420-105020

Ahuja, S., & Zaheer, S. (2025). Advancements in pathology: Digital transformation, precision medicine, and beyond. *Journal of Pathology Informatics, 16*, 100408. https://doi.org/10.1016/j.jpi.2024.100408

Almatroudi, A. (2024). Investigating biofilms: Advanced methods for comprehending microbial behavior and antibiotic resistance. *Frontiers in Bioscience-Landmark, 29*. https://doi.org/10.31083/j.fbl2904133

Aplakidou, E., Vergoulidis, N., Chasapi, M., et al. (2024). Visualizing metagenomic and metatranscriptomic data: A comprehensive review. *Computational and Structural Biotechnology Journal, 23*, 2011–2033. https://doi.org/10.1016/j.csbj.2024.04.060

Arnold, J. W., Roach, J., & Azcarate-Peril, M. A. (2016). Emerging technologies for gut microbiome research. *Trends in Microbiology, 24*, 887–901. https://doi.org/10.1016/j.tim.2016.06.008

Auslander, N., Gussow, A. B., & Koonin, E. V. (2021). Incorporating machine learning into established bioinformatics frameworks. *International Journal of Molecular Sciences, 22*, 2903. https://doi.org/10.3390/ijms22062903

Babu, M., & Snyder, M. (2023). Multi-omics profiling for health. *Molecular & Cellular Proteomics, 22*, 100561. https://doi.org/10.1016/j.mcpro.2023.100561

Balomenos, A. D., Stefanou, V., & Manolakos, E. S. (2021). Analytics and visualization tools to characterize single-cell stochasticity using bacterial single-cell movie cytometry data. *BMC Bioinformatics, 22*, 531. https://doi.org/10.1186/s12859-021-04409-9

Bang, C., Dagan, T., Deines, P., et al. (2018). Metaorganisms in extreme environments: Do microbes play a role in organismal adaptation? *Zoology, 127*, 1–19. https://doi.org/10.1016/j.zool.2018.02.004

Baquero, F., Coque, T. M., Galán, J. C., & Martinez, J. L. (2021). The origin of niches and species in the bacterial world. *Frontiers in Microbiology, 12*. https://doi.org/10.3389/fmicb.2021.657986

Barbosa, A., Miranda, S., Azevedo, N. F., et al. (2023). Imaging biofilms using fluorescence in situ hybridization: Seeing is believing. *Frontiers in Cellular and Infection Microbiology, 13*. https://doi.org/10.3389/fcimb.2023.1195803

Bedia, C. (2022). Metabolomics in environmental toxicology: Applications and challenges. *Trends in Environmental Analytical Chemistry, 34*, e00161. https://doi.org/10.1016/j.teac.2022.e00161

Berg, G., Rybakova, D., Fischer, D., et al. (2020). Microbiome definition re-visited: Old concepts and new challenges. *Microbiome, 8*, 103. https://doi.org/10.1186/s40168-020-00875-0

Beurel, E. (2024). Stress in the microbiome-immune crosstalk. *Gut Microbes, 16*. https://doi.org/10.1080/19490976.2024.2327409

Bhattacharyya, A., Mavrodi, O., Bhowmik, N., et al. (2023). Bacterial biofilms as an essential component of rhizosphere plant-microbe interactions. *Methods in Microbiology, 53*, 3–48.

Böhning, J., Ghrayeb, M., Pedebos, C., et al. (2022). Donor-strand exchange drives assembly of the TasA scaffold in Bacillus subtilis biofilms. *Nature Communications, 13*, 7082. https://doi.org/10.1038/s41467-022-34700-z

Boo, A., Ledesma Amaro, R., & Stan, G.-B. (2021). Quorum sensing in synthetic biology: A review. *Current Opinion in Systems Biology, 28*, 100378. https://doi.org/10.1016/j.coisb.2021.100378

Brehm-Stecher, B. F., & Johnson, E. A. (2004). Single-cell microbiology: Tools, technologies, and applications. *Microbiology and Molecular Biology Reviews, 68*, 538–559. https://doi.org/10.1128/MMBR.68.3.538-559.2004

Bubier, J. A., Chesler, E. J., & Weinstock, G. M. (2021). Host genetic control of gut microbiome composition. *Mammalian Genome, 32*, 263–281. https://doi.org/10.1007/s00335-021-09884-2

Buermans, H. P. J., & den Dunnen, J. T. (2014). Next generation sequencing technology: Advances and applications. *Biochimica et Biophysica Acta (BBA) – Molecular Basis of Disease, 1842*, 1932–1941. https://doi.org/10.1016/j.bbadis.2014.06.015

Cahana, I., & Iraqi, F. A. (2020). Impact of host genetics on gut microbiome: Take-home lessons from human and mouse studies. *Animal Models and Experimental Medicine, 3*, 229–236. https://doi.org/10.1002/ame2.12134

Carson, M. D., Westwater, C., & Novince, C. M. (2023). Adolescence and the microbiome. *The American Journal of Pathology, 193*, 1900–1909. https://doi.org/10.1016/j.ajpath.2023.07.004

Chakaroun, R., Massier, L., & Kovacs, P. (2020). Gut microbiome, intestinal permeability, and tissue bacteria in metabolic disease: Perpetrators or bystanders? *Nutrients, 12*, 1082. https://doi.org/10.3390/nu12041082

Chan, K.-G., Liu, Y.-C., & Chang, C.-Y. (2015). Inhibiting N-acyl-homoserine lactone synthesis and quenching pseudomonas quinolone quorum sensing to attenuate virulence. *Frontiers in Microbiology, 6*. https://doi.org/10.3389/fmicb.2015.01173

Chen, S., Zhang, K., Li, Z., et al. (2023). Hydrogen-bonded supramolecular adhesives: Synthesis, responsiveness, and application. *Supramolecular Materials, 2*, 100032. https://doi.org/10.1016/j.supmat.2023.100032

Chiquito-Contreras, C. J., Meza-Menchaca, T., Guzmán-López, O., et al. (2024). Molecular insights into plant–microbe interactions: A comprehensive review of key mechanisms. *Frontiers in Bioscience-Elite, 16*. https://doi.org/10.31083/j.fbe1601009

Coenye, T., Ahonen, M., Anderson, S., et al. (2024). Global challenges and microbial biofilms: Identification of priority questions in biofilm research, innovation and policy. *Biofilms, 8*, 100210. https://doi.org/10.1016/j.bioflm.2024.100210

Cremer, J., Melbinger, A., Wienand, K., et al. (2019). Cooperation in microbial populations: Theory and experimental model systems. *Journal of Molecular Biology, 431*, 4599–4644. https://doi.org/10.1016/j.jmb.2019.09.023

Culp, E. J., & Goodman, A. L. (2023). Cross-feeding in the gut microbiome: Ecology and mechanisms. *Cell Host & Microbe, 31*, 485–499. https://doi.org/10.1016/j.chom.2023.03.016

Daisley, B. A., Koenig, D., Engelbrecht, K., et al. (2021). Emerging connections between gut microbiome bioenergetics and chronic metabolic diseases. *Cell Reports, 37*, 110087. https://doi.org/10.1016/j.celrep.2021.110087

Dal Co, A., Ackermann, M., & van Vliet, S. (2023). Spatial self-organization of metabolism in microbial systems: A matter of enzymes and chemicals. *Cell Systems, 14*, 98–108. https://doi.org/10.1016/j.cels.2022.12.009

Dankovich, T. M., & Rizzoli, S. O. (2021). Challenges facing quantitative large-scale optical super-resolution, and some simple solutions. *iScience, 24*, 102134. https://doi.org/10.1016/j.isci.2021.102134

Das, N., & Chandran, P. (2011). Microbial degradation of petroleum hydrocarbon contaminants: An overview. *Biotechnology Research International, 2011*, 1–13. https://doi.org/10.4061/2011/941810

Dawan, J., & Ahn, J. (2022). Bacterial stress responses as potential targets in overcoming antibiotic resistance. *Microorganisms, 10*, 1385. https://doi.org/10.3390/microorganisms10071385

Dekaboruah, E., Suryavanshi, M. V., Chettri, D., & Verma, A. K. (2020). Human microbiome: An academic update on human body site specific surveillance and its possible role. *Archives of Microbiology, 202*, 2147–2167. https://doi.org/10.1007/s00203-020-01931-x

Deo, P., & Deshmukh, R. (2019). Oral microbiome: Unveiling the fundamentals. *Journal of Oral and Maxillofacial Pathology, 23*, 122. https://doi.org/10.4103/jomfp.JOMFP_304_18

Dey, S., Nayak, A. K., Rajaram, H., & Das, S. (2025). Exploitative stress within Bacillus subtilis biofilm determines the spatial distribution of pleomorphic cells. *Microbiological Research, 292*, 128034. https://doi.org/10.1016/j.micres.2024.128034

Donlan, R. M. (2002). Biofilms: Microbial life on surfaces. *Emerging Infectious Diseases, 8*, 881–890. https://doi.org/10.3201/eid0809.020063

Dutta, B., Chatterjee, D., Sarkar, N., et al. (2024). Multi-omics technology in detection of multi-species biofilm. *The Microbe, 4*, 100128. https://doi.org/10.1016/j.microb.2024.100128

Ejaz, M. R., Badr, K., Hassan, Z. U., et al. (2024). Metagenomic approaches and opportunities in arid soil research. *Science of the Total Environment, 953*, 176173. https://doi.org/10.1016/j.scitotenv.2024.176173

Etesami, H. (2024). Enhancing soil microbiome resilience: The mitigating role of silicon against environmental stresses. *Frontiers in Agronomy, 6*. https://doi.org/10.3389/fagro.2024.1465165

Foster, K. R., & Bell, T. (2012). Competition, not cooperation, dominates interactions among Culturable microbial species. *Current Biology, 22*, 1845–1850. https://doi.org/10.1016/j.cub.2012.08.005

Gallardo-Navarro, O., Aguilar-Salinas, B., Rocha, J., & Olmedo-Álvarez, G. (2024). Higher-order interactions and emergent properties of microbial communities: The power of synthetic ecology. *Heliyon, 10*, e33896. https://doi.org/10.1016/j.heliyon.2024.e33896

Ghannam, R. B., & Techtmann, S. M. (2021). Machine learning applications in microbial ecology, human microbiome studies, and environmental monitoring. *Computational and Structural Biotechnology Journal, 19*, 1092–1107. https://doi.org/10.1016/j.csbj.2021.01.028

Granato, E. T., Meiller-Legrand, T. A., & Foster, K. R. (2019). The evolution and ecology of bacterial warfare. *Current Biology, 29*, R521–R537. https://doi.org/10.1016/j.cub.2019.04.024

Groussin, M., Poyet, M., Sistiaga, A., et al. (2021). Elevated rates of horizontal gene transfer in the industrialized human microbiome. *Cell, 184*, 2053–2067.e18. https://doi.org/10.1016/j.cell.2021.02.052

Guillaume, O., Butnarasu, C., Visentin, S., & Reimhult, E. (2022). Interplay between biofilm microenvironment and pathogenicity of Pseudomonas aeruginosa in cystic fibrosis lung chronic infection. *Biofilms, 4*, 100089. https://doi.org/10.1016/j.bioflm.2022.100089

Gulliver, E. L., Young, R. B., Chonwerawong, M., et al. (2022). Review article: The future of microbiome-based therapeutics. *Alimentary Pharmacology & Therapeutics, 56*, 192–208. https://doi.org/10.1111/apt.17049

Hajiagha, M. N., & Kafil, H. S. (2023). Efflux pumps and microbial biofilm formation. *Infection, Genetics and Evolution, 112*, 105459. https://doi.org/10.1016/j.meegid.2023.105459

Hemmati, M. A., Monemi, M., Asli, S., et al. (2024). Using new technologies to analyze gut microbiota and predict cancer risk. *Cells, 13*, 1987. https://doi.org/10.3390/cells13231987

Hitch, T. C. A., Hall, L. J., Walsh, S. K., et al. (2022). Microbiome-based interventions to modulate gut ecology and the immune system. *Mucosal Immunology, 15*, 1095–1113. https://doi.org/10.1038/s41385-022-00564-1

Hou, K., Wu, Z.-X., Chen, X.-Y., et al. (2022). Microbiota in health and diseases. *Signal Transduction and Targeted Therapy, 7*, 135. https://doi.org/10.1038/s41392-022-00974-4

Hu, P., Zhang, W., Xin, H., & Deng, G. (2016). Single cell isolation and analysis. *Frontiers in Cell and Development Biology, 4*. https://doi.org/10.3389/fcell.2016.00116

Huang, G., Khan, R., Zheng, Y., et al. (2023). Exploring the role of gut microbiota in advancing personalized medicine. *Frontiers in Microbiology, 14*. https://doi.org/10.3389/fmicb.2023.1274925

Iqbal, B., Li, G., Alabbosh, K. F., et al. (2023). Advancing environmental sustainability through microbial reprogramming in growth improvement, stress alleviation, and phytoremediation. *Plant Stress, 10*, 100283. https://doi.org/10.1016/j.stress.2023.100283

Jain, A., Sarsaiya, S., Singh, R., et al. (2024). Omics approaches in understanding the benefits of plant-microbe interactions. *Frontiers in Microbiology, 15*. https://doi.org/10.3389/fmicb.2024.1391059

Jeong, D., & Kim, D. (2022). Recent developments in correlative super-resolution fluorescence microscopy and electron microscopy. *Molecules and Cells, 45*, 41–50. https://doi.org/10.14348/molcells.2021.5011

Jo, S. J., Kwon, J., Kim, S. G., & Lee, S.-J. (2023). The biotechnological application of bacteriophages: What to do and where to go in the middle of the post-antibiotic era. *Microorganisms, 11*, 2311. https://doi.org/10.3390/microorganisms11092311

Johnson, J. S., Spakowicz, D. J., Hong, B.-Y., et al. (2019). Evaluation of 16S rRNA gene sequencing for species and strain-level microbiome analysis. *Nature Communications, 10*, 5029. https://doi.org/10.1038/s41467-019-13036-1

Joseph, P. T., & Niddk. (2013). In 6th (Ed.), *Catalyst*. National Institutes of Health.

Juszczuk-Kubiak, E. (2024). Molecular aspects of the functioning of pathogenic bacteria biofilm based on quorum sensing (QS) signal-response system and innovative non-antibiotic strategies for their elimination. *International Journal of Molecular Sciences, 25*, 2655. https://doi.org/10.3390/ijms25052655

Kabir, A. H., Baki, M. Z. I., Ahmed, B., & Mostofa, M. G. (2024). Current, faltering, and future strategies for advancing microbiome-assisted sustainable agriculture and environmental resilience. *New Crops, 1*, 100013. https://doi.org/10.1016/j.ncrops.2024.100013

Kamel, M., Aleya, S., Alsubih, M., & Aleya, L. (2024). Microbiome dynamics: A paradigm shift in combatting infectious diseases. *Journal of Personalized Medicine, 14*, 217. https://doi.org/10.3390/jpm14020217

Kang, X., Yang, X., He, Y., et al. (2023). Strategies and materials for the prevention and treatment of biofilms. *Materials Today Bio, 23*, 100827. https://doi.org/10.1016/j.mtbio.2023.100827

Karygianni, L., Ren, Z., Koo, H., & Thurnheer, T. (2020). Biofilm Matrixome: Extracellular components in structured microbial communities. *Trends in Microbiology, 28*, 668–681. https://doi.org/10.1016/j.tim.2020.03.016

Kaur, N., & Dey, P. (2023). Bacterial exopolysaccharides as emerging bioactive macromolecules: From fundamentals to applications. *Research in Microbiology, 174*, 104024. https://doi.org/10.1016/j.resmic.2022.104024

Kennedy, M. S., & Chang, E. B. (2020). The microbiome: Composition and locations. *Progress in Molecular Biology and Translational Science, 176*, 1–42.

Kim, J. S., Chen, M.-H., Wang, H. E., et al. (2023). Inflammatory bowel disease and neurodegenerative diseases. *Gut Liver, 17*, 495–504. https://doi.org/10.5009/gnl220523

Kim, N., Ma, J., Kim, W., et al. (2024). Genome-resolved metagenomics: A game changer for microbiome medicine. *Experimental & Molecular Medicine, 56*, 1501–1512. https://doi.org/10.1038/s12276-024-01262-7

Kogut, M. H., Lee, A., & Santin, E. (2020). Microbiome and pathogen interaction with the immune system. *Poultry Science, 99*, 1906–1913. https://doi.org/10.1016/j.psj.2019.12.011

Kondakindi, V. R., Pabbati, R., Erukulla, P., et al. (2024). Bioremediation of heavy metals-contaminated sites by microbial extracellular polymeric substances – A critical view. *Environmental Chemistry and Ecotoxicology, 6*, 408–421. https://doi.org/10.1016/j.enceco.2024.05.002

Kreve, S., & Reis, A. C. D. (2021). Bacterial adhesion to biomaterials: What regulates this attachment? A review. *Japanese Dental Science Review, 57*, 85–96. https://doi.org/10.1016/j.jdsr.2021.05.003

Kuppan, N., Padman, M., Mahadeva, M., et al. (2024). A comprehensive review of sustainable bioremediation techniques: Eco friendly solutions for waste and pollution management. *Waste Management Bulletin, 2*, 154–171. https://doi.org/10.1016/j.wmb.2024.07.005

Kwa, W. T., Sundarajoo, S., Toh, K. Y., & Lee, J. (2023). Application of emerging technologies for gut microbiome research. *Singapore Medical Journal, 64*, 45–52. https://doi.org/10.4103/singaporemedj.SMJ-2021-432

Lago, A., Rocha, V., Barros, O., et al. (2024). Bacterial biofilm attachment to sustainable carriers as a clean-up strategy for wastewater treatment: A review. *Journal of Water Process Engineering, 63*, 105368. https://doi.org/10.1016/j.jwpe.2024.105368

Li, Y., Xin, M., Xie, D., et al. (2021). Variation in extracellular polymeric substances from *Enterobacter* sp. and their Pb^{2+} adsorption behaviors. *ACS Omega, 6*, 9617–9628. https://doi.org/10.1021/acsomega.1c00185

Li, Y., Narayanan, M., Shi, X., et al. (2024). Biofilms formation in plant growth-promoting bacteria for alleviating agro-environmental stress. *Science of the Total Environment, 907*, 167774. https://doi.org/10.1016/j.scitotenv.2023.167774

Liang, A., Kong, Y., Chen, Z., et al. (2024). Advancements and applications of single-cell multi-omics techniques in cancer research: Unveiling heterogeneity and paving the way for precision therapeutics. *Biochemistry and Biophysics Reports, 37*, 101589. https://doi.org/10.1016/j.bbrep.2023.101589

Ma, Z., Zuo, T., Frey, N., & Rangrez, A. Y. (2024). A systematic framework for understanding the microbiome in human health and disease: From basic principles to clinical translation. *Signal Transduction and Targeted Therapy, 9*, 237. https://doi.org/10.1038/s41392-024-01946-6

Madhogaria, B., Bhowmik, P., & Kundu, A. (2022). Correlation between human gut microbiome and diseases. *Infectious Medicine, 1*, 180–191. https://doi.org/10.1016/j.imj.2022.08.004

Mahmud, K., Missaoui, A., Lee, K., et al. (2021). Rhizosphere microbiome manipulation for sustainable crop production. *Current Plant Biology, 27*, 100210. https://doi.org/10.1016/j.cpb.2021.100210

Malard, F., Dore, J., Gaugler, B., & Mohty, M. (2021). Introduction to host microbiome symbiosis in health and disease. *Mucosal Immunology, 14*, 547–554. https://doi.org/10.1038/s41385-020-00365-4

Matijašić, M., Meštrović, T., Paljetak, H. Č., et al. (2020). Gut microbiota beyond bacteria-mycobiome, virome, archaeome, and eukaryotic parasites in IBD. *International Journal of Molecular Sciences, 21*. https://doi.org/10.3390/ijms21082668

Matthews, K., Cavagnaro, T., Weinstein, P., & Stanhope, J. (2024). Health by design; optimising our urban environmental microbiomes for human health. *Environmental Research, 257*, 119226. https://doi.org/10.1016/j.envres.2024.119226

Mazziotta, C., Tognon, M., Martini, F., et al. (2023). Probiotics mechanism of action on immune cells and beneficial effects on human health. *Cells, 12*, 184. https://doi.org/10.3390/cells12010184

Mcmeeking, A., Dieckmann, E., & Cheeseman, C. (2024). Production methods for bacterial biomaterials: A review. *Materials Today Sustainability, 25*, 100623. https://doi.org/10.1016/j.mtsust.2023.100623

Mhade, S., & Kaushik, K. S. (2023). Tools of the trade: Image analysis programs for confocal laser-scanning microscopy studies of biofilms and considerations for their use by experimental researchers. *ACS Omega, 8*, 20163–20177. https://doi.org/10.1021/acsomega.2c07255

Mishra, S., Gupta, A., Upadhye, V., et al. (2023). Therapeutic strategies against biofilm infections. *Life, 13*, 172. https://doi.org/10.3390/life13010172

Mohseni, P., & Ghorbani, A. (2024). Exploring the synergy of artificial intelligence in microbiology: Advancements, challenges, and future prospects. *Computational and Structural Biotechnology Reports, 1*, 100005. https://doi.org/10.1016/j.csbr.2024.100005

Muteeb, G., Rehman, M. T., Shahwan, M., & Aatif, M. (2023). Origin of antibiotics and antibiotic resistance, and their impacts on drug development: A narrative review. *Pharmaceuticals, 16*, 1615. https://doi.org/10.3390/ph16111615

Myhrstad, M. C. W., Tunsjø, H., Charnock, C., & Telle-Hansen, V. H. (2020). Dietary fiber, gut microbiota, and metabolic regulation—Current status in human randomized trials. *Nutrients, 12*, 859. https://doi.org/10.3390/nu12030859

Naureen, Z., Dautaj, A., Anpilogov, K., et al. (2020). Bacteriophages presence in nature and their role in the natural selection of bacterial populations. *Acta Biomed, 91*, e2020024. https://doi.org/10.23750/abm.v91i13-S.10819

Netrusov, A. I., Liyaskina, E. V., Kurgaeva, I. V., et al. (2023). Exopolysaccharides producing bacteria: A review. *Microorganisms, 11*, 1541. https://doi.org/10.3390/microorganisms11061541

Orcutt, B. N., Bradley, J. A., Brazelton, W. J., et al. (2020). Impacts of deep-sea mining on microbial ecosystem services. *Limnology and Oceanography, 65*, 1489–1510. https://doi.org/10.1002/lno.11403

Pandit, A., Adholeya, A., Cahill, D., et al. (2020). Microbial biofilms in nature: Unlocking their potential for agricultural applications. *Journal of Applied Microbiology, 129*, 199–211. https://doi.org/10.1111/jam.14609

Panlilio, H., & Rice, C. V. (2021). The role of extracellular DNA in the formation, architecture, stability, and treatment of bacterial biofilms. *Biotechnology and Bioengineering, 118*, 2129–2141. https://doi.org/10.1002/bit.27760

Park, S., & Sauer, K. (2022). Controlling biofilm development through cyclic di-GMP signaling. *Advances in Experimental Medicine and Biology, 1386*, 69–94.

Parrilli, E., Tutino, M. L., & Marino, G. (2022). Biofilm as an adaptation strategy to extreme conditions. *Rendiconti Lincei. Scienze Fisiche e Naturali, 33*, 527–536. https://doi.org/10.1007/s12210-022-01083-8

Peng, Q., Tang, X., Dong, W., et al. (2022). A review of biofilm formation of Staphylococcus aureus and its regulation mechanism. *Antibiotics, 12*, 12. https://doi.org/10.3390/antibiotics12010012

Perry, E. K., & Tan, M.-W. (2023). Bacterial biofilms in the human body: Prevalence and impacts on health and disease. *Frontiers in Cellular and Infection Microbiology, 13*. https://doi.org/10.3389/fcimb.2023.1237164

Pinto, R. M., Soares, F. A., Reis, S., et al. (2020). Innovative strategies toward the disassembly of the EPS matrix in bacterial biofilms. *Frontiers in Microbiology, 11*. https://doi.org/10.3389/fmicb.2020.00952

Poulin, M. B., & Kuperman, L. L. (2021). Regulation of biofilm exopolysaccharide production by cyclic Di-guanosine monophosphate. *Frontiers in Microbiology, 12*. https://doi.org/10.3389/fmicb.2021.730980

Preda, V. G., & Săndulescu, O. (2019). Communication is the key: Biofilms, quorum sensing, formation and prevention. *Discover, 7*, e10. https://doi.org/10.15190/d.2019.13

Prescott, R. D., & Decho, A. W. (2020). Flexibility and adaptability of quorum sensing in nature. *Trends in Microbiology, 28*, 436–444. https://doi.org/10.1016/j.tim.2019.12.004

Puig-Castellví, F., Pacheco-Tapia, R., Deslande, M., et al. (2023). Advances in the integration of metabolomics and metagenomics for human gut microbiome and their clinical applications. *TrAC Trends in Analytical Chemistry, 167*, 117248. https://doi.org/10.1016/j.trac.2023.117248

Rafeeq, H., Afsheen, N., Rafique, S., et al. (2023). Genetically engineered microorganisms for environmental remediation. *Chemosphere, 310*, 136751. https://doi.org/10.1016/j.chemosphere.2022.136751

Rahimian, M., & Panahi, B. (2024). Next generation sequencing-based transcriptome data mining for virus identification and characterization: Review on recent progress and prospects. *Journal of Clinical Virology Plus, 4*, 100194. https://doi.org/10.1016/j.jcvp.2024.100194

Rame, R., Purwanto, P., & Sudarno, S. (2024). Industry 5.0 and sustainability: An overview of emerging trends and challenges for a green future. *Innovation and Green Development, 3*, 100173. https://doi.org/10.1016/j.igd.2024.100173

Rather, M. A., Gupta, K., & Mandal, M. (2021). Microbial biofilm: Formation, architecture, antibiotic resistance, and control strategies. *Brazilian Journal of Microbiology, 52*, 1701–1718. https://doi.org/10.1007/s42770-021-00624-x

Rattray, J. B., Thomas, S. A., Wang, Y., et al. (2022). Bacterial quorum sensing allows graded and bimodal cellular responses to variations in population density. *MBio, 13*. https://doi.org/10.1128/mbio.00745-22

Reed, J., Bain, S., & Kanamarlapudi, V. (2024). The regulation of metabolic homeostasis by Incretins and the metabolic hormones produced by pancreatic islets. *Diabetes, Metabolic Syndrome and Obesity, 17*, 2419–2456. https://doi.org/10.2147/DMSO.S415934

Relucenti, M., Familiari, G., Donfrancesco, O., et al. (2021). Microscopy methods for biofilm imaging: Focus on SEM and VP-SEM pros and cons. *Biology (Basel), 10*, 51. https://doi.org/10.3390/biology10010051

Rizvi, S. M. D., Lila, A. S. A., Moin, A., et al. (2025). Bacteriophage resurrection: Innovative impacts in medicine, biotechnology, and environmental solutions. *Scientific African, 27*, e02506. https://doi.org/10.1016/j.sciaf.2024.e02506

Ronan, V., Yeasin, R., & Claud, E. C. (2021). Childhood development and the microbiome—The intestinal microbiota in maintenance of health and development of disease during childhood development. *Gastroenterology, 160*, 495–506. https://doi.org/10.1053/j.gastro.2020.08.065

Rowland, I., Gibson, G., Heinken, A., et al. (2018). Gut microbiota functions: Metabolism of nutrients and other food components. *European Journal of Nutrition, 57*, 1–24. https://doi.org/10.1007/s00394-017-1445-8

Rutherford, S. T., & Bassler, B. L. (2012). Bacterial quorum sensing: Its role in virulence and possibilities for its control. *Cold Spring Harbor Perspectives in Medicine, 2*, a012427. https://doi.org/10.1101/cshperspect.a012427

Samantaray, A., Chattaraj, S., Mitra, D., et al. (2024). Advances in microbial based bio-inoculum for amelioration of soil health and sustainable crop production. *Current Research in Microbial Sciences, 7*, 100251. https://doi.org/10.1016/j.crmicr.2024.100251

Samrot, A. V., Abubakar Mohamed, A., Faradjeva, E., et al. (2021). Mechanisms and impact of biofilms and targeting of biofilms using bioactive compounds—A review. *Medicina (B Aires), 57*, 839. https://doi.org/10.3390/medicina57080839

Sarkar, A., McInroy, C. J. A., Harty, S., et al. (2024). Microbial transmission in the social microbiome and host health and disease. *Cell, 187*, 17–43. https://doi.org/10.1016/j.cell.2023.12.014

Sartaj, K., Patel, A., Matsakas, L., & Prasad, R. (2022). Unravelling metagenomics approach for microbial biofuel production. *Genes (Basel), 13*, 1942. https://doi.org/10.3390/genes13111942

Sasso, J. M., Ammar, R. M., Tenchov, R., et al. (2023). Gut microbiome-brain Alliance: A landscape view into mental and gastrointestinal health and disorders. *ACS Chemical Neuroscience, 14*, 1717–1763. https://doi.org/10.1021/acschemneuro.3c00127

Satam, H., Joshi, K., Mangrolia, U., et al. (2023). Next-generation sequencing technology: Current trends and advancements. *Biology (Basel), 12*, 997. https://doi.org/10.3390/biology12070997

Sauer, K., Stoodley, P., Goeres, D. M., et al. (2022). The biofilm life cycle: Expanding the conceptual model of biofilm formation. *Nature Reviews. Microbiology, 20*, 608–620. https://doi.org/10.1038/s41579-022-00767-0

Sessitsch, A., Wakelin, S., Schloter, M., et al. (2023). Microbiome interconnectedness throughout environments with major consequences for healthy people and a healthy planet. *Microbiology and Molecular Biology Reviews, 87*, e0021222. https://doi.org/10.1128/mmbr.00212-22

Sharma, S., Mohler, J., Mahajan, S. D., et al. (2023). Microbial biofilm: A review on formation, infection, antibiotic resistance, control measures, and innovative treatment. *Microorganisms, 11*, 1614. https://doi.org/10.3390/microorganisms11061614

Shineh, G., Mobaraki, M., Perves Bappy, M. J., & Mills, D. K. (2023). Biofilm formation, and related impacts on healthcare, food processing and packaging, industrial manufacturing, marine industries, and sanitation–a review. *Applied Microbiology, 3*, 629–665. https://doi.org/10.3390/applmicrobiol3030044

Shree, P., Singh, C. K., Sodhi, K. K., et al. (2023). Biofilms: Understanding the structure and contribution towards bacterial resistance in antibiotics. *Medicine in Microecology, 16*, 100084. https://doi.org/10.1016/j.medmic.2023.100084

Shukla, V., Singh, S., Verma, S., et al. (2024). Targeting the microbiome to improve human health with the approach of personalized medicine: Latest aspects and current updates. *Clinical Nutrition ESPEN, 63*, 813–820. https://doi.org/10.1016/j.clnesp.2024.08.005

Sidhu, S. R. K., Kok, C. W., Kunasegaran, T., & Ramadas, A. (2023). Effect of plant-based diets on gut microbiota: A systematic review of interventional studies. *Nutrients, 15*, 1510. https://doi.org/10.3390/nu15061510

Simon, J.-C., Marchesi, J. R., Mougel, C., & Selosse, M.-A. (2019). Host-microbiota interactions: From holobiont theory to analysis. *Microbiome, 7*, 5. https://doi.org/10.1186/s40168-019-0619-4

Sousa, A., Phung, A. N., Škalko-Basnet, N., & Obuobi, S. (2023). Smart delivery systems for microbial biofilm therapy: Dissecting design, drug release and toxicological features. *Journal of Controlled Release, 354*, 394–416. https://doi.org/10.1016/j.jconrel.2023.01.003

Spacapan, M., Bez, C., & Venturi, V. (2023). Quorum sensing going wild. *iScience, 26*, 108000. https://doi.org/10.1016/j.isci.2023.108000

Su, Y., & Ding, T. (2023). Targeting microbial quorum sensing: The next frontier to hinder bacterial driven gastrointestinal infections. *Gut Microbes, 15*. https://doi.org/10.1080/19490976.2023.2252780

Suárez-Martínez, C., Santaella-Pascual, M., Yagüe-Guirao, G., & Martínez-Graciá, C. (2023). Infant gut microbiota colonization: Influence of prenatal and postnatal factors, focusing on diet. *Frontiers in Microbiology, 14*. https://doi.org/10.3389/fmicb.2023.1236254

Suman, J., Rakshit, A., Ogireddy, S. D., et al. (2022). Microbiome as a key player in sustainable agriculture and human health. *Frontiers in Soil Science, 2*. https://doi.org/10.3389/fsoil.2022.821589

Sun, W., Shahrajabian, M. H., & Lin, M. (2022). Research Progress of fermented functional foods and protein factory-microbial fermentation technology. *Fermentation, 8*, 688. https://doi.org/10.3390/fermentation8120688

Szydlowski, L. M., Bulbul, A. A., Simpson, A. C., et al. (2024). Adaptation to space conditions of novel bacterial species isolated from the international Space Station revealed by functional gene annotations and comparative genome analysis. *Microbiome, 12*, 190. https://doi.org/10.1186/s40168-024-01916-8

Tihlaříková, E., Neděla, V., & Shiojiri, M. (2013). *In situ* study of live specimens in an environmental scanning electron microscope. *Microscopy and Microanalysis, 19*, 914–918. https://doi.org/10.1017/S1431927613000603

Tomasulo, A., Simionati, B., & Facchin, S. (2024). Microbiome one health model for a healthy ecosystem. *Science in One Health, 3*, 100065. https://doi.org/10.1016/j.soh.2024.100065

Trivedi, P., Mattupalli, C., Eversole, K., & Leach, J. E. (2021). Enabling sustainable agriculture through understanding and enhancement of microbiomes. *New Phytologist, 230*, 2129–2147. https://doi.org/10.1111/nph.17319

Tuon, F. F., Suss, P. H., Telles, J. P., et al. (2023). Antimicrobial treatment of staphylococcus aureus biofilms. *Antibiotics, 12*, 87. https://doi.org/10.3390/antibiotics12010087

Vadakkan, K., Choudhury, A. A., Gunasekaran, R., et al. (2018). Quorum sensing intervened bacterial signaling: Pursuit of its cognizance and repression. *Journal of Genetic Engineering and Biotechnology, 16*, 239–252. https://doi.org/10.1016/j.jgeb.2018.07.001

Valiei, A., Dickson, A., Aminian-Dehkordi, J., & Mofrad, M. R. K. (2024). Metabolic interactions shape emergent biofilm structures in a conceptual model of gut mucosal bacterial communities. *npj Biofilms and Microbiomes, 10*, 99. https://doi.org/10.1038/s41522-024-00572-y

Vinogradova, E., Mukhanbetzhanov, N., Nurgaziyev, M., et al. (2024). Impact of urbanization on gut microbiome mosaics across geographic and dietary contexts. *mSystems, 9*. https://doi.org/10.1128/msystems.00585-24

Wang, X., Liu, M., Yu, C., et al. (2023). Biofilm formation: Mechanistic insights and therapeutic targets. *Molecular Biomedicine, 4*, 49. https://doi.org/10.1186/s43556-023-00164-w

Wang, X., Chi, Y., & Song, S. (2024). Important soil microbiota's effects on plants and soils: A comprehensive 30-year systematic literature review. *Frontiers in Microbiology, 15*. https://doi.org/10.3389/fmicb.2024.1347745

Weiland-Bräuer, N. (2021). Friends or foes—Microbial interactions in nature. *Biology (Basel), 10*, 496. https://doi.org/10.3390/biology10060496

Weiskopf, S. R., Rubenstein, M. A., Crozier, L. G., et al. (2020). Climate change effects on biodiversity, ecosystems, ecosystem services, and natural resource management in the United States. *Science of the Total Environment, 733*, 137782. https://doi.org/10.1016/j.scitotenv.2020.137782

Wensel, C. R., Pluznick, J. L., Salzberg, S. L., & Sears, C. L. (2022). Next-generation sequencing: Insights to advance clinical investigations of the microbiome. *Journal of Clinical Investigation, 132*. https://doi.org/10.1172/JCI154944

Wille, J., & Coenye, T. (2020). Biofilm dispersion: The key to biofilm eradication or opening Pandora's box? *Biofilms, 2*, 100027. https://doi.org/10.1016/j.bioflm.2020.100027

Wu, H., Cui, H., Fu, C., et al. (2024). Unveiling the crucial role of soil microorganisms in carbon cycling: A review. *Science of the Total Environment, 909*, 168627. https://doi.org/10.1016/j.scitotenv.2023.168627

Yiallouris, A., Pana, Z. D., Marangos, G., et al. (2024). Fungal diversity in the soil Mycobiome: Implications for ONE health. *One Health, 18*, 100720. https://doi.org/10.1016/j.onehlt.2024.100720

Young, G. R., Sherry, A., & Smith, D. L. (2023). Built environment microbiomes transition from outdoor to human-associated communities after construction and commissioning. *Scientific Reports, 13*, 15854. https://doi.org/10.1038/s41598-023-42427-0

Zampieri, G., Campanaro, S., Angione, C., & Treu, L. (2023). Metatranscriptomics-guided genome-scale metabolic modeling of microbial communities. *Cell Reports Methods, 3*, 100383. https://doi.org/10.1016/j.crmeth.2022.100383

Zhang, P. (2022). Influence of foods and nutrition on the gut microbiome and implications for intestinal health. *International Journal of Molecular Sciences, 23*, 9588. https://doi.org/10.3390/ijms23179588

Zhang, C., Li, L., Jin, B., et al. (2021). The effects of delivery mode on the gut microbiota and health: State of art. *Frontiers in Microbiology, 12*. https://doi.org/10.3389/fmicb.2021.724449

Zhang, Y., Chen, R., Zhang, D., et al. (2023). Metabolite interactions between host and microbiota during health and disease: Which feeds the other? *Biomedicine & Pharmacotherapy, 160*, 114295. https://doi.org/10.1016/j.biopha.2023.114295

Zhao, X., Yu, Z., & Ding, T. (2020). Quorum-sensing regulation of antimicrobial resistance in bacteria. *Microorganisms, 8*, 425. https://doi.org/10.3390/microorganisms8030425

Zhao, A., Sun, J., & Liu, Y. (2023). Understanding bacterial biofilms: From definition to treatment strategies. *Frontiers in Cellular and Infection Microbiology, 13*. https://doi.org/10.3389/fcimb.2023.1137947

Zheng, D., Liwinski, T., & Elinav, E. (2020). Interaction between microbiota and immunity in health and disease. *Cell Research, 30*, 492–506. https://doi.org/10.1038/s41422-020-0332-7

Part II

Omics in Biofilm Research

5 Genomic Insights into the Field of Biofilm Research

Zarin Taj, Seelam Sandhya Rani, Rohit Gundamaraju, and Indranil Chattopadhyay

5.1 Introduction

Biofilm production is an important adaptation strategy for microbes, allowing them to thrive in various and frequently difficult conditions. These complex microbial communities are enclosed in a matrix composed of extracellular polymeric substances (EPSs), providing structural stability, nutrition retention, and protection from external challenges like antimicrobial agents and immunological reactions (Pham et al., 2020). Understanding the genetic foundations of biofilm formation has been critical in revealing the molecular pathways that allow bacteria to change from a planktonic to a sessile lifestyle. Genomic studies have revealed a diverse repertoire of genes that regulate biofilm development's adhesion, growth, maturation, and dispersion phases, underscoring these microbial systems' complexity and adaptability.

Key regulatory pathways and genetic determinants vary across species, reflecting biofilm-forming microorganisms' diverse ecological roles and survival strategies. For instance, *Pseudomonas aeruginosa* relies on the *rhlI-rhlR* quorum sensing (QS) system and the *pelA* gene for EPS production, while *Staphylococcus aureus* utilizes the *icaADBC* operon for biofilm formation on biotic and abiotic surfaces (Abdelraheem et al., 2021; Kot et al., 2018). In oral biofilms, *Streptococcus mutans* employs *gtfs* and *gbps* genes to mediate adhesion and structural integrity, whereas *Escherichia coli* utilizes type I fimbriae encoded by the *fimABCDEFGH* gene cluster to establish initial surface attachment (Afrasiabi & Partoazar, 2024). These

Z. Taj · S. S. Rani · I. Chattopadhyay (✉)
Department of Biotechnology, School of Integrative Biology, Central University of Tamil Nadu, Neelakudi, Thiruvarur, Tamil Nadu, India
e-mail: indranil@cutn.ac.in

R. Gundamaraju
ER Stress and Mucosal Immunology Lab, School of Health Sciences, University of Tasmania, Launceston, TAS, Australia

S. Busi et al. (eds.), *Omics Approaches in Biofilm Research*,
https://doi.org/10.1007/978-3-031-91863-6_5

species-specific genetic pathways highlight the evolutionary adaptability of biofilm-forming microorganisms (Table 5.1).

The genomic insights gained through next-generation sequencing (NGS) and multi-omics approaches have significantly advanced our understanding of the biofilm dynamics (Fig. 5.1). Genomics has been instrumental in identifying genes involved in adhesion, EPS synthesis, and antimicrobial resistance. Concurrently, transcriptomics has provided valuable information about gene expression patterns during various stages of biofilm formation, revealing the interplay between environmental signals and genetic regulation (Chiu & Miller, 2019). Proteomic analyses have complemented these findings by elucidating protein expression profiles and interactions within biofilm communities, while metabolomics has shed light on the

Table 5.1 Genes involved in the biofilm formation in different microbes

Microorganism	Genes	References
MRSA	*hly*, *flu*, *icaA*, *icaD*	Hamida et al. (2020), Montazeri et al. (2020)
Pseudomonas putida	*csgA*, *alg8*	Thuptimdang et al. (2015)
Staphylococcus epidermidis	*icaA*, *icaB*, *icaC*, *icaR*	Swolana et al. (2022)
P. aeruginosa	*pslA*, *rhlI*, *lasI*, *lasA*, *lasB*, *rhlA*, *rhlB*, *rhlR*, *rhlI*, *pqsC*, *fimH*, *rmpA*, *mrkA* *lecA*, *lecB*	Bhargava et al. (2018), Elwakil et al. (2024)
Staphylococcus aureus	*icaA*, *icaR*	Moulavi et al. (2019)
S. mutans	*gtfB*, *gtfc*, *gtfd*, *spaP*, *gbpB*, *ldh*, *comD*, *comE*, *luxS*	Jin et al. (2017), Li et al. (2013a, b)
Porphyromonas gingivalis	*FimA*, *Mfa3*	Jin et al. (2017)
Klebsiella pneumoniae	*MrkA*, *MrkB*, *MrkF*	Foroohimanjili et al. (2020)

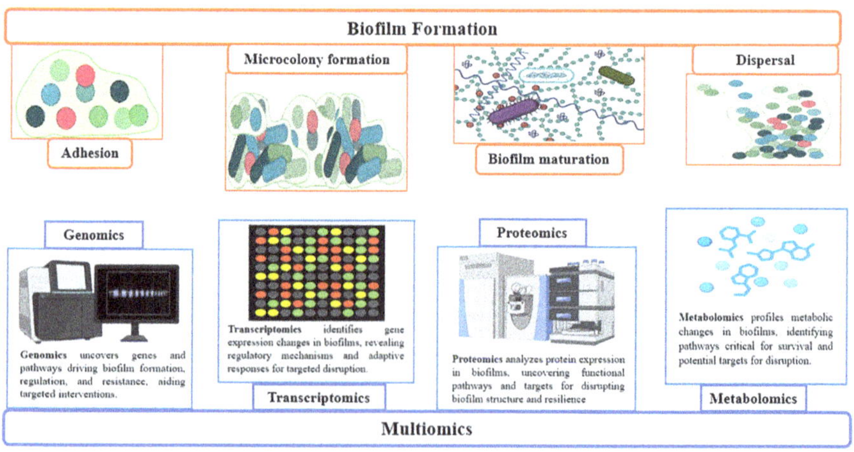

Fig. 5.1 Omics approaches as tools for elucidating the mechanisms of biofilm formation

metabolic adaptations that enable biofilm resilience under stress conditions (Soufi & Soufi, 2016; Wan et al., 2018).

These genomic advancements have also deepened our understanding of interspecies interactions within mixed-species biofilms, which often display enhanced antimicrobial resistance and functional synergy. For example, the exchange of genetic material by horizontal gene transfer (HGT) within biofilms promotes the development of antibiotic resistance genes, providing important issues in clinical and industrial situations (Santos et al., 2018). In addition, merging metagenomics with computational and machine learning methods has permitted the analysis of microbial communities at a systems level, bringing fresh perspectives on the molecular basis of biofilm development and its implications for host–pathogen interactions (Dutta et al., 2024).

Applying genomic insights into biofilm research has practical implications for developing novel therapeutic and preventive strategies. By targeting specific genetic pathways, such as quorum sensing or EPS synthesis, researchers can design interventions to disrupt biofilm formation and combat biofilm-associated infections. Additionally, genomic data can guide the identification of new antimicrobial targets and enhance microbial risk assessment in clinical and environmental settings (Alcock et al., 2023; Kean et al., 2018).

In summary, genomic insights have revolutionized the study of biofilm formation by uncovering the intricate genetic and molecular mechanisms that drive microbial adaptation and resilience. These findings address the challenges posed by biofilm-associated antimicrobial resistance and persistence and highlight biofilm research's ecological and biotechnological potential. As omics technologies and computational tools continue to advance, they promise to unlock new strategies for understanding and managing biofilms in diverse contexts.

5.2 Revolutionizing Biofilm Research Through Genomic Insights

Genomic insights have significantly advanced biofilm research by unveiling the genetic and molecular mechanisms underlying biofilm formation, behavior, and resistance. High-throughput sequencing and comparative genomics have facilitated decoding entire genomes of biofilm-forming microorganisms, shedding light on the genes and regulatory networks involved in biofilm development and maintenance. These studies have revealed how specific genes are activated during biofilm stages, from attachment and microcolony formation to maturation and dispersal. Key discoveries include the role of quorum sensing and other cell-to-cell communication pathways in coordinating microbial activities, ensuring biofilm integrity and resilience (Chiu & Miller, 2019).

Genomic research has highlighted the genetic diversity within biofilms, identifying subpopulations with specialized roles that enhance biofilm adaptability. Additionally, genes associated with antimicrobial resistance and stress responses have been uncovered, explaining biofilms' ability to withstand antibiotics and immune defenses. Metagenomics has further illuminated the complexity of multi-species biofilms, revealing synergistic interactions between microbial species that enhance biofilm stability and resistance (Dutta et al., 2024).

These advancements have paved the way for innovative strategies to disrupt biofilm formation, such as targeting specific genetic pathways, inhibiting quorum sensing, and designing antimicrobials to penetrate the EPS matrix. Moreover, genomics has improved predictions of biofilm behavior across medical and ecological contexts and informed precision medicine strategies for biofilm-associated infections.

This chapter will explore genomic technologies like next-generation sequencing and transcriptomics, emphasizing their transformative impact on biofilm research and future directions for managing biofilm-related challenges in medicine, industry, and the environment.

5.2.1 Overview of Biofilms

5.2.1.1 Definition, Structure, and Formation Process

Microorganisms generate biofilms with distinct structures and functions at different phases, including initial adhesion, aggregation, development, maturation, and dispersion (Sauer et al., 2022). Mature microbial biofilms perform a variety of critical activities. Certain microbes develop filamentous structures, such as hyphae or pseudohyphae, and their aggregation and proliferation play a part in biofilm creation (Wang et al. 2024a, b). The first phase in biofilm formation is the attachment of individual planktonic cells to a surface, followed by continual development, maturity, EPS synthesis, and eventual dispersion. However, new research suggests that surface contact is not necessarily required for biofilm formation. Many aggregated flocs in the planktonic state share traits with biofilms, particularly in their aggregation characteristics. As a result, new research perspectives necessitate updating and refining our understanding of microbial biofilm formation processes (Gao et al., 2024).

Additionally, internal genetic mechanisms closely regulate biofilm formation, but much of the research has focused on the model organisms *Saccharomyces cerevisiae* and *Candida albicans*. These studies emphasize several genes, complex regulatory mechanisms, and disparate conclusions (Peters et al., 2010). Systematizing the genes and metabolic pathways involved in biofilm formation in *Saccharomyces cerevisiae* and *Candida albicans* is critical for understanding the molecular mechanisms underlying biofilm formation in these species and providing theoretical insights into the regulatory mechanisms in other microorganisms (Gulati & Nobile, 2016).

Quorum sensing is also critical in regulating biofilm formation in microbial communities. Quorum sensing factors influence the behavior of microbial populations and regulate morphogenesis. Mixed microbial biofilms have numerous benefits, including substrate exchange, increased antimicrobial resistance, environmental adaptability, improved adhesion, and interspecies communication (Santos et al., 2018). Thus, it is essential to summarize research on mixed biofilms in order to clarify the internal structure, cellular connectivity, and functions of biofilms created by different bacteria as well as adhesion processes and signal transduction in multi-species interactions (Wang et al., 2024a, b).

5.2.2 Significance in Different Environments

5.2.2.1 Medical Significance

Human infections involving microbial adherence and biofilm formation account for approximately 65% of all infections. Microbial biofilms can form on tissues, including intact or damaged mucosa, teguments, teeth, and bones, leading to infections such as valvular endocarditis, otitis media, chronic tonsillitis, pulmonary infections in cystic fibrosis patients, bacterial prostatitis, necrotizing fasciitis, osteomyelitis, periodontitis, dental caries, genitourinary infections, biliary tract infections, and postoperative and post-traumatic infections (Lazăr & Chifiriuc, 2010). Biofilm-associated adherence to mucosal surfaces helps pathogens evade expulsion mechanisms like secretions, blood flow, peristalsis, coughing, and sneezing, allowing colonization despite host defenses (Mishra et al., 2024).

Medical devices provide surfaces for biofilm formation, affecting artificial heart valves, pacemakers, endotracheal tubes, dialysis equipment, prosthetic joints, shunts, catheters, intrauterine devices, and contact lenses (Wang et al. 2024a, b). These infections are characterized by nonspecific symptoms like fever and therapeutic failure with antibiotics targeting planktonic cells. Pathogenesis involves resistance to host defenses, endotoxin production (e.g., lipopolysaccharides (LPS) from Gram-negative pathogens), nutrient accumulation in biofilm matrices prolonging infection, and cell detachment promoting dissemination (Dsouza et al., 2024).

Gram-positive bacteria (*Staphylococcus epidermidis*, *S. aureus*), Gram-negative pathogens (*Pseudomonas aeruginosa*, *Escherichia coli*), and fungi like the *Candida* species—especially *C. albicans* and *C. parapsilosis*—are among the microorganisms that create biofilms and are linked to illnesses. *Candida* causes a large percentage of nosocomial infections, and biofilms from this infection cause considerable morbidity and mortality (Datta et al., 2024). Approximately 40% of catheterized patients develop *Candida*-related fungemia. Non-device-related infections, such as *Candida* endocarditis, also exhibit biofilm involvement. Despite the fact that lipid-based amphotericin B formulations and echinocandins such as caspofungin and micafungin exhibit encouraging anti-biofilm action, studies demonstrate that *Candida* biofilm cells are more resistant to antimicrobials than their planktonic counterparts (Lazăr & Chifiriuc, 2010).

5.2.2.2 Industrial Significance

Biofilm development has a major effect on human health and industrial productivity; in the United States, microbially induced corrosion results in billions of dollars' worth of losses yearly (Shineh et al., 2023). Construction, transportation, shipping, chemicals, oil, gas, textiles, and power are among the industries impacted by microbial proliferation and biofouling. Textile fibers are especially susceptible because of their molecular characteristics. Biofilm growth causes fabric damage, faded colors, and unpleasant odors in the clothing manufacturing industry. In knitting and dyeing areas, warm, humid conditions encourage the growth of microorganisms such as *Pseudomonas* sp., *Escherichia coli*, and *Staphylococcus aureus*. While cellulolytic bacteria use cellulolytic enzymes to break down cellulose fibers, natural fibers are particularly vulnerable to bacterial and fungal development. Microbial contamination can be effectively reduced by biocide substances such as chlorinated phenols and organo-coppers and nanotechnology-based solutions like titanium dioxide (TiO_2) and antifouling magnetic nanoparticles (Anghel et al., 2012).

In construction, microbial-induced corrosion causes severe damage to concrete structures. The *Thiobacillus* sp. reacts with sulfur compounds to produce sulfuric acid, which degrades concrete by forming gypsum and calcium composites, leading to microcracks. Acid-producing *Fusarium* sp. biofilms further reduce concrete pH and thickness (George et al., 2013). Incorporating ZnO, lignin, and nanoparticles like silicon dioxide and titanium dioxide enhances concrete's antibacterial properties and durability (Kumari et al., 2016).

The maritime, energy, and chemical industries are impacted by biofilm buildup in water and gas pipelines, condenser tubes, and storage tanks. Bacteria in water pipelines reduce flow rates, while biofilms in condenser tubes lower heat transfer and efficiency (Vishwakarma, 2020). Sulfate-reducing bacteria corrode pipelines, causing leaks and disasters (Enning & Garrelfs, 2014). Surface modifications, such as electroless plating and antibacterial nanofilms (e.g., copper and nickel), provide effective biofilm resistance. Advanced coatings using chemical or physical vapor deposition also enhance antibacterial properties, albeit at higher costs (Butler et al., 2023).

5.2.2.3 Natural Significance

Biofilms are integral to natural ecosystems, facilitating stability and nutrient cycling across diverse habitats. These structured microbial consortia are pivotal to biogeochemical processes, including carbon, nitrogen, sulfur, and phosphorus cycles. In aquatic environments, biofilms colonize substrates such as rocks, sediments, and plant roots, mediating the degradation of organic matter and recycling nutrients to sustain freshwater and marine productivity. Biofilms contribute to water quality by degrading contaminants and filtering pollutants, thereby enhancing natural water purification systems (Sukmarini et al., 2024).

In terrestrial ecosystems, biofilms play critical roles in soil fertility and plant–microbe interactions. Rhizosphere biofilms formed by bacteria and fungi colonize plant roots, promoting nutrient acquisition through mechanisms like nitrogen fixation and phosphorus solubilization. These biofilms improve plant health by suppressing root pathogens via resource competition or the secretion of antimicrobial compounds, thereby enhancing resilience in nutrient-deficient soils (Abdian et al., 2024).

In wetland systems, biofilms form microbial mats that facilitate organic matter degradation, nutrient cycling, and pollutant filtration from agricultural runoff. These mats stabilize sedimentary environments by aggregating particles, reducing erosion, and preserving habitat structure for diverse organisms. Marine biofilms, forming on submerged surfaces, initiate fouling communities and provide sustenance for larger organisms such as barnacles and mussels. Biofilms also influence the global carbon cycle by sequestering and degrading organic carbon, thus affecting carbon storage and climate regulation (Behzadnia et al., 2024).

Within anoxic environments, such as hydrothermal vents and wetlands, biofilm-associated microbial communities drive sulfur oxidation and reduction, enhancing nutrient availability and energy flow (Xu et al., 2024). The extracellular polymeric substance (EPS) matrix of biofilms concentrates nutrients and shields microorganisms from environmental stressors like ultraviolet (UV) radiation, temperature extremes, and desiccation, promoting ecosystem resilience. These biofilms serve as microhabitats, fostering biodiversity and sustaining ecological functions critical to environmental health and ecosystem services (do Socorro Costa et al., 2024).

5.3 Genomic Approaches in Biofilm Research

Over the past 15 years, advancements in multi-omics science have significantly influenced molecular biology, environmental biology, and microbial physiology (Thakur et al., 2008). The sciences of biochemistry, molecular biology, and microbiology have become whole-systems-level through the combination of large-scale omics techniques with computational and post-data analysis techniques. Multi-omics makes it possible to identify pathogen-specific proteins and offers comprehensive insights into host–pathogen interactions (Otto et al., 2012). Additionally, omics-based approaches aid in the systems-level evaluation of microbial physiology, including protein production and metabolic pathways related to carbon and carbohydrates (Dutta et al., 2024).

Important omics techniques like proteomics, metabolomics, transcriptomics, and genomes have proven crucial in improving microbial risk assessment models. For instance, genomics helps identify druggable viral genome features and facilitates the genomic study of viruses, including strains of the Viridae family. The identification of genes implicated in biofilm development and antibiotic resistance is made possible by transcriptomics (Chiu & Miller, 2019). Metabolomics has been used to study microbial activity in fecal matter using ^{13}C-labeled techniques (Date et al., 2010). Additionally, metabolomics has revealed biofilm formation and resistance mechanisms in *Candida auris* (Kean et al., 2018). Combining multi-omics approaches with metagenomics and microbial community analysis has improved our understanding of biofilm infections and contaminated site ecology. This integration holds promise for advancing therapies, antiviral targets, and microbial ecology.

5.3.1 Whole Genome Sequencing (WGS) (Genomics)

Finding the genes causing microbial biofilm development, antibiotic resistance, and virulence requires whole genome sequencing (WGS). Researchers can thoroughly evaluate the genetic components that contribute to these traits by sequencing the complete genomes of bacterial isolates (Zhou et al., 2010).

5.3.1.1 Biofilm Formation Genes

One of the main virulence factors of many microbes is the creation of biofilms, which helps them survive in harsh environments like industrial settings, human tissues, or food production systems. WGS is used to identify particular genes, such as the intercellular adhesion (ica) operon, involved in forming biofilms. For example, polysaccharide intercellular adhesin (PIA), which is synthesized by the icaADBC gene cluster, helps bacteria adhere to surfaces and create biofilms (Omidi et al., 2020). WGS enables the identification of these genes across a broad spectrum of biofilm-forming microorganisms, providing a deeper understanding of their genetic foundation and aiding in developing targeted strategies to combat biofilm-related infections (Atshan et al., 2013).

5.3.1.2 Antibiotic Resistance Genes

WGS makes it possible to identify the genes in microbial genomes that resist antibiotics. For instance, this study has used phenotypic techniques to evaluate antibiotic resistance patterns. WGS, on the other hand, offers a more thorough method by identifying the genes that cause antibiotic resistance. Sequenced genomes are analyzed using tools like the Resistance Gene Identifier (RGI) to look for known resistance genes that are listed in databases like the Comprehensive Antibiotic Resistance Database (CARD) (Alcock et al., 2023). This helps understand the mechanisms behind multidrug resistance (MDR) in biofilm-forming pathogens, which are often more difficult to treat due to their enhanced resistance (Sergelidis & Angelidis, 2017). By comparing the resistome of different strains, researchers can track the spread of resistance genes and identify novel targets for antimicrobial therapy.

5.3.1.3 Virulence Genes

Microorganisms' capacity to cause illness is influenced by virulence factors. By using WGS, scientists can find the genes linked to virulence that allow bacteria to infiltrate host organs, elude immunological reactions, or release poisons. Genomes are frequently analyzed for virulence determinants using the Virulence Factor Database (VFDB), which offers information on the pathogenic potential of various strains (Liu et al., 2022). Combining biofilm formation and virulence factors can significantly enhance pathogenicity in biofilm-forming microorganisms, making them more resistant to host defenses and antimicrobial treatments.

5.3.1.4 Mobile Genetic Elements (MGEs)

MGEs, such as plasmids, transposons, and prophages, are essential for the horizontal spread of virulence and antibiotic resistance genes. WGS enables identifying and characterizing these elements, which are often responsible for spreading resistance and virulence traits among bacterial populations (Johansson et al., 2021). Tools like MGEfinder and PHASTER help identify and analyze MGEs within sequenced genomes, providing insights into how these elements facilitate the spread of detrimental traits in microbial communities (Arndt et al., 2016).

5.3.1.5 Phylogenetic Analysis

WGS also aids in phylogenetic analysis, which helps understand the evolutionary relationships between strains. By comparing the genomes of biofilm-forming and antibiotic-resistant bacteria with other strains in public databases, researchers can determine the genetic lineage and track the transmission of resistant and virulent strains (Wattam et al., 2017). Phylogenetic trees generated using platforms like PATRIC or the Mash/MinHash algorithm can provide valuable insights into biofilm-forming microbes' genetic diversity and evolutionary patterns. This analysis is essential for identifying sources of infection, monitoring outbreaks, and developing effective control measures (Stamatakis, 2014).

In conclusion, whole genome sequencing effectively determines the genetic foundations of virulence, antibiotic resistance, and biofilm formation in microorganisms. By identifying the key genes involved in these processes, WGS enables a deeper understanding of microbial behavior, which is essential for developing new therapeutic strategies to combat biofilm-related infections and resistance in pathogenic microorganisms (Ajose et al., 2024).

5.4 Transcriptomics

5.4.1 Unravelling Biofilm Development Through Transcriptomics

Understanding the structure of the proteome and the dynamics of gene regulation requires understanding transcriptomics, which is the study of transcriptomes or the total amount of mRNA transcribed from a genome (Lowe et al., 2017). It sheds light on the molecular processes that control the development of drug resistance and the change from the planktonic state to biofilm formation (Pan et al., 2015). Transcriptome analysis relies heavily on well-established techniques like microarrays, quantitative polymerase chain reaction (qPCR), and RNA sequencing (RNA-Seq).

Among these, qPCR is very useful for monitoring changes in biofilms' gene expression over time, even during therapeutic interventions. For example, qPCR has been successfully used to assess the expression of the *esp* gene in *Enterococcus faecalis* biofilms following treatment with zinc oxide nanoparticles (Partoazar et al., 2019). Moreover, qPCR often validates results obtained from RNA-Seq and microarray studies (Shemesh et al., 2007). Despite their efficiency and cost-effectiveness, qPCR and related methods are typically target-specific, limiting their scope to preselected genes of interest.

DNA microarrays are another prevalent tool in transcriptomic studies. They combine targeted DNA sequences on microchips, enabling concurrent analysis of multiple gene expression levels. This method has proven effective in comparing immune responses to oral pathogenic bacteria in planktonic and biofilm states (Ebersole et al., 2019). However, the popularity of microarrays has decreased due to the emergence of next-generation sequencing (NGS). Because of its remarkable sensitivity, capacity to identify uncommon mutations, and single-base resolution for transcript quantification, RNA-Seq, an NGS-based technology, has become well-known. As transcriptomics develops, it becomes a vital tool for researching drug resistance mechanisms, host–biofilm interactions, and biofilm development, offering thorough insights into the behavior and reactions of microorganisms (Lowe et al., 2017).

Biofilm formation involves sequential steps, including irreversible adhesion, microcolony formation, and maturation. Transcriptomic approaches have been instrumental in analyzing differential gene expression throughout these stages. For example, comparative DNA microarray analysis has revealed that the expression of the WapA surface protein in *S. mutans* biofilms is upregulated in biofilms with greater thickness (200–400 μm) compared to thinner biofilms (100 μm) (Shemesh et al., 2008).

RNA-Seq has further clarified the biochemical characteristics and genetic or environmental factors affecting biofilm formation. A total of 27 metabolically active genera were found in 1 study using the transcriptome profiles of dental plaque biofilms; the most prevalent of these were the *Streptococcus* species, the *Veillonella* species, and the *Capnocytophaga* species (Peterson et al., 2014).

Transcriptomics has also been used to study interactions between species within biofilms. For example, mutagenesis and RNA-Seq investigations have demonstrated that the adhesion between *Porphyromonas gingivalis* and *Candida albicans* is mediated by the internalin family protein In1J. In *P. gingivalis* and *C. albicans* communities, this interaction has resulted in the upregulation of 57 genes linked to integrins and the type IX secretion machinery. These results suggest that type IX secretion systems both promote defense mechanisms against *Candida albicans* and increase the pathogenicity of *P. gingivalis* (Dutta et al., 2024).

Transcriptomics, as the initial stage of biofilm surveillance, offers valuable insights into the molecular mechanisms underlying biofilm development, drug resistance, and host–microbe interactions. While it provides a robust framework for understanding gene expression, it is crucial to integrate transcriptomic data with proteomic and metabolomic analyses to capture translational and posttranslational modifications. Careful interpretation of transcriptomic data, supported by advanced tools like machine learning, will further enhance our understanding of biofilm dynamics and pave the way for innovative therapeutic strategies (Dutta et al., 2024).

5.5 Proteomics

5.5.1 Exploring Biofilm Dynamics, Drug Resistance, and EPS Role in Proteomics

Research on microbial biofilms has been transformed by high-throughput proteomics, which has produced vast amounts of data that have significantly increased our comprehension of these intricate microbial communities. Specifically, gel-free and differential isotopic labeling methods have gained popularity in biofilm research, including tandem mass tag (TMT), isobaric tags for relative and absolute quantitation (iTRAQ), and Stable isotope labeling by amino acids in cell culture (SILAC). These methods make it possible to measure the proportion of proteins in biofilm communities, which is essential for comprehending the dynamics of biofilms (Dutta et al., 2024).

For instance, the proteome of *Neisseria gonorrhoeae's* planktonic and biofilm forms has been investigated using the SILAC labeling approach, which has shed light on the different protein profiles of these two states (Phillips et al., 2012). The protein expression of antibiotic-sensitive and antibiotic-resistant strains of the biofilm-forming *Aeromonas hydrophila* and *Candida* species has been compared using TMT labeling in conjunction with iTRAQ analysis (Ellepola et al., 2019). Biofilm research has also benefited from label-free techniques that use spectral intensity measurements. By using these techniques to examine subgingival biofilms made up of *Aggregatibacter actinomycetemcomitans* and 10 other bacterial species, important patterns of protein regulation in species such as *Prevotella intermedia* and *Campylobacter rectus* have been discovered (Bao et al., 2015). These proteins' pathway enrichment study has shed more light on the molecular mechanisms behind biofilm formation, including ferric iron binding and 5S RNA binding.

Another important element of biofilm formation that has been the focus of numerous proteomics investigations is the extracellular polymeric substance (EPS) matrix. Although studies on the precise processes by which EPSs contribute to biofilm drug resistance have produced conflicting findings, it is believed that EPSs play a significant role in this resistance. While some studies concentrate more on the function of EPSs in biofilm architecture and structural integrity, others contend that EPSs shield biofilm-forming bacteria from antimicrobial treatments (Dutta et al., 2024).

For instance, a proteomic analysis of *Salmonella enterica* under different flow conditions revealed that although the EPS content and the biofilm's shape changed somewhat, the overall protein expression patterns of the biofilm cells remained mostly unchanged (Gil et al., 2014). However, immuno evasive proteins found in the extracellular polymeric substances (EPSs) of methicillin-resistant and methicillin-sensitive *Staphylococcus aureus* biofilms indicate that these proteins may play a role in stress resistance and immune evasion (Gil et al., 2014). Similarly, studies on *Pseudomonas aeruginosa* biofilms at various phases have shown that EPSs contain type III secretion system effectors, emphasizing the matrix's function in pathogenesis, nutrition uptake, and stress tolerance (Zhang et al., 2015).

Proteomics has demonstrated a connection between antibiotic resistance and fatty acid synthesis in biofilms. Specifically, it has been demonstrated that *Aeromonas hydrophila* biofilms enhance the formation of fatty acids, which is linked to increased antibiotic resistance (Li et al., 2016). Furthermore, research on secondary metabolites, like carolacton from *Sorangium cellulosum*, has shown that it can compromise the integrity of biofilms by rupturing the cell wall and preventing the formation of peptidoglycans (Li et al. 2013a, b).

Proteomic analysis of biofilm persister populations has also shed light on the unique characteristics of these cells, which are believed to be responsible for chronic infections. In *P. aeruginosa* biofilms, persister cells have been shown to express higher levels of LasB (a protease), chitinase, and pyocyanin synthesis proteins, all of which contribute to the biofilm's ability to tolerate antibiotics. Type IV pili, critical for biofilm motility and quorum sensing, were also overexpressed in persister cells, suggesting that motility inhibitors could be used with antibiotics to prevent biofilm persistence (Chua et al., 2016).

Proteomics has become essential in studying microbial biofilms, providing valuable insights into their composition, structure, and behavior under various conditions. Through advanced techniques such as mass spectrometry, proteomics has allowed researchers to explore the complex molecular interactions within biofilms and between biofilms and their hosts (Soufi & Soufi, 2016). The findings from proteomic studies have enhanced our understanding of biofilm resistance mechanisms, including metabolic shifts, oxidative stress responses, and the role of the EPS matrix (Gil et al., 2014). These insights have opened new avenues for developing strategies to combat biofilm-related infections, which remain a significant challenge in clinical settings (Dutta et al., 2024).

5.6 Metabolomics

Biofilms are communities of microorganisms attached to surfaces and exhibit complex metabolic behaviors that differ significantly from planktonic (free-floating) bacteria (Wan et al., 2018). Metabolomics has become crucial in studying biofilm formation and its associated antimicrobial resistance. The development of biofilms is heavily influenced by both primary and secondary metabolites, which can affect microbial growth, attachment, and resistance to environmental stresses, including antibiotic treatment. Understanding the metabolomic changes during biofilm formation is essential for identifying potential therapeutic targets to combat biofilm-related infections (Dutta et al., 2024).

Research has shown that biofilms can alter their metabolic profiles in response to environmental changes and antibiotic exposure, leading to antimicrobial resistance. Scientists can identify the key metabolic pathways involved in biofilm formation and resistance mechanisms by analyzing these changes. For example, in the case of *Staphylococcus aureus*, metabolic changes have been shown to play a crucial role in biofilm development and antibiotic resistance (Antti et al., 2013).

Metabolomics can also aid in identifying new drugs to prevent or disrupt biofilm formation. A promising strategy involves screening for compounds that interfere with autoinducers (AIs), signaling molecules that regulate biofilm formation. Autoinducers are involved in quorum sensing, which enables bacteria to coordinate biofilm formation and other collective behaviors (Wilson et al., 2012). Researchers can develop new therapeutic agents that disrupt biofilm formation by using metabolomics to identify compounds that block or mimic autoinducers.

In addition to identifying potential therapeutic targets, metabolomics can help elucidate the metabolic heterogeneity within biofilm communities. Biofilms are spatially organized structures, and different regions within the biofilm may exhibit distinct metabolic activities. This heterogeneity is crucial for the survival and persistence of biofilms under varying environmental conditions. Metabolomics provides a means to study these complex metabolic profiles, offering insights into the metabolic states contributing to biofilm formation and persistence (Zhang & Powers, 2012).

Metabolomics, particularly when coupled with metabolic flux analysis, provides powerful tools for studying biofilm formation and its associated metabolic processes (Wan et al., 2018). By examining the metabolites produced during biofilm development, researchers can gain insights into the mechanisms underlying antimicrobial resistance and identify potential targets for therapeutic intervention. Moreover, using stable isotope labeling techniques in Metabolic flux analysis allows for the real-time tracking of metabolic fluxes, providing a dynamic view of the metabolic reprogramming that occurs during biofilm formation (Winter & Krömer, 2013).

As the field of biofilm metabolomics continues to evolve, these tools are likely to play an increasingly important role in developing new strategies to combat biofilm-associated infections. Metabolomics has the potential to completely change how we manage chronic infections and disorders linked to biofilms with additional developments in analytical methods and a better comprehension of biofilm metabolism (Dutta et al., 2024).

5.7 Genomic Insights into Biofilm Formation and Regulation

Microorganisms exhibit flexibility in a range of environmental situations by naturally aggregating at different surfaces to create permanent communities, such as single or mixed microbial biofilms (Rodrigues et al., 2019). A three-dimensional structure made up of microbial cells and their extracellular polymeric substances (EPSs) is commonly referred to as a microbial biofilm (Ras et al., 2011). EPSs are a complex matrix made up of both small and macromolecules that play important roles in adhesion, cell aggregation, defense against host defense mechanisms and antimicrobial agents, supplying nutrients, stabilizing cell communities, and facilitating the exchange of genetic information (Pham et al., 2020).

Different EPS components and ratios are secreted by cells as a result of changes in microbial biology, environmental factors, physical stress, and nutritional availability. As a result, the content, three-dimensional structure, and function of biofilm substrates might differ greatly (Lin et al., 2018). Biofilm-associated infections are difficult to treat because of EPSs' protective function against environmental elements. Additionally, the synthesis of EPSs promotes the sequestration and elimination of specific heavy metals and improves cell surface adherence. Therefore, studying the formation and function of biofilms requires an understanding of the crucial roles that EPS components play in biofilms (Grujić et al., 2018).

Microbial colonies that are organized and flourish on both biotic and abiotic surfaces are called biofilms. Cells, an extracellular matrix (ECM), and other polymeric materials make up these communities. Exopolysaccharides (EPS), extracellular DNA, proteins, and amyloidogenic proteins make up the majority of the extracellular matrix (ECM), a biopolymer matrix that supports the stability and functionality of biofilms (Gupta et al., 2016). Several genes are critical for biofilm development in different bacterial species.

In *Pseudomonas aeruginosa*, biofilm formation involves key genes such as *pqsR-pqsA*, *lasR*, *lecA*, *pelA*, *rhlI-rhlR*, and *rhlAB* (Abdelraheem et al., 2021). The genes icaADBC, *eno* (laminin-binding protein), *ebps* (elastin-binding protein), and *fib* (fibrinogen-binding protein) are critical for *Staphylococcus aureus* because they promote the production of biofilms (Kot et al., 2018). In *Streptococcus mutans*, the adhesion process relies on *gtfs* genes, while *gbps*, *smu630*, *relA*, and *comDE* contribute to biofilm development. Similarly, type I fimbriae, which are expressed by the *fim*ABCDEFGH gene cluster, mediate cell attachment in *Escherichia coli* (Afrasiabi & Partoazar, 2024).

These genes and their products underline the complexity and specificity of biofilm formation, highlighting the significance of genetic and biochemical pathways in microbial adhesion and persistence across various environments. Understanding these mechanisms can inform strategies for managing biofilm-associated infections and other biofilm-related challenges.

5.7.1 Process of Biofilm Formation

5.7.1.1 Adhesion

The first stage of biofilm formation is bacterial adhesion to surfaces, where the bacterium approaches the surface and permits the initial attachment. Hydrophobic, electrostatic, and Van der Waals forces are some of the forces that mediate this attachment. The liquid flow rate can readily separate the bacteria at this point since they are still moving in a Brownian motion (Afrasiabi & Partoazar, 2024). The bacteria produce extracellular polymeric substances (EPSs) and may develop filamentous appendages like pili, flagella, or non-fimbrial adhesins during the irreversible attachment phase that ensues after the initial encounter. Stronger physical or chemical pressures are now needed to separate the germs from the surface since these structures improve adhesion (Palmer et al., 2007).

The contact area between the bacterial cell and the surface, the equilibrium of repulsive and attractive forces between the two surfaces, surface hydrophobicity or hydrophilicity, roughness, surface charge, and the substrate's physicochemical and topographical characteristics are some of the variables that affect bacterial attachment (Renner & Weibel, 2011). The *atlE* gene mediates the synthesis of autolysin, which is essential for bacterial adherence in *Staphylococcus aureus*. *AtlE*-deficient strains exhibit a markedly decreased capacity for adhesion. Furthermore, fibrinogen-binding protein is produced due to the expression of the *fbe* and a *sap* gene, which helps *S. aureus* adhere to fibrin (Afrasiabi & Partoazar, 2024). The *iapABD* gene is essential for the synthesis of extracellular matrix (ECM) during the accumulation phase, which ultimately results in the development of a multilayer structure in mature biofilms that resembles a pileus (Zhao & Ashraf, 2016).

5.7.1.2 Synthesis of EPSs

During biofilm formation, the production of extracellular polymeric substances (EPSs) is essential for fostering irreversible adherence. Bacterial cells' quorum sensing (QS) systems strictly control this process (Afrasiabi & Partoazar, 2024). EPSs, essential components of the extracellular matrix (ECM) of a biofilm, mediate bacterial adhesion and cohesiveness through hydrophobic interactions and ion-bridging pressures. EPSs are essential for several biofilm-related functions, including attachment to microbial cell surfaces, cell-to-cell adhesion, biofilm architecture, water retention, cell communication, protection, nutrient trapping, and genetic adaptation (Czaczyk & Myszka, 2007).

Furthermore, it has been determined that the cyclic dinucleotide di-GMP (c-di-GMP) is a crucial initiator of the change from reversible to irreversible attachment, which promotes the synthesis of EPSs and the development of cell surface structures. EPSs are composed of several different substances, including enzymes, structural proteins, DNA, polysaccharides, phospholipids, and glycoproteins (Afrasiabi & Partoazar, 2024). Among these, polysaccharides are an essential part of the EPS matrix and are essential for the growth of bacteria and the formation of biofilms (Flemming, 2016). The production and composition of EPSs are thus central to biofilm formation and the persistence of bacterial communities in various environments.

5.7.1.3 Microcolony Formation

Microbial cells divide and proliferate after adhering to a biotic or abiotic surface, resulting in a stable attachment. Certain chemical signals inside the extracellular polymeric substance (EPS) matrix drive this process (Afrasiabi & Partoazar, 2024). The development of microcolonies is the next stage after solid adhesion. The bacterial cells in these microcolonies, which make up different micro-communities within the biofilm, interact in a coordinated manner. The formation and operation of biofilms depend on the interchange of substrates, the distribution of vital metabolic products, and the excretion of metabolic wastes, all of which are made easier by such coordination (Van Gestel et al., 2015).

5.7.1.4 Maturation of Biofilms

The bacteria embedded in the extracellular polymeric substance (EPS) matrix support the development of microcolonies and the eventual maturation of biofilms. EPSs serve as a biological "adhesive" that links the bacterial cells, with changes in gene expression playing a pivotal role in this process (Afrasiabi & Partoazar, 2024). The biofilm guarantees the delivery of nutrients to the bacterial communities and eliminates waste materials by forming water-filled pathways through matrix development. These biofilms frequently exhibit a distinctive multicellular structure resembling a mushroom. Bacterial movement inside the microcolonies is restricted during the maturation phase due to the inhibition of surface structure development.

Additionally, the gene expression profile of bacteria within the biofilm differs substantially from that of planktonic cells, reflecting the altered physiological state of the biofilm-associated bacteria (Muhammad et al., 2020).

5.7.1.5 Biofilm Dispersal

Bacterial cells use the dispersal process to dissolve biofilms and start a fresh biofilm life cycle. Effectors, signal transduction pathways, and environmental cues regulate this intricate process. During dispersal, the expression of the genes responsible for extracellular polymeric substance (EPS) production, initial attachment, and fimbriae synthesis is typically downregulated, while that of the genes involved in EPS degradation and flagella synthesis is upregulated (Afrasiabi & Partoazar, 2024). Inhibition of cyclic di-GMP (c-di-GMP) signaling pathways has been identified as an effective mechanism for promoting biofilm dispersion. Additionally, various environmental factors, including temperature fluctuations, nutrient availability (particularly glucose), pH changes, and oxygen depletion, can all contribute to the dispersal of biofilm-forming bacteria (Toyofuku et al., 2016).

5.7.1.6 Quorum Sensing

Bacteria can sense population density and control the synthesis of virulence factors and the creation of biofilms owing to a chemical communication mechanism called quorum sensing (QS). One possible tactic to counteract bacterial pathogenicity is to disrupt QS (Colino et al., 2018). By releasing tiny extracellular signaling molecules called autoinducers (AIs), which start genetic programming and synchronize group behaviors, bacteria carry out QS. The autoinducing peptide (AIP) system in Gram-positive bacteria, the acyl-homoserine lactone (AHL) system in Gram-negative bacteria, and the AI-2 system, which is shared by both Gram-positive and Gram-negative bacteria, are the three main QS systems that have been found. QS plays a role in controlling the formation of biofilm architecture and bacterial swarming behavior (Sadekuzzaman et al., 2015).

Any tactic that disrupts bacterial communication by interfering with microbial QS signals is known as quorum quenching (QQ). This interference can occur through two mechanisms: the inhibition or degradation of AIs and the disruption of AI–receptor interactions. It has been discovered that QQ chemicals reduce virulence without preventing bacterial growth by blocking bacterial communication (Afrasiabi & Partoazar, 2024). Depending on the regulatory context of QS (whether it enhances or suppresses virulence), QQ agents must either inhibit or stimulate QS-regulated gene expression (Colino et al., 2018). QQ enzymes are commonly used to suppress QS and, hence, lower bacterial pathogenicity; they frequently target AHLs (Hong et al., 2012).

5.7.1.7 Efflux Pumps

Membrane-bound proteins called efflux pumps are essential for the export of different compounds outside of the bacterial cell, such as heavy metals, medications, detergents, and quorum sensing (QS) signals, and for the development of antibiotic resistance. Bacterial chromosomes encode efflux pump genes, which can also be found on mobile genetic components like plasmids (Webber & Piddock, 2003). Efflux pumps may perform at least four functions in the production of biofilms. Initially, they aid in the export of quorum quenching (QQ) and/or extracellular polymeric substances (EPSs), which help regulate QS and create a biofilm matrix. Second, they might indirectly alter genes that are involved in the formation of biofilms. Third, they help expel harmful compounds like antibiotics and metabolic byproducts. Finally, by preventing adherence to surfaces and other cells, efflux pumps might affect bacterial aggregation (Alav et al., 2018).

Efflux pump genes such *emrK*, *araJ*, *gltK*, *ddpD*, *ycbO*, and *yhdX* have elevated expression during biofilm generation in *Escherichia coli* (May et al., 2009). *Pseudomonas aeruginosa* biofilm formation was unaffected by mutations in *nalB* and *nfxB*, which overexpress the MexAB-OprM and MexCD-OprJ efflux systems, respectively. In actuality, the biofilms of *nalB* mutant strains were noticeably denser than those of the wild type. On the other hand, the MexEF-OprN multidrug efflux pump overexpression decreased intracellular QS signal concentrations, which hindered the production of biofilms (Alav et al., 2018). Furthermore, *Proteus mirabilis*'s *bcr* efflux gene mutation led to decreased biofilm formation and abnormalities in swimming and swarming motility (Holling et al., 2014).

5.7.1.8 Intracellular Adhesion (ica) Operon

Enzymes involved in the manufacture of polysaccharide intercellular adhesin (PIA) are encoded by the intercellular adhesion (ica) operon, which is made up of the *genes icaA*, *icaD*, *icaB*, and *icaC*. The expression of *icaA* and *icaD* tightly regulates EPS synthesis. Simultaneously, by promoting its transport to the bacterial cell surface and deacetylating exopolysaccharide molecules, *icaB* and *icaC* aid in synthesizing poly-*N*-acetylglucosamine (PNAG). In addition, *N*-acetylglucosaminyltransferase and *icaD* interact with UDP-N-acetylglucosamine to produce EPSs (Afrasiabi & Partoazar, 2024). The final product is then carried by icaC across the cytoplasmic membrane. PIA must be deacetylated by *icaB* in order for it to stick to biotic and abiotic surfaces and encourage the production of biofilms. Although PIA is essential for biofilm development, some strains of biofilm-forming bacteria do not have the *ica* operon genes. Hence, it is not strictly necessary for biofilm formation (Swolana et al., 2022).

5.8 Conclusions

A crucial adaptation mechanism that allows bacteria to endure and flourish in various harsh conditions is biofilm formation. Genomic insights have been critical in uncovering the molecular mechanisms underpinning the transition from planktonic to sessile lifestyles. Studies have identified species-specific genetic determinants, such as the rhlI-rhlR quorum sensing system in *Pseudomonas aeruginosa* and the icaADBC operon in *Staphylococcus aureus*, which highlight the evolutionary adaptability of biofilm-forming microorganisms.

Next-generation sequencing (NGS) and multi-omics approaches have significantly advanced the understanding of biofilm dynamics, including genes involved in adhesion, EPS synthesis, and antimicrobial resistance. Transcriptomics has revealed gene expression patterns, while proteomics and metabolomics have illuminated protein interactions and metabolic adaptations critical for biofilm resilience. These tools have also shed light on interspecies interactions within mixed-species biofilms, emphasizing their enhanced antimicrobial resistance and the role of horizontal gene transfer in spreading resistance genes.

Integrating genomics with computational and machine learning tools has expanded the systems-level understanding of biofilms, offering insights into host–pathogen interactions and microbial community dynamics. These findings have practical implications for developing novel therapeutic strategies, including targeting specific genetic pathways to disrupt biofilm formation and identifying new antimicrobial targets.

Genomic research has revolutionized the study of biofilms, providing a comprehensive understanding of their genetic and molecular basis. These advancements promise to address biofilm-associated challenges and leverage their ecological and biotechnological potential.

References

Abdelraheem, W. M., Khairy, R. M., Zaki, A. I., & Zaki, S. H. (2021). Effect of ZnO nanoparticles on methicillin, vancomycin, linezolid resistance and biofilm formation in Staphylococcus aureus isolates. *Annals of Clinical Microbiology and Antimicrobials, 20*(1), 54.

Abdian, P. L., Cámara, M., Cassan, F. D., Castrillo, G., Díaz, M., Gonzalez Anta, G., et al. (2024). Creating pathways for collaboration between Argentina and the UK to utilise microbial biofilms in sustainable agriculture. *CABI Agriculture and Bioscience, 5*(1), 23.

Afrasiabi, S., & Partoazar, A. (2024). Targeting bacterial biofilm-related genes with nanoparticle-based strategies. *Frontiers in Microbiology, 15*, 1387114.

Ajose, D. J., Abolarinwa, T. O., Oluwarinde, B. O., et al. (2024). Whole genome sequence analysis of multi-drug resistant and biofilm-forming staphylococcus haemolyticus isolated from bovine milk. *BMC Microbiology, 24*, 426. https://doi.org/10.1186/s12866-024-03575-z

Alav, I., Sutton, J. M., & Rahman, K. M. (2018). Role of bacterial efflux pumps in biofilm formation. *Journal of Antimicrobial Chemotherapy, 73*(8), 2003–2020.

Alcock, B. P., Huynh, W., Chalil, R., Smith, K. W., Raphenya, A. R., Wlodarski, M. A., et al. (2023). CARD 2023: Expanded curation, support for machine learning, and resistome prediction at the comprehensive antibiotic resistance database. *Nucleic Acids Research, 51*(D1), D690–D699.

Anghel, I., Grumezescu, A. M., Andronescu, E., Anghel, A. G., Ficai, A., Saviuc, C., et al. (2012). Magnetite nanoparticles for functionalized textile dressing to prevent fungal biofilms development. *Nanoscale Research Letters, 7*, 1–6.

Antti, H., Fahlgren, A., Nässtrröm, E., Kouremenos, K., Sunden-Cullberg, J., Guo, Y., et al. (2013). Metabolic profiling for detection of Staphylococcus aureus infection and antibiotic resistance. *PLoS One, 8*(2), e56971.

Arndt, D., Grant, J. R., Marcu, A., Sajed, T., Pon, A., Liang, Y., & Wishart, D. S. (2016). PHASTER: A better, faster version of the PHAST phage search tool. *Nucleic Acids Research, 44*(W1), W16–W21.

Atshan, S. S., Shamsudin, M. N., Karunanidhi, A., van Belkum, A., Lung, L. T. T., Sekawi, Z., et al. (2013). Quantitative PCR analysis of genes expressed during biofilm development of methicillin resistant Staphylococcus aureus (MRSA). *Infection, Genetics and Evolution, 18*, 106–112.

Bao, K., Bostanci, N., Selevsek, N., Thurnheer, T., & Belibasakis, G. N. (2015). Quantitative proteomics reveal distinct protein regulations caused by Aggregatibacteractinomycetemcomitans within subgingival biofilms. *PLoS One, 10*(3), e0119222.

Behzadnia, A., Moosavi-Nasab, M., & Oliyaei, N. (2024). Anti-biofilm activity of marine algae-derived bioactive compounds. *Frontiers in Microbiology, 15*, 1270174.

Bhargava, A., Pareek, V., Roy Choudhury, S., Panwar, J., & Karmakar, S. (2018). Superior bactericidal efficacy of fucose-functionalized silver nanoparticles against Pseudomonas aeruginosa PAO1 and prevention of its colonization on urinary catheters. *ACS Applied Materials & Interfaces, 10*(35), 29325–29337.

Butler, J., Handy, R. D., Upton, M., & Besinis, A. (2023). Review of antimicrobial nanocoatings in medicine and dentistry: mechanisms of action, biocompatibility performance, safety, and benefits compared to antibiotics. *ACS Nano, 17*(8), 7064–7092. https://doi.org/10.1021/acsnano.2c12488. Epub 2023 Apr 7. PMID: 37027838; PMCID:PMC10134505.

Chiu, C. Y., & Miller, S. A. (2019). Clinical metagenomics. *Nature Reviews Genetics, 20*(6), 341–355.

Chua, S. L., Yam, J. K. H., Hao, P., Adav, S. S., Salido, M. M., Liu, Y., et al. (2016). Selective labelling and eradication of antibiotic-tolerant bacterial populations in Pseudomonas aeruginosa biofilms. *Nature Communications, 7*(1), 1–11.

Colino, C. I., Millán, C. G., & Lanao, J. M. (2018). Nanoparticles for signaling in biodiagnosis and treatment of infectious diseases. *International Journal of Molecular Sciences, 19*(6), 1627.

Czaczyk, K., & Myszka, K. (2007). Biosynthesis of extracellular polymeric substances (EPS) and its role in microbial biofilm formation. *Polish Journal of Environmental Studies, 16*(6), 799–806.

Date, Y., Nakanishi, Y., Fukuda, S., Kato, T., Tsuneda, S., Ohno, H., & Kikuchi, J. (2010). New monitoring approach for metabolic dynamics in microbial ecosystems using stable-isotope-labeling technologies. *Journal of Bioscience and Bioengineering, 110*(1), 87–93.

Datta, S., Nag, S., & Roy, D. N. (2024). Biofilm-producing antibiotic-resistant bacteria in Indian patients: A comprehensive review. *Current Medical Research and Opinion, 40*(3), 403–422.

do Socorro Costa, M., da Silva, A. R. P., Araújo, N. J. S., Paulo, C. L. R., Alencar, G. G., de Alencar, M. A. S., et al. (2024). Assessment of the potential susceptibility of planktonic cells and bacterial biofilms by diosgenin. *Food Bioscience, 62*, 105106.

Dsouza, F. P., Dinesh, S., & Sharma, S. (2024). Understanding the intricacies of microbial biofilm formation and its endurance in chronic infections: A key to advancing biofilm-targeted therapeutic strategies. *Archives of Microbiology, 206*(2), 85.

Dutta, B., Chatterjee, D., Sarkar, N., Lahiri, D., Nag, M., & Ray, R. R. (2024). Multi-omics technology in detection of multispecies biofilm. *The Microbe, 4*, 100128.

Ebersole, J. L., Peyyala, R., & Gonzalez, O. A. (2019). Biofilm-induced profiles of immune response gene expression by oral epithelial cells. *Molecular Oral Microbiology, 34*(1). https://doi.org/10.1111/omi.12251

Ellepola, K., Truong, T., Liu, Y., Lin, Q., Lim, T. K., Lee, Y. M., et al. (2019). Multi-omics analyses reveal synergistic carbohydrate metabolism in Streptococcus mutans-Candida albicans mixed-species biofilms. *Infection and Immunity, 87*(10), 10–1128.

Elwakil, B. H., Eldrieny, A. M., Almotairy, A. R. Z., & El-Khatib, M. (2024). Potent biological activity of newly fabricated silver nanoparticles coated by a carbon shell synthesized by electrical arc. *Scientific Reports, 14*(1), 5324.

Enning, D., & Garrelfs, J. (2014). Corrosion of iron by sulfate-reducing bacteria: New views of an old problem. *Applied and Environmental Microbiology, 80*(4), 1226–1236.

Flemming, H. C. (2016). EPS—Then and now. *Microorganisms, 4*(4), 41.

Foroohimanjili, F., Mirzaie, A., Hamdi, S. M. M., Noorbazargan, H., Hedayati Ch, M., Dolatabadi, A., et al. (2020). Antibacterial, antibiofilm, and antiquorum sensing activities of phytosynthesized silver nanoparticles fabricated from Mespilus germanica extract against multidrug resistance of Klebsiella pneumoniae clinical strains. *Journal of Basic Microbiology, 60*(3), 216–230.

Gao, Z., Chen, X., Wang, C., Song, J., Xu, J., Liu, X., et al. (2024). New strategies and mechanisms for targeting Streptococcus mutans biofilm formation to prevent dental caries: A review. *Microbiological Research, 278*, 127526.

George, R. P., Ramya, S., Ramachandran, D., & Mudali, U. K. (2013). Studies on biodegradation of normal concrete surfaces by fungus fusarium sp. *Cement and Concrete Research, 47*, 8–13.

Gil, C., Solano, C., Burgui, S., Latasa, C., García, B., Toledo-Arana, A., et al. (2014). Biofilm matrix exoproteins induce a protective immune response against Staphylococcus aureus biofilm infection. *Infection and Immunity, 82*(3), 1017–1029.

Grujić, S., Radojevic, I., Vasić, S., Čomić, L., & Ostojić, A. (2018). Heavy metal tolerance and removal efficiency of the Rhodotorula mucilaginosa and Saccharomyces boulardii planktonic cells and biofilm. *Kragujevac Journal of Science*. https://doi.org/10.5937/KgJSci1840217G

Gulati, M., & Nobile, C. J. (2016). Candida albicans biofilms: Development, regulation, and molecular mechanisms. *Microbes and Infection, 18*(5), 310–321.

Gupta, P., Sarkar, S., Das, B., Bhattacharjee, S., & Tribedi, P. (2016). Biofilm, pathogenesis and prevention—A journey to break the wall: A review. *Archives of Microbiology, 198*, 1–15.

Hamida, R. S., Ali, M. A., Goda, D. A., Khalil, M. I., & Al-Zaban, M. I. (2020). Novel biogenic silver nanoparticle-induced reactive oxygen species inhibit the biofilm formation and virulence activities of methicillin-resistant Staphylococcus aureus (MRSA) strain. *Frontiers in Bioengineering and Biotechnology, 8*, 433.

Holling, N., Lednor, D., Tsang, S., Bissell, A., Campbell, L., Nzakizwanayo, J., et al. (2014). Elucidating the genetic basis of crystalline biofilm formation in Proteus mirabilis. *Infection and Immunity, 82*(4), 1616–1626.

Hong, K. W., Koh, C. L., Sam, C. K., Yin, W. F., & Chan, K. G. (2012). Quorum quenching revisited—From signal decays to signalling confusion. *Sensors, 12*(4), 4661–4696.

Jin, J., Zhang, L., Shi, M., Zhang, Y., & Wang, Q. (2017). Ti-GO-Ag nanocomposite: The effect of content level on the antimicrobial activity and cytotoxicity. *International Journal of Nanomedicine, 12*, 4209–4224.

Johansson, M. H., Bortolaia, V., Tansirichaiya, S., Aarestrup, F. M., Roberts, A. P., & Petersen, T. N. (2021). Detection of mobile genetic elements associated with antibiotic resistance in Salmonella enterica using a newly developed web tool: MobileElementFinder. *Journal of Antimicrobial Chemotherapy, 76*(1), 101–109.

Kean, R., Delaney, C., Sherry, L., Borman, A., Johnson, E. M., Richardson, M. D., et al. (2018). Transcriptome assembly and profiling of Candida auris reveals novel insights into biofilm-mediated resistance. *Msphere, 3*(4), 10–1128.

Kot, B., Sytykiewicz, H., & Sprawka, I. (2018). Expression of the biofilm-associated genes in methicillin-resistant Staphylococcus aureus in biofilm and planktonic conditions. *International Journal of Molecular Sciences, 19*(11), 3487.

Kumari, K., Preetha, R., Ramachandran, D., Vishwakarma, V., George, R. P., Sundaramurthy, C., et al. (2016). Nanoparticles for enhancing mechanical properties of fly ash concrete. *Materials Today: Proceedings, 3*(6), 2387–2393.

Lazăr, V., & Chifiriuc, M. C. (2010). Medical significance and new therapeutical strategies for biofilm associated infections. *Roumanian Archives of Microbiology and Immunology, 69*(3), 125–138.

Li, J., Wang, W., Wang, Y., & Zeng, A. P. (2013a). Two-dimensional gel-based proteomic of the caries causative bacterium S treptococcusmutans UA 159 and insight into the inhibitory effect of carolacton. *Proteomics, 13*(23–24), 3470–3477.

Li, F., Weir, M. D., Chen, J., & Xu, H. H. (2013b). Comparison of quaternary ammonium-containing with nano-silver-containing adhesive in antibacterial properties and cytotoxicity. *Dental Materials, 29*(4), 450–461.

Li, W., Yao, Z., Sun, L., Hu, W., Cao, J., Lin, W., & Lin, X. (2016). Proteomics analysis reveals a potential antibiotic cocktail therapy strategy for Aeromonas hydrophila infection in biofilm. *Journal of Proteome Research, 15*(6), 1810–1820.

Lin, Y. C., Cornell, W. C., Jo, J., Price-Whelan, A., & Dietrich, L. E. (2018). The Pseudomonas aeruginosa complement of lactate dehydrogenases enables use of d-and l-lactate and metabolic cross-feeding. *MBio, 9*(5), 10–1128.

Liu, B., Zheng, D., Zhou, S., Chen, L., & Yang, J. (2022). VFDB 2022: A general classification scheme for bacterial virulence factors. *Nucleic Acids Research, 50*(D1), D912–D917.

Lowe, R., Shirley, N., Bleackley, M., Dolan, S., & Shafee, T. (2017). Transcriptomics technologies. *PLoS Computational Biology, 13*(5), e1005457.

May, T., Ito, A., & Okabe, S. (2009). Induction of multidrug resistance mechanism in Escherichia coli biofilms by interplay between tetracycline and ampicillin resistance genes. *Antimicrobial Agents and Chemotherapy, 53*(11), 4628–4639.

Mishra, A., Aggarwal, A., & Khan, F. (2024). Medical device-associated infections caused by biofilm-forming microbial pathogens and controlling strategies. *Antibiotics, 13*(7), 623.

Montazeri, A., Salehzadeh, A., & Zamani, H. (2020). Effect of silver nanoparticles conjugated to thiosemicarbazide on biofilm formation and expression of intercellular adhesion molecule genes, icaAD, in Staphylococcus aureus. *Folia Microbiologica, 65*, 153–160.

Moulavi, P., Noorbazargan, H., Dolatabadi, A., Foroohimanjili, F., Tavakoli, Z., Mirzazadeh, S., et al. (2019). Antibiofilm effect of green engineered silver nanoparticles fabricated from Artemisia scoporia extract on the expression of icaA and icaR genes against multidrug-resistant Staphylococcus aureus. *Journal of Basic Microbiology, 59*(7), 701–712.

Muhammad, M. H., Idris, A. L., Fan, X., Guo, Y., Yu, Y., Jin, X., et al. (2020). Beyond risk: Bacterial biofilms and their regulating approaches. *Frontiers in Microbiology, 11*, 928.

Omidi, M., Firoozeh, F., Saffari, M., Sedaghat, H., Zibaei, M., & Khaledi, A. (2020). Ability of biofilm production and molecular analysis of spa and ica genes among clinical isolates of methicillin-resistant Staphylococcus aureus. *BMC Research Notes, 13*, 1–7.

Otto, A., Bernhardt, J., Hecker, M., & Becher, D. (2012). Global relative and absolute quantitation in microbial proteomics. *Current Opinion in Microbiology, 15*(3), 364–372.

Palmer, J., Flint, S., & Brooks, J. (2007). Bacterial cell attachment, the beginning of a biofilm. *Journal of Industrial Microbiology and Biotechnology, 34*(9), 577–588.

Pan, W., Fan, M., Wu, H., Melander, C., & Liu, C. (2015). A new small molecule inhibits Streptococcus mutans biofilms in vitro and in vivo. *Journal of Applied Microbiology, 119*(5), 1403–1411.

Partoazar, A., Talaei, N., Bahador, A., Pourhajibagher, M., Dehpour, S., Sadati, M., & Bakhtiarian, A. (2019). Antibiofilm activity of natural zeolite supported NanoZnO. Inhibition of Esp gene expression of enterococcus faecalis. *Nanomedicine, 14*(6), 675–687.

Peters, B. M., Jabra-Rizk, M. A., Scheper, M. A., Leid, J. G., Costerton, J. W., & Shirtliff, M. E. (2010). Microbial interactions and differential protein expression in Staphylococcus aureus–Candida albicans dual-species biofilms. *FEMS Immunology & Medical Microbiology, 59*(3), 493–503.

Peterson, S. N., Meissner, T., Su, A. I., Snesrud, E., Ong, A. C., Schork, N. J., & Bretz, W. A. (2014). Functional expression of dental plaque microbiota. *Frontiers in Cellular and Infection Microbiology, 4*, 108.

Pham, D. Q., Bryant, S. J., Cheeseman, S., Huang, L. Z., Bryant, G., Dupont, M. F., et al. (2020). Micro-to nano-scale chemical and mechanical mapping of antimicrobial-resistant fungal biofilms. *Nanoscale, 12*(38), 19888–19904.

Phillips, N. J., Steichen, C. T., Schilling, B., Post, D. M., Niles, R. K., Bair, T. B., et al. (2012). Proteomic analysis of Neisseria gonorrhoeae biofilms shows shift to anaerobic respiration and changes in nutrient transport and outermembrane proteins. *PLoS One, 7*(6), e38303.

Ras, M., Lefebvre, D., Derlon, N., Paul, E., & Girbal-Neuhauser, E. (2011). Extracellular polymeric substances diversity of biofilms grown under contrasted environmental conditions. *Water Research, 45*(4), 1529–1538.

Renner, L. D., & Weibel, D. B. (2011). Physicochemical regulation of biofilm formation. *MRS Bulletin, 36*(5), 347–355.

Rodrigues, M. E., Gomes, F., & Rodrigues, C. F. (2019). Candida spp./bacteria mixed biofilms. *Journal of Fungi, 6*(1), 5.

Sadekuzzaman, M., Yang, S., Mizan, M. F. R., & Ha, S. D. (2015). Current and recent advanced strategies for combating biofilms. *Comprehensive Reviews in Food Science and Food Safety, 14*(4), 491–509.

Santos, A. L. S. D., Galdino, A. C. M., Mello, T. P. D., Ramos, L. D. S., Branquinha, M. H., Bolognese, A. M., et al. (2018). What are the advantages of living in a community? A microbial biofilm perspective! *Memórias do Instituto Oswaldo Cruz, 113*(9), e180212.

Sauer, K., Stoodley, P., Goeres, D. M., Hall-Stoodley, L., Burmølle, M., Stewart, P. S., & Bjarnsholt, T. (2022). The biofilm life cycle: Expanding the conceptual model of biofilm formation. *Nature Reviews Microbiology, 20*(10), 608–620.

Sergelidis, D., & Angelidis, A. S. (2017). Methicillin-resistant Staphylococcus aureus: A controversial food-borne pathogen. *Letters in Applied Microbiology, 64*(6), 409–418.

Shemesh, M., Tam, A., & Steinberg, D. (2007). Differential gene expression profiling of Streptococcus mutans cultured under biofilm and planktonic conditions. *Microbiology, 153*(5), 1307–1317.

Shemesh, M., Tam, A., Kott-Gutkowski, M., Feldman, M., & Steinberg, D. (2008). DNA-microarrays identification of Streptococcus mutans genes associated with biofilm thickness. *BMC Microbiology, 8*, 1–12.

Shineh, G., Mobaraki, M., Perves Bappy, M. J., & Mills, D. K. (2023). Biofilm formation, and related impacts on healthcare, food processing and packaging, industrial manufacturing, marine industries, and sanitation–a review. *Applied Microbiology, 3*(3), 629–665.

Soufi, Y., & Soufi, B. (2016). Mass spectrometry-based bacterial proteomics: Focus on dermatologic microbial pathogens. *Frontiers in Microbiology, 7*, 181.

Stamatakis, A. (2014). RAxML version 8: A tool for phylogenetic analysis and post-analysis of large phylogenies. *Bioinformatics, 30*(9), 1312–1313.

Sukmarini, L., Atikana, A., & Hertiani, T. (2024). Antibiofilm activity of marine microbial natural products: Potential peptide-and polyketide-derived molecules from marine microbes toward targeting biofilm-forming pathogens. *Journal of Natural Medicines, 78*(1), 1–20.

Swolana, D., Kępa, M., Kruszniewska-Rajs, C., & Wojtyczka, R. D. (2022). Antibiofilm effect of silver nanoparticles in changing the biofilm-related gene expression of Staphylococcus epidermidis. *International Journal of Molecular Sciences, 23*(16), 9257.

Thakur, N. L., Jain, R., Natalio, F., Hamer, B., Thakur, A. N., & Müller, W. E. (2008). Marine molecular biology: An emerging field of biological sciences. *Biotechnology Advances, 26*(3), 233–245.

Thuptimdang, P., Limpiyakorn, T., McEvoy, J., Prüß, B. M., & Khan, E. (2015). Effect of silver nanoparticles on Pseudomonas putida biofilms at different stages of maturity. *Journal of Hazardous Materials, 290*, 127–133.

Toyofuku, M., Inaba, T., Kiyokawa, T., Obana, N., Yawata, Y., & Nomura, N. (2016). Environmental factors that shape biofilm formation. *Bioscience, Biotechnology, and Biochemistry, 80*(1), 7–12.

Van Gestel, J., Vlamakis, H., & Kolter, R. (2015). Division of labor in biofilms: The ecology of cell differentiation. *Microbiology Spectrum, 3*, MB-0002-2014.

Vishwakarma, V. (2020). Impact of environmental biofilms: Industrial components and its remediation. *Journal of Basic Microbiology, 60*(3), 198–206.

Wan, N., Wang, H., Ng, C. K., Mukherjee, M., Ren, D., Cao, B., & Tang, Y. J. (2018). Bacterial metabolism during biofilm growth investigated by 13C tracing. *Frontiers in Microbiology, 9*, 2657.

Wang, X., Chen, C., Hu, J., Liu, C., Ning, Y., & Lu, F. (2024a). Current strategies for monitoring and controlling bacterial biofilm formation on medical surfaces. *Ecotoxicology and Environmental Safety, 282*, 116709.

Wang, D., Zeng, N., Li, C., Li, Z., Zhang, N., & Li, B. (2024b). Fungal biofilm formation and its regulatory mechanism. *Heliyon, 10*(12), e32766.

Wattam, A. R., Davis, J. J., Assaf, R., Boisvert, S., Brettin, T., Bun, C., et al. (2017). Improvements to PATRIC, the all-bacterial bioinformatics database and analysis resource center. *Nucleic Acids Research, 45*(D1), D535–D542.

Webber, M. A., & Piddock, L. J. (2003). The importance of efflux pumps in bacterial antibiotic resistance. *Journal of Antimicrobial Chemotherapy, 51*(1), 9–11.

Wilson, C. M., Aggio, R. B., O'Toole, P. W., Villas-Boas, S., & Tannock, G. W. (2012). Transcriptional and metabolomic consequences of LuxS inactivation reveal a metabolic rather than quorum-sensing role for LuxS in Lactobacillus reuteri 100-23. *Journal of Bacteriology, 194*(7), 1743–1746.

Winter, G., & Krömer, J. O. (2013). Fluxomics–connecting 'omics analysis and phenotypes. *Environmental Microbiology, 15*(7), 1901–1916.

Xu, H., Zhu, D., Zhong, M., Li, C., Wen, C., Zhu, S., et al. (2024). Source-oriented risks of heavy metals and their effects on resistance genes in natural biofilms. *Journal of Hazardous Materials, 484*, 136735.

Zhang, B., & Powers, R. (2012). Analysis of bacterial biofilms using NMR-based metabolomics. *Future Medicinal Chemistry, 4*(10), 1273–1306.

Zhang, W., Sun, J., Ding, W., Lin, J., Tian, R., Lu, L., et al. (2015). Extracellular matrix-associated proteins form an integral and dynamic system during Pseudomonas aeruginosa biofilm development. *Frontiers in Cellular and Infection Microbiology, 5*, 40.

Zhao, L., & Ashraf, M. A. (2016). Influence of silver-hydroxyapatite nanocomposite coating on biofilm formation of joint prosthesis and its mechanism. *The West Indian Medical Journal, 64*(5), 506.

Zhou, X., Ren, L., Li, Y., Zhang, M., Yu, Y., & Yu, J. (2010). The next-generation sequencing technology: A technology review and future perspective. *Science China. Life Sciences, 53*, 44–57.

Proteomics and its Applications in Biofilm Research

6

Z. H. Tawfeeq Ahmed, Prabhakar Singh, Ekramul Haque, Deepika Vasudevan, Vijayavarsine Rajesh, Showparnickaa Subburao Ravichandran, Dharshini Srinivasan, Mohsin Maqbool, and Saqib Hassan

6.1 Introduction

Biofilms are aggregations of bacterial microcolonies that stick to an organic or inorganic substrate and are surrounded by a wet extracellular matrix (ECM) made of extracellular polymeric substances (EPS) with a specific structure. The ECM protects the biofilm from the action of antimicrobials, desiccation, ultraviolet (UV) radiation, and immune responses, which makes it recalcitrant. A phenotypic switch that changes the morphology of biofilms between a planktonic state and its whole state via biological and chemical fluctuations in the environment is quintessential to the functioning and survival of biofilms. This conditional switching allows it to evade host immune responses and raises antibacterial tolerance (Koo et al., 2017). Biofilms are the main reason for nosocomial infections because they stick to medical tools and cause endocarditis, prosthetic infections, osteomyelitis, and wound infections in hospital patients (Divya et al., 2019; Fathima et al., 2023). On average, around 65–85% of all bacterial infections are caused by microbial biofilms, according to the Centers for Disease Control and Prevention (CDC), while they are also associated with plant diseases in farming environments and food spoilage (Rahman et al., 2022; Sadiq et al., 2017). Biofilms are also attracting investors in technologies that exploit their properties in the agricultural landscape, sewage treatment plants, the production of renewable bioenergy, and bioremediation to break down imperishable pollutants (Herschend et al., 2017).

Z. H. T. Ahmed · P. Singh · D. Vasudevan · V. Rajesh · S. S. Ravichandran · D. Srinivasan · S. Hassan (✉)
Department of Biotechnology, School of Bio and Chemical Engineering, Sathyabama Institute of Science and Technology, Chennai, Tamil Nadu, India

E. Haque
Department of Botany, Prabhat Kumar College, Purba Medinipur, Contai, West Bengal, India

M. Maqbool
Lerner Research Institute, Cleveland Clinic, Cleveland, OH, USA

S. Busi et al. (eds.), *Omics Approaches in Biofilm Research*,
https://doi.org/10.1007/978-3-031-91863-6_6

These microbial biofilms are the most primitive mode of growth in the natural environment, which gives them several benefits. High levels of population density within these biofilms help them perform specialized functions that unicellular organisms cannot perform. They are receptive and adapt to changes in climate and extreme environmental changes, such as scarce nutrition and radical deviations in pH and temperature. Biofilms also exhibit a high tolerance towards antimicrobials and efficiently evade host defense systems through horizontal gene transfer (HGT), which helps in bacterial proliferation and dissemination (Khemiri et al., 2016). Properties intrinsic to biofilms, such as quorum sensing (QS), help facilitate cell-to-cell mediation in Gram-positive and Gram-negative bacteria. However, both categories possess diverse mechanisms to carry out this function. For instance, Gram-negative *Escherichia coli* synthesizes N-acyl homoserine lactones (AHLs) for cell-to-cell communication; whereas, Gram-positive *Staphylococcus aureus* possesses a complicated two-part system that synthesizes autoinducer peptides (AIPs) that vary between different virulent strains (Karathanasi et al., 2018; Yin et al., 2019).

A "proteome" refers to a complete set of all proteins an organism expresses, which can be set off by environmental stresses and triggers allowing for its expression. Proteomics in biofilms involves the separation of proteins based on size, charge, or phase using techniques such as two-dimensional polyacrylamide gel electrophoresis (2D-PAGE) and liquid chromatography (Oliveira et al., 2014). There are limits to current technology for separating proteins due to low resolution and limited detection of proteins; however, they are quintessential tools for studying the transcriptome. Quantitative and qualitative proteomic approaches have recently evolved from gel-based techniques to mass spectrometry (MS)-based approaches, which can easily measure low-abundance proteins and membrane proteins in a relatively small amount of time (Sauer, 2003; Soufi & Soufi, 2016). These proteins are further identified by MS and quantified using labelling techniques, after which they are sequenced and collected into databases based on functions, such as antibiotic resistance, biofilm formation, and matrix production. Several proteomic studies are undertaken to analyze the various factors that influence their development, such as biofilm formation, bacterial pathogenicity in virulent strains, metabolic pathways, regulatory mechanisms, and tolerance to extreme conditions (Herschend et al., 2017; Santos et al., 2019; Wang et al., 2019; Wang et al., 2020; Yang et al., 2018). For certain bacterial biofilms like those formed by clinically relevant *Listeria monocytogenes*, *Mycobacterium tuberculosis*, and *Staphylococcus aureus*, there is a pronounced lack of studies that detail the proteomics and the functional mechanisms that underscore its pathogenicity and virulence (Wang et al., 2019). Antibiotic resistance mechanisms such as dormant persister cells, enzyme production, and the overexpression of multidrug efflux pumps in biofilms formed by *Acinetobacter baumannii* and *Staphylococcus aureus* make them a potent target for the study of multidrug effective antibiotics. The prevalence of "omics" tools, such as proteomics, metabolomics, and transcriptomics, as well as the overall multiomics framework, are crucial to studying biofilms and their regulatory mechanisms. The polymicrobial nature of biofilms and synergized antibiotic resistance mechanisms call for a multiomics approach towards studying biofilms rather than narrowly defined research (Seneviratne et al., 2020; Wuyts et al., 2018).

Planktonic cells, which are the precursors to biofilms, are highly susceptible to antibiotics during the initial stage, and those bacterial cells that persist on the bottom layer of the biofilm are deprived of oxygen and nutrition, making them equally vulnerable to antibiotics. Despite these vulnerabilities, the EPS effectively renders the biofilm invulnerable to antibiotic penetration, making the study of proteomics essential to solving medical infections involving virulent biofilms. Protein quantification and tagging through tandem mass tag (TMT) label-free quantitative analysis, stable isotope labelling by amino acids in cell culture (SILAC) label-based quantitative analysis, isotope-coded affinity tags (ICAT), and tags for relative and absolute quantitation (iTRAQ) are tools used to quantify proteins in biofilms for later sequencing, alternative to gel-based techniques (Chua et al., 2016; Ellepola et al., 2019; Pham et al., 2010). For instance, TMT is used in studies involving the development of antibiofilm drugs, such as broad-spectrum silver nanoparticles (AgNPs) and their influence on the expression of biofilm proteins. AgNPs promise to eradicate planktonic cells and the overall biofilm by releasing silver particles that corrode and disrupt the impenetrable EPS and biofilm matrix. The penetration of the matrix effectively disrupts cell communication, intercepts QS, destroys disulfide bonds, disintegrates cell membranes, and produces reactive oxygen species (ROS), ultimately inhibiting biofilm formation (Zhang et al., 2020). In lieu of the above, this chapter will provide an overview of the proteomics approaches used in protein identification and analysis while highlighting novel technologies increasingly used in biofilm proteomics. The following text will also delve into biofilm-specific profiles and pathogenicity, including a key summary of pathogenic biofilms' antibiotic resistance mechanisms and virulence factors. Additional discussion about the integration of proteomic techniques into the environmental and industrial sectors and therapeutic interventions for combating the dissemination of clinically relevant strains of biofilms and drug development will be provided in full detail.

6.2 Biofilm Formation and Structure

6.2.1 Stages of Biofilm Development

Biofilms are complex arrangements of cellular aggregations enclosed in an EPS matrix that provide protection while increasing chances of survival in extreme environments. The growth of biofilms follows a well-defined developmental cycle containing five distinct stages: initial attachment, irreversible adhesion, microcolony formation, maturation, and dispersion. The different stages of biofilm formation are discussed below:

Initial Attachment In the first stage, planktonic (free-floating) microbial cells attach to an abiotic or biotic surface through weak, reversible interactions such as hydrophobic and van der Waals forces. Surface appendages like flagella, pili, and fimbriae are vital in cell attachment. Proteins like adhesins and surface-associated enzymes also contribute to this stage.

Irreversible Adhesion Once microbial cells strongly adhere to the surface, they release EPS, anchoring themselves irreversibly. This step is often facilitated by specific exopolysaccharides, adhesins, and outer membrane proteins. Transitioning from reversible to irreversible adhesion is an important step in biofilm formation, leading to structural stability.

Microcolony Formation The attached bacterial cells multiply, creating three-dimensional microcolonies. Intercellular communication through QS is crucial in coordinating and regulating biofilm maturation and gene expression. During this stage, metabolic cooperation and nutrient gradients form within the biofilm.

Maturation The biofilm develops into a well-ordered structure with water channels facilitating water and nutrient exchange. The ECM undergoes extensive modification, incorporating extracellular DNA (eDNA), proteins, and lipids, further enhancing structural integrity. Proteomic analysis has identified key metabolic and structural proteins in biofilm maturation.

Dispersion Environmental factors, such as nutrient exhaustion, shear forces, or QS signaling, activate the liberation of microbial planktonic cells from the parent microcolony, allowing the colonization of new niches. Proteins, such as dispersin-B and hydrolases, destroy the components of the EPS matrix, aiding cell detachment. The entirety of the lifecycle of the biofilm and the influence of QS is illustrated in Fig. 6.1.

6.2.2 Extracellular Polymeric Substance (EPS) and Matrix Composition

The structural integrity provided by the EPS is crucial to biofilms, contributing to antibiotic resistance, structural stability, and stress tolerance. It mainly comprises polysaccharides, lipids, proteins, and nucleic acids, which interact to form a gel-like, hydrated matrix. The different components of the cellular matrix are discussed below:

Polysaccharides Exopolysaccharides are the significant components of the EPS matrix, providing adhesion and mechanical stability. Species-specific

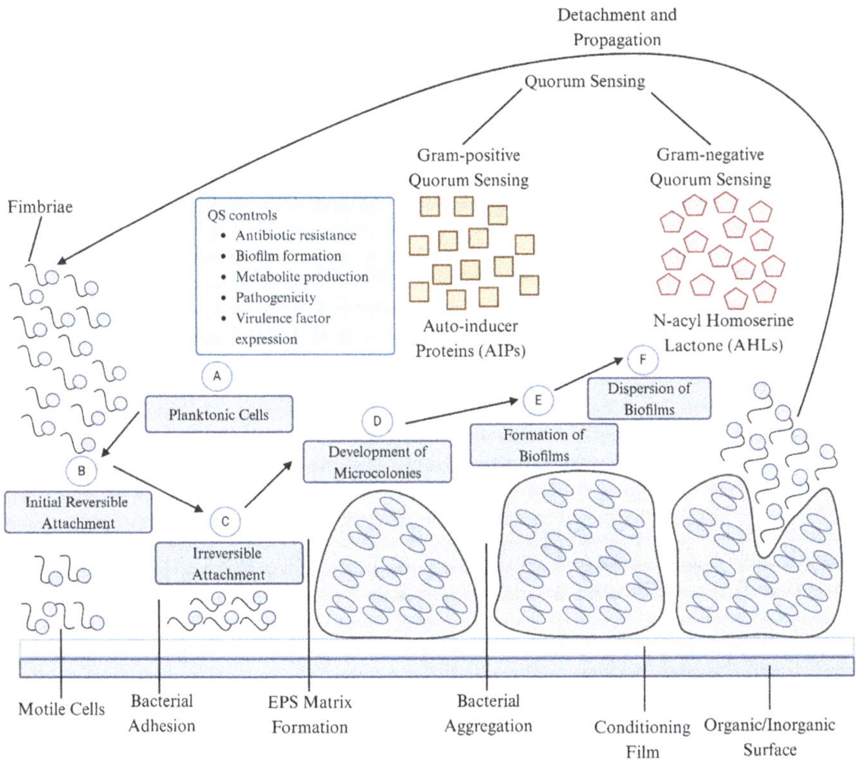

Fig. 6.1 An illustration of the lifecycle and development of biofilms on abiotic/biotic surfaces and the influence of QS in biofilm regulation

polysaccharides, such as alginate in *Pseudomonas aeruginosa*, PIA (poly-β-1,6-N-acetylglucosamine) in *Staphylococcus epidermidis*, and cellulose in *E. coli*, are critical for biofilm development.

Lipids Lipid parts, including membrane vesicles, contribute to biofilm heterogeneity and mediate HGT. Current proteomic studies have shown the presence of lipid-binding proteins that impact biofilm construction.

Proteins Structural and functional proteins inside the EPS matrix include enzymes, adhesins, and amyloid fibrils that improve cohesion. Functional proteomics has identified biofilm-associated proteins such as Bap (biofilm-associated protein) in *S. aureus* and CdrA in *P. aeruginosa*, which contribute to EPS matrix integrity and stability.

Extracellular DNA (eDNA) eDNA fulfills the structural and regulatory functions in biofilm creation, helping neighboring cells communicate and facilitate genetic material exchange. It boosts resistance to antimicrobial compounds by binding cationic antibiotics, reducing their penetration.

6.3 Proteomics Approaches in Biofilm Studies

6.3.1 Overview of Proteomics Techniques

Various proteomics methods for protein extraction and separation include gel-based methods derivatives of 2D-PAGE, such as two-dimensional electrophoresis (2-DE), laser capture microdissection (LCM), and two-dimensional difference gel electrophoresis (2D-DIGE). There are gel-free techniques with high processing power like ICAT and MS. An MS-based method like the shotgun proteomics method, is utilized to recognize and measure a vast array of proteins in a complex biological sample (Chandramouli & Qian, 2009). High-throughput proteomics has generated considerable data that are now widely utilized to study microbial biofilms. Numerous recent studies on microbial biofilms have utilized many proteomics techniques to quantify the proteins. A remarkably adaptable and widely used technique for examining an organism's proteome is 2D-PAGE. In particular, a flexible variant of 2D-PAGE called 2D-DIGE uses up to three fluorescent tags to compartmentalize large protein samples. This method offers better reactivity and replication than traditional gel electrophoresis methods to perform quantitative comparisons of protein abundance in samples. Each method offers unique benefits for separating and identifying numerous proteins, including isoforms and modifications that occur after translation (Meleady, 2018). Differing gel-free labelling techniques, such as TMT, iTRAQ, and SILAC, are commonly used for protein quantification. The SILAC labeling method can identify the proteomes of biofilms and their precursor planktonic bacterial cells (Dutta et al., 2024; Phillips et al., 2012).

Progress in proteomics methods during the last 10 years, closely fused with genomic and physicochemical strategies, has significantly contributed to understanding bacterial biofilm development. Although the versatility of the bacterial proteome is notable, numerous studies have shown that proteomics methods can effectively clarify the biofilm phenotype (Rani & Babu, 2018).

2D-PAGE has been employed to show phenotypic distinctions between biofilm cells of *Pseudomonas aeruginosa* and their planktonic counterparts under specific culture conditions. In a study conducted in 2001, Steyn et al. (2001) utilized 2D-PAGE to examine phenotypic variations between biofilm and planktonic cells of *P. aeruginosa*. This approach allowed the researchers to discover 41 proteins with notable expression differences throughout various growth stages, providing insights into the molecular processes of biofilm development. Using 2-DE, it was found that 15 peptides were uniquely produced in the biofilms of 2-h Bacillus cereus and 7 proteins in 18-h biofilms in a different study by Oosthuizen et al. (2002).

2D-DIGE, an enhanced form of 2D-PAGE, employs as many as three fluorescent labels for differential tagging, providing improved sensitivity, reproducibility, and more precise quantitative evaluation of variations in protein expression. Numerous investigations of biofilm proteomes have used the 2D-DIGE method, which was first created to examine the biofilms. *Pseudomonas aeruginosa* biofilm production is a developmental process, according to Sauer et al.'s analysis of the changes in proteome patterns using 2D-DIGE. Additionally, genetically dissimilar strains in proteome profiles and the added pressure of climatic stresses on the proteomes were analyzed using 2D-DIGE. *P. aeruginosa* was studied using the 2D-DIGE as it transitions from planktonic (free-swimming) to biofilm (surface-attached) growth modes and discovered proteins exhibiting different expression levels throughout biofilm development, providing better comprehension of the biological processes involved in the biofilm lifecycle and maintenance *in P. aeruginosa* (Sauer et al., 2002). A study in 2006 using a 2D-DIGE examination of *Campylobacter jejuni* 11,168 biofilms showed the involvement of the motility complex in biofilm development. This study discovered that two flagellins (FlaA/FlaB), the filament cap (FliD), and other basal proteins (FlgG/FlgG2) showcased an increased level of expression in the biofilm form compared to planktonic cells. Moreover, proteins that play roles in stress responses, adhesins (Peb1, FlaC), and numerous biosynthetic and energy-producing functions exhibited enhanced biofilm expression (Kalmokoff et al., 2006). A summary of all the proteomic techniques discussed above is illustrated in Fig. 6.2.

6.3.2 Emerging Technologies in Biofilm Proteomics

Developing technologies in biofilm proteomics have greatly enhanced the comprehension of biofilm development, upkeep, and resistance processes. These developments enable a more thorough examination of protein expressions, interactions, and molecular pathways in biofilms. The emergence of better yielding technologies (including Illumina, Roche 454, SOLiD, and Pacific Biosciences) recently enabled researchers to acquire genome-wide data on biofilms. The novel sequencing techniques have enabled the acquisition of the entire set of genome sequences for swathes of microbes involved in biofilm formation (Franklin et al., 2015).

6.3.2.1 Transcriptomics Approaches for Biofilm Studies

The RNA generated under various conditions is extracted, and every RNA transcript is measured. Transcriptomics is utilized for studying different biofilms to extract data on gene activity and patterns specifically attuned to the development of biofilms. The transcriptomics methods used to analyze the biofilms include quantitative reverse transcription polymerase chain reaction (RT-qPCR) and microarrays. The complete transcriptome of these microbial colonies is also analyzed using next-generation sequencing technologies such as RNAseq, which sequences cDNA libraries made from RNA pools (Folsom et al., 2010).

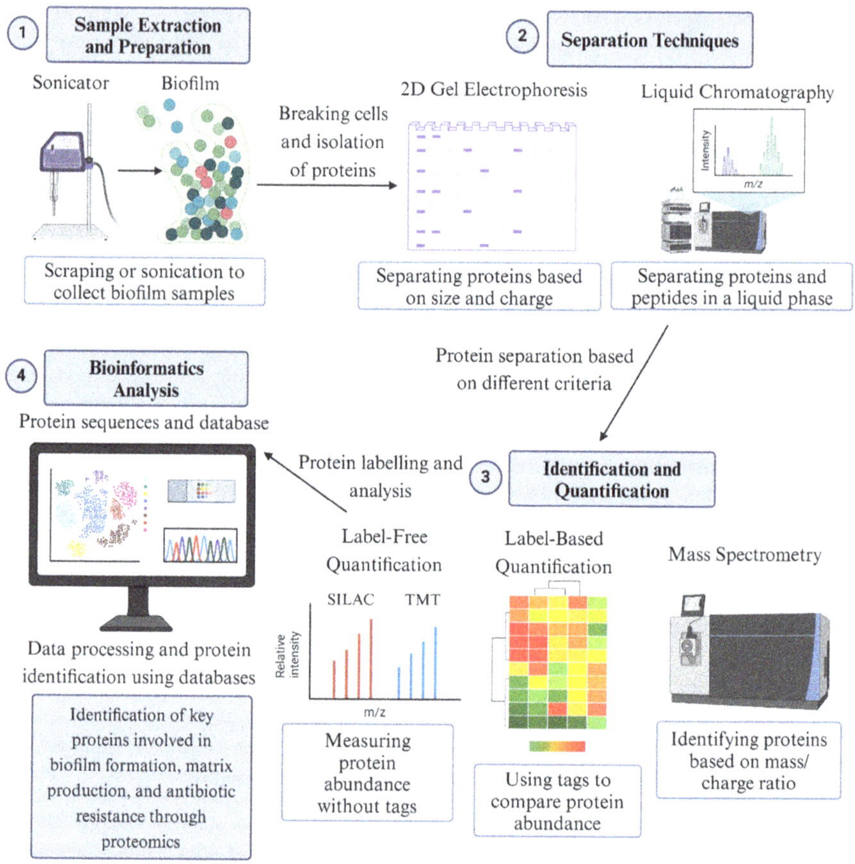

Fig. 6.2 An overview of the various proteomic techniques involved in the extraction, separation, identification, quantification, and analysis of the key proteins involved in processes of biofilm regulation

6.3.2.2 Proteomics and Metabolomics Approaches for Biofilm Studies

2D-DIGE and shotgun proteomics are the two primary techniques used in proteomics. Every strategy has its own variants. Since its creation, 2D-DIGE has been applied to numerous biofilm proteome investigations. Metabolomics is the newest tool in the omics toolbox used to study biofilms. Meta proteomics studies of extracellular and cell membrane proteins in biofilm specimens have been conducted using shotgun proteomics (Van Oudenhove & Devreese, 2013).

6.3.2.3 Biofilm Imaging

The ability to view aqueous active biofilms in three dimensions as they develop using a confocal scanning laser microscope (CSLM) has been a significant breakthrough in our understanding of biofilm organization, operation, and behavior.

6.4 Biofilm-Specific Proteomic Profiles

6.4.1 Cellular Response and Gene Expression

Gene expression alterations during mixed biofilm formation offer a more thorough picture of interspecies interactions. The structural integrity and behavior of the polymicrobial biofilms are shaped by a combination of environmental signals and antagonistic and synergistic interactions, which may be reflected in altered gene expression patterns. Fluctuations in gene expression from a single-species to polymicrobial biofilms have been observed in several investigations. As previously mentioned, when *V. parvula* and *S. mutans*, two oral bacteria, underwent simultaneous culturing, which made *V. parvula* modify the structure of *S. mutans*, *S. mutans* showed more excellent resistance to antimicrobial drugs. Coculturing changed the manifestation of numerous *S. mutans* genes, such as those related to protein formation and intracellular polysaccharide, purine, and amino acid metabolism (Luppens et al., 2008). As previously stated, another dental biofilm model revealed gene expression fluctuations linked to metabolism during bacterial coaggregation. Genes responsible for arginine synthesis were expressed in *Streptococcus gordonii* during coaggregation with *Actinomyces naeslundii*. Furthermore, the development of coaggregates with *A. naeslundii* was necessary for the growth of *S. gordonii*, whether arginine was present or absent at low concentrations (Jakubovics et al., 2008). Similarly, it was demonstrated that *P. aeruginosa* produced much more exotoxin A when *S. aureus* was present (Goldsworthy, 2008). In this case, fluctuations in gene expression brought about by the synthesis of polymicrobial biofilms led to the production of a virulence gene. Notably, from a therapeutic standpoint, the rise in bacterial pathogenicity that results from these two microorganisms mediating different diseases is a troubling phenotypic shift. It is crucial to stress that *P. aeruginosa* and *S. aureus* rarely ever modulate the synergistic functions and that *P. aeruginosa* can occasionally outcompete *S. aureus* when cocultured (Biswas et al., 2009).

In some situations, activity among polymicrobial biofilms might result in the emergence of species variations. As mentioned, Hansen and associates employed a cocultured community of *P. putida* and *Acinetobacter* sp., using benzyl alcohol (C_7H_8O) as a carbon source. During the investigation, it was found that the two strains relied on each other for carbon and did not need benzyl alcohol. After observing the relationship between the two bacteria, the researchers noticed that a descendent of *P. putida*, known to commonly produce rough colony variants, had appeared. Furthermore, a particular gene was discovered to be mutated, and this trait was heritable (Hansen et al., 2007b).

Compared to the ancestral strain, these mutants were better at covering the *Acinetobacter* biofilm colonies, which made it easier for them to connect and create a mixed biofilm. Furthermore, in response to oxygen shortage, the rough versions showed decreased dispersion and increased formation of a polymer bearing a striking resemblance to cellulose, which is responsible for the non-dispersal phenotype and their ability to surround *Acinetobacter*. The evolutionary distinctions between the novel *P. putida* varieties occurred as a result of introducing environmental stimulus, specifically the type of development and prevalence of *Acinetobacter*, which eventually results in a more stable and productive community. Spatial structure is crucial in determining species interaction, as evidenced by evolutionary studies (Hansen et al., 2007a). The interaction between *S. aureus* and *P. aeruginosa* illustrates the development of variant species. Earlier, this connection was discussed about the synergy that allows for survival against biocides. Because *P. aeruginosa* secretes 2-heptyl-4-hydroxyquinoline N-oxide (HQNO), a by-product that halts the proliferation of Gram-positive bacteria (Machan et al., 1992), the prevalence of *P. aeruginosa* causes small colony variants (SCVs) produced by *S. aureus* to emerge. When the quinolone triggered alternate σ-B factors in the bacteria, the SCV phenotype manifested more effectively. Numerous virulence variables, such as those governing the ability to attach, infiltrate, and stay inside cells that are being attacked, were altered in expression as a result (Mitchell et al., 2010). The presence of both *S. aureus* HQNO and SCVs in cystic fibrosis (CF) lungs suggests that these variations evolve inside the human host. According to Biswas and associates, *S. aureus* SCV production is a survival tactic to resist *P. aeruginosa* competition (Biswas et al., 2009). This last example of a mixed biofilm supports the claim that species interactions within biofilms should be considered when developing and selecting treatments (Table 6.1).

6.4.2 Metabolic Adaptability and Stress Response of Biofilms

Low-molecular-weight compounds known as oxidizing agents take electrons from other substances to oxidize them. They include ROS like HO^-, $O2^-$, H_2O_2, and hydroxyl radicals (HO^-), as well as reactive chlorine species (RCS) like NaOCl and HOCl. These microbes come across oxidizing chemicals in a wide range of settings and from various sources. They can be created exogenously by the mammalian immune system, internally by respiration, or by using cleaning products and disinfectants. Oxidants are also produced by natural processes like photosynthesis and

Table 6.1 Functional categorization of proteins linked to biofilms (Latasa et al., 2006)

Microorganism	Protein	Function
Staphylococcus aureus	Bap	Primary attachment and biofilm development are involved in persistent infections on abiotic surfaces
Salmonella enteritidis	BapA	Biofilm formation and host colonization
Pseudomonas putida	Mus 20	Initial seed colonization
Enterococcus faecium	Espfm	Adhesion to eukaryotic cells
Burkholderia cepacia	Bap	Late stages of biofilm formation
Pseudomonas fluorescens	LapA	Transition to the formation of biofilms and permanent adhesion to quartz sand
Enterococcus faecalis	Esp	Affixation on inorganic substrates and subsequent biofilm creation
Vibrio parahaemolyticus	VP1443	Biofilm maturation
Escherichia coli	YeeJ	Adhesion

volcanic eruptions, as well as human activities like runoff from agriculture and pollution from industrial effluent. Despite their antimicrobial properties, bacteria have evolved several defenses against the harm these harmful substances may inflict. According to earlier studies, growing as a biofilm improves bacterial survivability against oxidizing chemicals. Knowledge of the intricacies of bacterial reactions to oxidative stress can help develop effective treatment plans and biofilm management techniques for various bacterial species (da Cruz Nizer et al., 2024).

Studies have used transcriptome analysis to ascertain gene expression in biofilm-dwelling cells but less often than in planktonic cells. This confirmed previously described processes and provided insight into novel molecules and pathways implicated in resistance. DNA microarrays revealed that the outer membrane protein YcfR in *E. coli* was activated by several stressors, including H_2O_2 stress, indicating that it functions as a molecule that resists stress. Subsequent investigation revealed that ycfR inhibits the development of biofilms, aggregation, and hydrophobicity on the cell surface (Zhang et al., 2007).

Using RNA-sequencing, Lipus et al. (2019) exposed the biofilm of *Pseudomonas fluorescens* and demonstrated that the most expressed genes play a role in regulating transcription (e.g., *area, tetR, lysR, iscR,* and *arsR*), multidrug efflux, transport, membrane-related genes (e.g., *mexE, term, ssuF, copZ, potAB, kdpA,* and *araJ*), and oxidative stress response (e.g., *Ohr, ahpC, ahpF, trxB, yedY,* and *katA*). According to these findings, the biofilm of *P. fluorescens* release sodium hypochlorite and hazardous byproducts by increasing the production of major facilitator superfamily (MFS) transporters and efflux pumps, such as MexEF-OprN (Lipus et al., 2019). Along with the increased production of genetic proteins that protect against oxidative stress (*apps, ahpF*, catalase, thioredoxin reductase (*trxB* and *trxA*), *ohr*, and *OsmC*-like protein), it was also noted that catalase *katB* and *recA* are essential for the biofilm of *Burkholderia cenocepacia* survive exposure to H_2O_2 and NaOCl. An extracytoplasmic function (ECF) σ70 factor *OrbS (BCAL1688)*, a putative σ factor *(BCAL3478)*, an ECF σ70 factor *FecI (BCAL1369)*, the iron-sulfur cluster (*iscSUA-hscAB-fdx-iscX* and *ice*), and genes related to flagella are among the other genes that have been upregulated (Peeters et al., 2010).

Compared to planktonic cells, *S. epidermidis* biofilms exposed to H_2O_2 or NaOCl generated substantially greater σB-dependent alkaline shock protein 23 (asp23). According to genomic studies, redox homeostasis disruption and the lack of thiol metabolic recycling genes affect NTHi biofilms' ability to survive under H_2O_2 stress (Hunt et al., 2022). Protein studies reveal that *S. enteritidis* biofilms exposed to H_2O_2 and benzalkonium chloride (BAC) had higher levels of the genes invA and avrA observed to oversee the invading of cells. Researchers conclude that the virus may be able to live in its host due to the increased ability of biofilm cells to invade (Romeu et al., 2020). It was also demonstrated that *Salmonella enterica* biofilms treated with BAC and H_2O_2 expressed avrA (Rodrigues et al., 2011).

The synthesis of matrix genes has been demonstrated to be adversely affected by the DNA damage response (DDR), which is triggered in the fully grown biofilm of *B. subtilis* as a result of elevated ROS (Gozzi et al., 2017). H_2O_2 stimulated the production of *nrdJ*, a ribonucleotide reductase (RNR) with a DNA mending function, in *P. aeruginosa* biofilms (Rubio-Canalejas et al., 2023). When the stringent response (SR) of these biofilms were triggered, other defense mechanisms used by biofilms in response to environmental triggers were deactivated, hydroxyl radical levels rose, and biofilms became vulnerable to paraquat and phenazine methosulfate because catalase and superoxide dismutase activity decreased (Nguyen et al., 2011).

6.4.3 Role of Quorum Sensing and Regulatory Proteins

QS is a cell communication process that synchronizes protein expression as a result of fluctuating cell densities. Biofilm development and QS are linked processes. The biofilm forms due to the activation of the QS gene, which also controls its development and disintegration. Only when a given volume contains a negligible number of bacteria is the phenomena of QS possible. The number of bacteria in each volume may be ascertained by counting the autoinducer signaling molecules the bacteria release in a microcolony (Sharma et al., 2023). However, other researchers disagree, claiming that autoinducer signaling molecules should not be classified as signaling molecules because they are only metabolic byproducts (Winzer et al., 2002). Several mathematical and modeling methods have been proposed to understand bacteria's QS mechanism while forming biofilms. Researchers think that blocking QS might be a valuable tool in the fight against biofilms since it is essential for the growth and maturation of biofilms. In order to identify goods and materials that may "quorum quench," a new field of study known as "quorum quenching" is being established (Seghal Kiran et al., 2017). Finally, studies on two different bacterial species associated with biofilms, *P. aeruginosa* and *B. cepacian*, have shown that quorum quenching products decrease antibiotic resistance in bacteria (Brackman et al., 2011).

Bacteria use a communication mechanism called QS to control protein expression caused by consistent fluctuation in cell densities. By identifying signaling molecules (autoinducers) and modifying target gene transcription, regulatory proteins are essential to this process. LuxR-type proteins, like LasR in *P. aeruginosa*, activate genes in Gram-negative bacteria involved in virulence and biofilm development. Two-component systems regulate toxin production and bacterial adhesion in Gram-positive bacteria, observed in agr mechanisms of *S. aureus*. Furthermore, interspecies communication is made possible via AI-2-based QS, which LsrR controls in *E. coli*. Knowing the QS regulatory proteins gives possible targets for antimicrobial treatments and insights into the pathophysiology of bacteria. One potential tactic to treat infections resistant to antibiotics is to use quorum sensing inhibitors (QSIs) to target these proteins (Schuster & Greenberg, 2006).

6.5 Proteomic Insights into Biofilm-Associated Pathogenicity

Graf et al. (2019) cultured *Staphylococcus aureus* biofilms in flow systems instead of planktonic cultures and fully characterized the intracellular and extracellular proteomes, including the extracellular metabolome, ECM, and flow-through (FT) fractions. The high concentration of proteins involved in capsule production and other virulence factors, such as hemolysins, leukotoxins, and lipases, demonstrate the expression of several pathogenicity factors in the ECM. The virulence factors and their influence were verified using blood assays and a virulent strain of *Galleria mellonella*. It was found that ribosomal proteins and ECM-trapped virulence agents play a hitherto unidentified role in biofilm stability. This stabilizing effect was assumed to be due to the substantial positive ions in these proteins within an acidic ECM, which is brought on by the liberation of formate, lactate, and acetate by-products under O_2-limited conditions. It was proposed that electrostatic interactions with anionic elements, such as metabolites, eDNA, and cell surfaces, could foster cell aggregation and stabilize biofilms. These results propose possible targets for the creation of antimicrobial treatments and offer significant fresh insights into the biofilm formation of *S. aureus*. Zheng et al. (2017) examined the virulence parameters linked to biofilm development in 265 clinical isolates of *Enterococcus faecalis* from patients in Shenzhen, China. Biofilms were found using crystal violet staining, and polymerase chain reaction (PCR) was applied to verify the pathogenicity-inducing genes. Compared to ST179 isolates, ST16 isolates had a higher likelihood of forming strong/medium biofilms ($p < 0.001$). Isolates that were resistant to linezolid ($p = 0.001$) and those that screened positive for virulence genes, such as asa1, cylA, or agg, were more likely to form biofilms over time in comparison to biofilm-negative samples, and gene expression analysis revealed that mild biofilm isolates had an increase in cycle of 4.02–6.00 times, while potent biofilm isolates had an increase in esp of 11.39 to 134.08 times. Stronger biofilms were associated with ST16 and linezolid resistance, while weak and strong biofilm development has been associated with cells. Antibiotics and disinfectants frequently fail to eradicate clinically relevant biofilms on organic and inorganic substrates, which presents a significant hurdle to the healthcare industry. Using *S. aureus, E. coli, P. aeruginosa,* and *Chromobacterium violaceum* as models, this review demonstrates that the inhibition of biofilms, reduction in virulence, and eradication of biofilms made possible by essential oils and their constituents. It has been observed that essential oils alter genes related to virulence factors and biofilm development. Anti-QS activity was shown in *C. violaceum*, where enantiomeric monoterpenes impacted violacein synthesis, controlled by QS. Carvone, limonene, and borneol's (+) enantiomers promoted violacein synthesis, while their (−) analogues prevented it. Lipophilic volatile chemicals must enter biofilm channels to eradicate biofilms (Reichling, 2020).

Antibiotic resistance is mainly shaped by the genetic diversification of bacteria in biofilm. Because of the closeness of cells within this multicellular structure, research has indicated that biofilm plays a significant role in transmitting conjugative plasmids. Various physical, physiological, and gene-related variables influence biofilm resistance to antibiotics. Therefore, the use of traditional antimicrobials is severely hampered by the complex relationship between drug tolerance and biofilm formation. In conclusion, bacterial biofilm plays a significant role in the emergence of antibiotic resistance (Abebe, 2020). Antibiotic resistance in clinical settings is primarily contributed by bacterial biofilms being up to 1000 times more resistant to antimicrobial drugs than planktonic cells. As an outcome, biofilm-induced infections like the lung infection produced by *Pseudomonas aeruginosa* in people with CF are notoriously hard to cure. The Gastrointestinal tract, nasal tract, and blood reservoirs are the primary proliferation targets of nosocomial illnesses because internal medical devices are frequently planted there. Therefore, biofilms are a serious health concern for victims who require medical implants, particularly for individuals with weakened immune systems. Once biofilms have formed on medical implants, they become extremely hard to eliminate (Luo et al., 2020). Surgical debridement is an efficient and helpful method of overcoming antibiotic resistance in biofilms, which breaks down biofilms and uncovers bacteria in their planktonic, antibiotic-sensitive state (Vuotto & Donelli, 2019). Since biofilm forms on medical devices inside the living system and is implicated in numerous acute chronic infections, it is a significant issue in healthcare. The food and food processing surface becomes the perfect place for biofilm formation since it has enough nutrients for microorganisms to thrive and adhere. As an outcome, biofilm growth on these surfaces, particularly on surfaces used in food processing, becomes problematic for human health and food safety (Abebe, 2020). In 2024, Liu et al. observed that, regardless of thwarting antibiotic access, eDNA in biofilms is vital for preserving the structure of the biofilm. Traditionally believed to originate solely from lysed cells, eDNA is now understood to have an active role in the stability and creation of biofilms. It is frequently linked to outer membrane vesicles in bacteria such as *P. aeruginosa* and can be produced actively. DNase has been shown to interfere with the production of biofilms and even disintegrate existing ones in bacteria like *E. coli* and *Micrococcus luteus*. The action of external cellular responses in biofilm resistance is further highlighted by the fact that cellulase, which breaks down cellulose in the biofilm matrix, can aid in removing biofilms, especially from species like *E. coli* and *P. aeruginosa*.

Synergistic interactions originating from biofilm-forming microorganisms and the microbial communities of four well-known and highly biofilm-forming pathogens, specifically, *P. aeruginosa, E. coli, Staphylococcus* sp., and *Candida* sp., were reviewed by Cangui-Panchi et al. (2023). Their investigation included the innate and adaptive immune responses from the infection with these microbes and how they evade the human immune system by forming biofilms. When comparing biofilm-associated infections to planktonic infections, the significance of fluctuations in the expression of specific virulence factors and the immune response is typically explained by how the biofilm architecture protects the pathogen and alters the host immune response mechanisms. Plenty of investigations into species that form biofilms have found that specific virulence factors can favorably influence the biofilm formation of these microbial communities and cause significant issues with current antibiotic resistance and the ability of pathogenic microorganisms to survive on abiotic surfaces.

6.6 Environmental and Industrial Biofilm Proteomics

Extracellular proteins are critical in wastewater treatment by facilitating microbial attachment, pollutant degradation, and stress resistance, primarily through their presence in EPS. These proteins support biofilm formation and microbial aggregation, enhancing treatment efficiency, while adhesion proteins and pili enable microbial attachment to surfaces. However, some extracellular proteins contribute to membrane fouling, posing operational challenges. Extracellular enzymes, such as cellulases, proteases, and glycosidases, aid in breaking down complex organic pollutants, making them more accessible for microbial metabolism. Their expression varies with environmental conditions, with microorganisms like *Clostridium cellulovorans* and *Aspergillus fumigatus*, adjusting enzyme production for optimal degradation. Additionally, extracellular proteins enhance microbial resilience against environmental stressors by upregulating stress-related proteins, assisting microbes in coping with temperature shifts, pH fluctuations, and pollutant loads. Beyond their biological roles, these proteins also influence wastewater treatment operations, particularly in biofilms found in reclaimed water distribution systems, which can lead to bacterial regrowth and impact water quality. Metaproteomics has been instrumental in studying active bacterial communities, but protein extraction methods remain inconsistent and untested for reclaimed water biofilms (Zhang et al., 2019). A recent study used a scoring system to determine effectiveness to compare five extraction methods—radioimmuno precipitation assay (RIPA), PreOmics, sodium dodecyl sulfate (SDS), AllPrep, and urea-B-PER—RIPA, PreOmics, and SDS emerged as the most effective, with RIPA yielding the highest protein amount and SDS and PreOmics extracting the most diverse range of proteins, demonstrating efficiency in lysing both Gram-positive and Gram-negative bacteria. PreOmics ranked highest overall, making it a promising method for extracting proteins from biofilms, and this study marked the first systematic evaluation of these methods for reclaimed water research. Another study employed shotgun proteomics to analyze extracellular proteins in activated sludge from wastewater treatment plants, identifying 130, 108, and 114 proteins in anaerobic, anoxic, and aerobic sludge samples, which came from microbial cells and were primarily involved in catalytic and binding activities. Anaerobic sludge contained a higher proportion of catalytic proteins, contributing to structural differences compared to aerobic sludge, and many of these proteins played key roles in metal ion and organic molecule binding, further aiding pollutant removal. However, despite advances in proteomics, identifying extracellular proteins remains a challenge due to database limitations and the presence of low-abundance proteins, which often evade detection (Almulhim & Hong, 2023). Expanding proteomic databases and refining detection techniques, such as single-cell imaging MS, are necessary steps toward a more comprehensive understanding of extracellular protein functions in wastewater treatment. Protein extraction methods must also be standardized to ensure reproducibility and accuracy in wastewater treatment research, with future studies focusing on refining existing methods and incorporating novel approaches to improve protein recovery. Integrating metaproteomics with metagenomics and metabolomics can offer a more holistic view of microbial communities, revealing deeper insights into extracellular protein roles in pollutant degradation, biofilm formation, and microbial adaptation to environmental stressors. Expanding reference databases is crucial, as many microbial species within wastewater treatment plants are underrepresented in current proteomic repositories, hindering accurate protein identification and functional analysis. Understanding extracellular proteins can also drive innovations in wastewater treatment, such as engineering microbial consortia with enhanced enzyme expression to improve the breakdown of complex pollutants or developing antifouling strategies to mitigate membrane fouling (Zhang et al., 2015). Furthermore, targeted modifications to filtration membranes and biofilm-disrupting agents could address the challenges associated with extracellular proteins contributing to fouling, enhancing overall treatment efficiency. Shotgun proteomics continues to provide valuable insights into extracellular protein composition and function, offering critical data on differences between anaerobic, anoxic, and aerobic sludge environments, yet challenges remain in fully characterizing these proteins due to analytical limitations. By refining extraction techniques, expanding proteomic databases, and integrating multiomics approaches, researchers can advance our understanding of extracellular proteins and optimize wastewater treatment processes. This deeper knowledge will facilitate the development of more effective, sustainable, and cost-efficient treatment strategies, ensuring improved water quality and environmental safety. Addressing current challenges in extracellular protein research will not only enhance wastewater treatment efficiency but also provide a foundation for future innovations in microbial community engineering and biofilm management, making wastewater treatment systems more resilient, adaptable, and efficient in pollutant removal (Carrascal et al., 2020).

6.6.1 Industrial Biofilms and Biocorrosion

The biocorrosion of mild steel (MS) and brass (BR) in cooling water systems is examined about bacterial communities. This study finds that Bacillus cereus and *Achromobacter xylosoxidans* are the main bacterial species that promote corrosion using electrochemical impedance spectroscopy, Scanning Electron Microscopy, and X-ray diffraction. According to the weight loss approach, MS has higher corrosion rates (0.7 mm/y) than BR (0.17 mm/y). SEM pictures confirm that XRD detects biofilm production and corrosion products such as iron and copper oxides. According to electrochemical research, BR resists corrosion better. This study identifies the role of bacteria in corrosion and offers methods for controlling corrosion in industrial systems (Narenkumar et al., 2019). Biofilm-forming bacteria's effects on the biocorrosion of carbon steel submerged in actual wastewater are investigated in this study, with particular attention paid to pH (3, 7, and 9) and chloride concentrations

(0.5 g/L and 1.5 g/L). Under acidic conditions, Pseudomonas aeruginosa and *Enterobacter aerogenes* in tannery wastewater developed robust biofilms that promoted corrosion and caused irreversible adhesion. *Brevundimonas diminuta,* on the other hand, showed reversible attachment in industrial wastewater, which led to decreased biofilm stability. By changing the surface electrochemical characteristics, biofilm growth substantially impacts corrosion rates, according to electrochemical impedance spectroscopy (EIS). Excessive levels of chlorine encouraged biofilm separation, which decreased corrosion resistance. The study emphasizes how metal degradation and biofilm stability are influenced by microbial hydrophobicity. The results indicate that regulating pH and chloride levels can help reduce biocorrosion in industrial environments by limiting microbial adhesion and biofilm formation. This study offers insightful information about the behavior of biofilms and possible methods for preventing corrosion in infrastructure exposed to wastewater (El-Bassi et al., 2020) Biocorrosion also referred to as microbiologically influenced corrosion, or MIC occurs when metals in diverse settings are impacted by microbial biofilms. Direct MIC occurs when microorganisms take electrons from metals to breathe, while indirect MIC occurs when protective oxide layers are damaged, speeding up corrosion. There are substantial financial losses due to MIC in sectors like biomedical applications, water utilities, and oil and gas. Acid-producing bacteria (APB) and sulfate-reducing bacteria (SRB) are important because they produce caustic compounds. The work emphasizes extracellular electron transport (EET), one of the most recent developments in MIC mechanisms. MIC mitigation techniques include corrosion-resistant materials, biocides, and biofilm dispersion agents. Microbial interactions in mixed biofilms and novel mitigation strategies should be the main topics of future studies. Effective MIC prevention requires understanding microbial synergy (Dou et al., 2021). Under different pH levels (3, 7, and 9) and chloride concentrations (0.5 g/L and 1.5 g/L), this study examines the effects of native bacteria in tannery (TWW) and industrial wastewaters (IWW) on the biocorrosion of metals. In TWW at pH 3, *Pseudomonas aeruginosa* (MN788665) and *Enterobacter aerogenes* (MN788688) have hydrophobic characteristics that help them adhere irreversibly to carbon steel surfaces in the early phases of biofilm formation. Alternatively, in IWW at the same pH, *Enterobacter aerogenes* (MN788689) and *Brevundimonas diminuta* (MN788672) exhibit reversible adhesion, which results in restricted biofilm formation (El-Bassi et al., 2020). Significant operational failures and environmental harm are caused by petroleum pipeline corrosion worldwide, and most of these incidents may have been exacerbated by microbial activity. The motivation behind this work is the paucity of scholarly articles that have examined the impact of seawater containing native microorganisms on the internal corrosion of welded joints in long-distance pipelines (Liduino et al., 2019). In this study, seawater from Guanabara Bay (Rio de Janeiro, Brazil) was used to measure the microbial abundance, diversity, and composition of biofilms formed on two corroded weldments (gas tungsten arc, or GTAW, and shielded metal arc, or SMAW) of an API 5L X65 microalloyed steel under laminar and turbulent flow regimes using a combination of culture-based and molecular microbiological techniques. Compared to laminar flow, turbulent flow did not, on average, decrease the biomass of sessile microorganisms on welded joints. According to phylogenetic analysis, the predominant phyla in biofilms affixed to GTAW and SMAW joints were Proteobacteria, Firmicutes, and Bacteroidetes. All samples displayed a comparable microbial makeup at the genus level, with *Desulfovibrio* predominating. Nevertheless, OTU-level analysis showed that several microbes were present in every corrosive sample despite their association with sulfate reducers. Therefore, the type of weld, flow regime, and biofilm age all impacted the makeup of the microbial communities.

6.6.2 Role of Proteomics in Biofouling Management

According to proteomic analysis, the parietal matrix was enriched with cuticle-associated proteins, while the base matrix had a greater diversity of enzymes, pheromones, and protease inhibitors. These differences in protein composition between the two regions helped to understand barnacle adhesion mechanisms. The notable differences in cement proteins in the base matrix highlighted its function in adhesion. Given the complexity of barnacle adherence and possible antifouling targets, the results imply that cuticle production, cement deposition, and biomineralization at the substrate interface are synchronized (Schultzhaus et al., 2024). The proteome profile of marine biofilms produced by the filamentous cyanobacterium *Toxifilum* sp. LEGE 06021 is examined in this research study about hydrodynamic circumstances. Two shear rates (4 s^{-1} and 40 s^{-1}) were used to culture biofilms on glass and Perspex; the lower shear rate resulted in more remarkable biofilm development. Proteomic analysis found 1877 proteins, which showed that hydrodynamic circumstances affected protein expression more than surface type. Enzymes, transporter proteins, and stress response factors were examples of differentially expressed proteins (DEPs). These results emphasize the function of proteomics in biofouling management and marine biotechnology, offering insights into the regulation of cyanobacterial biofilms and possible antibiofilm tactics (Romeu et al., 2022). A proteomic method is used in this study to examine coaggregation processes in drinking water biofilms. Scientists examined two strains of *Delftia acidovorans*, one that could coaggregate (C+) and the other that could not (C−). Greater mobility, increased production of type IV pilus proteins, and variations in the flagellar structure were all observed in the aggregating strain. Proteomic analysis distinguished the two strains by identifying important proteins linked to adhesion and biofilm development. These discoveries highlight the molecular underpinnings of bacterial interactions in aquatic environments. Through the identification of protein biomarkers for coaggregation, this study offers valuable information for enhancing microbial control, water system maintenance, and biofilm control (Afonso et al., 2024) Ascidians are common fouling organisms that have a significant detrimental ecological and economic impact on coastal habitats, especially those that are highly invasive. Some isolated Ascidians have a special structure called a stolon that allows them to finish the necessary adhesion mechanism and have a very high tolerance to environmental stressors during invasions and biofouling. Nevertheless, little is known about the mechanisms underpinning environmental tolerance. Here, we examined the molecular response to environmental stressors (temperature and salinity) in the stolon of a very invasive fouling ascidian, *Ciona robusta*, using the quantitative proteomics technique known as isobaric tags for relative and absolute quantification (iTRAQ). Under low salinity, high salinity, low temperature, and high-temperature stress, 75, 86, 123, and 83 differential abundance proteins were found compared to the control.

The "cytoskeleton," "signal transduction," and "posttranslational modification" were identified as the main pathways under both temperature and salinity stressors by bioinformatic analysis. These pathways were implicated in protein synthesis, stolon structure stability, and stress response activation. By controlling protein synthesis and cell signal transmission, the "extracellular matrix" pathway was found to be essential under low salinity stress. By triggering the "carbohydrate/lipid transport" and "catabolism" pathways, the stolon may store more energy to cope with the high salinity stress (Li et al., 2021). Although copper is a necessary component of living cells, it can be hazardous to many creatures due to its high abundance in specific marine settings. Marine creatures may have particular adaptive strategies to avoid cell harm in polluted regions. To examine how copper affects a single organism's metabolism, the bacterial strain *Pseudoalteromonas lipolytica* TC8 that forms biofilms was the subject of a dual strategy that combined proteomics and metabolomics. The response of *P. lipolytica* TC8 to copper stress was investigated in planktonic and biofilm culture modes under growth inhibitory copper concentrations to highlight differential adaptation based on the phenotypic. As anticipated, exposure to copper triggered detoxification and defense mechanisms. As a result, distinct protein and metabolite profiles were noted for each condition (control versus copper-treated cultures and planktonic versus biofilm). Exposure to copper appeared to cause significant alterations in the lipid content of the bacterial cell membrane and alter the number of proteins in both planktonic and biofilm growth types known to play a functional role in maintaining copper cell homeostasis. Biofilms showed significantly more proteins that were differently expressed following copper treatment than planktonic cells, suggesting that the biofilm cells' reaction to this metallic stress was more varied (Favre et al., 2018).

6.7 Clinical and Therapeutic Applications

6.7.1 Biomarker Discovery for Biofilm-Related Infections

Biomarkers are crucial for diagnosing biofilm-associated infections and monitoring treatment responses. Proteomics ensures the identification of differentially expressed proteins in biofilms compared to planktonic bacteria, providing insights into biofilm formation and antibiotic resistance mechanisms. Proteins such as e-DNA-binding proteins, lipoproteins, and enzymes associated with polysaccharides are potential indicators for biofilm detection. The components of EPS, including these DNA-binding proteins, lipoproteins, and enzymes that break down polysaccharides, are essential for biofilm formation, maintaining structural integrity, and enabling surface adherence. These proteins contribute significantly to the stability and persistence of biofilms, particularly in industrial environments, where biofilms can accumulate on surfaces like pipelines, machinery, and equipment. In manufacturing contexts, manipulating or targeting these proteins can either promote biofilm formation for specific applications or inhibit the growth of undesirable biofilms, depending on the intended use.

QS Regulators: Proteins involved in QS pathways, such as LuxS and the Agr system, play a crucial role in biofilm development and regulation. These signaling systems enable bacteria to coordinate collective behaviors, including biofilm formation, virulence expression, and resistance mechanisms.

To counteract biofilm formation, small molecules like furanone and analogues of AHLs can effectively disrupt bacterial communication signals. By interfering with QS pathways, these compounds inhibit the regulatory mechanisms that sustain biofilm integrity, making bacterial communities more vulnerable to antimicrobial treatments.

Stress Response Proteins: Elevated levels of chaperones, heat shock proteins, and oxidative stress proteins can indicate biofilm-associated infections.

6.7.2 Targeting Biofilm-Associated Proteins for Drug Development

Proteomics has enabled the identification of proteins associated with biofilms essential for biofilm creation, resistance, and adhesiveness. These proteins represent potential drug targets for the development of innovative antibiofilm therapeutics.

Potential Drug Targets

- *Adhesins and Surface Proteins*—Inhibiting key surface proteins like FimH in *E. coli* and Spa in *S. aureus* can prevent biofilm formation.
- *EPS Matrix Disruptors*—Targeting matrix-associated enzymes such as Dispersin B and DNase aids in biofilm degradation.
- *Quorum Sensing Inhibitors*—Small molecules like furanone and AHL analogs can interfere with QS signaling, disrupting biofilm communication and formation.
- *Efflux Pump Inhibitors*—Blocking efflux pumps such as MexAB-OprM in *P. aeruginosa* can enhance the effectiveness of conventional antibiotics.

6.7.3 Proteomic Insights for Novel Antibiofilm Strategies

Advancements in proteomics have facilitated the production of innovative antibiofilm techniques aimed at either inhibiting biofilm creation or promoting its removal. These approaches focus on targeting biofilm-specific metabolic pathways and stress response mechanisms to disrupt bacterial persistence. In manufacturing industries, biofilm formation poses significant challenges, leading to operational inefficiencies and potential product contamination. Addressing this issue requires strategic proteomic interventions to either suppress biofilm development or enhance its eradication, ensuring improved process efficiency and product safety. Enzyme-based strategies, including dispersin-B, DNase I, and proteases, have demonstrated significant potential in degrading biofilm structures. These enzymes effectively break down biofilm matrix components, facilitating improved antimicrobial penetration and removal. Their application in industrial cleaning, particularly in water treatment systems, food production equipment, and biofilm reactors, offers an efficient method for minimizing biofilm buildup while preserving the integrity of sensitive surfaces and machinery. In addition to enzymatic approaches, nanotechnology-driven strategies have leveraged proteomics to design nanoparticle-based drug delivery systems capable of penetrating biofilms and enhancing antimicrobial efficacy. Nanoparticles hold promise for targeted biofilm interventions, including their use in biofilm-based bioreactors and bioprocessing applications. Furthermore, phage therapy has emerged as an innovative biofilm control strategy. Phage-derived enzymes, such as endolysins and depolymerases, effectively degrade biofilm matrices, making biofilm-associated bacteria more susceptible to antimicrobial treatments. This approach offers a potential future for eradicating biofilm infections in hospitals and food industries.

6.8 Conclusion

Proteomics has become an effective technique in biofilm research that provides a profound understanding of the molecular processes behind biofilm growth, resistance, and creation. The full range of proteins produced in biofilms may be analyzed to help researchers find important regulatory pathways, stress responses, and adaptive processes contributing to biofilm resilience. Mass spectrometry-based methods and other advanced proteomic techniques have made it easier to identify biomarkers linked to biofilms and possible targets for innovative antibacterial tactics. Industrial biotechnology, environmental science, and clinical microbiology are just a few domains where proteomics is used in biofilm research. Proteomic research aids in developing more potent treatments in clinical settings by revealing the processes behind pathogenic biofilms' resistance to antibiotics. Proteomics supports biofilm management tactics, wastewater treatment, and bioremediation procedures in the industrial and environmental sectors.

References

Abebe, G. M. (2020). The role of bacterial biofilm in antibiotic resistance and food contamination. *International Journal of Microbiology, 2020*(1), 1705814.

Afonso, A. C., Simões, M., Saavedra, M. J., Simões, L., Lema, J. M., & Trueba-Santiso, A. (2024). Exploring coaggregation mechanisms involved in biofilm formation in drinking water through a proteomic-based approach. *Journal of Applied Microbiology, 135*, lxae143.

Almulhim, F., & Hong, P. Y. (2023). Evaluation of protein extraction methods to improve metaproteomics analysis of treated wastewater biofilms. *Proteomics, 23*(20), 2300191.

Biswas, L., Biswas, R., Schlag, M., Bertram, R., & Gotz, F. (2009). Small-colony variant selection as a survival strategy for Staphylococcus aureus in the presence of Pseudomonas aeruginosa. *Applied and Environmental Microbiology, 75*, 6910–6912.

Brackman, G., Cos, P., Maes, L., Nelis, H. J., & Coenye, T. (2011). Quorum sensing inhibitors increase the susceptibility of bacterial biofilms to antibiotics in vitro and in vivo. *Antimicrobial Agents and Chemotherapy, 55*(6), 2655–2661.

Cangui-Panchi, S. P., Ñacato-Toapanta, A. L., Enríquez-Martínez, L. J., Salinas-Delgado, G. A., Reyes, J., Garzon-Chavez, D., & Machado, A. (2023). Battle royale: immune response on biofilms–host-pathogen interactions. *Current Research in Immunology, 4*, 100057.

Carrascal, M., Abián, J., Ginebreda, A., & Barceló, D. (2020). Discovery of large molecules as new biomarkers in wastewater using environmental proteomics and suitable polymer probes. *Science of the Total Environment, 747*, 141145.

Chandramouli, K., & Qian, P. Y. (2009). Proteomics: challenges, techniques and possibilities to overcome biological sample complexity. *Human Genomics and Proteomics, 2009*, 239204. https://doi.org/10.4061/2009/239204. PMID: 20948568; PMCID: PMC2950283.

Chua, S. L., Yam, J. K. H., Hao, P., Adav, S. S., Salido, M. M., Liu, Y., Givskov, M., Sze, S. K., Tolker-Nielsen, T., & Yang, L. (2016). Selective labelling and eradication of antibiotic-tolerant bacterial populations in Pseudomonas aeruginosa biofilms. *Nature Communications, 7*(1), 1–11.

da Cruz Nizer, W. S., Adams, M. E., Allison, K. N., Montgomery, M. C., Mosher, H., Cassol, E., & Overhage, J. (2024). Oxidative stress responses in biofilms. *Biofilms, 7*, 100203.

Divya, M., Kiran, G. S., Hassan, S., & Selvin, J. (2019). Biogenic synthesis and effect of silver nanoparticles (AgNPs) to combat catheter-related urinary tract infections. *Biocatalysis and Agricultural Biotechnology, 18*, 101037.

Dou, W., Xu, D., & Gu, T. (2021). Biocorrosion caused by microbial biofilms is ubiquitous around us. *Microbial Biotechnology, 14*(3), 803–805.

Dutta, B., Chatterjee, B., Sarkar, N., Lahiri, D., & Ray, R. R. (2024). Multi-omics technology in detection of multispecies biofilm. *The Microbe, 4*, 100128. https://doi.org/10.1016/j.microb.2024.100128

El-Bassi, L., Ziadi, I., Belgacem, S., Bousselmi, L., & Akrout, H. (2020). Investigations on biofilm forming bacteria involved in biocorrosion of carbon steel immerged in real wastewaters. *International Biodeterioration & Biodegradation, 150*, 104960.

Ellepola, K., Truong, T., Liu, Y., Lin, Q., Lim, T., Lee, Y., Cao, T., Koo, H., & Seneviratne, C. (2019). Multi-omics analyses reveal synergistic carbohydrate metabolism in Streptococcus mutans-Candida albicans mixed-species biofilms. *Infection and Immunity, 87*(10), 00339-00319. https://doi.org/10.1128/iai

Fathima, A., Arafath, Y., Hassan, S., Prathiviraj, R., Kiran, G. S., & Selvin, J. (2023). Microbial biofilms: A persisting public health challenge. In *Understanding microbial biofilms* (pp. 291–314). Academic Press.

Favre, L., Ortalo-Magné, A., Pichereaux, C., Gargaros, A., Burlet-Schiltz, O., Cotelle, V., & Culioli, G. (2018). Metabolome and proteome changes between biofilm and planktonic phenotypes of the marine bacterium Pseudoalteromonas lipolytica TC8. *Biofouling, 34*(2), 132–148.

Folsom, J. P., Richards, L., Pitts, B., Roe, F., Ehrlich, G. D., Parker, A., Mazurie, A., & Stewart, P. S. (2010). Physiology of Pseudomonas aeruginosa in biofilms as revealed by transcriptome analysis. *BMC Microbiology, 10*, 294. https://doi.org/10.1186/1471-218010-294

Franklin, M. J., Chang, C., Akiyama, T., & Bothner, B. (2015). New technologies for studying biofilms. *Microbiology Spectrum, 3*(4). https://doi.org/10.1128/microbiolspec.MB-0016-2014. PMID: 26350329; PMCID: PMC4821632.

Goldsworthy, M. J. H. (2008). Gene expression of Pseudomonas aeruginosa and MRSA within a catheter-associated urinary tract infection biofilm model. *Bioscience Horizon, 1*, 28–37.

Gozzi, K., Ching, C., Paruthiyil, S., Zhao, Y., Godoy-Carter, V., & Chai, Y. (2017). Bacillus subtilis utilizes the DNA damage response to manage multicellular development. *Npj Biofilms Microbiome, 3*, 1–7.

Graf, A. C., Leonard, A., Schäuble, M., Rieckmann, L. M., Hoyer, J., Maass, S., Lalk, M., Becher, D., Pane-Farre, J., & Riedel, K. (2019). Virulence factors produced by Staphylococcus aureus biofilms have a moonlighting function contributing to biofilm integrity*[S]. *Molecular & Cellular Proteomics, 18*(6), 1036–1053.

Hansen, S. K., Rainey, P. B., Haagensen, J. A. J., & Molin, S. (2007a). Evolution of species interactions in a biofilm community. *Nature, 445*, 533–536.

Hansen, S. K., Haagensen, J. A. J., Gjermansen, M., Jorgensen, T. M., Tolker-Nielsen, T., & Molin, S. (2007b). Characterization of a Pseudomonas putida rough variant evolved in a mixed-species biofilm with Acinetobacter sp. strain C6. *Journal of Bacteriology, 189*, 4932–4943.

Herschend, J., Damholt, Z. B., Marquard, A. M., Svensson, B., Sørensen, S. J., Hägglund, P., & Burmølle, M. (2017). A meta-proteomics approach to study the interspecies interactions affecting microbial biofilm development in a model community. *Scientific Reports, 7*(1), 16483.

Hunt, B. C., Xu, X., Gaggar, A., & Swords, W. E. (2022). Nontypeable Haemophilus influenzae redox recycling of protein thiols promotes resistance to oxidative killing and bacterial survival in biofilms in a smoke-related infection model. *mSphere, 7*, e0084721.

Jakubovics, N. S., Gill, S. R., Iobst, S. E., Vickerman, M. M., & Kolenbrander, P. E. (2008). Regulation of gene expression in a mixed-genus community: Stabilized arginine biosynthesis in Streptococcus gordonii by coaggregation with Actinomyces naeslundii. *Journal of Bacteriology, 190*, 3646–3657.

Kalmokoff, M., Lanthier, P., Tremblay, T. L., Foss, M., Lau, P. C., Sanders, G., Austin, J., Kelly, J., & Szymanski, C. M. (2006). Proteomic analysis of Campylobacter jejuni 11168 biofilms reveals a role for the motility complex in biofilm formation. *Journal of Bacteriology, 188*(12), 4312–4320. https://doi.org/10.1128/JB.01975-05. PMID: 16740937; PMCID: PMC1482957.

Karathanasi, G., Bojer, M. S., Baldry, M., Johannessen, B. A., Wolff, S., Greco, I., Kilstrup, M., Hansen, P. R., & Ingmer, H. (2018). Linear peptidomimetics as potent antagonists of Staphylococcus aureus agr quorum sensing. *Scientific Reports, 8*(1), 3562.

Khemiri, A., Jouenne, T., & Cosette, P. (2016). Proteomics dedicated to biofilmology: What have we learned from a decade of research? *Medical Microbiology and Immunology, 205*(1), 1–19.

Koo, H., Allan, R. N., Howlin, R. P., Stoodley, P., & Hall-Stoodley, L. (2017). Targeting microbial biofilms: current and prospective therapeutic strategies. *Nature Reviews Microbiology, 15*(12), 740–755.

Latasa, C., Solano, C., Penadés, J. R., & Lasa, I. (2006). Biofilm-associated proteins. *Comptes Rendus Biologies, 329*(11), 849–857.

Li, X., Li, S., Cheng, J., Fu, R., & Zhan, A. (2021). Proteomic response to environmental stresses in the stolon of a highly invasive fouling ascidian. *Frontiers in Marine Science, 8*, 761628.

Liduino, V. S., Cravo-Laureau, C., Noel, C., Carbon, A., Duran, R., Lutterbach, M. T., & Sérvulo, E. F. C. (2019). Comparison of flow regimes on biocorrosion of steel pipe weldments: community composition and diversity of biofilms. *International Biodeterioration & Biodegradation, 143*, 104717.

Lipus, D., Vikram, A., Gulliver, D., & Bibby, K. (2019). Upregulation of peroxide scavenging enzymes and multidrug efflux proteins highlight an active sodium hypochlorite response in Pseudomonas fluorescens biofilms. *Biofouling, 35*, 329–339.

Liu, H. Y., Prentice, E. L., & Webber, M. A. (2024). Mechanisms of antimicrobial resistance in biofilms. *npj Antimicrobials and Resistance, 2*(1), 27.

Luo, Y., Yang, Q., Zhang, D., & Yan, W. (2020). Mechanisms and control strategies of antibiotic resistance in pathological biofilms. *Journal of Microbiology and Biotechnology, 31*(1), 1.

Luppens, S. B. I., Kara, D., Bandounas, L., et al. (2008). Effect of Veillonella parvula on the antimicrobial resistance and gene expression of Streptococcus mutans grown in a dual-species biofilm. *Oral Microbiology and Immunology, 23*, 183–189.

Machan, Z. A., Taylor, G. W., Pitt, T. L., Cole, P. J., & Wilson, R. (1992). 2-Heptyl-4-hydroxyquinoline N-oxide, an antistaphylococcal agent produced by Pseudomonas aeruginosa. *Journal of Antimicrobial Chemotherapy, 30*, 615–623.

Meleady, P. (2018). Two-Dimensional Gel Electrophoresis and 2D-DIGE. *Methods in Molecular Biology, 1664*, 3–14. https://doi.org/10.1007/978-1-4939-7268-5_1

Mitchell, G., Seguin, D. L., Asselin, A.-E., et al. (2010). Staphylococcus aureus sigma B-dependent emergence of small-colony variants and biofilm production following exposure to Pseudomonas aeruginosa 4-hydroxy-2-heptylquinoline-N-oxide. *BMC Microbiology, 10*, 33.

Narenkumar, J., AlSalhi, M. S., Arul Prakash, A., Abilaji, S., Devanesan, S., Rajasekar, A., & Alfuraydi, A. A. (2019). Impact and role of bacterial communities on biocorrosion of metals used in the processing industry. *ACS Omega, 4*(25), 21353–21360.

Nguyen, D., Joshi-Datar, A., Lepine, F., Bauerle, E., Olakanmi, O., Beer, K., McKay, G., Siehnel, R., Schafhauser, J., Wang, Y., Britigan, B. E., & Singh, P. K. (2011). Active starvation responses mediate antibiotic tolerance in biofilms and nutrient-limited bacteria. *Science, 334*, 982–986.

Oliveira, B. M., Coorssen, J. R., & Martins-de-Souza, D. (2014). 2DE: The phoenix of proteomics. *Journal of Proteomics, 104*, 140–150.

Oosthuizen, M. C., Steyn, B., Theron, J., Cosette, P., Lindsay, D., Von Holy, A., & Brözel, V. S. (2002). Proteomic analysis reveals differential protein expression by Bacillus cereus during biofilm formation. *Applied and Environmental Microbiology, 68*(6), 2770–2780. https://doi.org/10.1128/AEM.68.6.2770-2780.2002. PMID: 12039732; PMCID: PMC123966.

Peeters, E., Sass, A., Mahenthiralingam, E., Nelis, H., & Coenye, T. (2010). Transcriptional response of Burkholderia cenocepacia J2315 sessile cells to treatments with high doses of hydrogen peroxide and sodium hypochlorite. *BMC Genomics, 11*, 90.

Pham, T. K., Roy, S., Noirel, J., Douglas, I., Wright, P. C., & Stafford, G. P. (2010). A quantitative proteomic analysis of biofilm adaptation by the periodontal pathogen Tannerella forsythia. *Proteomics, 10*(17), 3130–3141.

Phillips, N. J., Steichen, C. T., Schilling, B., Post, D. M., Niles, R. K., Bair, T. B., Falsetta, M. L., Apicella, M. A., & Gibson, B. W. (2012). Proteomic analysis of Neisseria gonorrhoeae biofilms shows shift to anaerobic respiration and changes in nutrient transport and outer membrane proteins. *PLoS One, 7*(6), e38303. https://doi.org/10.1371/journal.pone.0038303. Epub 2012 Jun 6. PMID: 22701624; PMCID: PMC3368942.

Rahman, M. A., Amirkhani, A., Chowdhury, D., Mempin, M., Molloy, M. P., Deva, A. K., Vickery, K., & Hu, H. (2022). Proteome of Staphylococcus aureus biofilm changes significantly with aging. *International Journal of Molecular Sciences, 23*(12), 6415.

Rani, A., & Babu, S. (2018). Environmental proteomic studies: closer step to understand bacterial biofilms. *World Journal of Microbiology and Biotechnology, 34*(8), 120. https://doi.org/10.1007/s11274-018-2504-x

Reichling, J. (2020). Anti-biofilm and virulence factor-reducing activities of essential oils and oil components as a possible option for bacterial infection control. *Planta Medica, 86*(08), 520–537.

Rodrigues, D., Cerca, N., Teixeira, P., Oliveira, R., Ceri, H., & Azeredo, J. (2011). Listeria monocytogenes and Salmonella enterica enteritidis biofilms susceptibility to different disinfectants and stress-response and virulence gene expression of surviving cells. *Microbial Drug Resistance (Larchmont, N.Y.), 17*, 181–189.

Romeu, M. J., Rodrigues, D., & Azeredo, J. (2020). Effect of sub-lethal chemical disinfection on the biofilm forming ability, resistance to antibiotics and expression of virulence genes of Salmonella Enteritidis biofilm-surviving cells. *Biofouling, 36*, 101–112.

Romeu, M. J., Domínguez-Pérez, D., Almeida, D., Morais, J., Araújo, M. J., Osório, H., et al. (2022). Hydrodynamic conditions affect the proteomic profile of marine biofilms formed by filamentous cyanobacterium. *npj Biofilms and Microbiomes, 8*(1), 80.

Rubio-Canalejas, A., Admella, J., Pedraz, L., & Torrents, E. (2023). Pseudomonas aeruginosa non-phosphorylated AlgR induces ribonucleotide reductase expression under oxidative stress infectious conditions. *mSystems, 8*, e0100522.

Sadiq, F. A., Flint, S., Yuan, L., Li, Y., Liu, T., & He, G. (2017). Propensity for biofilm formation by aerobic mesophilic and thermophilic spore forming bacteria isolated from Chinese milk powders. *International Journal of Food Microbiology, 262*, 89–98.

Santos, T., Viala, D., Chambon, C., Esbelin, J., & Hébraud, M. (2019). Listeria monocytogenes biofilm adaptation to different temperatures seen through shotgun proteomics. *Frontiers in Nutrition, 6*, 89.

Sauer, K. (2003). The genomics and proteomics of biofilm formation. *Genome Biology, 4*, 1–5.

Sauer, K., Camper, A. K., Ehrlich, G. D., Costerton, J. W., & Davies, D. G. (2002). Pseudomonas aeruginosa displays multiple phenotypes during development as a biofilm. *Journal of Bacteriology, 184*(4), 1140–1154. https://doi.org/10.1128/jb.184.4.1140-1154.2002. PMID: 11807075; PMCID:PMC134825.

Schultzhaus, J., Hervey, J., Fears, K., & Spillmann, C. (2024). Proteomic comparison of the organic matrices from parietal and base plates of the acorn barnacle Amphibalanus amphitrite. *Open Biology, 14*(5), 230246.

Schuster, M., & Greenberg, E. P. (2006). A network of networks: quorum-sensing gene regulation in Pseudomonas aeruginosa. *International Journal of Medical Microbiology, 296*(2–3), 73–81.

Seghal Kiran, G., Hassan, S., Sajayan, A., & Selvin, J. (2017). Quorum quenching compounds from natural sources. In *Bioresources and bioprocess in biotechnology: Volume 2* (Exploring potential biomolecules) (pp. 351–364).

Seneviratne, C. J., Suriyanarayanan, T., Widyarman, A. S., Lee, L. S., Lau, M., Ching, J., Delaney, C., & Ramage, G. (2020). Multi-omics tools for studying microbial biofilms: current perspectives and future directions. *Critical Reviews in Microbiology, 46*(6), 759–778.

Sharma, S., Mohler, J., Mahajan, S. D., Schwartz, S. A., Bruggemann, L., & Aalinkeel, R. (2023). Microbial biofilm: a review on formation, infection, antibiotic resistance, control measures, and innovative treatment. *Microorganisms, 11*(6), 1614.

Soufi, Y., & Soufi, B. (2016). Mass spectrometry-based bacterial proteomics: focus on dermatologic microbial pathogens. *Frontiers in Microbiology, 7*, 181.

Steyn, B., Oosthuizen, M. C., MacDonald, R., Theron, J., & Brözel, V. S. (2001). The use of glass wool as an attachment surface for studying phenotypic changes in Pseudomonas aeruginosa biofilms by two-dimensional gel electrophoresis. *Proteomics, 1*(7), 871.

Van Oudenhove, L., & Devreese, B. (2013). A review on recent developments in mass spectrometry instrumentation and quantitative tools advancing bacterial proteomics. *Applied Microbiology and Biotechnology, 97*, 4749–4762. https://doi.org/10.1007/s00253-013-4897-7

Vuotto, C., & Donelli, G. (2019). Novel treatment strategies for biofilm-based infections. *Drugs, 79*, 1635–1655.

Wang, C., Zhang, Q., Wang, Y., Tang, X., An, Y., Li, S., Xu, H., Li, Y., Luan, W., & Wang, X. (2019). Comparative proteomics analysis between biofilm and planktonic cells of Mycobacterium tuberculosis. *Electrophoresis, 40*(20), 2736–2746.

Wang, N., Sadiq, F. A., Li, S., He, G., & Yuan, L. (2020). Tandem mass tag-based quantitative proteomics reveals the regulators in biofilm formation and biofilm control of Bacillus licheniformis. *Food Control, 110*, 107029.

Winzer, K., Hardie, K. R., & Williams, P. (2002). Bacterial cell-to-cell communication: sorry, can't talk now–Gone to lunch! *Current Opinion in Microbiology, 5*(2), 216–222.

Wuyts, J., Van Dijck, P., & Holtappels, M. (2018). Fungal persister cells: The basis for recalcitrant infections? *PLoS Pathogens, 14*(10), e1007301.

Yang, X., Wang, H., He, A., & Tran, F. (2018). Biofilm formation and susceptibility to biocides of recurring and transient Escherichia coli isolated from meat fabrication equipment. *Food Control, 90*, 205–211.

Yin, L., Li, Q., Xue, M., Wang, Z., Tu, J., Song, X., Shao, Y., Han, X., Xue, T., & Liu, H. (2019). The role of the phoP transcriptional regulator on biofilm formation of avian pathogenic Escherichia coli. *Avian Pathology, 48*(4), 362–370.

Zhang, X.-S., García-Contreras, R., & Wood, T. K. (2007). YcfR (BhsA) influences Escherichia coli biofilm formation through stress response and surface hydrophobicity. *Journal of Bacteriology, 189*, 3051–3062.

Zhang, P., Shen, Y., Guo, J. S., Li, C., Wang, H., Chen, Y. P., & Fang, F. (2015). Extracellular protein analysis of activated sludge and their functions in wastewater treatment plant by shotgun proteomics. *Scientific Reports, 5*(1), 12041.

Zhang, P., Zhu, J., Xu, X. Y., Qing, T. P., Dai, Y. Z., & Feng, B. (2019). Identification and function of extracellular protein in wastewater treatment using proteomic approaches: A minireview. *Journal of Environmental Management, 233*, 24–29.

Zhang, Y., Pan, X., Liao, S., Jiang, C., Wang, L., Tang, Y., Wu, G., Dai, G., & Chen, L. (2020). Quantitative proteomics reveals the mechanism of silver nanoparticles against multidrug-resistant Pseudomonas aeruginosa biofilms. *Journal of Proteome Research, 19*(8), 3109–3122.

Zheng, J. X., Wu, Y., Lin, Z. W., Pu, Z. Y., Yao, W. M., Chen, Z., Li, D. Y., Deng, Q. W., Qu, D., & Yu, Z. J. (2017). Characteristics of and virulence factors associated with biofilm formation in clinical Enterococcus faecalis isolates in China. *Frontiers in Microbiology, 8*, 2338.

7 Transcriptomics in Understanding the Biofilm Dynamics and Inhibition

Subhaswaraj Pattnaik, Siddhardha Busi, and Rajaneesh Anupam

7.1 Introduction

The pathogenic microorganisms have the inherent ability to form self-produced aggregates of polygenic materials called "Biofilms," which enable the encased microbial communities to protect from the shear stress conditions. An important aspect of microbial biofilms is their ability to cause serious health consequences and other socioeconomic hazards (Armbruster & Parsek, 2018; Gautam et al., 2013). The extracellular polymeric framework of biofilms not only exhibits healthcare hazards but also promotes strategic survival skills and persistence of embedded microbial communities over the therapeutic modules. In addition, the highly variable microenvironment associated with biofilms limit the use of spatially designed therapeutic interventions as compared to nonbiofilm forming counterparts by virtue of their enhanced viscoelastic properties as evident from the mechanistic interpretation of microbial biofilms (Gloag et al., 2020). The surface-based dynamics of microbial biofilms could be attributed to the typical composition of the polymer-enriched matrix including extracellular proteins, polysaccharides, and nucleic acid materials (eDNA). In majority of cases, the composition of the biofilm matrix determines the course of life-threatening chronic infections and resistance to antimicrobial agents

S. Pattnaik (✉)
PG Department of Zoology, Government Women's College, Sambalpur University, Sambalpur, Odisha, India

S. Busi
Department of Microbiology, School of Life Sciences, Pondicherry University, Kalapet, Puducherry, India

R. Anupam
Department of Biochemistry and Molecular Biology, School of Life Sciences, Pondicherry University, Puducherry, India
e-mail: ranupam@pondiuni.ac.in

S. Busi et al. (eds.), *Omics Approaches in Biofilm Research*,
https://doi.org/10.1007/978-3-031-91863-6_7

(van Hoogstraten et al., 2024). The extracellular polymeric substances (EPS) not only promote adhesion to different biotic and abiotic surfaces but also allow interspecies recognition and polymicrobial interactions within the matrix (Karygianni et al., 2020).

The formation of biofilms is highly dynamic and multistep process including initial attachment to biotic or abiotic surfaces, followed by establishment of micro-colony formation, maturation, and dispersion (Misra et al., 2022). The biofilm-associated infections (BAIs) are highly recalcitrant in nature and contribute to approximately 80% of chronic infections across the globe. As a result of which, the chronic microbial biofilms have managed to sustain the effect of antimicrobial agents and create a socioeconomic impact (Mirzaei et al., 2024). Owing to the importance of biofilms in microbial growth and infection severity, it is important to understand the insights into the biofilm microenvironment.

7.1.1 An Insight into Biofilm Microenvironment

As per the evidences, 40–80% of prokaryotic cells with particular reference to bacteria, predominantly reside in EPS-based biofilm matrices, depicting their prevalence and widespread effects. Among the biofilm-mediated chronic health implications, the healthcare implant-based infections account for the major concern to the existing healthcare facilities, particularly in developing countries across the globe (Mirzaei et al., 2024). Within the biofilm microenvironment, the encased microbial community exhibits high cell density patterns, reduced rate of diffusion process, changes in growth kinetics, differential metabolic activities, and the appearance of physicochemical heterogeneity, which are responsible for the transformative physiological behavior, inleading to extensive adaptation to the environmental stress including aided protection against antibiotic treatment (Bjarnsholt et al., 2022; Létoffé et al., 2017). The healthcare implications of emergent biofilms could be attributed to the composition and functional aspects of the EPS. The composition of EPS, structural diversity, and functional attributes of the biomolecules associated with the assemblage of biofilm matrices (including flagellar appendages, adhesion proteins, exopolysaccharides, eDNAs, etc.) as well as its pathophysiological characteristics are termed as "Matrixome" (Karygianni et al., 2020).

The cell density within the biofilm matrices and the maintenance of an acidic environment critically inhibit the responsive antibiotic treatment and the actions of antimicrobial macrophages, thereby promoting biofilm-associated chronic infections (Mei et al., 2023). As discussed earlier, the formation and development of biofilms in pathogenic microorganisms is an adaptive survival strategy to tackle environmental stress conditions and increased virulence profiles. In this regard, the pathophysiological interplay between biofilm mechanics and the associated virulence profiles could provide an epidemiological analysis of biofilm dynamics (Davignon et al., 2024).

7.1.2 Effects of Microbial Biofilms

The complex microbial biofilm community has several potential effects with special reference to healthcare sectors, food processing industries, aquatic environment, industrial sectors, and socio-economic values. The biofilm associated infections (BAIs) are considered to decipher several health consequences in humans affecting majority of body systems such as cardiovascular systems (infective endocarditis), respiratory system (cystic fibrosis), urinary system [urinary tract infections (UTIs)], reproductive system (bacterial vaginosis), integumentary system (chronic wound infections), auditory system (otitis media), and digestive system (dental caries) (Vestby et al., 2020). Biofilms are inherently associated with severe chronic infections by several mechanisms, such as the biofilm dispersal (generally involved in blood-related and urinary infections), horizontal transfer of resistant genes, and decreased susceptibility towards the therapeutic interventions with particular reference to antimicrobial agents. Besides, the biofilms are also involved in the production of diverse range of endotoxins and have the ability to counteract the host immune system leading to increased infection severity (Sharma et al., 2023).

As per the recent trends, the development of biofilms in the gastrointestinal tract and periodontal disease condition could lead to severe carcinogenic incidences, suggesting the correlation of biofilm and cancer leading to severe health hazards (Perry & Tan, 2022). Besides, biofilms are considered to be the major drivers of recalcitrant chronic wound infections and cystic fibrosis. Since biofilms are recalcitrant in nature, the global expenditure on healthcare settings with particular reference to biofilm-mediated healthcare effects are considered to be more than $350 billion in the USA. The global economic impact of biofilm-associated infections is also a major concern across the globe with special emphasis on the low and middle income countries (Cámara et al., 2022). The ability of biofilms to clinical recurrence of chronic infections and persistent severity of the infections have further aggravated the serious challenges to the infrastructure associated with global healthcare settings (Ciofu et al., 2022).

In addition, several biomedical implants associated with healthcare settings are also considered to be critically affected by the risk of biofilm growth, resulting in implant failure, chronic infections, and increased rate of mortality (Kadirvelu et al., 2024). The development of recurrent biofilms on the surfaces of medical devices in several hospital-acquired settings is considered to be the major contributing factor to the high rate of mortality, morbidity, and chronic infections, such as endocarditis, catheter-associated urinary tract infections (CAUTIs), chronic sinusitis, ventilator-associated pneumonia (VAP), central line-associated bloodstream infections (CLABSIs), and other device-associated nosocomial infections (Brigmon & Brigmon, 2023; Mishra et al., 2024). Since biofilms can inherently tolerate antimicrobial agents, their growth in several food products in different food processing industries critically affects the quality of the food products and reduces the shelf life of the products, resulting in significant economic loss (Carrascosa et al., 2021).

The large-scale food production units and food processing industries are associated with comparatively long production cycles and complex processing facilities, thereby providing a suitable conducive environment for the growth and development of microbial biofilms on food processing lines and contact surfaces. In addition, as the biofilms are inherently resistant to antimicrobial agents, exploration of these antimicrobial agents in food processing industries against biofilm-forming pathogens remains ineffective, leading to inappropriate food quality maintenance and severe economic loss (Sharan et al., 2022). As per the recent trends, the inherent ability of pathogenic microorganisms to form recalcitrant biofilms has critically transformed their ability to withstand antimicrobial agents and create severe health hazards, and socioeconomic implications. Hence, it is imperative to look for advanced molecular tools and omics-based approaches to improve the diagnostic values and therapeutic interventions of biofilm-mediated infections.

7.2 OMICS-Based Approaches for Understanding Biofilm Mechanics

Since biofilm pathophysiology critically affects the healthcare settings and socioeconomic values, it is imperative to understand the biology associated with the recalcitrant biofilms. The advent of multifaceted OMICS-based approaches provides novel avenues to decipher in-depth analysis of several biological components, such as nucleic acids, proteins, and other metabolites of interest with improved efficacy. The OMICS-based technologies are referred to as the high-throughput molecular tools to study the differential expression of genes (Genomics), profiling of transcripts (transcriptomics), regulated expression profiles of proteins of interest (Proteomics), and metabolite profiling (Metabolomics) in a controlled manner, depicting the overall aspects of biological macromolecules of interest (Narad & Kirthanashri, 2018; Nuñez-Belmar et al., 2023). Since the OMICS-based approaches of system biology are considered to be non-biased and non-targeted in nature, they are of particular interest in molecular biology research with particular reference to advanced screening, improved diagnostic values, increased prognosis of health ailments, improved therapeutic interventions with aided optimization, and overall biomarker discovery prospects (Narad & Kirthanashri, 2018). The integration of several OMICS-based approaches conceptualized as "meta-omics" tools is being developed to understand the regulative expression profiles of genes, transcripts, proteins, and metabolites of interest at the systems level (Panicker, 2024).

The OMICS-based molecular tools are spatially employed to determine the differential-based tools, expression profiles of genes, phenotypic elucidation of cells or tissues during development, disease progression, and chronic infection severity. Besides, the integrative OMICS-based tools also comprehensively detect the underlying mechanisms associated with cancer and could provide effective therapeutic modules for the efficient management of cancer development and progression (Chakraborty et al., 2024; Yako et al., 2023). OMICS-based technologies are explored for the discovery of specific biomolecular markers of pathophysiological relevance and are used for theranostic applications. Besides, these OMICS-based approaches inherently promote the current theranostic strategies with precision and provide elemental support and holistic infrastructure towards the development of personalized medicines for efficient biological system investigation (Dai & Shen, 2022; Magro et al., 2024).

In this regard, the advent of molecular-based approaches, particularly the OMICS-based methods could provide novel avenues to decipher an insight into the biofilm microenvironment, particularly the identification of antibiotic resistance genes (ARGs), uncovering the regulation of target genes expression, regulatory networks related to the expression of biofilm-related genes, complex pathophysiological pathways related to biofilm development, complex signaling mechanisms including quorum sensing (QS), and the transformation in the metabolic activities. The OMICS-based tools not only provide rapid detection methods for biofilm pathophysiology but also deliver new avenues to spatial design and development of strategic therapeutic modules to control the biofilm pathophysiology, drug resistance profiles, and infection severity. Based on the information provided by OMICS-based technologies, the differential profile of biological macromolecules of interest could be derived with a stringent response system (An et al., 2021; Dutta et al., 2024; Vohra et al., 2024).

In addition, integrated OMICS approaches also focus on detecting the differential biofilm signatures under different physiological, environmental, and mutational factors and how their interactions facilitate the biofilm pathophysiology. The OMICS approaches also provide novel avenues for efficient detection of marine biofouling and the interactions of the participating biofouling microorganisms with the relevant microenvironment, associated with high economic values suggesting their widespread relevance (Azevedo et al., 2009; Dobretsov & Rittschof, 2023). As the advance OMICS technologies provide an integrative approach to understand the intriguing physiological behavior of biofilm microenvironment towards several environmental stress, these technologies could be implemented towards the early biofilm detection and improved therapeutic management of recalcitrant biofilm infections.

7.2.1 Transcriptomics: An Insight into Molecular Tool

Among the several OMICS-based approaches, the study of transcriptomes received considerable attention. The transcriptomes refer to the entire set of the total mRNA transcripts produced by the genetic material. The study of transcriptomes refers to "transcriptomics." The ultimate advantage of transcriptomics is that it evaluates the RNA levels throughout the genome both qualitatively and quantitatively depicting the functional attributes of transcriptome studies. Since 80% of the genome are transcribed but only 3% of the genome encode proteins, it is important that the study of transcriptomics also enables the scientific community to understand several pathophysiological processes associated with noncoding RNAs (ncRNAs) including ribosomal RNA (rRNA), transfer RNA (tRNA), micro-RNA (miRNA), small interfering RNA (siRNA), small nuclear RNA (snRNA), small nucleolar RNA (snoRNA), and long noncoding RNA (lncRNA), which is otherwise not able to

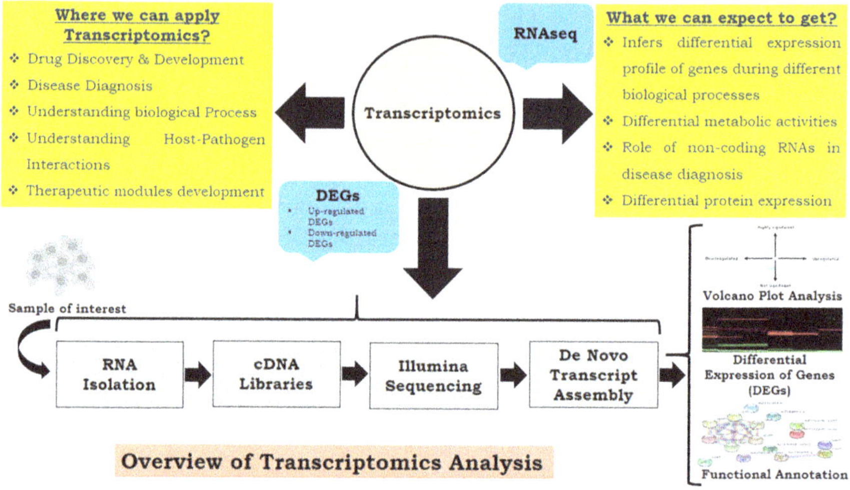

Fig. 7.1 Schematic representation of the transcriptomic analysis for diagnosis and therapeutics

decipher in proteomics studies (Hasin et al., 2017; Supplitt et al., 2021). Hence, the transcriptomic analysis provides regulated profiles of both coding sequences (standard RNA sequence) and the noncoding sequences (strand-specific RNA sequence) and thereby explores the functional attributes of the genes of interest (Maghembe et al., 2020). The overall procedures employed for the transcriptomics analysis for several applications are depicted in Fig. 7.1.

As the transcriptomes consist of the coding sequences responsible for the composition of the proteome, understanding the transcriptomics could provide novel avenues to decipher the first level biochemical profiles of gene regulation under different environmental conditions. Since the transcriptomics approaches provide the intriguing biological phenomena associated with several pathophysiological conditions, these tools could be explored for efficient diagnostic, prognostic, and therapeutic management of disease of interest by understanding the differential expression profiles of several hundreds of genes of pathophysiological interest (Kittleson et al., 2009; Melouane et al., 2019; Seneviratne et al., 2020).

The real-time monitoring and phenotypic characterization of transcript could be analyzed by high throughput molecular techniques such as quantitative polymerase chain reaction (qPCR), DNA microarray, and advanced next generation RNA-sequencing depicting a precised detection and therapies. The molecular tools employed in transcriptomics have their inherent advantages over the conventional approaches as the parallel quantification and differential expression profiles of thousands of genes of pathophysiological relevance could be generated at a short span of time with high precision and specificity (Klopfleisch & Gruber, 2012). The transcriptomics-based molecular tools, such as the express sequence tag (EST), are actively engaged in the spatial detection for sepsis; whereas, the advanced Serial Gene Expression Analysis (SAGE) is explored towards the real-time detection of chronic urinary tract infections (UTIs), which is associated with nosocomial pathogenic infections. Meanwhile, microarray-based tools are often employed for efficient detection of polymicrobial infections, and RNA-Seq-based tools are used for detection of meningitis. In this regard, the high-throughput transcriptomic tools could be efficiently used for disease diagnosis and facilitate the current therapeutic modules (Dutta et al., 2024). Among the several molecular tools, the qPCR-based technologies are actively engaged in the detection of differentially expressed genes (DEGs) associated with pathophysiological targets of interest owing to its simplicity in handling of samples and the real-time monitoring of up-regulated and down-regulated DEGs, which enable efficient disease diagnosis and treatment (Vlaeminck et al., 2022).

7.2.2 Transcriptomics: A Molecular Tool to Understand Biofilm Microenvironment

Among the several OMICS-based molecular approaches in healthcare settings, the transcriptomics approach received widespread recognition as there is adequate availability of standard transcriptomic tools. The exploration of these transcriptomic tools could be implemented towards the understanding of biofilm pathophysiology with special reference to the formation of biofilms on several abiotic and biotic surfaces, biofilm development, biofilm-host interactions, biofilm maturation, biofilm-mediated drug resistance phenomena, and improved therapeutic modules for the management of microbial risks (Dutta et al., 2024; Seneviratne et al., 2020). Owing to the complex signaling network in biofilm pathophysiology and the involvement of several genes in the overall expression of the biofilm phenotype in pathogenic microorganisms, it is imperative to utilize the large dataset generated by transcriptome profiling depicting the functional attributes of several DEGs involved in biofilm formation and its progression pathways. Based on the data generated, spatial therapeutic modules could be designed to specifically target the downregulation of overexpressed DEGs of interest involved in biofilm pathophysiology (Xu et al., 2022).

7.3 Transcriptomics: Management of Biofilm Infections

As evident from biofilm pathophysiological studies, approximately 65–80% of chronic infection are associated with biofilm-forming pathogenic bacteria and hence are considered to be a serious public health issue. In this regard, it is important to decipher the detailed information on the several genes associated with biofilm pathology. As several groups of pathogens are involved in biofilm mechanics under different pathophysiological conditions, the DEGs associated with biofilms also significantly diverse from each other. In this regard, the expression profiles of the selective DEGs could be evaluated for their stringent role in biofilm mechanics, and transcriptomics could provide new dimensions to understand the same and decipher spatial therapeutic modules to tackle the public health issue (Silva et al., 2021)

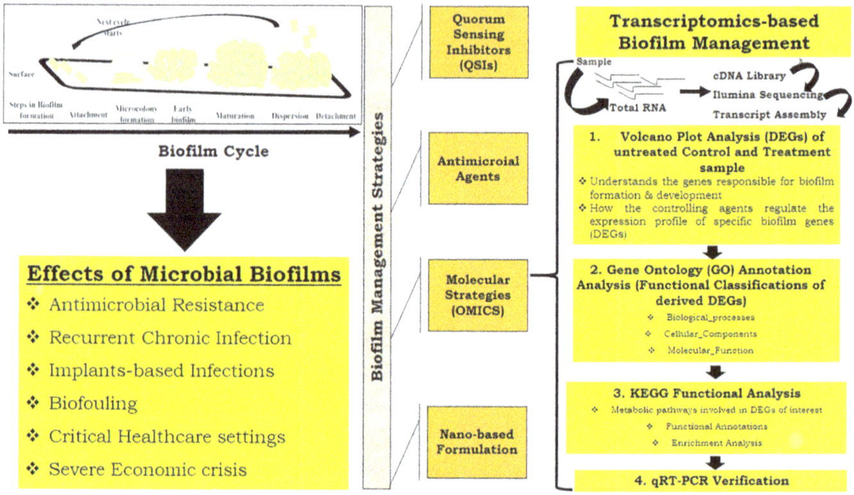

Fig. 7.2 Schematic overview of the diagnostic and therapeutic management of bacterial biofilms

(Fig. 7.2). The transcriptomic data analysis tools have the ability to provide a comprehensive transcriptome profile of thousands of DEGs responsible for several physiological attributes. Based on their expression profiles and the ability to facilitate the biofilm dynamics including c-di-GMP signaling and QS regulatory network, the DEGs could be annotated. Based on the dataset generated, these specific DEGs could be spatially targeted for the management of biofilm dynamics (Tan et al., 2015). In *Vibrio vulnificus*, the biofilm dynamics are critically regulated by two important DEGs such as *cabH* and *brpN,* which are exclusively associated with biofilm matrix calcium-binding genes and putative capsular polysaccharides, respectively. During biofilm formation, these two DEGs are critically overexpressed, resulting in the production of typical exopolysaccharides, which are not only associated with adhesion but also infer increased virulence properties (Lee et al., 2023).

Besides, the transcriptome profile also exhibits the information on the DEGs associated with the resistance profiles and could be specifically targeted in the therapeutic management of drug resistance phenomena in pathogenic microorganisms (Chattagul et al., 2021). Among the several advantages, the transcriptomic data analysis also critically extrapolate the biofilm heterogeneity with decreased metabolic profiling of specific DEGs coupled with qRT-PCR analysis, which is influential in biofilm dynamics and could be instrumental for diagnosis and therapies (Korshoj & Kielian, 2024). The mechanistic evaluation of the transcriptome profile of nosocomial pathogen, *P. aeruginosa*, also revealed the inherent role of efflux pump in mediating biofilm dynamics, promotion of antibiotic resistance, and adaptive stress management (Rajput et al., 2022). As evident from transcriptome profiling, a direct correlation exist between the QS regulatory signaling network and biofilm formation. In absence of an autoinducer synthase gene, the QS-associated communication is downregulated and critically influence the downregulation of biofilm forming DEGs as evident from the transcriptomic analysis (Xiong et al., 2022).

7.3.1 Transcriptomics for Biofilm Diagnostics

Since biofilms are an effective survival strategy of pathogenic microorganisms against several environmental stress conditions and promote increased tolerance and resistance towards the available antibiotics, they are involved in serious public health issues across the globe and infer selective pressure on healthcare settings. Owing to the pathophysiological importance of biofilms in healthcare sectors, it is imperative to improve the current understanding of detection and selective diagnosis of the progression of biofilm dynamics. Hence, detecting the clinical manifestation of biofilm mechanics with high-throughput OMICS-based technologies with special reference to transcriptomics analysis could be taken into consideration (Cleaver & Garnett, 2023). Among the several tools employed to detect and diagnose the biofilm progression, OMICS-based tools are considered to be instrumental for diagnosis owing to their specificity. The transcriptomic profiling has the added advantage over other conventional strategies in differentiating the DEGs involved in recurrent biofilm infections over non-recurrent infections, suggesting their specificity in biofilm diagnosis (Rubio-Mendoza et al., 2023; Silva et al., 2021).

bvnGPS, a spatially designed diagnostic tool, constructed based upon the large dataset obtained from integrative host transcriptomic data with relatively diverse gene pair signatures (GPS), could be employed to detect the relative expression profiles of several genes associated with bacterial biofilm infections (Li et al. 2023a, b). The recent trends in the use of advanced molecular tools also promote the development of sequence-based approaches to differentiate the virulence profile of different pathogenic microorganisms with special reference to distinguish the clinical isolates from the commensal isolates based on their relative expression profiles. Since the RNA-sequence-based transcriptomic tools critically complement whole genome sequencing (WGS), these transcriptomic tools could be developed for improved diagnosis of several rare diseases including microbial infections (Brás & França, 2022; Curry et al., 2021).

7.3.2 Transcriptomics: Therapeutic Approaches Against Biofilm Infections

Among the different pathogenic bacteria, methicillin-resistant *Staphylococcus aureus* (MRSA) has the inherent tendency to exhibit promising resistance against several conventional antibiotics owing to the ability to form recalcitrant biofilms. The ability of MRSA to form biofilm could be contributed to the overexpression of several DEGs, such as clumping factors, surface-associated proteins (SasG), adhesion proteins like polysaccharide intercellular adhesins-PIA, immunoglobulin binding protein (SpA), and iron-regulated surface determinants as evident from the transcriptomic profiling, which was further corroborated by qPCR analysis. Based on the information obtained from transcriptomic profiling, the biofilm pathophysiology could be explained and thereby therapeutic modules could be spatially designed (Liu et al., 2018; Tomlinson et al., 2021). Loratadine showed promising anti-infective properties based on its regulation of transcript levels with contrasting effects on relative expression of DEGs involved in a plethora of pathophysiological responses including biofilm development, biofilm heterogeneity through persister inhibition, virulence profiles, drug resistance patterns, and metabolic activities in MRSA depicting its prospective role in the regulation of bacterial virulence and biofilms (Viering et al., 2024).

Besides, the biofilm regulation modules could be designed against the biofilm formation and development in opportunistic pathogen, *Haemophilus parasuis*, based on the relative expression profiles of several DEGs, such as artM/artQ (ABC transporter permease), ssrS (noncoding sequence of RNA associated with RNA degradation), pflA, and HutX (involved in biofilm metabolism), which are associated with different stages of biofilm development (Jiang et al., 2021). Among the most pathogenic bacteria with the severe drug resistance profile as revealed by WHO report, the ESKAPE group of pathogens with particular reference to nosocomial pathogen, *Pseudomonas aeruginosa*, have the tendency to form recalcitrant biofilms owing to their ability control and coordinate the overexpression of several DEGs including alginate biosynthetic gene, phenazine biosynthesis proteins, proteases (*LasA*), efflux transporters, two-component response regulator (*PmrA*), and other transcription regulatory proteins (D'Arpa et al., 2021; Harrington et al., 2022).

As the transcriptome profile of thousands of DEGs are considered to be inevitable in regulation of biofilm mechanics, it is important to develop strategic therapeutic modules based on the transcriptomic data. According to a recent report, Glabridin, an extract of licorice with reported antimicrobial properties exhibited promising downregulation of several DEGs, such as *SarA* (QS regulated transcription regulator), clumping factors (*ClfA*), adhesion proteins (*icaD* and *icaR*), fibronectin binding protein A (*FnbA*), etc. Besides, it also regulated the transcript levels of biofilm-related genes linked with the two-component regulatory system which was confirmed by the qPCR analysis suggesting its candidature as potential therapeutic agent against chronic biofilm infections (Ma et al., 2024). A spatially designed chemical moiety, Y0-C10-HSL, exhibited promising anti-biofilm activities against *P. aeruginosa* biofilms as evident from in vitro studies as well as transcriptomic analysis. As per the transcriptome analysis, on treatment with Y0-C10-HSL, a total of 585 DEGs were observed of which 254 and 331 DEGs are associated with upregulation and downregulation, respectively. Among the downregulated DEGs, majority of DEGs are associated with QS-regulated pathways, adhesion properties, and biofilm development, suggesting its ability to control the resistance shown by *P. aeruginosa* (Tang et al., 2024).

The presence of zinc sulfate in combination with antibiotics critically downregulated the expression profiles of several DEGs including *wbbM* (encoding galactosyltransferase), *mrkA* (encoding fimbriae proteins), and *pbaA* (adhesion polysaccharides) correlated with biofilm mechanics in nosocomial pathogen,

Klebsiella pneumonia (Shebl et al., 2023). The combinatorial approach where naturally derived therapeutic bioactive secondary metabolites could potentially improve the sensitivity of conventional antibiotics against pathogenic microorganisms. In an earlier study, carvacrol significantly improved the sensitivity of β-lactam antibiotics by downregulation of specific DEGs involved in two-component signaling system in *S. aureus*-based infections with particular reference to the downregulation of transcriptional regulator, sarA, and promotes regulation of the β-lactam resistance profile (Li et al. 2023a, b). Table 7.1 depicts the role of several therapeutic agents (natural and synthetic) in the downregulation of several DEGs associated with biofilm dynamics as evident from transcriptomic studies in pathogenic microorganisms (Table 7.1).

7.4 Current Trends and Future Perspectives of Transcriptomics in Biofilm Research

The formation of biofilms by pathogenic microorganisms has critical implications on healthcare settings and economical implications in several food-based industries with particular reference to dairy industries. Besides, the biofilms have provided an adaptive strategy to counteract the actions of several anti-infective agents including antibiotics leading to resistance mechanisms. In addition, the inherent ability of bacterial biofilms in recurrence of infections further aggravate the chronic infection state. In this regard, it is important to look for the state-of-an-art approach to understand the pathophysiological responses of biofilms, their behavioral patterns under different environmental conditions, and the therapeutic management of these recalcitrant biofilms. The omics-based tools are actively engaged in determining the key aspects of several DEGs linked with biofilm formation, development, and drug resistance (Yuan et al., 2023).

As per the recent trends, the transcriptomics approach could be integrated with other omics-based approaches and other molecular tools for an improved understanding of biofilm pathophysiology. No doubt, multiomics approaches provide large dataset on the pathophysiological responses of biofilms, the interspecific signaling, and other factors, affecting the data analysis leading to its limitations in the accurate analysis of data integration (Yuan et al., 2023). As discussed earlier, among the different omics-based approaches, the transcriptomics received widespread recognition based on their usability, ease of operations, interpretation of both qualitative as well as quantitative data analysis, and promotion of analysis of both coding as well as noncoding sequence of the RNA transcripts of interest, which is not observed in other omics-based tools. Besides, the quantitative qPCR and the RNA-sequencing tools effectively provide the relative expression profiles of several DEGs associated with biofilm dynamics with real-time monitoring (Meng et al., 2024). The high-throughput RNA-sequencing tools have the inherent tendency to provide large data that could derive the relative expression profile of millions of nucleotide sequences from an individual transcriptome profile with relative information on the

Table 7.1 Role of several therapeutic agents (natural and synthetic) in the downregulation of several DEGs associated with biofilm dynamics as evident from transcriptomic studies in pathogenic microorganisms

Sl. no.	Therapeutic agents	Source	Target microorganism	Down-regulated DEGs	Inference on biofilm management	References
1.	Kojic acid	*Aspergillus flavus* A7	*Acinetobacter baumannii*	Two-component system response regulator (OmpR), type IV pilus secretin (PilQ), membrane protein, phospholipase C (Plc), multidrug efflux MFS transporter (AbaQ), and LuxR family transcriptional regulator	Transcriptional attenuation of quorum sensing (QS)-mediated virulence factors and biofilm formation	Li et al. (2022)
2.	2-phenyl ethyl methyl ether (PEME)	Kewda essential oil	*Staphylococcus aureus*	Transcriptional regulator (SarA), transport protein (norA), and two-component system (AgrA)	Regulation of QS-mediated biofilm dynamics by controlling expression profiles of genes of interest	Cheruvanachari et al. (2023)
3.	Terpinen-4-ol	Plant-derived essential oil	Methicillin resistant *S. aureus* (MRSA)	Biosynthesis of branched chain amino acids (BCAAs), ABC transporters, and metabolism of 2-oxocarboxylic acid	Attenuate the biosynthesis of bacterial amino acids and nucleotide metabolism and regulate the biofilm	Cheng et al. (2021)
4.	Combination of p-anisaldehyde and epigallocatechin gallate (EGCG)	Plant-derived essential oil	*Pseudomonas aeruginosa*	Cell envelope proteins, membrane transport, drug efflux, and energy metabolism	Promote antimicrobial activities by regulating biofilm response	Adewunmi et al. (2020)
5.	Rhodethrin with chloramphenicol	Antibiotics	*Enterococcus faecalis*	Biofilm formation responsible genes (Ace, AtpB, lepA, bopD, and typA), QS-associated genes (sylA, fsrC and came), and resistant genes (liaX, typA, EfrA, and lepA)	Attenuation of QS communication and regulation of virulence genes associated with biofilm development	Tatta et al. (2023)
6.	*Coleus amboinicus* essential oil (EO-CA)	Plant-derived essential oil	*Salmonella typhimurium* ATCC 14028	Motility factors (flhD, fljB, and fimD), fimbriae proteins (csgD), and invasion factor (hilA)	Regulate bacterial virulence and biofilm forming ability	Leesombun et al. (2023)

7.	Erythritol	Plants	*Staphylococcus pseudintermedius*	ABC transporter protein	Controlled biofilm formation by downregulation of biofilm specific genes	Fujii et al. (2023)
8.	Alpha-mangostin (α-MG)	Plant polyphenol	*Staphylococcus epidermidis* RP62A	Cytoplasmic membrane integrity (yidC2, secA, and mscL), cell division (ftsY and divlB), teichoic acid biosynthesis (tagG and dltA), fatty-acid biosynthesis (accB, accC, fabD, fabH, fabI, and fabZ), and biofilm formation (icaA)	Attenuated the biofilm mechanics by the loss of cytoplasmic membrane integrity	Sivaranjani et al. (2019)
9.	Isobavachalcone	Naturally occurring flavonoid	*Cryptococcus neoformans*	Biological pathways, including oxidative phosphorylation, pentose and glucuronate interconversions, pyruvate metabolism, lysine degradation, and biosynthesis of cofactors.	Downregulation of bacterial virulence and resistance mechanism	Qian et al. (2024)
10.	Curcumin in combination with the antibiotic, tigecycline	*Curcuma* sp.	*Escherichia coli*	Pathways such as cationic antimicrobial peptide (CAMP) resistance, two-component systems, QS regulatory network, and lipopolysaccharide (LPS) biosynthesis	Downregulation of bacterial virulence and biofilm dynamics by effectively downregulation of LPS biosynthesis, multidrug efflux pumps, CAMP resistance, and the electron transport chain	Zheng et al. (2025)

identification of specific DEGs, which are actively transcribed during biofilm mechanics (Deshpande et al., 2023).

The recent advances in the applications of multifaceted nanoparticles and nanocomposites could also be integrated with transcriptomics for improved biosensing and enhanced drug delivery platforms for a clear understanding of bacterial biofilms in the majority of pathogenic microorganisms (Vagena et al., 2024). The integrative approach further corroborates in understanding the polymicrobial biofilms with special reference to interspecific interactions among the participating microbial community, highly specific intraspecific signaling, and stress response mechanisms. Based on the information obtained, strategic therapeutic modules could be spatially designed to control the menace of polymicrobial infections (Jeong et al., 2024). The advent of a machine learning approach in combination with transcriptomics data analysis is considered next-generation high-throughput data analysis tools for providing the in-depth analysis of several aspects of bacterial biofilm pathophysiology, relative expression profiles of DEGs, and co-expression profiles of several genes under different environmental conditions (Zhang et al., 2024a).

The advent of state-of-the-art computational tools to derive the large datasets obtained from transcriptomics analysis further improved our efficacy to understand the pathophysiology of bacterial biofilms with an added advantage to predict the occurrence of specific genes and/or proteins of interest. Recently, an open resource platform, BBSdb, was developed to predict the relative expression profiles of bacterial biofilm-associated proteins with a relatively high number of DEGs, enabling a user-friendly approach to predict the relative expression profile patterns of several biofilm-associated proteins (Zhang et al., 2024b). In this regard, it is imperative to integrate the transcriptomics tools with advanced computational approaches for effective interpretation of data and could revolutionize the differential expression profiles of DEGs, and active participation of those specific genes during biofilm formation could also be derived.

7.5 Conclusion

The advent of transcriptomic data analysis, no doubt, is considered to be an effective and potential data generation tool, providing holistic insights into the relative expression profiles of several DEGs associated with several cellular and biological systems, thereby providing a comprehensive approach to understand the underlying mechanisms. In this regard, the transcriptome profiling of biofilm-forming pathogenic microorganisms could provide large datasets depicting the relative expression profiles of several DEGs intriguingly associated with biofilm formation, development, maturation, and resistance profile of biofilms. With the evolution of advanced molecular tools, such as qPCR, RNA-sequencing, and microarray-based technologies further improve the interpretation of the transcriptome profile of pathogenic microorganisms associated with recalcitrant biofilm formation. Besides, the integrative omics approaches further improved the relative gene expression patterns of specific genes to understand their intricate regulatory mechanisms in biofilm dynamics. In addition, advanced molecular tools could also be used in combination with transcriptomics for further interpretation of data analysis linked with biofilms. Based on the interpretation of data generated from transcriptomics analysis, integrative omics-based tools could be actively engaged towards the development of regulative therapeutic modules for the active regulation of biofilm-associated chronic infections and drug resistance phenomena.

References

Adewunmi, Y., Namjilsuren, S., Walker, W. D., Amato, D. N., Amato, D. V., Mavrodi, O. V., Patton, D. L., & Mavrodi, D. V. (2020). Antimicrobial activity of, and cellular pathways targeted by, p-anisaldehyde and epigallocatechin gallate in the opportunistic human pathogen *Pseudomonas aeruginosa*. *Applied and Environmental Microbiology, 86*(4), e02482–e02419. https://doi.org/10.1128/AEM.02482-19

An, A. Y., Choi, K. G., Baghela, A. S., & Hancock, R. E. W. (2021). An overview of biological and computational methods for designing mechanism-informed anti-biofilm agents. *Frontiers in Microbiology, 12*, 640787. https://doi.org/10.3389/fmicb.2021.640787

Armbruster, C. R., & Parsek, M. R. (2018). New insight into the early stages of biofilm formation. *The Proceedings of the National Academy of Sciences, 115*(17), 4317–4319. https://doi.org/10.1073/pnas.1804084115

Azevedo, N. F., Lopes, S. P., Keevil, C. W., Pereira, M. O., & Vieira, M. J. (2009). Time to "go large" on biofilm research: Advantages of an omics approach. *Biotechnology Letters, 31*, 477–485. https://doi.org/10.1007/s10529-008-9901-4

Bjarnsholt, T., Whiteley, M., Rumbaugh, K. P., Stewart, P. S., Jensen, P. Ø., & Frimodt-Møller, N. (2022). The importance of understanding the infectious microenvironment. *The Lancet Infectious Diseases, 22*(3), e88–e92. https://doi.org/10.1016/S1473-3099(21)00122-5

Brás, S., & França, A. (2022). Transcriptome mining to identify molecular markers for the diagnosis of *Staphylococcus epidermidis* bloodstream infections. *Antibiotics, 11*(11), 1596. https://doi.org/10.3390/antibiotics11111596

Brigmon, M. M., & Brigmon, R. L. (2023). Infectious diseases impact on biomedical devices and materials. *Biomedical Materials & Devices, 1*, 74–81. https://doi.org/10.1007/s44174-022-00035-y

Cámara, M., Green, W., MacPhee, C. E., Rakowska, P. D., Raval, R., Richardson, M. C., et al. (2022). Economic significance of biofilms: A multidisciplinary and cross-sectoral challenge. *npj Biofilms and Microbiomes, 8*, 42. https://doi.org/10.1038/s41522-022-00306-y

Carrascosa, C., Raheem, D., Ramos, F., Saraiva, A., & Raposo, A. (2021). Microbial biofilms in the food industry-A comprehensive review. *International Journal of Environmental Research and Public Health, 18*(4), 2014. https://doi.org/10.3390/ijerph18042014

Chakraborty, S., Sharma, G., Karmakar, S., & Banerjee, S. (2024). Multi-OMICS approaches in cancer biology: New era in cancer therapy. *Biochimica et Biophysica Acta – Molecular Basis of Disease, 1870*(5), 167120. https://doi.org/10.1016/j.bbadis.2024.167120

Chattagul, S., Khan, M. M., Scott, A. J., Nita-Lazar, A., Ernst, R. K., Goodlett, D. R., & Sermswan, R. W. (2021). Transcriptomics analysis uncovers transient ceftazidime tolerance in *Burkholderia* biofilms. *ACS Infectious Diseases, 7*(8), 2324–2336. https://doi.org/10.1021/acsinfecdis.1c00003

Cheng, F., Mo, Y., Chen, K., Shang, X., Yang, Z., Hao, B., et al. (2021). Integration of metabolomics and transcriptomics indicates changes in MRSA exposed to terpinen-4-ol. *BMC Microbiology, 21*, 305. https://doi.org/10.1186/s12866-021-02348-2

Cheruvanachari, P., Pattnaik, S., Mishra, M., Pragyandipta, P., Pattnaik, A., & Naik, P. K. (2023). Deciphering the antibiofilm potential of 2-Phenylethyl methyl ether (PEME), a bioactive compound of Kewda essential oil against *Staphylococcus aureus*. *Microbial Pathogenesis, 179*, 106093. https://doi.org/10.1016/j.micpath.2023.106093

Ciofu, O., Moser, C., Jensen, P. Ø., & Høiby, N. (2022). Tolerance and resistance of microbial biofilms. *Nature Reviews. Microbiology, 20*, 621–635. https://doi.org/10.1038/s41579-022-00682-4

Cleaver, L., & Garnett, J. A. (2023). How to study biofilms: Technological advancements in clinical biofilm research. *Frontiers in Cellular and Infection Microbiology, 13*, 1335389. https://doi.org/10.3389/fcimb.2023.1335389

Curry, P. D. K., Broda, K. L., & Carroll, C. J. (2021). The role of RNA-sequencing as a new genetic diagnosis tool. *Current Genetic Medicine Reports, 9*, 13–21. https://doi.org/10.1007/s40142-021-00199-x

D'Arpa, P., Karna, S. L. R., Chen, T., & Leung, K. P. (2021). *Pseudomonas aeruginosa* transcriptome adaptations from colonization to biofilm infection of skin wounds. *Scientific Reports, 11*, 20632. https://doi.org/10.1038/s41598-021-00073-4

Dai, X., & Shen, L. (2022). Advances and trends in omics technology development. *Frontiers in Medicine (Lausanne), 9*, 911861. https://doi.org/10.3389/fmed.2022.911861

Davignon, G., Pietrosemoli, N., Benaroudj, N., Soupé-Gilbert, M. E., Cagliero, J., Turc, É., et al. (2024). *Leptospira interrogans* biofilm transcriptome highlights adaption to starvation and general stress while maintaining virulence. *npj Biofilms and Microbiomes, 10*, 95. https://doi.org/10.1038/s41522-024-00570-0

Deshpande, D., Chhugani, K., Chang, Y., Karlsberg, A., Loeffler, C., Zhang, J., et al. (2023). RNA-seq data science: From raw data to effective interpretation. *Frontiers in Genetics, 14*, 997383. https://doi.org/10.3389/fgene.2023.997383

Dobretsov, S., & Rittschof, D. (2023). "Omics" techniques used in marine biofouling studies. *International Journal of Molecular Sciences, 24*(13), 10518. https://doi.org/10.3390/ijms241310518

Dutta, B., Chatterjee, D., Sarkar, N., Lahiri, D., Nag, M., & Ray, R. R. (2024). Multi-omics technology in detection of multispecies biofilm. *The Microbe, 4*, 100128. https://doi.org/10.1016/j.microb.2024.100128

Fujii, T., Tochio, T., & Nishifuji, K. (2023). Erythritol alters gene transcriptome signatures, cell growth, and biofilm formation in *Staphylococcus pseudintermedius*. *BMC Veterinary Research, 19*, 146. https://doi.org/10.1186/s12917-023-03711-3

Gautam, C. K., Srivastav, A. K., Bind, S., Madhav, M., & Shanthi, V. (2013). An insight into biofilm ecology and its applied aspects. *International Journal of Pharmacy and Pharmaceutical Sciences, 5*(4), 69–73.

Gloag, E. S., Fabbri, S., Wozniak, D. J., & Stoodley, P. (2020). Biofilm mechanics: Implications in infection and survival. *Biofilms, 2*, 100017. https://doi.org/10.1016/j.bioflm.2019.100017

Harrington, N. E., Littler, J. L., & Harrison, F. (2022). Transcriptome analysis of *Pseudomonas aeruginosa* biofilm infection in an ex vivo pig model of the cystic fibrosis lung. *Applied and Environmental Microbiology, 88*, e01789-21. https://doi.org/10.1128/aem.01789-21

Hasin, Y., Seldin, M., & Lusis, A. (2017). Multi-omics approaches to disease. *Genome Biology, 18*, 83. https://doi.org/10.1186/s13059-017-1215-1

Jeong, G. J., Khan, F., Tabassum, N., Cho, K. J., & Kim, Y. M. (2024). Strategies for controlling polymicrobial biofilms: A focus on antibiofilm agents. *International Journal of Antimicrobial Agents, 64*(2), 107243. https://doi.org/10.1016/j.ijantimicag.2024.107243

Jiang, R., Xiang, M., Chen, W., Zhang, P., Wu, X., Zhu, G., et al. (2021). Biofilm characteristics and transcriptomic analysis of *Haemophilus parasuis*. *Veterinary Microbiology, 258*, 109073. https://doi.org/10.1016/j.vetmic.2021.109073

Kadirvelu, L., Sivaramalingam, S. S., Jothivel, D., Chithiraiselvan, D. D., Govindarajan, D. K., & Kandaswamy, K. (2024). A review on antimicrobial strategies in mitigating biofilm-associated infections on medical implants. *Current Research in Microbial Sciences, 6*, 100231. https://doi.org/10.1016/j.crmicr.2024.100231

Karygianni, L., Ren, Z., Koo, H., & Thurnheer, T. (2020). Biofilm matrixome: Extracellular components in structured microbial communities. *Trends in Microbiology, 28*(8), 668–681. https://doi.org/10.1016/j.tim.2020.03.016

Kittleson, M. M., Irizarry, R., Heidecker, B., & Hare, J. M. (2009). Transcriptomics: Translation of global expression analysis to genomic medicine. In H. F. Willard & G. S. Ginsburg (Eds.), *Genomic and personalized medicine* (pp. 143–156). Academic. https://doi.org/10.1016/B978-0-12-369420-1.00012-3

Klopfleisch, R., & Gruber, A. D. (2012). Transcriptome and proteome research in veterinary science: What is possible and what questions can be asked? *The Scientific World Journal, 2012*, 254962. https://doi.org/10.1100/2012/254962

Korshoj, L. E., & Kielian, T. (2024). Bacterial single-cell RNA sequencing captures biofilm transcriptional heterogeneity and differential responses to immune pressure. *Nature Communications, 15*, 10184. https://doi.org/10.1038/s41467-024-54581-8

Lee, H., Im, H., Hwang, S. H., Ko, D., & Choi, S. H. (2023). Two novel genes identified by large-scale transcriptomic analysis are essential for biofilm and rugose colony development of *Vibrio vulnificus*. *PLoS Pathogens, 19*(1), e1011064. https://doi.org/10.1371/journal.ppat.1011064

Leesombun, A., Sungpradit, S., Sariya, L., Taowan, J., & Boonmasawai, S. (2023). Transcriptional profiling of the effect of *Coleus amboinicus* L. essential oil against *Salmonella typhimurium* biofilm formation. *Antibiotics, 12*(11), 1598. https://doi.org/10.3390/antibiotics12111598

Létoffé, S., Chalabaev, S., Dugay, J., Stressmann, F., Audrain, B., Portais, J. C., et al. (2017). Biofilm microenvironment induces a widespread adaptive amino-acid fermentation pathway conferring strong fitness advantage in *Escherichia coli*. *PLoS Genetics, 13*(5), e1006800. https://doi.org/10.1371/journal.pgen.1006800

Li, H., Du, X., Chen, C., Qi, J., & Wang, Y. (2022). Integrating transcriptomics and metabolomics analysis on kojic acid combating *Acinetobacter baumannii* biofilm and its potential roles. *Microbiological Research, 254*, 126911. https://doi.org/10.1016/j.micres.2021.126911

Li, J. G., Chen, X. F., Lu, T. Y., Zhang, J., Dai, S. H., Sun, J., et al. (2023a). Increased activity of β-Lactam antibiotics in combination with carvacrol against MRSA bacteremia and catheter-associated biofilm infections. *ACS Infectious Diseases, 9*(12), 2482–2493. https://doi.org/10.1021/acsinfecdis.3c00338

Li, Q., Zheng, X., Xie, J., Wang, R., Li, M., Wong, M. H., et al. (2023b). bvnGPS: A generalizable diagnostic model for acute bacterial and viral infection using integrative host transcriptomics and pretrained neural networks. *Bioinformatics, 39*(3), btad109. https://doi.org/10.1093/bioinformatics/btad109

Liu, J., Yang, L., Hou, Y., Soteyome, T., Zeng, B., Su, J., et al. (2018). Transcriptomics study on *Staphylococcus aureus* biofilm under low concentration of ampicillin. *Frontiers in Microbiology, 9*, 2413. https://doi.org/10.3389/fmicb.2018.02413

Ma, Y., Mao, Y., Kang, X., Zhang, B., Wang, J., Wang, G., et al. (2024). Transcriptomic analysis of the effect of glabridin on biofilm formation in *Staphylococcus aureus*. *Foodborne Pathogens and Disease*. https://doi.org/10.1089/fpd.2024.0038

Maghembe, R., Damian, D., Makaranga, A., Nyandoro, S. S., Lyantagaye, S. L., Kusari, S., et al. (2020). Omics for bioprospecting and drug discovery from bacteria and microalgae. *Antibiotics, 9*(5), 229. https://doi.org/10.3390/antibiotics9050229

Magro, D., Venezia, M., & Balistreri, C. R. (2024). The omics technologies and liquid biopsies: Advantages, limitations, applications. *Medicine in Omics, 11*, 100039. https://doi.org/10.1016/j.meomic.2024.100039

Mei, J., Xu, D., Wang, L., Kong, L., Liu, Q., Li, Q., et al. (2023). Biofilm microenvironment-responsive self-assembly nanoreactors for all-stage biofilm associated infection through bacterial cuproptosis-like death and macrophage re-rousing. *Advanced Materials, 35*(36), e2303432. https://doi.org/10.1002/adma.202303432

Melouane, A., Ghanemi, A., Yoshioka, M., & St-Amand, J. (2019). Functional genomics applications and therapeutic implications in sarcopenia. *Mutation Research-Reviews in Mutation Research, 781*, 175–185. https://doi.org/10.1016/j.mrrev.2019.04.003

Meng, F., Lyu, F., Bie, X., Lu, Y., & Lu, Z. (2024). Advances in transcriptomic analysis of *Salmonella* biofilms and their correlation with food safety. *Current Opinion in Food Science, 55*, 101110. https://doi.org/10.1016/j.cofs.2023.101110

Mirzaei, R., Campoccia, D., Ravaioli, S., & Arciola, C. R. (2024). Emerging issues and initial insights into bacterial biofilms: From orthopedic infection to metabolomics. *Antibiotics (Basel), 13*(2), 184. https://doi.org/10.3390/antibiotics13020184

Mishra, A., Aggarwal, A., & Khan, F. (2024). Medical device-associated infections caused by biofilm-forming microbial pathogens and controlling strategies. *Antibiotics (Basel), 13*(7), 623. https://doi.org/10.3390/antibiotics13070623

Misra, T., Tare, M., & Jha, P. N. (2022). Insights into the dynamics and composition of biofilm formed by environmental isolate of *Enterobacter cloacae*. *Frontiers in Microbiology, 13*, 877060. https://doi.org/10.3389/fmicb.2022.877060

Narad, P., & Kirthanashri, S. V. (2018). Introduction to omics. In P. Arivaradarajan & G. Misra (Eds.), *Omics approaches, technologies and applications* (pp. 1–10). Springer. https://doi.org/10.1007/978-981-13-2925-8_1

Nuñez-Belmar, J., Morales-Olavarria, M., Vicencio, E., Vernal, R., Cárdenas, J. P., & Cortez, C. (2023). Contribution of –Omics Technologies in the Study of *Porphyromonas gingivalis* during periodontitis pathogenesis: A Minireview. *International Journal of Molecular Sciences, 24*(1), 620. https://doi.org/10.3390/ijms24010620

Panicker, S. (2024). Meta-omics to understand the behavior of multispecies microbial communities. In J. A. Parray (Ed.), *Progress in soil microbiome research. Progress in soil science* (pp. 303–317). Springer. https://doi.org/10.1007/978-3-031-71487-0_13

Perry, E. K., & Tan, M. W. (2022). Bacterial biofilms in the human body: Prevalence and impacts on health and disease. *Frontiers in Cellular and Infection Microbiology, 13*, 1237164. https://doi.org/10.3389/fcimb.2023.1237164

Qian, W., Lu, J., Gao, C., Liu, Q., Li, Y., Zeng, Q., et al. (2024). Deciphering antifungal and antibiofilm mechanisms of isobavachalcone against *Cryptococcus neoformans* through RNA-seq and functional analyses. *Microbial Cell Factories, 23*, 107. https://doi.org/10.1186/s12934-024-02369-2

Rajput, A., Tsunemoto, H., Sastry, A. V., Szubin, R., Rychel, K., Chauhan, S. M., Pogliano, J., & Palsson, B. O. (2022). Advanced transcriptomic analysis reveals the role of efflux pumps and media composition in antibiotic responses of *Pseudomonas aeruginosa*. *Nucleic Acids Research, 50*(17), 9675–9688. https://doi.org/10.1093/nar/gkac743

Rubio-Mendoza, D., Córdova-Fletes, C., Martínez-Meléndez, A., Morfín-Otero, R., Maldonado-Garza, H. J., & Garza-González, E. (2023). Transcriptomic analysis of biofilm formation in strains of *Clostridioides difficile* associated with recurrent and non-recurrent infection reveals potential candidate markers for recurrence. *PLoS One, 18*(8), e0289593. https://doi.org/10.1371/journal.pone.0289593

Seneviratne, C. J., Suriyanarayanan, T., Widyarman, A. S., Lee, L. S., Lau, M., Ching, J., Delaney, C., & Ramage, G. (2020). Multi-omics tools for studying microbial biofilms: Current perspectives and future directions. *Critical Reviews in Microbiology, 46*(6), 759–778. https://doi.org/10.1080/1040841X.2020.1828817

Sharan, M., Vijay, D., Dhaka, P., Bedi, J. S., & Gill, J. P. S. (2022). Biofilms as a microbial hazard in the food industry: A scoping review. *Journal of Applied Microbiology, 133*(4), 2210–2234. https://doi.org/10.1111/jam.15766

Sharma, S., Mohler, J., Mahajan, S. D., Schwartz, S. A., Bruggemann, L., & Aalinkeel, R. (2023). Microbial biofilm: A review on formation, infection, antibiotic resistance, control measures, and innovative treatment. *Microorganisms, 11*(6), 1614. https://doi.org/10.3390/microorganisms11061614

Shebl, R. I., Elkhatib, W. F., & Badawy, M. S. E. M. (2023). Modulating the transcriptomic profile of multidrug-resistant *Klebsiella pneumoniae* biofilm formation by antibiotics in combination with zinc sulfate. *Annals of Clinical Microbiology and Antimicrobials, 22*, 84. https://doi.org/10.1186/s12941-023-00634-7

Silva, N. B. S., Marques, L. A., & Röder, D. D. B. (2021). Diagnosis of biofilm infections: Current methods used, challenges and perspectives for the future. *Journal of Applied Microbiology, 131*(5), 2148–2160. https://doi.org/10.1111/jam.15049

Sivaranjani, M., Leskinen, K., Aravindraja, C., Saavalainen, P., Pandian, S. K., Skurnik, M., et al. (2019). Deciphering the antibacterial mode of action of Alpha-Mangostin on *Staphylococcus epidermidis* RP62A through an integrated transcriptomic and proteomic approach. *Frontiers in Microbiology, 10*, 150. https://doi.org/10.3389/fmicb.2019.00150

Supplitt, S., Karpinski, P., Sasiadek, M., & Laczmanska, I. (2021). Current achievements and applications of transcriptomics in personalized cancer medicine. *International Journal of Molecular Sciences, 22*(3), 1422. https://doi.org/10.3390/ijms22031422

Tan, X., Qin, N., Wu, C., Sheng, J., Yang, R., Zheng, B., et al. (2015). Transcriptome analysis of the biofilm formed by methicillin-susceptible *Staphylococcus aureus*. *Scientific Reports, 5*, 11997. https://doi.org/10.1038/srep11997

Tang, D., Liu, Y., Yao, H., Lin, Y., Xi, Y., Li, M., et al. (2024). Transcriptome analysis reveals the mechanism of y0-c10-hsl on biofilm formation and motility of *Pseudomonas aeruginosa*. *Pharmaceuticals, 17*(12), 1719. https://doi.org/10.3390/ph17121719

Tatta, E. R., Paul, S., & Kumavath, R. (2023). Transcriptome analysis revealed the synergism of novel rhodethrin inhibition on biofilm architecture, antibiotic resistance and quorum sensing in *Enterococcus faecalis*. *Gene, 871*, 147436. https://doi.org/10.1016/j.gene.2023.147436

Tomlinson, B. R., Malof, M. E., & Shaw, L. N. (2021). A global transcriptomic analysis of *Staphylococcus aureus* biofilm formation across diverse clonal lineages. *Microbial Genomics, 7*(7), 000598. https://doi.org/10.1099/mgen.0.000598

Vagena, I.-A., Gatou, M.-A., Theocharous, G., Pantelis, P., Gazouli, M., Pippa, N., et al. (2024). Functionalized ZnO-based nanocomposites for diverse biological applications: Current trends and future perspectives. *Nanomaterials, 14*(5), 397. https://doi.org/10.3390/nano14050397

van Hoogstraten, S. W. G., Kuik, C., Arts, J. J. C., & Cillero-Pastor, B. (2024). Molecular imaging of bacterial biofilms–A systematic review. *Critical Reviews in Microbiology, 50*(6), 971–992. https://doi.org/10.1080/1040841X.2023.2223704

Vestby, L. K., Grønseth, T., Simm, R., & Nesse, L. L. (2020). Bacterial biofilm and its role in the pathogenesis of disease. *Antibiotics (Basel), 9*(2), 59. https://doi.org/10.3390/antibiotics9020059

Viering, B. L., Balogh, H., Cox, C. F., Kirpekar, O. K., Akers, A. L., Federico, V. A., et al. (2024). Loratadine combats methicillin-resistant *Staphylococcus aureus* by modulating virulence, antibiotic resistance, and biofilm genes. *ACS Infectious Diseases, 10*(1), 232–250. https://doi.org/10.1021/acsinfecdis.3c00616

Vlaeminck, J., Lin, Q., Xavier, B. B., De Backer, S., Berkell, M., De Greve, H., et al. (2022). The dynamic transcriptome during maturation of biofilms formed by methicillin-resistant *Staphylococcus aureus*. *Frontiers in Microbiology, 13*, 882346. https://doi.org/10.3389/fmicb.2022.882346

Vohra, M., Kour, A., Kalia, N. P., Kumar, M., Sharma, S., Jaglan, S., et al. (2024). A comprehensive review of genomics, transcriptomics, proteomics, and metabolomic insights into the differentiation of *Pseudomonas aeruginosa* from the planktonic to biofilm state: A multi-omics approach. *International Journal of Biological Macromolecules, 257*(Pt 1), 128563. https://doi.org/10.1016/j.ijbiomac.2023.128563

Xiong, L., Yi, F., Yu, Q., Huang, X., Ao, K., Wang, Y., et al. (2022). Transcriptomic analysis reveals the regulatory role of quorum sensing in the *Acinetobacter baumannii* ATCC 19606 via RNA-seq. *BMC Microbiology, 22*, 198. https://doi.org/10.1186/s12866-022-02612-z

Xu, S., Cao, Q., Liu, Z., Chen, J., Yan, P., Li, B., et al. (2022). Transcriptomic analysis reveals the role of tmRNA on biofilm formation in *Bacillus subtilis*. *Microorganisms, 10*(7), 1338. https://doi.org/10.3390/microorganisms10071338

Yako, H., Niimi, N., Takaku, S., & Sango, K. (2023). Advantages of omics approaches for elucidating metabolic changes in diabetic peripheral neuropathy. *Frontiers in Endocrinology (Lausanne), 14*, 1208441. https://doi.org/10.3389/fendo.2023.1208441

Yuan, L., Dai, H., He, G., Yang, Z., & Jiao, X. (2023). Current perspectives for analyzing the dairy biofilms by integrated multiomics. *Journal of Dairy Science, 106*(12), 8181–8192. https://doi.org/10.3168/jds.2023-23306

Zhang, Z., Chen, G., Hussain, W., Pan, Y., Yang, Z., Liu, Y., et al. (2024a). Machine learning and network analysis with focus on the biofilm in *Staphylococcus aureus*. *Computational and Structural Biotechnology Journal, 23*, 4148–4160. https://doi.org/10.1016/j.csbj.2024.11.011

Zhang, Z., Pan, Y., Hussain, W., Chen, G., & Li, E. (2024b). BBSdb, an open resource for bacterial biofilm-associated proteins. *Frontiers in Cellular and Infection Microbiology, 14*, 1428784. https://doi.org/10.3389/fcimb.2024.1428784

Zheng, M., Wu, X., Xu, Y., Ma, S., Shen, J., Li, T., et al. (2025). Curcumin reverses high-level tigecycline resistance mediated by different mechanisms in Gram-negative bacteria. *Phytomedicine, 136*, 156319. https://doi.org/10.1016/j.phymed.2024.156319

8 Metabolomics Approaches in Understanding the Biofilm Formation and Its Inhibitors

Barath Gajendran, Hemavathy Nagarajan, Busi Siddhartha, Sharada Ramasubramanyan, Dinakara Rao Ampasala, Ra mya Chandar Charles Mariasoosai, and Sampathkumar Ranganathan

8.1 Introduction

Biofilms are structured communities of microbial cells adhering to the surface and embedded in a self-producing matrix of extracellular polymeric substances (EPS). This matrix comprises polysaccharides, proteins, and nucleic acids, which are essential for the structural support of microbial survival and interaction with the environment (Awaness et al., 2024) (Kang et al., 2022). The importance of biofilms in healthcare is evident from the fact that they cause medical device-related infections where the microorganisms can settle on medical devices, such as catheters, prosthetic joints, and contact lenses, thus posing serious health problems (Jastaniah et al., 2022).

B. Gajendran
Central Research Instrumentation Facility, Kamalnayan Bajaj Institute for Research in Vision and Ophthalmology, Vision Research Foundation, Chennai, Tamil Nadu, India

H. Nagarajan · S. Ranganathan (✉)
Centre for Bioinformatics, Kamalnayan Bajaj Institute for Research in Vision and Ophthalmology, Vision Research Foundation, Chennai, Tamil Nadu, India
e-mail: drsampath@snmail.org

B. Siddhartha
Department of Microbiology, School of Life Sciences, Pondicherry University, Puducherry, India

S. Ramasubramanyan
R.S. Mehta Jain Department of Biochemistry and Cell Biology, KBIRVO, Vision Research Foundation, Chennai, Tamil Nadu, India

D. R. Ampasala
Department of Bioinformatics, School of Life Sciences, Pondicherry University, Puducherry, India

R. C. C. Mariasoosai
Department of Chemistry and Biochemistry, University of Texas at Dallas, Austin, TX, USA

S. Busi et al. (eds.), *Omics Approaches in Biofilm Research*,
https://doi.org/10.1007/978-3-031-91863-6_8

The formation of a biofilm begins with the reversible adhesion of planktonic bacteria to a surface, followed by the production of EPS that allows for irreversible adhesion (Muhammad et al., 2020). This will lead to a three-dimensional structure inhabited by heterogeneous microbial populations that include bacteria, fungi, and protozoa (Liang et al., 2010; Yin et al., 2019). The EPS matrix physically protects and controls nutrient and waste material interchange through its canal-like structure and is vital in maintaining microbial survival within the biofilm. One of the major problems faced by biofilms is their innate resistance to the action of antimicrobial agents. Bacteria living in biofilms can tolerate high levels of antibiotics. This results in the need for extremely high doses to kill the counterparts (Zhang et al., 2018). Several factors are associated with this resistance, such as limited penetration of antimicrobials through the EPS matrix, the presence of persister cells, and the development of adaptive stress responses (Somasundar et al., 2023). Moreover, biofilms can induce host inflammation, resulting in chronic infections and complications, such as implant failure (Jastaniah et al., 2022).

Biofilms are associated with a spectrum of device-related infections in clinical environments, mainly *Staphylococcus aureus* and *Staphylococcus epidermidis*, and multiresistant Gram-negative bacteria, including *Escherichia coli* and *Pseudomonas auroginosa*, according to Jastaniah et al. (2022). Such infections can be attributed to microorganisms that may arise from the patient or the environment, which implies the necessity for strategies that effectively inhibit biofilm formation on medical devices (Zhao et al., 2018). Biofilm infections are complicated, thus requiring multifaceted treatments, such as developing novel antimicrobial agents and applying biofilm-disrupting therapies (Zhang et al., 2022). Outside healthcare, industrial settings, especially food processing and wastewater treatment, are also affected and faced various challenges. Biofilms contribute to contamination and spoilage in these contexts; therefore, good research-based management strategies are needed (Ogawa et al., 2018; Wang et al., 2022). Understanding biofilm formation dynamics and EPS's role can inform the development of materials and coating that inhibit biofilm adhesion and growth. Bacterial biofilm represents a significant challenge in healthcare and industrial contexts due to its complex structure, resistance to antimicrobials, and ability to cause chronic infections. Continued research into the mechanism of biofilm formation, the role of EPS, and the development of effective treatment strategies are essential in mitigating the impact of biofilm on public health and industry.

8.2 Metabolites and Need to Understand Biofilm Dynamics

Metabolomics studies of all small molecules, or metabolites, within a biological system. It is essential for understanding the physiological and phenotypical state of organisms. These metabolites include amino acids, carbohydrates, lipids, peptides, and nucleotides, integral parts of the cellular process. Metabolomics provides a dynamic view of cellular functions and responses to environmental changes, thus offering valuable insights into the functional status of biological systems. Recent advancements in analytical techniques, such as mass spectrometry (MS) and nuclear magnetic resonance (NMR), have enabled researchers to analyze complex biological samples with unprecedented precision and depth, solidifying metabolomics as an essential component of systems biology, medicine, and microbial research (Karygianni et al., 2020). Biofilms are structured communities of microorganisms that adhere to surfaces and are encased in a self-produced extracellular matrix (ECM).

These communities are strikingly resilient and plastic, which poses an immense challenge in healthcare or industrial realms. Understanding their metabolic adaptations governing biofilm growth, development, and resistance is essential when devising and creating effective intervention strategies to perturb biofilm formation. Metabolomics will provide detailed high-resolution imaging to unravel some metabolic intricacies, unraveling biochemical pathways, sustaining biofilms under various environmental perturbations. The specific pathways and shifts in metabolites can travel from analyzing the metabolite profiles of the biofilm-forming organisms that facilitate growth. For example, research indicates that alterations in nutrient utilization and energy production are crucial for the synergistic interactions of dual-species biofilm, such as *Staphylococcus aureus* and *Klebsiella oxytoca*, as observed in their metabolic profiles (Liu et al., 2022). Moreover, through metabolomics, specific biofilm-specific biomarkers are identified, which serve as markers for biofilm presence and activity and advance the understanding of the microbial interaction in these communities (Jastaniah et al., 2022). Lastly, metabolites enable the discovery of novel intervention opportunities through targeted metabolic pathways involved with biofilm. For instance, research has recently demonstrated that specific metabolites produced by marine microorganisms could inhibit biofilm formations, offering a new platform for biofilm control in industrial applications (Muhammad et al., 2020). This integrative perspective underscores the critical role of metabolomics in advancing our understanding of biofilm dynamics and developing effective strategies for managing biofilm-related challenges in both healthcare and industrial setups.

8.3 Biofilm Formation and Metabolic Adaptations

Biofilm formation has five developing steps as described below:

I. Primary adhesion: Biofilm formation is a process that transfers free-floating into organized community structures. This process begins with the primary adhesion of planktonic bacteria to surfaces, where they first establish loose and reversible attachments. During this early stage, microorganisms are usually found in a polar orientation on the surface. As they stick around, the bacteria fight through diverse physical states that otherwise inhibit their ability to establish a biofilm. Bis-3′,5′-cyclic dimeric guanosine monophosphate (C-Di-GMP) is a significant intracellular signaling molecule for this initiation. C-Di-GMP increased production allows them to continue production of the biofilm matrix; simultaneously, it impairs the production of flagella-mediated motility, locking them down onto a substratum (Nobile et al., 2012).

The Pil-chp surface-sensing mechanism in *Pseudomonas aeruginosa* causes an increase in the C-Di-GMP concentration with each attachment or detachment event, facilitating the transition from the surface-naïve planktonic cells characterized by low C-Di-GMP levels to surface sentient planktonic cells with high C-Di-GMP levels. Once firmly attached, these microorganisms multiply and aggregate within the self-produced EPS, forming microcolonies. The motility of flagella and type IV pili is essential for microorganism-surface contacts and cell-cell aggregation, respectively (Ramos et al., 2010).

II. Maturation: The biofilm maturation significantly depends on EPS, which plays various roles. It stabilizes the three-dimensional structure of the biofilm, aids in the attachment of microbes to a surface, aggregates cells, and protects against several stresses, including oxidative damage, host immune responses, and antimicrobial agents.

In addition, EPS captures signaling molecules important for metabolic functions, enzyme activity, and quorum sensing (QS) (Sasaki-Imamura et al., 2010). With the maturation of biofilms, they may assume unique morphologies, including "Mushroom or Tower" structures that depend on the metabolic rates and aerotolerance of the bacteria inside them. A fully grown biofilm often has three layers: the external layer is said to house planktonic bacteria poised to disperse, an internal regulatory layer, and a core microbial foundation middle layer (Zheng et al., 2015).

III. Dispersion: Dispersion of a fully grown biofilm initiates a cycle of new biofilm development. Dispersion can be carried out passively because of physical properties, such as fluid dynamics, or active through motility and degradation of EPS. Biofilm dispersion can be driven by the overgrowth of populations, competition for available resources, nutrient depletion, and alterations in environmental parameters, such as a change in temperature and insufficient oxygen. Furthermore, the upregulation of genes related to motility in addition to EPS degradation, while the downregulation of genes responsible for polysaccharide and fimbriae production, is of prime significance in this process.

In conclusion, metabolomics is a powerful tool for understanding complex biochemical networks involved in biofilm formation. Identifying specific metabolic pathways associated with biofilm development and persistence is possible by analyzing the profiles of metabolites from biofilm-forming organisms (Fig. 8.1). For example, Wu et al. (2024) studied various bacterial species and demonstrated the impact of indole on biofilm formation through the modulation of quorum-sensing signals. Additionally, metabolite accumulation could reflect alterations in the biofilm's structure and heterogeneity, for instance, through changes in *Pseudomonas aeruginosa*, in which differential metabolism between biofilm and planktonic states has been described. The elucidation of such metabolic dynamics will thus be pivotal to design targeted approaches aimed at interrupting biofilm development and maximizing therapeutic success against biofilm-related infections.

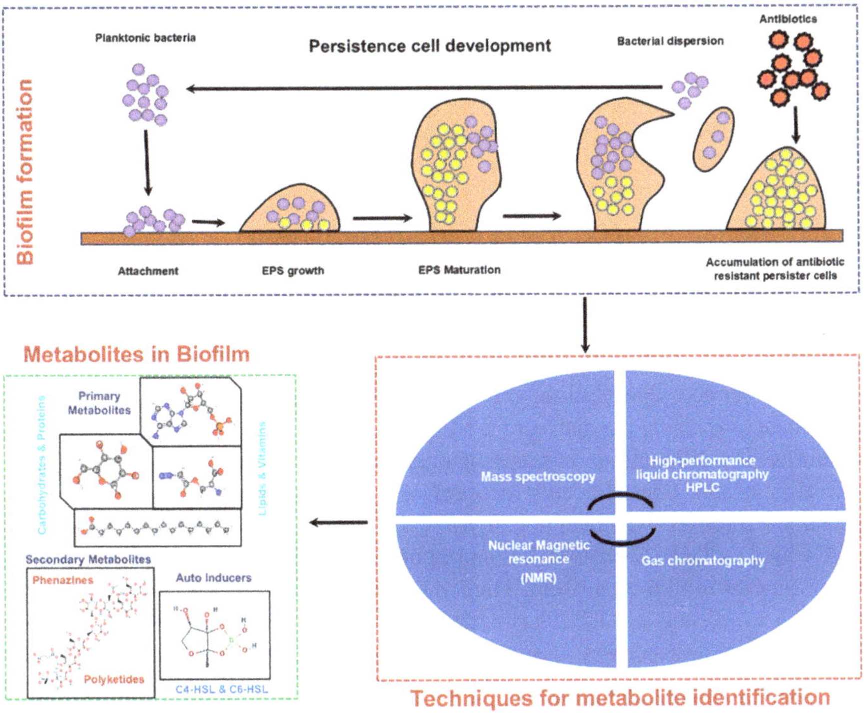

Fig. 8.1 Analytical strategies for biofilm metabolite profiling

8.3.1 Metabolomics Shift During Biofilm Formation

It is a very complex and dynamic metabolic shift that characterizes biofilms. Because of this process, microbial communities can thrive in various environments. Shifts of this nature are essential to biofilms' establishment, maintenance, and survival. Metametabolomics is an invaluable tool for illuminating these shifts that provide a global view of biochemical networks and pathways that underpin biofilm dynamics.

8.3.2 Shifts in Metabolic Pathways

Significant metabolic alterations, mostly associated with the carbon metabolism cycle, take place in the early phases of the biofilm mode of development. Although mature biofilms often belong to the tricarboxylic acid (TCA) cycle, glycolysis is thought to be the primary energy route for microbial production. For example, cells in anaerobic zones depend on fermentation or other electron acceptors like nitrate, whereas cells in aerobic conditions need aerobic respiration (Kernan et al., 2015).

The second critical regulation is redox balance, which maintains redox equilibrium. In this case, it has to endure oxidative stress due to ROS (reactive oxygen species). Secondary metabolites, such as phenazines, are highly essential as they assist in maintaining redox equilibrium by acting as an electron shuttle to enhance the biofilm's structural stability and robustness (Costerton et al., 1999).

8.3.3 Metabolism Critical to Biofilm Formation

Several metabolites are needed to form a biofilm that provides structural strength, stress resistance, and intercellular communication. The amino acids glutamate, arginine, and proline are essential in maintaining osmotic balance and acting as a precursor to secondary metabolites (Zhu et al., 2007). Lipids and polysaccharides also contribute to forming the matrix of the EPS that surrounds and shields biofilm communities. The EPS matrix consists of macromolecules including polysaccharides like alginate and poly-N-acetylglucosamine (PNAG), which confer structural support and hydration (Nobile et al., 2012).

Cyclic Di-GMP (C-Di-GMP) is an important signaling molecule involved in the planktonic-to-biofilm transition. High intracellular concentrations of C-Di-GMP promote adherence and EPS accumulation; whereas, low concentrations promote detachment and dispersion. In biofilms, metabolic substrates, including secret and pyruvate, also accumulate as cross-feeding substrates for microbial populations to promote cooperation and increase the resistance to biofilms' nutrient-poor conditions (Xu et al., 2017).

8.3.4 Role of Quorum Sensing and Secondary Metabolites

Quorum sensing (QS), alternatively called cell-cell communications is an essential mechanism that controls biofilm and maintenance. Microbial cells communicate with each other by signaling molecules to coordinate collective behaviors, such as adhesion and EPS production. In Gram-negative bacteria, acyl-homoserine lactones (AHL) regulate biofilm maturation, while autoinducing peptides (AIPs) mediate QS in Gram-positive bacteria (He et al., 2015). The secondary metabolites produced in response to QS play diverse roles in biofilm dynamics. For instance, phenazines contribute to redox balance, while rhamnolipid enhances biofilm expansion by promoting cell motility and adhesion (Table 8.1) (Zhao et al., 2022).

In addition to the complexity arising from variations in metabolite production among cells, QS systems add to the biofilm's heterogeneity because differences in these production levels generate various microenvironments where diverse populations can exist and evolve. As such, the biofilm will resist external stressors better, such as antibiotics and the host's immune response.

Table 8.1 Key microbial metabolites and their functions in biofilm development and regulation

S. no	Metabolite type	Specific metabolites	Uses
1.	Primary metabolites	Extracellular polymeric substance (EPS)	It forms an extracellular polymer, which promotes adhesion and serves as a structural matrix that leads to the formation of microbial biofilm (Donlan, 2001)
		Amino acids	It involved in the signaling of metabolism within the biofilm formation (Karygianni et al., 2020)
		Lipids	It acts as a signaling molecule and involved in the cell membrane structure and integrity (Karygianni et al., 2020)
4.	Secondary metabolites	Quorum sensing(QS)	It is a signaling molecule produced by bacteria to interact with each other to promote the formation of biofilms (Solano et al., 2014)
		Antimicrobial peptides (AMPs)	Depending on the bacterial type and concentration, it will either encourage or prevent the production of biofilms (Brogden, 2005)
		Pigments	It helps in maintaining the stability and structure of biofilms and provides defense against environmental stress (López et al., 2010)
		Autoinducers (AIs)	It plays a major role in gene regulation, intracellular interactions, and biofilm formation (López et al., 2010)

8.3.5 Application of Metabolomics in Biofilm Research

Metabolomics is one of the most effective methods for dealing with metabolic shifts and biomolecular interactions between biofilm communities. It is a driving force for scientists to understand and even detect new specific markers affecting the biofilm. Moreover, it can diagnose established metabolic weaknesses to implement targeted interventions. For example, inhibiting the synthesis of the secondary messenger molecule cyclic diguanosine monophosphate (C di GMP) or the Quorum-Sensing (QS) pathways may be a good way to destroy biofilms (Goytia et al., 2013). With metabolomics, the methods of detecting new targets for antimicrobials, e.g., identifying some pathways or metabolites vital for the biofilm's survival, will be developed (Liu et al., 2015).

Furthermore, the health sector will majorly use biofilm metabolism to combat biofilm infections like *Pseudomonas aeruginosa*, *Staphylococcus aureus*, and ESKAPE, among other pathogens. Thus, it is crucial to make new substances to kill them because they are resistant to common antibiotics and mainly cause problems in chronic wound treatment and implant infections (Paino et al., 2011). Applying metabolomics to biofilm development will be able to neutralize biofilm pathogens with anti-adhesion coatings or related to metabolite inhibitors (Erickson et al., 2010).

Besides medicine, metabolomics will also find applications in the industrial and environmental fields. In fact, the metabolic behavior of biofilms, in particular, governs fouling, corrosion, and contamination of equipment and appliances in the food processing industries, water treatment plants, and pipes. More detailed information gained during the experiments with biofilms at the metabolic level has permitted scientists to work out effective control methods for biofilms by utilizing targeted biocides and metabolic inhibitors. This penetration of other metabolic pathways is significant proof of microbial community adaptability. The sequence of biofilm formation is tightly linked to carbon catabolism and oxidative stress balance by the nucleotides and quorum molecules. Metabolomics would thus aid in a better understanding of this issue. On the other hand, this knowledge would be mandatory to develop novel control strategies as new findings reveal more and more aspects of the metabolic background of biofilms and thus might fulfill its implementation in the application of targeted interventions in healthcare, industry, and the environment (Erickson et al., 2010).

8.4 Metabolomics Approaches in Studying Biofilm

8.4.1 Analytical Techniques for Metabolic in Biofilm Formation

8.4.1.1 Mass Spectrometry (MS) and Nuclear Magnetic Resonance (NMR) in Biofilm Metabolomics

Mass spectrometry (MS) and nuclear magnetic resonance (NMR) spectroscopy are the main analytical tools applied in metabolomics. MS is very effective for analyzing various metabolites by determining their mass-to-charge ratio. It is usually combined with separation methods, such as gas chromatography (GC), high-performance liquid chromatography (HPLC), and capillary electrophoresis (CE) for the profiling of metabolic biofilm constituents. GC-mass spectrometry (GC-MS) is exceptionally well adapted for analyzing volatile and thermally stable compounds, such as volatile small organic acids and alcohols, released during biofilm growth. These metabolic by products provides information regarding the biochemical routes that microbes employ to facilitate biofilm growth and also acquire resistance to environmental stress (Lu et al., 2019).

Conversely, liquid chromatography-mass spectrometry (LC-MS) can analyze various biomolecules, including heat-sensitive and nonvolatile compounds, such as lipids, polysaccharides, and proteins. Its diversity makes LC-MS a go-to method for analyzing intricate biofilm matrices, in which metabolites from diverse pathways co-occur and interact. For instance, LC-MS has been used effectively to determine the prominent biofilm extracellular matrix molecules, including biofilm-specific lipids and polysaccharides, responsible for adhesion and immune evasion (Pisithkul et al., 2019). Despite its advantages, MS-based methodologies are subject to limitations like molecular ionization inefficiencies and the risk of sample loss or modification during chromatographic separation. Nevertheless, MS is still an important instrument for biofilm metabolomics due to its sensitivity and specificity.

Conversely, NMR spectroscopy allows metabolite identification without previous separation, keeping sample loss to a minimum. It delivers detailed structural data through chemical shifts, peak splitting (J-coupling), and signal intensity, which suits accurate metabolite quantification. Despite its low sensitivity compared to MS,

NMR's potential to provide precise structural and quantitative data renders it an excellent complement for biofilm research (Awadh et al., 2021). Both MS and NMR, collectively, deliver an all-around knowledge of biofilm metabolomics, which encapsulates metabolic diversity and complexity.

8.4.1.2 Gas Chromatography-Mass Spectrometry (GC-MS) and Liquid Chromatography (LC-MS) in Biofilm Analysis

GC-MS and LC-MS are indispensable for examining the metabolic adaptations of biofilms and discovering pathways that may be targeted for inhibition. GC-MS is beneficial for reviewing volatile metabolites generated by biofilms, including short-chain fatty acids and alcohols, which are vital for microbial signaling and matrix formation in biofilms. These volatilities usually indicate active metabolism in biofilms and can give insights into the environmental conditions and the availability of nutrients that dictate biofilm formation (Wong et al., 2018; Zhang et al., 2021).

LC-MS expands the analytic scope by enabling the detection of nonvolatile and thermally labile biomolecules. This technique has proven effective for analyzing key components of the biofilm extracellular matrix, such as lipids, polysaccharides, and proteins. By profiling these metabolites, researchers can gain insights into how biofilm maintains its structural integrity and adapts to stress. For example, LC-MS studies have identified biofilm-specific lipids and polysaccharides critical for adhesion and immune evasion, highlighting potential targets for biofilm inhibitors (Nothias et al., 2018). Additionally, LC-MS can be used to study secondary metabolites, such as quorum-sensing molecules and antibiotics, which regulate biofilm behavior and interspecies interactions (Kriswandini et al., 2024).

8.5 Metabolomics in Identifying Biofilm Inhibitors

Biofilm metabolomics researchers have highly focused to discover and develop transformative agents, blocking biofilm establishment and maintain research profile of biofilms metabolic reactions due to impact of environmental stress and chemical to identify survival critical metabolic pathways. As demonstrated by metabolomics research, biofilms utilize glycolysis, the TCA cycle, and other specific carbon metabolism pathways to obtain their necessary energetic requirements. Biofilm energy production would decrease, further the growth and resilience would also be reduced if researchers used targeted inhibitors on these critical pathways (Parga et al., 2023).

Metabolomics successively played a crucial role in identifying substances that impede quorum sensing mechanisms. Through quorum sensing mechanisms, Gram-negative bacteria execute intercellular communication using AHLs, and Gram-positive bacteria employ AIPs to control biofilm processes. Metabolic research investigations have identified molecules that alter these signaling pathways and efficiently accomplish biofilm formation reduction alongside virulence decrease. The fundamental understanding of the biofilm inhibitor mechanism of action comes from metabolomics research, which revealed the potential of natural compounds, such as furanone combined with synthetic quorum-sensing analogs. Identified secondary metabolites show potential in weakening biofilm structures while simultaneously boosting the drug-enterable vulnerability of such structures. Using metabolomics enables the researchers to identify compounds that degrade EPS matrix structures and inhibit EPS synthesis processes. During the disruption of biofilm protective barriers, these inhibitors enhance biofilm exposure to antibiotics and host defense mechanisms (Parga et al., 2023).

Metabolomics has provided considerable benefits in solving the questions of biofilms or finding their inhibitors; nevertheless, numerous challenges persist. The complexity and heterogeneity of biofilm metabolism pose obstacles to applying exhaustive analyses. Besides, robust models are needed that accurately represent minimum *in vivo* conditions to expand the application of metabolomics findings; such models would also pave the way for Multidisciplinary (MD)-biofilm research integrated with other omics approaches and newly formulated emerging technologies like high-resolution MS, multidimensional NMR, etc. (Parga et al., 2023).

Modern technologies, such as MS, NMR, etc., can lead to fundamental discoveries by unravelling the biochemical process that contributes to the biofouling resilience. Therapeutic approaches seek to transfigure the minimum biofilm conditions. Metabolomics plays a crucial role not only in identifying trends within biofilm population but also in predicting future challenges that make biofilm so persistent. These insights will enhance our understanding of microbial communities and support strategic interaction to combat biofilm formation across healthcare, industry, and environmental settings. (Parga et al., 2023).

8.6 Sample Preparation and Extraction Methods from Biofilm Matrix

Mass spectrometry methods used for sample preparation and extraction within biofilm environments require focus. Several essential processing procedures exist for metabolite extraction from biofilm matrices, starting with cell disruption through metabolite extraction and then cell quenching, followed by harvesting. The main strength of NMR-based metabolomics is its essential and easy sample preparation requirement. Biofilms that develop on various habitat surfaces create conditions that necessitate substantial discrepancies in culturing and harvesting protocols (Gaubert-Boussarie et al., 2020). The accuracy of metabolome detection requires rapid biological and enzymatic stoppage to obtain precise snapshots during sample prediction. The rapid metabolomic turnover rates of fructose-6-phosphate and other crucial metabolites, such as fumarate, glutamate, phosphoenolpyruvate, and pyruvate, demands immediate attention (Santiago et al., 2016). To ensure experimental reliability, studies should focus on minimizing cellular delays and stress responses, as these factors can introduce variability and impactful. Experimental rapid cooldown of cell temperatures through rapid quenching proves an effective method to reduce cellular enzyme activity. Central to this approach is using methanol because it provides quick freezing capabilities while being less dangerous than other organic freezing agents. Gram-positive bacteria and fungi represent the only microorganisms that methanol can penetrate successfully because cell leakage and metabolite loss occur during quench-and-wash procedures (Lee et al., 2018).

Two principal strategies exist for medium elimination prior to metabolite harvesting: filtration and centrifugation. Sample preparation using centrifugation produces uniform results at the cost of additional time, potentially harming the metabolome through induced stress. One can achieve rapid processing through filtration, while cells function smoothly when placed on filter membranes. The practical process of removing all frozen cells from filter paper consistently presents technical obstacles (Papaianni et al., 2020). Direct *E. coli* cultivation on filter paper offers a potential solution for efficient cell quenching procedures and metabolomics extraction methods. The chosen washing methods, quenching techniques, and separation protocols are essential to achieve valid metabolomics research results (Bryce et al., 2020). Dual extraction approaches for metabolites and cellular liquids exist in specific situations. The Fast-Prep system and organic solvent methods have shown notable success in metabolomic studies. The traditional use of trichloroacetic acid as a cell lysis agent from filter paper leads to filter paper degradation that produces excessive background metabolite activity (Wong et al., 2015). The preferred extraction solution must handle metabolites without chemical or physical alteration. The extraction method must recover all metabolites effectively while showing no selectivity toward specific physical characteristics or vice versa chemical properties. Extracting lipid metabolites typically require a 1 v/v ratio combination between methanol and chloroform or water and chloroform (Stipetic et al., 2016).

8.7 Data Processing and Statistical Analysis

This challenging step starts with raw data curation through noise elimination and systematic error correction, which could utilize baseline adjustments, instrument response analysis, and peak alignment methods (González-Domínguez et al., 2017). The univariate analysis methods include t-tests, ANOVA (Analysis of Variance), and multivariate analyses involving PCA (Principal Component Analysis) and PLS-DA (Partial Least Squares Discriminant Analysis), which allow researchers to visualize differences in metabolite profiles between experimental groups following preprocessing. Pattern interpretation between treatments and cluster formation reveals that these analytical methods can achieve biofilm-based metabolic alterations (Ząbek et al., 2017).

Research shows that deploying modern machine learning (ML) algorithms as sophisticated statistical methods boosts their forecast capabilities. Machine learning algorithms help to detect essential biomarkers for monitoring antimicrobial resistance and biofilm presence, guiding proper therapeutic choice (Vyas et al., 2016). New understandings of biofilm dynamics emerge when researchers combine metabolomics with genomic and transcriptomic approaches (Rockstroh et al., 2024).

8.8 Metabolomics Profiling of Biofilm-Forming Microorganism

Biofilms form complex microbial communities enclosed by an exopolymeric substance (EPS) matrix, which offers stability and involves nutrient accessibility while defending against environmental threats. Metabolomics profiling is a critical investigative method to disclose the biochemical pathways in biofilm development and continuous metabolic transformations that help biofilms survive in different environments. The metabolic profiles obtained from biofilm-forming microorganisms, including bacteria, fungi, and mixed species communities, highlight their resistance, adaptable capabilities, and complex biological relationships. Due to their essential industrial and clinical roles, the scientific community has focused primarily on bacterial biofilms to examine *Pseudomonas aeruginosa* and *Staphylococcus aureus* species. Due to its infection of medical devices and role in cystic fibrosis pathogenesis, *P. aeruginosa* is a prominent model organism in biofilm investigation. *Pseudomonas aeruginosa* and *Staphylococcus aureus* both form biofilms that contain thick EPS matrices, which include polysaccharides alongside extracellular DNA (eDNA) and proteins, making these species resistant to antibiotics while causing immune evasion (Nett et al., 2010). As *S. aureus* forms its biofilm structures, it uses surface-based proteins and teichoic acids to maintain attachment and stability. The formation of biofilms by Gram-positive and Gram-negative bacteria occurs across different environments, including medical devices and industrial pipelines. It continues to be among the principal contributors to biofouling (Vediyappan et al., 2010).

8.8.1 Unique Metabolic Features of Bacterial Biofilms

Bacterial biofilms' metabolic behavior stands apart from planktonic bacteria's because environmental variations within the biofilm structure produce these differences. Biofilms exhibit varied metabolic distribution patterns according to location, as metabolomics studies indicate that the outer oxygen-rich zones use oxidative phosphorylation. At the same time, oxygen-poor inner sections operate via fermentation or anaerobic respiration with alternative electron acceptors like nitrate or fumarate (Deorukhkar et al., 2014). The survival of biofilms depends on secondary metabolites, which enhance their endurance ability. *Pseudomonas aeruginosa* biofilms generate phenazines as redox-active molecules, enabling low-oxygen electron transfer processes to help maintain biological activity (Marta Setiabudy et al., 2023). Biofilm maturation co-occurs with an accumulation of signaling molecules such as cyclic di-GMP (C-di-GMP), enabling both EPS production and biofilm maturation processes. Lipid metabolism changes occur within the biofilm. Through these adaptations, biofilm cells can produce enhanced levels of fatty acids, strengthening membrane stability while providing antimicrobial resistance (Mohandas & Ballal, 2011). Metabolic adaptations demonstrate complex biofilm characteristics, making them desirable pharmacological intervention targets.

8.9 Fungal and Mixed Biofilms

8.9.1 Fungal Biofilms

Patients with weak immune function represent the main groups who develop severe biofilm-associated *Candida albicans* infections. Fungal planktonic cells demonstrate lower antifungal medication resistance, but biofilms exhibit improved resistance due to their ECM structure. Fungal biofilm extracellular matrices (ECM) contains glucans and mannans alongside other polysaccharides establish a physical obstacle for drug molecules while simultaneously acting as a protective layer against environmental stresses (Nerurkar et al., 2012). *C. albicans* pathway mediated through CDR1 and MDR1 genes upregulates efflux pump activity, causing antifungal agent removal from fungal cells and improving resistance against antifungal agents (Sida et al., 2016). Fungal biofilms achieve effectiveness across various surfaces, including medical devices and mucosal tissues, through metabolic plasticity, which drives their involvement in infections, such as candidiasis (Tiwari et al., 2019).

Metabolomics research findings establish that carbohydrate metabolic pathways like glycolysis and pentose phosphate cycle gather bioenergies while simultaneously boosting extracellular matrix component production in fungal biofilms. The metabolic activity within biofilms displays heterogeneity since distinct biofilm locations harbor cells at different metabolic states. Diverse cell states help sustain biofilm stability and create an environment that allows the formation of persister cells that resist antifungal treatment (Bruder-Nascimento et al., 2014). The fungal signaling molecule farnesol plays a crucial regulatory role in biofilm establishment and microorganism communication, proving that metabolomics is essential in fungal biofilm research (Andriana et al., 2024).

8.9.2 Metabolic Interactions in Mixed Microbial Biofilms

Interspecies bacterial and fungal communities in mixed biofilms produce complex ecosystems, potentially creating stronger resilience than mono-species biofilms. Microorganisms succeed together through metabolic supportive relationships, leading to increased chances of survival. The lactate produced by bacterial cells in *Candida albicans* and *Staphylococcus aureus* biofilms functions as a carbon source for fungal metabolism. Barophilic fungal cells build protected zones that shield bacterial cells from the effects of antibiotics, according to Wolber et al. (2017). Independent shared metabolic outputs from different species show how mixed biofilm populations build synergistic dynamics.

Integrated metabolomics analyses reveal the exclusive presence of combined quorum-sensing molecules responsible for regulating interspecies dialogue and maintaining biofilm resilience in mixed community structures. Analyses indicate that mixed biofilms contain increased stable forms due to incorporating bacterial lipopolysaccharides (LPS) and fungal glucans within their extracellular matrix framework (Iman et al., 2013).

8.10 Differences in Metabolomic Signatures Between Monomicrobial and Polymicrobial Biofilms

Different species interaction within biofilms creates distinct metabolomic profiles that differ between mono-species and polymicrobial systems. Management through restricted metabolic pathways fits mono-species biofilms, but polymicrobial biofilms achieve metabolic flexibility through multiple substrate options. Bacterial and fungal species within polymicrobial biofilms employ different carbon sources during metabolism to generate a metabolic cooperative framework supporting biofilm growth under limited nutritional conditions (de Cremer et al., 2016). Metabolomic data reveal that mixed biofilms contain elevated stress-adaptation metabolites, including antioxidants and secondary metabolites, which perform better than mono-species biofilms under adverse conditions (Lorin et al., 2017). The metabolic profiles obtained through metabolomic analysis revealed physiological differences in bacterial, fungal, and biofilm communities. Metabolomics research demonstrates how biochemical processes change, and microbial groups interact to help biofilms survival.

8.11 Biofilm Inhibitors: A Metabolomics Perspective

8.11.1 Natural and Synthetic Inhibitors

The inhibition of biofilms depends on antibiotics alongside phytochemicals and synthetic molecules with separate anti-biofilm mechanisms. The bacterial processes of DNA replication and protein synthesis succumb to attack by the antibiotic drugs, ciprofloxacin and rifampin, frequently prescribed to fight biofilm producing microorganisms. The low success rate of these antibiotics against thick biofilms drives scientists to search for different inhibiting agents. Plant-derived compounds, carnosol from rosemary and terpinene-4-ol from tea tree oil, prevent biofilm structural collapse through changes in signaling pathways while suppressing the metabolic actions of bacteria. When applied to *Staphylococcus aureus*, carnosol disrupts bacterial amino acid processes, resulting in altered bacterial biofilm structures, according to Zhang et al. (2021). Trehalose analogs and peptide-based synthetic agents target quorum-sensing systems and extracellular glucans to restrict biofilm development while disabling its structural integrity (Pisithkul et al., 2019). Metabolomic approaches identify diverse inhibitor capabilities, demonstrating their potential for discovering and assessing biofilm metabolic changes. According to their study, Kang et al. (2022) found that metabolic inhibitors successfully disrupted the biofilm-forming C-di-GMP biosynthesis pathway to minimize biofilm mass (Kang et al., 2022). Such findings spotlight the need for precise knowledge of the pathways that the inhibitors target.

8.11.2 Metabolomics Changes in Biofilms Upon Inhibitor Treatment

Biofilm inhibitor administration triggers transformative changes to vital metabolic pathways that sustain biofilms throughout their developmental stages. The regulatory molecules that control biofilm formation show noticeable reductions due to these metabolic changes. Biofilm biomass levels decrease substantially when cells change from biofilm affiliation to a planktonic state because of C-di-GMP inhibitor administration, which decreases extracellular polymeric substance production and alters cellular energy consumption patterns (Kang et al., 2022). The application of metabolomics helps researchers to detect specific biomarkers that track inhibitor-induced changes in addition to producing quantitative data about inhibitor performance.

Metabolite concentration variations function as fundamental markers that help to identify biofilm inhibitory effects. The energy-related metabolites, ATP and NADPH, demonstrate substantial decreases in inhibitor-treated biofilms, indicating disrupted energy metabolism throughout these systems. A biofilm dysfunction caused by sodium hypochlorite sublethal disinfection led Lu and colleagues to notice substantial metabolic decreases of glycolytic and Krebs cycle metabolites in *Klebsiella pneumoniae* biofilms (Lu et al., 2019). Carnosol treatment caused reductions in two essential amino acids required for biofilm stability: glycine and serine, according to Zhang et al. (2021). Research shows how biofilm inhibitors impact three fundamental metabolic pathways responsible for energy creation, amino acid synthesis, and stress response functions, providing detailed information about biofilm disrupting mechanisms.

8.11.2.1 Case Studies: Metabolomic Profiles Before and After Inhibitor Treatment

Case studies demonstrate biofilms' responsive metabolic changes during inhibitor exposure by illustrating the molecular comprehension made possible through metabolomics analysis. Research by Wolber and colleagues tested trehalose analogs on *Mycobacterium smegmatis* to determine direct suppression links between trehalose metabolic breakdown and diminished biofilm growth (Wolber et al., 2017). Metabolomic data from these experiments illustrated that inhibitors interfered with glucan extracellular production, preventing biofilm matrix function and resulting in biofilm biomass and structural framework degradation.

During *Bacillus subtilis* biofilm growth, Pisithkul et al. (2019) observed metabolic shifts and showed that proper inhibitors could restore normal metabolism and decrease biofilm survival (Pisithkul et al., 2019). The research conducted by Hobson et al. (2022) revealed that terpinen-4-ol created substantial disruptions to genes and metabolites linked with ATP, impacting methicillin-resistant *Staphylococcus aureus* (MRSA) biofilms. Biofilm development suffered inhibition due to multiple metabolic deviations, including lower cellular energy production and impaired stress-response functions (Lu et al., 2019). Through case study analyses, we demonstrate how metabolomics reveals the mechanisms by which inhibitors affect biofilms, which provides vital information for developing targeted antimicrobial approaches.

8.12 Key Metabolite Markers of Biofilm Inhibition

Metabolomics revealed multiple biochemical indicators to detect biofilm regulatory effects. Exposed biofilms show a decrease in both glycolytic intermediates and tricarboxylic acid (TCA) cycle metabolites, while levels of amino acids glycine and serine become reduced. Core biofilm support processes like energy production, EPS synthesis, and quorum sensing, have inhibited, leading to observed metabolic changes. The metabolic profile of *Klebsiella pneumoniae* underwent significant modifications after applying sodium hypochlorite, according to Zhang et al. (2021), who reported that this procedure lowered biofilm-associated metabolic activity (Lu et al., 2019). Demonstrated carnosol's disruption of *S. aureus* biofilms by identifying disrupted amino acid metabolic pathways through metabolomics methodology. Combining multiple omics approaches helps scientists to gain better insights into inhibitor-triggered metabolic reprogramming events. Research evaluations of biofilm inhibition achieve complete comprehension when metabolomics works alongside transcriptomics and proteomics approaches. Cheng et al. (2021) showed that terpinen-4-ol treatment altered both MRSA metabolites and ATP-related gene expression, demonstrating the interlinked nature of changes in biofilm inhibition according to Zhang et al. (2021).

Biofilm inhibitors impact the metabolic processes that disturb biofilm growth and stability by modifying measurable biomarkers in metabolite profiles, which indicate inhibition severity. Metabolomics offers a robust analytical basis to examine these biological changes and allows researchers to detect disturbed pathways, leading to the discovery of new inhibitors. Research through case studies demonstrates how inhibitors modify biofilm structure and metabolic processes to reveal their mode of operation. The fusion of metabolomics with other omics technologies enables researchers to refine their understanding of biofilm inhibition, creating new avenues for innovative solutions to tackle clinical and industrial biofilm problems.

8.13 Targeting Biofilm Metabolism for Drug Discovery

8.13.1 Metabolic Vulnerabilities in Biofilms

The essential metabolic pathways that bacteria activate to form biofilm structures are life-sustaining mechanisms that build biofilm strength and prevent antibiotic interruption. These modifications encompass alterations in lipid, nucleotide, amino acid, and carbohydrate metabolism, critical aspects of biofilm physiology. Metabolic research has revealed key biofilm architecture cycles while uncovering ways to block their formation mechanisms, which sustain biofilm integrity through destined therapeutic targets (Rieusset et al., 2020).

8.13.2 Lipid and Nucleotide Metabolism in Biofilms

Biofilm formation and maintenance depend on lipid metabolism, which fulfills two critical functions: maintaining cellular membrane stability and developing extracellular matrix (ECM). Biofilm-forming bacteria exhibit noticeable alterations to their lipid metabolism process that creates metabolites controlling the biofilm defenses against stress exposures and antibiotic effects (Zhang et al., 2023). Treating lipid metabolic pathways through key enzyme inhibition demonstrates great promise in biofilm maintenance since it leads to biofilm deterioration. Biofilm matrix vulnerability increases due to phospholipid biosynthesis inhibitors, which enhance the susceptibility of bacterial cells to existing antibiotics, according to Roy et al. (2023). Nucleotide metabolism maintains vital roles during biofilm growth because it enables both DNA formation and cellular energy flow along with regulatory sequences that support cell division and intercellular communication. The biofilm formation of particular bacteria leads to purine and pyrimidine biosynthesis pathway activation for creating the abundant extracellular DNA (eDNA) that constitutes a prominent element of the ECM. Research shows that disruption of nucleotide metabolism leads to degraded biofilm stability, thus making the nucleotide pathways attractive for therapeutic development (Vega-Dominguez et al., 2020). The mutually dependent biofilm process demonstrates the interconnectedness of these pathways yet establishes their importance as therapeutic targets.

8.13.3 Amino Acid and Carbohydrate Metabolism in Biofilm Formation

The organic compound from carbohydrate and amino acid metabolism used in biofilms comprises ECM. *Bacillus subtilis*, the known biofilm model, alters its production of amino acids during biofilm formation, specifically, amino acid synthesis pathways. For example, glutamate and glutamine are the most critical players in ECM synthesis; blocking their pathways which significantly decreases biofilm biomass and the biofilm matrix's mechanical stability (Rieusset et al., 2020). Amino acids also function as both precursors of the molecules and proteins that regulate biofilm maturation, such that those metabolic pathways are the most attractive ones for therapy. The biosynthesis of the polysaccharides, the components of the ECM, is a process aided by carbohydrate metabolism. The upregulation of glycolysis and the pentose phosphate pathway together provide the exact precursors needed for the production of polysaccharides, which are necessary for the precise 3D structure of the biofilm matrix (Vega-Dominguez et al., 2020). Using the inhibitors of carbohydrate metabolic enzymes that may be involved in glucan or cellulose, biosynthesis can interrupt the ECM-producing process and thus weaken the biofilms. The data confirm that the metabolic pathways are the base of the biofilm matrix and are potential drug targets.

8.13.4 Identification of Key Metabolic Bottlenecks

Biofilms present significant therapeutic challenges due to their inherent resistance to antibiotics and immune responses. This resistance is attributed to the protective extracellular matrix (ECM), metabolic heterogeneity, and dormant persister cells. Identifying metabolic bottlenecks within biofilms offers a promising avenue for overcoming these challenges. Recent metabolomics studies have highlighted critical vulnerabilities in biofilm metabolism, such as dependencies on specific amino acids, lipids, and energy production pathways (Rieusset et al., 2020). These bottlenecks are potential targets for anti-biofilm therapies, as their disruption can impair biofilm formation and stability.

For instance, glutamine and glutamate metabolisms are essential for biofilm stability in several bacterial species. Modulating these pathways has been found to significantly reduce biofilm biomass and compromise structural integrity, positioning them as key metabolic vulnerabilities (Hobson et al., 2022). Additionally, targeting pathways involved in synthesizing secondary metabolites, such as quorum sensing molecules and redox-active compounds, can disrupt biofilm communication and resilience. These insights underscore the power of metabolomics in identifying and characterizing metabolic targets for the development of effective anti-biofilm strategies.

8.14 Potential Drug Targets Based on Metabolomics Data

Integrating metabolomics with transcriptomics and proteomics has become instrumental in enabling scientists to refer to the metabolic reprogramming that permits biofilm formation, thereby providing a systems-level understanding of biochemical changes resulting in biofilm architecture. This comprehensive tool offers a glimpse at some metabolites governing biofilm formation and persistence among lipid, amino acid, and nucleotide metabolism, most of which are available in tandem with the help of systems biology (Vega-Dominguez et al., 2020). These metabolites may serve as markers for further identification of pathways capable of being therapeutically targeted to disrupt biofilms.

Metabolic inhibitors targeting specific pathways have shown promise in reducing biofilm formation. Glutamate metabolism involves a small molecule inhibitor detrimental to the extracellular matrix (ECM) and destabilizes the biofilm (Hobson et al., 2022). Similarly, newly developed compounds targeting enzymes involved in carbohydrate metabolism have been shown to decrease the structural integrity of the biofilm. Knowing these metabolic vulnerabilities, combination therapy could have been developed. By combining metabolic inhibitors with antibiotics, the synergistic action increases treatment effectiveness over standard antibiotic therapy.

8.15 Targeting Biofilm Specific Pathways

The metabolic pathways, which were changed during biofilm formation, are essential for the bacteria's survival and resistance to the treatment. Dealing with the paths connected to lipid, nucleotide, amino acid, and carbohydrate metabolism provides the most beneficial therapeutic outcomes. Metabolomics makes sense of these pathways and identifies the most critical areas of the metabolic process, and directs the development of antibiotics against biofilm.

There are ways to integrate metabolomics with other omics, e.g., transcriptomics and proteomics, to get deeper insights into the metabolic complexity of the biofilm. This omics approaches and understanding is to fight against biofilm-related infections in clinical and industrial settings. Integration of systems biology and targeted therapeutic strategies may significantly promote the management of biofilm related challenges.

8.16 Applications of AI/ML in Metabolomic Data Analysis

Biofilm inoculums mounted on EPI-HA-coated slides were analyzed by liquid chromatography-mass spectrometry (LC-MS) using both artificial intelligence (AI) and machine learning (ML). The technology of metabolomics encompasses an acceleration in biofilm formation assays. It has become an essential to have this technique to discern the intercellular steps of biofilm production and maintenance. Data-mining tools yielded several promising ideas predicting inhibitor effectiveness, but the latest studies have shown progress in developing a more predictive approach in biofilm metabolomics. Therefore, current methodologies towards a better understanding of metabolic pathways associated with biofilm formation could pave the way for the rapid exploration of promising anti-biofilm agents. Subsequently, a study elaboration is introduced by integrating AI for enhanced prediction of complex biofilm metabolomic dynamics regarding multilevel molecular targets. AI applications conveniently detect the key metabolic markers connected to biofilm formation. Artificial intelligence algorithms like deep learning, increase the efficiency of allowing metabolite identification and classification, thus giving global insights into the metabolic shifts happening throughout the growth of biofilms (Roy et al., 2023). AI algorithms have another unique feature that distinguishes them from statistics: the capability to trace complex patterns and relationships in large datasets generated through metabolomics studies. In the case of biofilms, there are substantially interconnected metabolic pathways and significant heterogeneity. These AI methods can aid in isolating the most critical metabolic indicators.

The fascinating domain of biofilm biology has benefited substantially from integrating metabolomics with other omics technologies like genomics, transcriptomics, and proteomics. Such an approach is well-suited for obtaining a detailed look at the molecular mechanism of biofilm formation, maintenance, and resistance. AI tools help create a synthesis by generating numerous intricate connections between metabolic pathways, transcriptome participation, and protein activity. For example, the neural network-based MiMeNet interacts with the microbiome and metabolome to permit the identification of gene-metabolite associations, key to biofilm growth, as related by Huang et al. (2024). Thus, integrative approaches provide a holistic understanding of biofilm dynamics as they clarify the interactions between metabolic pathways and proteomic and genetic layers. Integrating AI applications with multiple omics techniques enhances the ability to properly understand the metabolic pathways leading to biofilm resilience and the novel drug targets for their challenging treatment. For instance, metabolomics profiles lead to identifying energy metabolism and biosynthesis profiles disrupted during biofilm formation, while transcriptional data unravel corresponding genetic changes. Proteomic analyses tend to provide the above findings by highlighting proteins' possible functioning during matrix biogenesis and stress coping (Gao et al., 2022).

8.17 Challenges and Future Perspectives in Biofilm Metabolomics

Although it has the potential to enhance our knowledge of microbial communities and biofilm-associated processes greatly, biofilm metabolomics is confronted with several technical issues that need to be overcome to exploit its potential fully. One of the key challenges is the intricate nature of the biofilm matrix, which can present obstacles to the efficient extraction of metabolites. Biofilms comprise a compact extracellular polymeric substance (EPS) surrounding microbial cells, forming barriers to metabolite recovery. This intricacy can result in variability in metabolite profiles based on the extraction method employed, as noted by Zarnowski et al. (2021), who underscored the importance of optimized approaches to reach the biofilm exometabolome, the total of the metabolic output of the microbial community (Seneviratne et al., 2020).

Efficient metabolomic extraction protocols that penetrate complex matrices and procedure representative metabolite profiles are crucial for precise metabolomic analysis. Moreover, the sheer diversity of metabolites synthesized by various microbial species in a biofilm makes it challenging to interpret data. Intersecting metabolic pathways and complex interspecies interactions present serious challenges in identifying key metabolites that play pivotal roles in the dynamics of biofilms. Advanced computational tools are required to interpret these complex datasets and metabolite-microbial species relationships. Combining metabolomics with other omics technologies, including transcriptomics, proteomics, and genomics, further exacerbates these challenges. Although multiomics techniques offer a complete perspective of biofilm biology, they demand sophisticated data integration strategies to correlate findings across different biological layers. Creusot et al. (2022) emphasized network-based integration's role in revealing predictive subnetworks that elucidate biofilm behavior (Chen et al., 2023).

Although data dimensionality, scale, and quality variations can cause inconsistencies, rendering seamless integration an ongoing challenge. Moreover, preanalytical variables like sample handling, collection times, and metabolite stability during processing, can introduce variability affecting metabolomic analysis's reliability. Standardization of sample collection and processing protocols is critical to reduce variability and improve the reproducibility of metabolomic studies. Resolution of these technical issues is key to develop biofilm metabolomics and realizing its potential to reveal the complex dynamics of biofilm formation and inhibition. Optimization of metabolite extraction protocols, the development of robust computational tools for data interpretation, and the standardization of pre-analytical workflows will allow researchers to produce more accurate and reproducible results. Overcoming these obstacles will enable biofilm metabolomics to provide deeper insights into microbial community interactions, identify new therapeutic targets, and design effective anti-biofilm strategies.

8.18 Prospects for Personalized Anti-biofilm Treatments Based on Metabolic Profiles

The potential for personalized anti-biofilm treatments based on metabolic profiles is particularly promising. By analyzing metabolomic data from biofilm samples, researchers can identify specific metabolites or metabolic pathways over represented in pathogenic biofilms, paving the way for tailored therapeutic interventions. For example, Gaubert-Boussarie et al. (2020) highlighted differences in the metabolomic profiles of *Helicobacter pylori* strains with varying capacities for biofilm formation, demonstrating that these profiles could inform the development of customized treatments (Gao et al., 2022). Additionally, identifying metabolic markers associated with biofilm production offers the potential for diagnostic tools that predict patient outcomes linked to biofilm-related infections. Integrating advanced metabolomic technologies into biofilm research, such as real-time metabolomics and single-cell metabolomics, will further enhance our ability to monitor metabolic changes dynamically and at high resolution. This capability could revolutionize our understanding of biofilm biology and support the development of individualized anti-biofilm therapies, ultimately improving therapeutic outcomes. As metabolomics research progresses, its application in clinical and industrial settings is likely to yield innovative strategies for addressing biofilm-associated challenges, such as the development of specific inhibitors targeting metabolic pathways that are crucial for biofilm formation and maintenance (Huang et al., 2024; Zarnowski et al., 2021).

8.19 Conclusion

Using metabolomics in biofilm research has considerably expanded our knowledge on the metabolic processes that promote biofilm growth and the likely pathways that may be targeted to inhibit them. Biofilm growth is a dynamic process controlled by multiple metabolic and environmental factors. The recent evolution of metabolomics has given great insight into the metabolic changes involved during biofilm development and has suggested ways of disturbing these communities. Understanding these changes is crucial in creating effective treatment strategies for biofilm-based infections. Metabolomics, as an excellent tool to study metabolites in biological systems, allows researchers to investigate both endometabolome (intracellular metabolites) and exometabolome (extracellular metabolites) of biofilms. This technique has shown how environmental conditions and microbial interactions impact biofilm metabolic responses. For example, Creusot et al. (2022) demonstrated the relevance of examining the metabolomic response of particular biofilm components to uncover the metabolic exchange influencing biofilm diversity and resilience (Creusot et al., 2022).

In addition, noted significant metabolic changes as bacteria move from planktonic to biofilm lifestyles, such as the upregulation of pathways for extracellular polymeric substance (EPS) production, which confers structural integrity and environmental robustness to biofilms. These observations point out the importance of metabolism in biofilm formation. The application of metabolomics has also revealed potential targets for biofilm inhibition. For instance, Nishanth and Prasanna (2022) illustrated that certain strains have distinct metabolomic profiles associated with biofilm-forming ability, highlighting the significance of specific metabolic pathways in biofilm formation (Nishanth et al., 2022). In addition,Cheng Feng et al. (2021) revealed that manipulating nucleotide metabolism with terpinen-4-ol treatment significantly decreased biofilm signaling and blocked methicillin-resistant *Staphylococcus aureus* (MRSA) biofilm formation. Such research illustrates the promise of targeted metabolic intervention in biofilm regulation. Coupling integrative metabolomics with other omics tools, including transcriptomics and proteomics, can further advance our knowledge of biofilm biology. Combining these technologies enables researchers to investigate the interplay between gene expression, protein function, and metabolic pathways, gaining new insights into biofilm dynamics and identifying novel therapeutic targets. For example, Xu et al. (2023) illustrated the utility of integrative omics in determining antimicrobial peptide-sensitive metabolic pathways, providing new strategies for biofilm dispersion and inhibition (Xu et al., 2023).

In summary, the integration of metabolomics in biofilm studies has revolutionized our understanding of the metabolic basis of biofilm formation and inhibition. By dissecting the biochemical pathways that maintain biofilms, metabolomics offers a basis for developing targeted interventions to inhibit biofilm formation and increase treatment effectiveness. Future research must continue to explore the complex interplay between microbial metabolism and biofilm dynamics, using integrative omics tools to identify novel solutions for managing biofilm-related infections. As the field advances, metabolomics will be at the forefront of biofilm studies, driving the discovery of novel inhibitors and our capacity to overcome biofilm-related issues in clinical and industrial applications.

References

Andriana, Y., Widodo, A. D. W., & Arfijanto, M. V. (2024). Synergistic Interactions between Pseudomonas aeruginosa and Candida albicans, Candida glabrata, Candida krusei, Candida parapsilosis as well as Candida tropicalis in the Formation of Polymicrobial Biofilms. *Journal of Pure and Applied Microbiology, 18*(1), 219–228. https://doi.org/10.22207/jpam.18.1.05

Awadh, A. A., Le Gresley, A., Forster-Wilkins, G., Kelly, A. F., & Fielder, M. D. (2021). Determination of metabolic activity in planktonic and biofilm cells of Mycoplasma fermentans and Mycoplasma pneumoniae by nuclear magnetic resonance. *Scientific Reports, 11*(1), 5650. https://doi.org/10.1038/s41598-021-84326-2

Awaness, A., Elkeeb, R., Afshari, S., & Atef, E. (2024). The pharmacokinetic changes in cystic fibrosis patients population: Narrative review. *Medicines (Basel, Switzerland), 12*(1). https://doi.org/10.7287/peerj.9020v0.3/reviews/1

Brogden, K. A. (2005). Antimicrobial peptides: pore formers or metabolic inhibitors in bacteria? *Nature Reviews. Microbiology, 3*(3), 238–250. https://doi.org/10.1038/nrmicro1098

Bruder-Nascimento, A., Camargo, C. H., Mondelli, A. L., Sugizaki, M. F., Sadatsune, T., & Bagagli, E. (2014). Candida species biofilm and Candida albicans ALS3 polymorphisms in clinical isolates. *Brazilian Journal of Microbiology, 45*(4), 1371–1377. https://doi.org/10.1590/s1517-83822014000400030

Bryce, G., Pratten, J., Ready, D., Ng, Y.-L., & Gulabivala, K. (2020). Disruption and bactericidal indices depicted in polygonal graphs to show multiple outcome effects of root canal irrigant supplements on single- and dual-species biofilms. *Clinical Oral Investigations, 24*(9), 3255–3264. https://doi.org/10.1007/s00784-020-03202-x

Chen, C., Wang, J., Pan, D., Wang, X., Xu, Y., Yan, J., et al. (2023). Applications of multi-omics analysis in human diseases. *MedComm, 4*(4), e315. https://doi.org/10.1002/mco2.315

Cheng, F., Mo, Y., Chen, K., et al. (2021). Integration of metabolomics and transcriptomics indicates changes in MRSA exposed to terpinen-4-ol. *BMC Microbiol, 21*, 305. https://doi.org/10.1186/s12866-021-02348-2

Costerton, J. W., Stewart, P. S., & Greenberg, E. P. (1999). Bacterial biofilms: a common cause of persistent infections. *Science (New York, N.Y.), 284*(5418), 1318–1322. https://doi.org/10.1126/science.284.5418.1318

Creusot, N., Chaumet, B., Eon, M., Mazzella, N., Moreira, A., & Morin, S. (2022). Metabolomics insight into the influence of environmental factors in responses of freshwater biofilms to the model herbicide diuron. *Environmental Science and Pollution Research International, 29*(20), 29332–29347. https://doi.org/10.1007/s11356-021-17072-7

de Cremer, K., de Brucker, K., Staes, I., Peeters, A., van den Driessche, F., Coenye, T., et al. (2016). Stimulation of superoxide production increases fungicidal action of miconazole against Candida albicans biofilms. *Scientific Reports, 6*, 27463. https://doi.org/10.1038/srep27463

Deorukhkar, S. C., Saini, S., & Mathew, S. (2014). Non-albicans Candida Infection: An Emerging Threat. *Interdisciplinary Perspectives on Infectious Diseases, 2014*, 615958. https://doi.org/10.1155/2014/615958

Donlan, R. M. (2001). Biofilms and device-associated infections. *Emerging Infectious Diseases, 7*(2), 277–281. https://doi.org/10.3201/eid0702.010226

Erickson, B. K., Mueller, R. S., VerBerkmoes, N. C., Shah, M., Singer, S. W., Thelen, M. P., et al. (2010). Computational prediction and experimental validation of signal peptide cleavages in the extracellular proteome of a natural microbial community. *Journal of Proteome Research, 9*(5), 2148–2159. https://doi.org/10.1021/pr900877a

Gao, K., Su, B., Dai, J., Li, P., Wang, R., & Yang, X. (2022). Anti-biofilm and anti-hemolysis activities of 10-Hydroxy-2-decenoic acid against staphylococcus aureus. *Molecules (Basel, Switzerland), 27*(5). https://doi.org/10.3390/molecules27051485

Gaubert-Boussarie, J., Prado, S., & Hubas, C. (2020). An untargeted metabolomic approach for microphytobenthic biofilms in intertidal mudflats. *Frontiers in Marine Science, 7*, 2641. https://doi.org/10.3389/fmars.2020.00250

González-Domínguez, R., Mateos, R. M., Lechuga-Sancho, A. M., González-Cortés, J. J., Corrales-Cuevas, M., Rojas-Cots, J. A., et al. (2017). Synergic effects of sugar and caffeine on insulin-mediated metabolomic alterations after an acute consumption of soft drinks. *Electrophoresis, 38*(18), 2313–2322. https://doi.org/10.1002/elps.201700044

Goytia, M., Dhulipala, V. L., & Shafer, W. M. (2013). Spermine impairs biofilm formation by Neisseria gonorrhoeae. *FEMS Microbiology Letters, 343*(1), 64–69. https://doi.org/10.1111/1574-6968.12130

He, X., Lu, F., Yuan, F., Jiang, D., Zhao, P., Zhu, J., et al. (2015). Biofilm Formation Caused by Clinical Acinetobacter baumannii Isolates Is Associated with Overexpression of the AdeFGH Efflux Pump. *Antimicrobial Agents and Chemotherapy, 59*(8), 4817–4825. https://doi.org/10.1128/aac.00877-15

Hobson, K. A., Kuwae, T., Drever, M. C., Easton, W. E., & Elner, R. W. (2022). Biofilm and invertebrate consumption by western sandpipers (Calidris mauri) and dunlin (Calidris alpina) during spring migratory stopover: insights from tissue and breath CO2 isotopic (δ13C, δ15N) analyses. *Conservation Physiology, 10*(1), coac006. https://doi.org/10.1093/conphys/coac006

Huang, M.-B., Brena, D., Wu, J. Y., Shelton, M., & Bond, V. C. (2024). SMR peptide antagonizes Staphylococcus aureus biofilm formation. *Microbiology Spectrum, 12*(2), e0258323. https://doi.org/10.1128/spectrum.02583-23

Iman, E. W., Rasha, H. B., Eman, A. E. S., Amira, E., Rahman, A., Razek, A. A., Amina, A. S., & Reem, F. (2013). Resistant virulent Candida species colonizing preterm neonates and in vitro promising prospect of chlorhexidine gluconate. *African Journal of Microbiology Research, 7*(26), 3421–3428. https://doi.org/10.5897/ajmr2013.5622

Jastaniah, S. D., Jamal, T. Y., Amashah, R. H., & Aly, M. M. (2022). New Strategies for Inhibition of Listeria monocytogenes and Klebsiella pneumoniae Biofilm Formation and Persistence. *Journal of Contemporary Medical Sciences, 8*(6). https://doi.org/10.22317/jcms.v8i6.1302

Kang, D., Liu, W., Kakahi, F. B., & Delvigne, F. (2022). Combined utilization of metabolic inhibitors to prevent synergistic multi-species biofilm formation. *AMB Express, 12*(1), 32. https://doi.org/10.1186/s13568-022-01363-4

Karygianni, L., Ren, Z., Koo, H., & Thurnheer, T. (2020). Biofilm matrixome: Extracellular components in structured microbial communities. *Trends in Microbiology, 28*(8), 668–681. https://doi.org/10.1016/j.tim.2020.03.016

Kernan, C., Chow, P. P., Christianson, R. J., & Huang, J. (2015). Experimental and computational investigation of biofilm formation by rhodopseudomonas palustris growth under two metabolic modes. *PLoS One, 10*(6), e0129354. https://doi.org/10.1371/journal.pone.0129354

Kriswandini, I. L., Sidarningsih, S., Hermanto, A. C., Tyas, P. R., & Aljunaid, M. A. (2024). The Influence of Streptococcus mutans biofilm formation in a polymicrobial environment (Streptococcus gordonii & Porphyromonas gingivalis). *European Journal of Dentistry, 18*(4), 1085–1089. https://doi.org/10.1055/s-0044-1782215

Lee, S. W., Gu, H., Kilberg, J. B., & Ren, D. (2018). Sensitizing bacterial cells to antibiotics by shape recovery triggered biofilm dispersion. *Acta Biomaterialia, 81*, 93–102. https://doi.org/10.1016/j.actbio.2018.09.042

Liang, R.-m., Cao, Y.-b., Zhou, Y.-j., Xu, Y., Gao, P.-h., Dai, B.-d., et al. (2010). Transcriptional response of Candida albicans biofilms following exposure to 2-amino-nonyl-6-methoxyl--tetralin muriate. *Acta Pharmacologica Sinica, 31*(5), 616–628. https://doi.org/10.1038/aps.2010.33

Liu, J., Prindle, A., Humphries, J., Gabalda-Sagarra, M., Asally, M., Lee, D.-y. D., et al. (2015). Metabolic co-dependence gives rise to collective oscillations within biofilms. *Nature, 523*(7562), 550–554. https://doi.org/10.1038/nature14660

Liu, S., Lu, H., Zhang, S., Shi, Y., & Chen, Q. (2022). Phages against pathogenic bacterial biofilms and biofilm-based infections: A review. *Pharmaceutics, 14*(2). https://doi.org/10.3390/pharmaceutics14020427

López, D., Vlamakis, H., & Kolter, R. (2010). Biofilms. *Cold Spring Harbor Perspectives in Biology, 2*(7), a000398. https://doi.org/10.1101/cshperspect.a000398

Lorin, D., Teusdea, V., Mitranescu, E., Muselin, F., Dumitrescu, E., Stancu, A., et al. (2017). The hazardous activity of yeasts embedded in biofilm and planktonic estimated through the effectiveness of four commonly used biocidal conditionings. *Materiale Plastice, 54*(2), 239–243. https://doi.org/10.37358/mp.17.2.4824

Lu, H., Que, Y., Wu, X., Guan, T., & Guo, H. (2019). Metabolomics Deciphered Metabolic Reprogramming Required for Biofilm Formation. *Scientific Reports, 9*(1), 13160. https://doi.org/10.1038/s41598-019-49603-1

Marta Setiabudy, Dewa Ayu Putri Sri Masyeni, Anak Agung Gede Indraningrat, Kadek Suryawan, I Ketut Agus Indra Adhiputra, & Muhammad Amirul bin Abdul Rahman. (2023). Biofilm formation in Staphylococcus aureus and coagulase-negative staphylococcus. *olia Medica Indonesiana, 59*(3), 222–228. https://doi.org/10.20473/fmi.v59i3.44598

Mohandas, V., & Ballal, M. (2011). Distribution of Candida species in different clinical samples and their virulence: Biofilm formation, proteinase and phospholipase production: A study on hospitalized patients in southern India. *Journal of Global Infectious Diseases, 3*(1), 4–8. https://doi.org/10.4103/0974-777x.77288

Muhammad, M. H., Idris, A. L., Fan, X., Guo, Y., Yu, Y., Jin, X., et al. (2020). Beyond risk: Bacterial biofilms and their regulating approaches. *Frontiers in Microbiology, 11*, 928. https://doi.org/10.3389/fmicb.2020.00928

Nerurkar, A., Solanky, P., Chavda, N., Baria, H., & Desai, B. (2012). Isolation of Candida Species in clinical specimens and its virulence factor. The biofilm. *International Journal of Medical Science and Public Health, 1*(2), 97. https://doi.org/10.5455/ijmsph.2012.1.97-100

Nett, J. E., Sanchez, H., Cain, M. T., & Andes, D. R. (2010). Genetic basis of Candida biofilm resistance due to drug-sequestering matrix glucan. *The Journal of Infectious Diseases, 202*(1), 171–175. https://doi.org/10.1086/651200

Nishanth, S., & Prasanna, R. (2022). Untargeted GC-MS reveals differential regulation of metabolic pathways in cyanobacterium Anabaena and its biofilms with Trichoderma viride and Providencia sp. *Current Research in Microbial Sciences, 3*, 100174. https://doi.org/10.1016/j.crmicr.2022.100174

Nobile, C. J., Fox, E. P., Nett, J. E., Sorrells, T. R., Mitrovich, Q. M., Hernday, A. D., et al. (2012). A recently evolved transcriptional network controls biofilm development in Candida albicans. *Cell, 148*(1–2), 126–138. https://doi.org/10.1016/j.cell.2011.10.048

Nothias, L.-F., Nothias-Esposito, M., da Silva, R., Wang, M., Protsyuk, I., Zhang, Z., et al. (2018). Bioactivity-based molecular networking for the discovery of drug leads in natural product bioassay-guided fractionation. *Journal of Natural Products, 81*(4), 758–767. https://doi.org/10.1021/acs.jnatprod.7b00737

Ogawa, A., Takakura, K., Sano, K., Kanematsu, H., Yamano, T., Saishin, T., & Terada, S. (2018). Microbiome analysis of biofilms of silver nanoparticle-dispersed silane-based coated carbon steel using a next-generation sequencing technique. *Antibiotics (Basel, Switzerland), 7*(4). https://doi.org/10.3390/antibiotics7040091

Paino, A., Tuominen, H., Jääskeläinen, M., Alanko, J., Nuutila, J., Asikainen, S. E., et al. (2011). Trimeric form of intracellular ATP synthase subunit β of Aggregatibacter actinomycetemcomitans binds human interleukin-1β. *PLoS One, 6*(4), e18929. https://doi.org/10.1371/journal.pone.0018929

Papaianni, M., Cuomo, P., Fulgione, A., Albanese, D., Gallo, M., Paris, D., et al. (2020). Bacteriophages promote metabolic changes in bacteria biofilm. *Microorganisms, 8*(4). https://doi.org/10.3390/microorganisms8040480

Parga, A., Muras, A., Otero-Casal, P., Arredondo, A., Soler-Ollé, A., Àlvarez, G., et al. (2023). The quorum quenching enzyme Aii20J modifies in vitro periodontal biofilm formation. *Frontiers in Cellular and Infection Microbiology, 13*, 1118630. https://doi.org/10.3389/fcimb.2023.1118630

Pisithkul, T., Schroeder, J. W., Trujillo, E. A., Yeesin, P., Stevenson, D. M., Chaiamarit, T., et al. (2019). Metabolic remodeling during biofilm development of Bacillus subtilis. *MBio, 10*, 3. https://doi.org/10.1128/mbio.00623-19

Ramos, I., Dietrich, L. E. P., Price-Whelan, A., & Newman, D. K. (2010). Phenazines affect biofilm formation by Pseudomonas aeruginosa in similar ways at various scales. *Research in Microbiology, 161*(3), 187–191. https://doi.org/10.1016/j.resmic.2010.01.003

Rieusset, L., Rey, M., Muller, D., Vacheron, J., Gerin, F., Dubost, A., et al. (2020). Secondary metabolites from plant-associated Pseudomonas are overproduced in biofilm. *Microbial Biotechnology, 13*(5), 1562–1580. https://doi.org/10.1111/1751-7915.13598

Rockstroh, J., Gerschler, S., Schultze, N., Schulze, C., Wurster, M., Methling, K., et al. (2024). A targeted approach for the metabolome analysis of E. coli biofilms. *Helvetica Chimica Acta, 107*(6), 4. https://doi.org/10.1002/hlca.202300240

Roy, S., Acosta, J. A. M., Karak, M., Ramirez-Velez, I., Torikai, K., Ren, D., & Barbosa, L. C. A. (2023). Effects of synthetic tetronamides and methylated denigrins on bacterial quorum sensing and biofilm formation. *ACS Omega, 8*(41), 37798–37807. https://doi.org/10.1021/acsomega.3c01729

Santiago, A. J., Ahmed, M. N. A., Wang, S.-L., Damera, K., Wang, B., Tai, P. C., et al. (2016). Inhibition and dispersal of Pseudomonas aeruginosa biofilms by combination treatment with Escapin intermediate products and hydrogen peroxide. *Antimicrobial Agents and Chemotherapy, 60*(9), 5554–5562. https://doi.org/10.1128/aac.02984-15

Sasaki-Imamura, T., Yano, A., & Yoshida, Y. (2010). Production of indole from L-tryptophan and effects of these compounds on biofilm formation by Fusobacterium nucleatum ATCC 25586. *Applied and Environmental Microbiology, 76*(13), 4260–4268. https://doi.org/10.1128/aem.00166-10

Seneviratne, C. J., Suriyanarayanan, T., Widyarman, A. S., Lee, L. S., Lau, M., Ching, J., et al. (2020). Multi-omics tools for studying microbial biofilms: Current perspectives and future directions. *Critical Reviews in Microbiology, 46*(6), 759–778. https://doi.org/10.1080/1040841x.2020.1828817

Sida, H., Shah, P., Pethani, J., Patel, L., & Shah, H. (2016). Study of biofilm formation as a virulence marker in Candida species isolated from various clinical specimens. *International Journal of Medical Science and Public Health, 5*(5), 842. https://doi.org/10.5455/ijmsph.2016.24082015139

Solano, C., Echeverz, M., & Lasa, I. (2014). Biofilm dispersion and quorum sensing. *Current Opinion in Microbiology, 18*, 96–104. https://doi.org/10.1016/j.mib.2014.02.008

Somasundar, A., Qin, B., Shim, S., Bassler, B. L., & Stone, H. A. (2023). Diffusiophoretic particle penetration into bacterial biofilms. *ACS Applied Materials & Interfaces, 15*(28), 33263–33272. https://doi.org/10.1021/acsami.3c03190

Stipetic, L. H., Dalby, M. J., Davies, R. L., Morton, F. R., Ramage, G., & Burgess, K. E. V. (2016). A novel metabolomic approach used for the comparison of Staphylococcus aureus planktonic cells and biofilm samples. *Metabolomics, 12*, 75. https://doi.org/10.1007/s11306-016-1002-0

Tiwari, G., Mishra, B., Dogra, V., Sood Loomba, P., & Arora, D. R. (2019). Detection of biofilm production and its quantification in Candida Isolates in a Tertiary Care Hospital. *Saudi Journal of Pathology and Microbiology, 4*(10), 755–760. https://doi.org/10.36348/sjpm.2019.v04i10.011

Vediyappan, G., Rossignol, T., & d'Enfert, C. (2010). Interaction of Candida albicans biofilms with antifungals: Transcriptional response and binding of antifungals to beta-glucans. *Antimicrobial Agents and Chemotherapy, 54*(5), 2096–2111. https://doi.org/10.1128/aac.01638-09

Vega-Dominguez, P., Peterson, E., Pan, M., Di Maio, A., Singh, S., Umapathy, S., et al. (2020). Biofilms of the non-tuberculous Mycobacterium chelonae form an extracellular matrix and display distinct expression patterns. *Cell Surface (Amsterdam, Netherlands), 6*, 100043. https://doi.org/10.1016/j.tcsw.2020.100043

Vyas, N., Sammons, R. L., Addison, O., Dehghani, H., & Walmsley, A. D. (2016). A quantitative method to measure biofilm removal efficiency from complex biomaterial surfaces using SEM and image analysis. *Scientific Reports, 6*, 32694. https://doi.org/10.1038/srep32694

Wang, Y., Zhang, R., Duan, J., Shi, X., Zhang, Y., Guan, F., et al. (2022). Extracellular polymeric substances and biocorrosion/biofouling: recent advances and future perspectives. *International Journal of Molecular Sciences, 23*(10). https://doi.org/10.3390/ijms23105566

Wolber, J. M., Urbanek, B. L., Meints, L. M., Piligian, B. F., Lopez-Casillas, I. C., Zochowski, K. M., et al. (2017). The trehalose-specific transporter LpqY-SugABC is required for antimicrobial and anti-biofilm activity of trehalose analogues in Mycobacterium smegmatis. *Carbohydrate Research, 450*, 60–66. https://doi.org/10.1016/j.carres.2017.08.003

Wong, H. S., Maker, G. L., Trengove, R. D., & O'Handley, R. M. (2015). Gas chromatography-mass spectrometry-based metabolite profiling of Salmonella enterica serovar Typhimurium differentiates between biofilm and planktonic phenotypes. *Applied and Environmental Microbiology, 81*(8), 2660–2666. https://doi.org/10.1128/aem.03658-14

Wong, E. H. J., Ng, C. G., Goh, K. L., Vadivelu, J., Ho, B., & Loke, M. F. (2018). Metabolomic analysis of low and high biofilm-forming Helicobacter pylori strains. *Scientific Reports, 8*(1), 1409. https://doi.org/10.1038/s41598-018-19697-0

Wu, W., Jing, J., Chen, Z., Guo, X., & Qu, Y. (2024). Indole signaling enhances biofilm formation and quorum sensing in sequencing biofilm batch reactors. *Journal of Environmental Chemical Engineering, 12*(3), 112494. https://doi.org/10.1016/j.jece.2024.112494

Xu, C.-G., Yang, Y.-B., Zhou, Y.-H., Hao, M.-Q., Ren, Y.-Z., Wang, X.-T., et al. (2017). Comparative proteomic analysis provides insight into the key proteins as possible targets involved in aspirin inhibiting biofilm formation of Staphylococcus xylosus. *Frontiers in Pharmacology, 8*, 543. https://doi.org/10.3389/fphar.2017.00543

Xu, T., Xiao, Y., Wang, H., Zhu, J., Lu, W., & Chen, W. (2023). Multiomics reveals the mechanism of B. longum in promoting the formation of mixed-species biofilms. *Food & Function, 14*(18), 8276–8290. https://doi.org/10.1039/d3fo01751f

Yin, W., Wang, Y., Liu, L., & He, J. (2019). Biofilms: The microbial "protective clothing" in extreme environments. *International Journal of Molecular Sciences, 20*(14). https://doi.org/10.3390/ijms20143423

Ząbek, A., Junka, A., Szymczyk, P., Wojtowicz, W., Klimek-Ochab, M., & Młynarz, P. (2017). Metabolomics analysis of fungal biofilm development and of arachidonic acid-based quorum sensing mechanism. *Journal of Basic Microbiology, 57*(5), 428–439. https://doi.org/10.1002/jobm.201600636

Zarnowski, R., Noll, A., Chevrette, M. G., Sanchez, H., Jones, R., Anhalt, H., et al. (2021). Coordination of fungal biofilm development by extracellular vesicle cargo. *Nature Communications, 12*(1), 6235. https://doi.org/10.1038/s41467-021-26525-z

Zhang, J.-M., Liu, J., Wang, K., Zhang, X., Zhao, T., & Luo, H.-M. (2018). Observations of bacterial biofilm on ureteral stent and studies on the distribution of pathogenic bacteria and drug resistance. *Urologia Internationalis, 101*(3), 320–326. https://doi.org/10.1159/000490621

Zhang, Q., Li, J., Lu, W., Zhao, J., Zhang, H., & Chen, W. (2021). Multi-Omics Reveals the Inhibition of Lactiplantibacillus plantarum CCFM8724 in Streptococcus mutans-Candida albicans Mixed-Species Biofilms. *Microorganisms, 9*(11). https://doi.org/10.3390/microorganisms9112368

Zhang, Y., Yu, H., Xie, Y., Guo, Y., Cheng, Y., & Yao, W. (2022). Quorum sensing inhibitory effect of hexanal on Autoinducer-2 (AI-2) and corresponding impacts on biofilm formation and enzyme activity in Erwinia carotovora and Pseudomonas fluorescens isolated from vegetables. *Food Processing Preservation, 46*(2). https://doi.org/10.1111/jfpp.16293

Zhang, Y., Cai, Y., Wang, T., & Wu, Q. (2023). A common tolerance mechanism of bacterial biofilms to antibiotics. *bioRxiv*. https://doi.org/10.1101/2023.01.30.526163

Zhao, X., Zhao, F., & Zhong, N. (2018). Quorum sensing inhibition and anti-biofilm activity of traditional Chinese medicines. In Y. El-Samragy (Ed.), *Food safety - Some global trends*. InTech.

Zhao, D., Li, H., Cui, Y., Tang, S., Wang, C., Du, B., & Ding, Y. (2022). MsmR1, a global transcription factor, regulates polymyxin synthesis and carbohydrate metabolism in Paenibacillus polymyxa SC2. *Frontiers in Microbiology, 13*, 1039806. https://doi.org/10.3389/fmicb.2022.1039806

Zheng, H., Kim, J., Liew, M., Yan, J. K., Herrera, O., Bok, J. W., et al. (2015). Redox metabolites signal polymicrobial biofilm development via the NapA oxidative stress cascade in Aspergillus. *Current Biology: CB, 25*(1), 29–37. https://doi.org/10.1016/j.cub.2014.11.018

Zhu, Y., Weiss, E. C., Otto, M., Fey, P. D., Smeltzer, M. S., & Somerville, G. A. (2007). Staphylococcus aureus biofilm metabolism and the influence of arginine on polysaccharide intercellular adhesin synthesis, biofilm formation, and pathogenesis. *Infection and Immunity, 75*(9), 4219–4226. https://doi.org/10.1128/iai.00509-07

Metagenomics in Understanding the Microbial Biofilms and Therapeutic Intervention

9

V. T. Anju

9.1 Introduction

Microbial biofilms are highly organized colonies of fungi, bacteria, and algae that stick to surfaces and are encased in a self-produced matrix of extracellular polymeric substance (EPS) (Sharma et al., 2023). According to Flemming and Wingender (2010), the EPS gives structural protection to the cells against environmental stressors, such as desiccation, antibiotic therapy, and the immune system in the host. Biofilms can develop on both biotic and abiotic surfaces, such as industrial settings, medical equipment, and human tissues (Flemming & Wingender, 2010). Compared to free-floating microbial cells, biofilm distinct structure allows them to exhibit greater functioning and resilience (Sharma et al., 2023).

Initial reversible attachment, irreversible attachment, maturation, and dispersion are the key stages that involve the biofilm formation. Microbial cells stick to surfaces in the early development stage via hydrophobic or weak van der Waals forces. The cells then produce EPS, which causes irreversible attachment as they firmly adhere to the substrate (Zhao et al., 2023). Microbiological communities expand and create intricate three-dimensional structures with pathways that promote waste and nutrient exchange as the biofilm ages (Peschel & Otto, 2013). The cycle is complete when some cells disperse from the biofilm and start colonizing new surfaces. Biofilm development is greatly affected by environmental variables, such as temperature, pH, and nutrient availability (Lee & Yoon, 2017).

Biofilms possess significant clinical and environmental consequences. They also contribute to bioremediation and nutrient recycling in natural settings by decomposing contaminants (Muhammad et al., 2020). However, they cause persistent infections as well as antibiotic resistance patterns and pose a serious threat in therapeutic settings. For example, chronic infections caused by biofilms formed on medical

V. T. Anju (✉)
Department of Biology, St. Joseph's College (Autonomous), Irinjalakuda, Thrissur, Kerala, India

S. Busi et al. (eds.), *Omics Approaches in Biofilm Research*,
https://doi.org/10.1007/978-3-031-91863-6_9

equipment, such as catheters and prosthetic implants, are difficult to control. Understanding the biofilm morphology is significant to explore various beneficial roles and to develop different control measures in industries and healthcare (Sahoo & Meshram, 2024).

Metagenomics helps to study microbial populations from different environments without the need of culturing and maintaining in the laboratory. The advancements made in DNA sequencing technologies in the late twentieth century made it possible to examine the genetic material of different microbial communities present in one sample at the same time (Garza & Dutilh, 2015). Microbial ecology and diversity have been explored widely even though many of them are difficult to isolate and cultivate in a laboratory setting. Metagenomics offers a glimpse of the microbial community by sequencing environmental DNA (eDNA) and provides information about the ecological roles and functional possibilities of unculturable microbes (Sharma et al., 2005).

A significant benefit of using metagenomic approaches over conventional microbial analysis is its capacity to identify and examine the entire range of microbial life, including unculturable organisms. Conventional techniques depend on separating and cultivating individual organisms, which restricts the range of microbial diversity that can be investigated. On the other hand, metagenomics provides a more comprehensive image of microbial communities and their interactions by capturing both key and abundant species (Vieites et al., 2009). Additionally, metagenomics makes it possible to investigate the functional genes present in the communities, offering important insights on resistance mechanisms, metabolic pathways, and other genetic characteristics, which facilitate the general functioning of the microbial ecosystem (Warnecke et al., 2007; Dinsdale et al., 2008).

Metagenomics has transformed as an important tool in biofilm research, especially for deciphering the intricate microbial communities that grow into biofilms. Metagenomics allows researchers to thoroughly examine complex microbial interactions in biofilms. Metagenomic studies have provided insight into the biology of biofilm development, resistance to antimicrobial therapies, and biofilm-associated infections by examining the microbial diversity and functional genes within biofilms (Imchen et al., 2022). This method is also crucial for discovering novel therapeutic targets for the control of biofilm-related infections as it reveals the genetic pathways involved in biofilm resistance. Thus, metagenomics has unlocked the possibilities for novel treatments to prevent the formation of biofilms or to improve traditional strategies (Kneis et al., 2022).

9.2 Microbial Biofilms: Challenges and Significance

Microbial biofilms are composed of groups of microorganisms that adhere to surfaces via the matrix of EPS such as polysaccharides, proteins, lipids, and extracellular DNA. The EPS makes up most of the extracellular matrix, which gives the cell structural integrity and serves as a barrier that blocks external stimuli (Rather et al., 2021). Cell-to-cell interactions in the biofilm structure create an ecosystem that promotes microbial cooperation and survival in harsh conditions. Initial attachment, irreversible adhesion, maturation into complex three-dimensional structures, and final cell dispersion to colonize other surfaces are the separate stages involved in the creation of biofilms (Muhammad et al., 2020).

The genetic diversity of biofilms is one of the distinguishing characteristics of biofilms. According to Lee et al. (2022), biofilms are frequently made up of several microbial species that work together amicably and exchange genetic materials through horizontal gene transfer (HGT). Genetic diversity also contributes to biofilm resistance, which accounts to different functions, such as food absorption, EPS generation, and resistance to external challenges. Due to varying heterogeneity among biofilms, they are biologically intricate and difficult to control (Lee et al., 2022).

Chronic infections are mainly caused by antibiotic-resistant biofilms. Biofilms formed on living tissues and medical equipment (dental implants, prosthetic joints, and catheters) protect themselves from host immune responses and antibiotic treatments (Ali et al., 2023). Endocarditis, cystic fibrosis lung infections, and chronic wounds are among the many chronic and recurring conditions related to biofilms. The protective nature of EPS and the altered metabolic state of cells make them to be more difficult to eradicate as they are resistant to conventional antibiotics (Mitra, 2024).

Biofilm communities are characterized by their resistance to antimicrobial agents. Antibiotics and other antimicrobial agents cannot pass through the EPS matrix because it serves as a physical barrier. Cells inside the biofilm may display phenotypic variability, with some cells showing dormant state, reducing their susceptibility to antimicrobial therapies. Additionally, genetic mechanisms, such as efflux pumps and HGT, improve the resistance profiles by dispersing more resistance genes among biofilm populations. This comprehensive resistance profile is an important roadblock in healthcare, demanding alternative, and effective biofilm management methods (Shree et al., 2023).

Biofilms also have an impact on several industries where they can lead to biofouling, equipment corrosion, and water system contamination in addition to their adverse healthcare effects. Biofilms on surfaces and equipment in sectors like food processing can cause product spoiling and pose health risks. Biofouling caused by microbial biofilms can lower productivity and enhance the maintenance expenses in the oil and maritime industries. These widespread challenges associated with biofilms draw attention to the financial and ecological costs of biofilms (Shineh et al., 2023).

In conclusion, the structural complexity, genetic diversity, and resistance mechanisms of microbial biofilms constitute serious challenges. They have significant implications in industrial settings and cause chronic illnesses, which have an impact on both public health and the economy. Thus, it is important to understand the biofilm ecosystem to develop more efficient strategies to manage biofilm infections (Abdallah et al., 2014).

9.3 Metagenomics: Tools and Approaches

Metagenomics is one of the promising approaches to investigate the genetic material directly from multiple environmental samples, enabling an easy method to study microbial communities. This technique avoids the shortcomings of conventional culturing technique, which mostly depends on growing specific species in their nutrient media. Whole genome sequencing (WGS), 16S rRNA gene sequencing, and shotgun metagenomics are the three widely used techniques in metagenomic research to study microbial communities. These techniques are the essential tools in microbial ecology because they offer varying degrees of resolution and functional understanding of microbial communities.

9.3.1 Shotgun Metagenomics

In shotgun metagenomics, DNA in the sample is randomly fragmented, and the fragments are then sequenced and computationally assembled. This method provides both taxonomic and functional information by capturing whole genetic makeup of microbial community. It enables scientists to forecast metabolic pathways throughout the community, identify microbial species, evaluate genetic diversity and abundant species. Nevertheless, it requires a lot of computational resources and could be influenced by host DNA, particularly human or animal samples (Quince et al., 2017). Despite these difficulties, shotgun metagenomics is widely used in different domains, such as environmental microbiology and human microbiome studies.

9.3.2 16S rRNA Gene Sequencing

In 16S rRNA sequencing methods, the 16S ribosomal RNA gene, which is one of the highly conserved genetic markers in bacteria and archaea, is targeted. In this method, microorganisms are taxonomically classified particularly into the genus level by amplifying and sequencing specific genomic regions. It is an easy and affordable method of sequencing microbial samples of even larger sizes (Janda & Abbott, 2007). However, this technique fails to provide functional roles and explores species or different taxa (such as fungi or viruses). Despite these drawbacks, 16S rRNA sequencing remains one of the commonly employed methods of characterizing microbial communities in environmental and clinical samples.

9.3.3 Whole Genome Sequencing (WGS)

Whole genome sequencing (WGS) method is used to determine genetic information of single microbial species or dominant species present within a community. WGS provides a high-resolution genetic composition, which includes information about gene roles, virulence factors, and evolutionary relationships. WGS is frequently used to investigate genome sequencing of single or dominant species present in intricate microbial ecosystems (Bergholz et al., 2014). However, the process requires several resources and efficient isolation of organisms from the sample mixture, which can be difficult in a variety of microbial populations.

9.4 Insights from Metagenomics in Biofilm Research

Metagenomics revolutionized the field of biofilm research, which enabled scientists to explore the whole microbiome present in unusual environments. The varying functional capabilities of each species and species richness were uncovered by employing these techniques. Different applications of metagenomics are summarized in Fig. 9.1.

Skipper et al. (2022) studied the microbiome component of limestone present in damaged and undamaged buildings. In this study, Illumina MiSeq next-generation sequencing system was employed to analyze the genetic makeup of microbial samples present on limestone. The study provided interesting findings of varied microbiome analysis on both surfaces indicating a core microbiome being unique to both places. Additionally, few unique species were present on damaged surfaces, such as *Bacillus licheniformis*, which was thought to be linked with the biodeterioration of stone. It was observed that strong biofilm formation and EPS matrix formation occurred at damaged surfaces leading to the biomechanical weathering of the stone (Skipper et al., 2022).

Multiomics approach is always a significant method that reveals different taxa present in varied samples. A similar approach was adapted by s group of researchers to examine the taxonomic and metabolic diversity present on Roman stuccoes of a Roman archaeological site. Both bacterial and eukaryotic diversity that are involved in the biodeterioration of the decorative element was studied using untargeted metabolomics and metabarcoding approaches (molecular markers used, 16S and 18S for bacterial eukaryotic communities, respectively). In most of the samples,

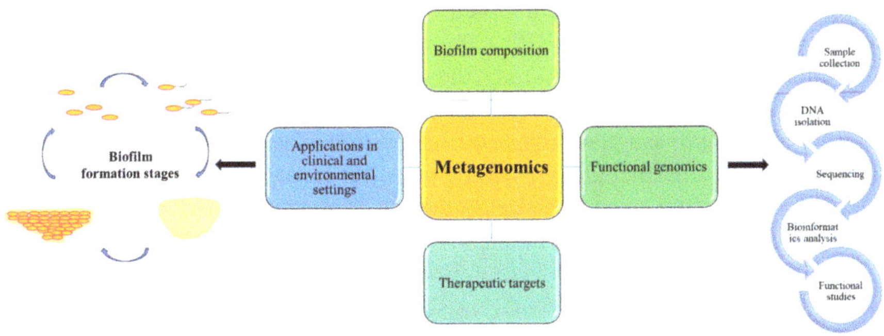

Fig. 9.1 Summarized illustration of metagenomic approaches in biofilm research including the study of biofilm composition to clinical applications

different bacterial taxa observed, and in few samples, eukaryotic organisms, such as fungi and chlorophytes, were noted. Around 95% of bacterial isolates were identified at the phylum and 59% at the family level in all samples. Approximately, 14 bacterial families were abundantly found in most of the samples; whereas, the *Euzebyaceae* family was notably abundant in all taxa. Most of the 162 metabolites observed in the study were attributed to different metabolic pathways, such as secondary metabolites, amino acids, protein, carbon, and microbial metabolism. They also conducted an anti-biofilm approach to remove the bacterial biofilms formed on tiles using essential oils (De Luca et al., 2024).

In another interesting study, metagenomic analysis of zinc oxide (ZnO) nanoparticle-contaminated soil provided information on biofilm genesis and transfer of antibiotic resistance elements. Biofilm formation was encountered in bacterial communities present in the environment exposed to ZnO nanoparticles. The study compared the effect of nano and bulk forms of ZnO on biofilm formation and transfer of multidrug resistance genes and mobile genetic elements. Both nano and bulk forms of nanoparticle exposure in soil enabled the formation of biofilms and resistance genes. They contributed to the production of EPS and bacterial cell attachment, formation of capsules, and biofilm dispersal mechanisms. Thus, the study suggested the proper disposal of nanoparticles to reduce the biofilm formation in natural ecosystems (Dinesh et al., 2023a). Similarly, another study from the same group of researchers observed a decrease in alpha microbial diversity with increasing nano zinc. A dose-dependent changes were observed in bacterial communities including a decrease of taxa Firmicutes, Actinobacteria, and *Chloroflexi* at higher bulk and nano ZnO. Also, increased stress mechanisms were observed under elevated nano zinc, such as component and secretion systems (Dinesh et al., 2023b).

Busi et al. (2022) investigated the biofilm ecology of microbial communities present in glacier-fed streams (GFS). The study discussed the adaptations of biofilm cells in harsh environments and their functional roles and interactions in biofilms. Biofilm samples from different habitats were analyzed using 16 and 18S rRNA sequencing methods. It was observed that a diverse group of communities, such as bacteria, photoautotrophs, and viruses, present in biofilm interactions. Different protective adaptations and energy-gaining methods, such as photosynthesis (via rhodopsins) and nutrient recycling, help biofilms to survive in harsh conditions, such as GFS (Busi et al., 2022).

Metagenomics approach to study the clinical settings is much significant to delineate the relationship between pathogens and host. A metagenomics study showed that sputum samples collected from cystic fibrosis sample contained many different pathogens, including *Pseudomonas aeruginosa*, *Rothia mucilaginosa*, *Streptococcus pneumoniae, Candida albicans,* and *Streptococcus pneumonia* single species biofilm (Ribeiro, 2017) (Table 9.1).

Oral microbiome analysis showed periodontal disease and associated pathogens. Deep sequencing (WGS) and 16S rDNA sequencing methods revealed microbiome analysis of healthy and patient samples. Around, 55 and 58 distinct genera were abundant in 0.1% or higher in the 16S rDNA and WGS methods, respectively. As per the Shannon diversity index, a more varied number of organisms was found in

Table 9.1 Metagenomic approaches to study different biofilm physiologies

Sl. No:	Techniques	Microbial diversity	Study details	References
1	Metagenome of activated sludge	*Mostly proteobacteria* *Nitrospira* *Rhodobacter capsulatus* *Nitrobacter*	Quorum sensing (QS) signaling systems of luxI/R and EPS production	Tan et al. (2021)
2	Metagenome of activated sludge and membrane bioreactor	*Proteobacteria* *Bacteroidetes* *Xanthomonadaceae*	Quorum sensing pathways, such as acyl-homoserine lactones (AHLs), AI-2, DSF, PQS, AIP, and cyclic diguanylate monophosphate (c-di-GMP)	Li et al. (2019)
3	Metagenome of Alaskan soil	*Gammaproteobacteria*	Novel LuxI/LuxR-type quorum sensing systems	Williamson et al. (2005)
4	Metagenome of gypsy moth gut	α, β, γ proteobacteria	Novel LuxI/LuxR-type quorum sensing systems	Hao et al. (2010)
5	Metagenome of activated sludge	α, β, γ proteobacteria	AHL-like quorum sensing systems	Valle et al. (2004)
6	Metagenome of membrane-aerated biofilm reactors	Pseudomonas Achromobacter Rhodococcus Granulicoccus Thauera	Acyl-homoserine lactones (AHLs) and diguanylate monophosphate	Hu et al. (2022)
7	Metagenome of activated sludge	Enriched growth of indole degrader (Burkholderiales)	Pathogens (*Giardia*, *Plasmodium*, and *Besnoitia*) growth inhibited by added indole. Enrichment of antibiotic resistance genes	Chunxiao et al. (2023)

diseased samples. Interestingly, the diseased samples exhibited a similar kind of microbial richness compared to the healthy ones. The most abundantly observed genera were *Actinomyces*, *Leptotrichia*, *Streptococcus*, *Veillonella*, etc. In addition to the disease etiology and taxonomic details, functional analysis is also provided through metagenomics. The metabolic analysis showed that diseased samples were enriched with pathways and genes associated with stress mechanisms, immune escape responses, virulence factor biosynthesis, and antibiotic resistance. A healthy poly-microbial biofilms were observed in healthy individuals exhibiting the homoserine lactone pathway. This allows healthy interaction among biofilms present in non-periodontal samples (Liu et al., 2012).

Metagenomics analysis provided valuable insights into the microplastics as a reservoir of antibiotic resistance genes. Few studies observed that antibiotic resistant genes (ARGs) are more abundant in microplastics than in surrounding environments. Biofilms of pathogenic and plastic-degrading bacteria covered the microplastics. Mobile genetic elements and ARGs were enriched on polystyrene microplastics revealing different pathogenic bacteria. The most abundant resistance genes, such as aminoglycoside and beta-lactam resistance genes were found to be enriched in the biofilms attached to the microplastics (Shi et al., 2021). In a similar research work, the enrichment of ARGs in freshwater bodies was determined through genotypic assays. The study revealed an increasing risk of infections to animals and humans owing to the uncontrolled release of antibiotics to the water bodies enriching the antibiotic resistance genes and bacteria. The sequencing analysis provided valuable insights into the abundance of tetracycline resistance genes and other antibiotic resistance genes coding for multidrug efflux pumps, sulfonamide, and macrolides (Lundström et al., 2016).

9.5 Metagenomics and Therapeutic Strategies

Understanding the biofilm physiology and function is an integral part of the design of novel therapeutics. There are several computational-based programs and approaches, which integrate large data generated through genome sequencing to develop databases containing potential anti-biofilm agents (Sharma et al., 2016). For instance, BaAMPs is a web server that contains information on numerous anti-biofilm peptides. The server was prepared by collecting the anti-microbial action of different peptides against biofilms and built using the Content Management System (CMS) Joomla. The website used different programming languages to enter the collected data and analysis (Di Luca et al., 2015).

With the advent of metagenomics, novel avenues were opened for the discovery of novel anti-microbial strategies including anti-biofilm methods. There are different types of subpopulations present in biofilms providing structural integrity and stability to the cells. Emerging technologies target these biofilm subpopulations to identify their metabolic functions and protection mechanisms. Eventually, this knowledge can be best applied to the design of novel therapeutics. In particular, targeting persister cells or metabolically inactive/dormant cells present in the biofilms is one of the promising therapeutic strategies to control biofilms (Bisht & Wakeman, 2019). Understanding the metabolic pathways and microbial interactions among biofilms helps to target biofilm matrix production and dispersal (Wang et al., 2023).

Functional analysis of metagenomic library prepared from slaughterhouse biofilms revealed the presence of putative and novel enzymes and biosurfactants. Environmental DNA from samples was isolated and prepared clones using the *Escherichia coli* expression system. The sequencing method revealed an abundance of the family Flavobacteriaceae. Several lipases, proteases, and biosurfactant and hemolysin-producing strains were screened through functional analysis. Biosurfactant, N-acyltyrosine exhibited good anti-bacterial activity against several bacteria (Thies et al., 2016).

Inhibition of quorum sensing (QS)/bacterial cell-to-cell communication is one of the effective methodologies to target biofilm formation among different bacteria. Several compounds and methods targeting QS were explored to target biofilm formation and associated infections in different pathogenic bacteria. The first study revealing the metagenome analysis screened QS quenching protein with oxidoreductase activity against different Gram-negative and positive bacteria. Similarly,

Table 9.2 Metagenomic approaches to study in the development of biofilm therapeutics

Sl. No:	Techniques	Study details	Activity	References
1	Soil metagenome	Short-chain dehydrogenase/ reductase named as bpiB09	Inhibition of biofilm in *Pseudomonas aeruginosa* and virulence in *Caenorhabditis elegans*	Bijtenhoorn et al. (2011)
2	Soil metagenome	Novel lactonase bpiB07	Inhibition of biofilm in *P. aeruginosa*	Schipper et al. (2009)
3	Marine metagenome	Lactonase activity in isolates	Acyl homoserine lactone quorum sensing inhibition	Romero et al. (2012)
4	Marine metagenome	Quorum quenching lactonase	Biofilm inhibition in food spoilage bacterium Pseudomonas fluorescens	Haramati et al. (2022)
5	Metagenomics of salt marsh	Quorum quenching proteins, QQ-5 and QQ-7	Biofilm inhibition in *Candida albicans* and *Staphylococcus epidermidis* Acyl homoserine lactone and autoinducer-2 QS inhibition	Weiland-Bräuer et al. (2019)
6	Metagenome of activated sludge and membrane bioreactor	Acylase from *Pseudoduganella, Actinoplanes, Cupriavidu,* and *Lactonases*	Inhibition of QS	Li et al. (2019)
7	Metagenome of pasture soil	Lactonase activity from members of the *Acidobacteria*	Inhibition of QS in *Pectobacterium carotovorum*	Riaz et al. (2008)
8	Metagenomics of Red Sea	Proteobacterial compounds	Biofilm and QS inhibition in *P. aeruginosa* PAO1 and A. baumannii	Yaniv et al. (2017)

lactonases were identified from soil metagenomes, which inhibited quorum sensing and biofilm formation in several bacteria (Weiland Bräucr ct al., 2016) (Table 9.2).

In an interesting research work, bacterial isolates were screened from marine eukaryotes and assessed their ability to inhibit and quench the QS. Most bacterial isolates found on marine algae and invertebrates exhibited quorum quenching activity against acyl homoserine lactone and autoinducer peptide-mediated QS in bacteria. The identified quorum quenching proteins from bacterial isolates, namely *Pseudoalteromonas* and *Vibrio,* inhibited biofilm formation on Gram-negative and Gram-positive bacteria (Weiland-Bräuer et al., 2015).

9.6 Conclusions

Metagenomics is one of the promising tools in modern biofilm research, focusing on the structural, genetic, and functional diversity. The taxonomic and functional profiling and antibiotic resistance of microbial biofilms can be studied through metagenomic approaches. Researchers explored several pathogenic elements and therapeutic potential of microbial biofilms using cutting-edge sequencing methods like whole-genome sequencing, 16S rRNA, and shotgun metagenomics. Quorum sensing inhibitors and anti-biofilm peptides are the two examples of tailored therapeutic approaches made possible by these discoveries. Despite obstacles such as computational requirements and unfinished databases, metagenomics keeps revealing new approaches to control biofilms in both healthcare and industries.

References

Abdallah, M., Benoliel, C., Drider, D., et al. (2014). Biofilm formation and persistence on abiotic surfaces in the context of food and medical environments. *Archives of Microbiology, 196*, 453–472. https://doi.org/10.1007/s00203-014-0983-1

Ali, A., Zahra, A., Kamthan, M., et al. (2023). Microbial biofilms: Applications, clinical consequences, and alternative therapies. *Microorganisms, 11*(8), 1934.

Bergholz, T. M., Moreno Switt, A. I., & Wiedmann, M. (2014). Omics approaches in food safety: Fulfilling the promise? *Trends in Microbiology, 22*, 275–281. https://doi.org/10.1016/j.tim.2014.01.006

Bijtenhoorn, P., Mayerhofer, H., Müller-Dieckmann, J., et al. (2011). A novel metagenomic Short-Chain dehydrogenase/reductase attenuates *Pseudomonas aeruginosa* biofilm formation and virulence on Caenorhabditis elegans. *PLoS One, 6*, e26278. https://doi.org/10.1371/journal.pone.0026278

Bisht, K., & Wakeman, C. A. (2019). Discovery and therapeutic targeting of differentiated biofilm subpopulations. *Frontiers in Microbiology, 10*. https://doi.org/10.3389/fmicb.2019.01908

Busi, S. B., Bourquin, M., Fodelianakis, S., et al. (2022). Genomic and metabolic adaptations of biofilms to ecological windows of opportunity in glacier-fed streams. *Nature Communications, 13*, 2168. https://doi.org/10.1038/s41467-022-29914-0

Chunxiao, D., Ma, F., Wu, W., et al. (2023). Metagenomic analysis reveals indole signaling effect on microbial community in sequencing batch reactors: Quorum sensing inhibition and antibiotic resistance enrichment. *Environmental Research, 229*, 115897. https://doi.org/10.1016/j.envres.2023.115897

De Luca, D., Piredda, R., Scamardella, S., et al. (2024). Taxonomic and metabolic characterisation of biofilms colonising Roman stuccoes at Baia's thermal baths and restoration strategies. *Scientific Reports, 14*, 26290. https://doi.org/10.1038/s41598-024-76637-x

Di Luca, M., Maccari, G., Maisetta, G., & Batoni, G. (2015). BaAMPs: The database of biofilm-active antimicrobial peptides. *Biofouling, 31*, 193–199. https://doi.org/10.1080/08927014.2015.1021340

Dinesh, R., Sreena, C. P., Sheeja, T. E., et al. (2023a). Metagenomics indicates abundance of biofilm related genes and horizontal transfer of multidrug resistant genes among bacterial communities in nano zinc oxide polluted soil. *Science of the Total Environment, 859*, 160032. https://doi.org/10.1016/j.scitotenv.2022.160032

Dinesh, R., Sreena, C. P., Sheeja, T. E., et al. (2023b). Soil polluted with nano ZnO reveals unstable bacterial communities and decoupling of taxonomic and functional diversities. *Science of the Total Environment, 889*, 164285. https://doi.org/10.1016/j.scitotenv.2023.164285

Dinsdale, E. A., Edwards, R. A., Hall, D., et al. (2008). Functional metagenomic profiling of nine biomes. *Nature, 452*, 629. https://doi.org/10.1038/nature06810

Flemming, H. C., & Wingender, J. (2010). The biofilm matrix. *Nature Reviews Microbiology, 8*, 623.

Garza, D. R., & Dutilh, B. E. (2015). From cultured to uncultured genome sequences: Metagenomics and modeling microbial ecosystems. *Cellular and Molecular Life Sciences, 72*, 4287.

Hao, Y., Winans, S. C., Glick, B. R., & Charles, T. C. (2010). Identification and characterization of new LuxR/LuxI-type quorum sensing systems from metagenomic libraries. *Environmental Microbiology, 12*, 105. https://doi.org/10.1111/j.1462-2920.2009.02049.x

Haramati, R., Dor, S., Gurevich, D., et al. (2022). Mining marine metagenomes revealed a quorum-quenching lactonase with improved biochemical properties that inhibits the food spoilage bacterium *Pseudomonas fluorescens*. *Applied and Environmental Microbiology, 88*. https://doi.org/10.1128/aem.01680-21

Hu, Y., Hu, Y., Li, Y., et al. (2022). Metagenomic insights into quorum sensing in membrane-aerated biofilm reactors for phenolic wastewater treatment. *Environmental Technology (United Kingdom), 43*, 1318. https://doi.org/10.1080/09593330.2020.1829084

Imchen, M., Anju, V. T., Busi, S., et al. (2022). Metagenomic insights into taxonomic, functional diversity and inhibitors of microbial biofilms. *Microbiological Research, 265*, 127207.

Janda, J. M., & Abbott, S. L. (2007). 16S rRNA gene sequencing for bacterial identification in the diagnostic laboratory: Pluses, perils, and pitfalls. *Journal of Clinical Microbiology, 45*, 2761–2764. https://doi.org/10.1128/JCM.01228-07

Kneis, D., Berendonk, T. U., Forslund, S. K., & Hess, S. (2022). Antibiotic resistance genes in river biofilms: A metagenomic approach toward the identification of sources and candidate hosts. *Environmental Science & Technology, 56*, 14913. https://doi.org/10.1021/acs.est.2c00370

Lee, K., & Yoon, S. S. (2017). Pseudomonas aeruginosa biofilm, a programmed bacterial life for fitness. *Journal of Microbiology and Biotechnology, 27*, 1053.

Lee, I. P. A., Eldakar, O. T., Gogarten, J. P., & Andam, C. P. (2022). Bacterial cooperation through horizontal gene transfer. *Trends in Ecology & Evolution, 37*, 223–232. https://doi.org/10.1016/j.tree.2021.11.006

Li, T., Guo, F., Lin, Y., et al. (2019). Metagenomic analysis of quorum sensing systems in activated sludge and membrane biofilm of a full-scale membrane bioreactor. *Journal of Water Process Engineering, 32*, 100952. https://doi.org/10.1016/j.jwpe.2019.100952

Liu, B., Faller, L. L., Klitgord, N., et al. (2012). Deep sequencing of the oral microbiome reveals signatures of periodontal disease. *PLoS One, 7*. https://doi.org/10.1371/journal.pone.0037919

Lundström, S. V., Östman, M., Bengtsson-Palme, J., et al. (2016). Minimal selective concentrations of tetracycline in complex aquatic bacterial biofilms. *Science of the Total Environment, 553*, 587. https://doi.org/10.1016/j.scitotenv.2016.02.103

Mitra, A. (2024). Combatting biofilm-mediated infections in clinical settings by targeting quorum sensing. *Cell Surface, 12*, 100133. https://doi.org/10.1016/j.tcsw.2024.100133

Muhammad, M. H., Idris, A. L., Fan, X., et al. (2020). Beyond risk: Bacterial biofilms and their regulating approaches. *Frontiers in Microbiology, 11*, 928.

Peschel, A., & Otto, M. (2013). Phenol-soluble modulins and staphylococcal infection. *Nature Reviews Microbiology, 11*, 667.

Quince, C., Walker, A. W., Simpson, J. T., Loman, N. J., & Segata, N. (2017). Shotgun metagenomics, from sampling to analysis. *Nature Biotechnology, 35*, 833–844. https://doi.org/10.1038/nbt.3935

Rather, M. A., Gupta, K., & Mandal, M. (2021). Microbial biofilm: Formation, architecture, antibiotic resistance, and control strategies. *Brazilian Journal of Microbiology, 52*, 1701–1718. https://doi.org/10.1007/s42770-021-00624-x

Riaz, K., Elmerich, C., Moreira, D., et al. (2008). A metagenomic analysis of soil bacteria extends the diversity of quorum-quenching lactonases. *Environmental Microbiology, 10*, 560. https://doi.org/10.1111/j.1462-2920.2007.01475.x

Ribeiro, S. (2017). Metagenomic approach to study biofilm in medical context. *Medical Research Archives, 5*. https://doi.org/10.18103/mra.v5i5.1222

Romero, M., Martin-Cuadrado, A.-B., & Otero, A. (2012). Determination of whether quorum quenching is a common activity in marine bacteria by analysis of cultivable bacteria and metagenomic sequences. *Applied and Environmental Microbiology, 78*, 6345–6348. https://doi.org/10.1128/AEM.01266-12

Sahoo, K., & Meshram, S. (2024). Biofilm formation in chronic infections: A comprehensive review of pathogenesis, clinical implications, and novel therapeutic approaches. *Cureus, 16*, 1–8. https://doi.org/10.7759/cureus.70629

Schipper, C., Hornung, C., Bijtenhoorn, P., et al. (2009). Metagenome-derived clones encoding two novel lactonase family proteins involved in biofilm inhibition in *Pseudomonas aeruginosa*. *Applied and Environmental Microbiology, 75*, 224. https://doi.org/10.1128/AEM.01389-08

Sharma, R., Ranjan, R., Kapardar, R. K., & Grover, A. (2005). "Unculturable" bacterial diversity: An untapped resource. *Current Science, 89*, 72–77.

Sharma, A., Gupta, P., Kumar, R., & Bhardwaj, A. (2016). DPABBs: A novel in silico approach for predicting and designing anti-biofilm peptides. *Scientific Reports, 6*. https://doi.org/10.1038/srep21839

Sharma, S., Mohler, J., Mahajan, S. D., et al. (2023). Microbial biofilm: A review on formation, infection, antibiotic resistance, control measures, and innovative treatment. *Microorganisms, 11*, 1614. https://doi.org/10.3390/microorganisms11061614

Shi, J., Wu, D., Su, Y., & Xie, B. (2021). Selective enrichment of antibiotic resistance genes and pathogens on polystyrene microplastics in landfill leachate. *Science of the Total Environment, 765*, 142775. https://doi.org/10.1016/j.scitotenv.2020.142775

Shineh, G., Mobaraki, M., Perves Bappy, M. J., & Mills, D. K. (2023). Biofilm formation, and related impacts on healthcare, food processing and packaging, industrial manufacturing, marine industries, and sanitation–a review. *Applied Microbiology, 3*, 629–665. https://doi.org/10.3390/applmicrobiol3030044

Shree, P., Singh, C. K., Sodhi, K. K., et al. (2023). *Biofilms: Understanding the structure and contribution towards bacterial resistance in antibiotics*. Med.

Skipper, P. J. A., Skipper, L. K., & Dixon, R. A. (2022). A metagenomic analysis of the bacterial microbiome of limestone, and the role of associated biofilms in the biodeterioration of heritage stone surfaces. *Scientific Reports, 12*, 4877. https://doi.org/10.1038/s41598-022-08851-4

Tan, C. H., Yeo, Y. P., Hafiz, M., et al. (2021). Functional metagenomic analysis of quorum sensing signaling in a nitrifying community. *npj Biofilms and Microbiomes, 7*, 79. https://doi.org/10.1038/s41522-021-00250-3

Thies, S., Rausch, S. C., Kovacic, F., et al. (2016). Metagenomic discovery of novel enzymes and biosurfactants in a slaughterhouse biofilm microbial community. *Scientific Reports, 6*. https://doi.org/10.1038/srep27035

Valle, A., Bailey, M. J., Whiteley, A. S., & Manefield, M. (2004). N-acyl-L-homoserine lactones (AHLs) affect microbial community composition and function in activated sludge. *Environmental Microbiology, 6*, 424. https://doi.org/10.1111/j.1462-2920.2004.00581.x

Vieites, J. M., Guazzaroni, M. E., Beloqui, A., et al. (2009). Metagenomics approaches in systems microbiology. *FEMS Microbiology Reviews, 33*(1), 236–255.

Wang, X., Liu, M., Yu, C., et al. (2023). Biofilm formation: Mechanistic insights and therapeutic targets. *Molecular Biomedicine, 4*(1), 49.

Warnecke, F., Luginbühl, P., Ivanova, N., et al. (2007). Metagenomic and functional analysis of hindgut microbiota of a wood-feeding higher termite. *Nature, 450*, 560. https://doi.org/10.1038/nature06269

Weiland-Bräuer, N., Pinnow, N., & Schmitz, R. A. (2015). Novel reporter for identification of interference with Acyl homoserine lactone and autoinducer-2 quorum sensing. *Applied and Environmental Microbiology, 81*, 1477. https://doi.org/10.1128/AEM.03290-14

Weiland-Bräuer, N., Kisch, M. J., Pinnow, N., et al. (2016). Highly effective inhibition of biofilm formation by the first metagenome-derived AI-2 quenching enzyme. *Frontiers in Microbiology, 7*. https://doi.org/10.3389/fmicb.2016.01098

Weiland-Bräuer, N., Malek, I., & Schmitz, R. A. (2019). Metagenomic quorum quenching enzymes affect biofilm formation of *Candida albicans* and *Staphylococcus epidermidis*. *PLoS One, 14*, e0211366. https://doi.org/10.1371/journal.pone.0211366

Williamson, L. L., Borlee, B. R., Schloss, P. D., et al. (2005). Intracellular screen to identify metagenomic clones that induce or inhibit a quorum-sensing biosensor. *Applied and Environmental Microbiology, 71*, 6335. https://doi.org/10.1128/AEM.71.10.6335-6344.2005

Yaniv, K., Golberg, K., Kramarsky-Winter, E., et al. (2017). Functional marine metagenomic screening for anti-quorum sensing and anti-biofilm activity. *Biofouling, 33*, 1. https://doi.org/10.1080/08927014.2016.1253684

Zhao, A., Sun, J., & Liu, Y. (2023). Understanding bacterial biofilms: From definition to treatment strategies. *Frontiers in Cellular and Infection Microbiology, 13*, 1137947.

Phenomics and Microbial Biofilms 10

G. Ganga

10.1 Introduction

Biofilms are highly organized communities of microorganisms that attach to surfaces and are embedded in a self-produced extracellular matrix consisting of polysaccharides, proteins, and nucleic acids (Costerton et al., 1999). These microbial structures are found in virtually all natural and artificial environments and are implicated in numerous challenges, such as chronic infections, industrial biofouling, and biocorrosion (Donlan & Costerton, 2002).

Traditional biofilm research methods, including microscopy and culture techniques, often fall short of capturing the intricate molecular interactions and complexity within these communities (Davies & Geesey, 1995). Biofilms contribute to a variety of clinical, environmental, and industrial issues, such as persistent infections in cystic fibrosis and implant-related complications, as well as problems like biocorrosion and fouling in water treatment systems. Understanding these complexities requires innovative approaches that go beyond traditional methodologies to address the multifaceted nature of biofilms effectively (Flemming & Wingender, 2010).

The emergence of high-throughput "omics" technologies, including genomics, transcriptomics, proteomics, and metabolomics, has transformed the field of biofilm research, ushering in a new era of molecular understanding (Hall-Stoodley & Stoodley, 2009). These cutting-edge methodologies offer a holistic view of the molecular processes that govern biofilm formation, structure, and function, enabling researchers to decode the intricate dynamics of these microbial communities.

This review highlights the pivotal role of omics-based approaches in biofilm characterization, focusing on recent advances and the promising future of this research domain. Studies by Noto and Miller (2017) and Watnick and Kolter (2000) have demonstrated how these methodologies uncover critical insights into the

G. Ganga (✉)
Sree Ayyappa College Eramallikkara Chengannur, Alappuzha, Kerala, India

S. Busi et al. (eds.), *Omics Approaches in Biofilm Research*,
https://doi.org/10.1007/978-3-031-91863-6_10

genetic, transcriptomic, proteomic, and metabolic networks that underpin biofilm behavior. By integrating these multiomics datasets, researchers aim to achieve a comprehensive understanding of biofilm systems, paving the way for novel therapeutic and industrial applications.

Omics approaches, including genomics, transcriptomics, proteomics, and metabolomics, are revolutionizing biofilm research by providing comprehensive insights into the molecular mechanisms underlying biofilm formation, maintenance, and resistance. Genomics, particularly whole genome sequencing (WGS), allows researchers to identify genes involved in biofilm formation and antibiotic resistance. By comparing the genomes of biofilm forming strains to planktonic strains, researchers can pinpoint specific genetic factors that contribute to biofilm-specific characteristics, such as the production of extracellular polymeric substances (EPS), adhesion molecules, and virulence factors. Transcriptomics, through RNA sequencing (RNA-Seq), enables the analysis of gene expression changes during biofilm formation, helping to identify key regulatory pathways that control biofilm development. These studies have highlighted the importance of quorum sensing, stress response mechanisms, and metabolic pathways in biofilm growth and persistence. Additionally, transcriptomic analysis can reveal the differential expression of genes associated with antibiotic resistance and immune evasion in biofilm-associated bacteria, which is crucial for understanding chronic infections (Flemming et al., 2016; Assisi et al., 2021; Dutta et al., 2024).

For instance, transcriptomic analyses have highlighted the activation of stress-response genes under hostile conditions, which enhances biofilm endurance against antibiotics and immune attacks. Similarly, genomics provides information on mutations and gene variations that confer resistance and adaptability in biofilm environments. By leveraging these advanced tools, researchers can better understand biofilm dynamics, leading to the development of targeted strategies to combat biofilm-related challenges, particularly in medical and industrial contexts. Various omics-based approaches in biofilm-based research is given in Fig. 10.1.

10.2 Biofilm Structure and Composition

The extracellular polymeric substances (EPS) matrix is a defining feature of biofilms, serving as a multifunctional scaffold that provides structural support, shields microbial cells, and maintains a hydrated environment essential for biofilm persistence. Composed primarily of polysaccharides, proteins, nucleic acids, and lipids, the EPS matrix anchors cells to surfaces and to one another traps nutrients to sustain microbial metabolism, and protects the biofilm from environmental stressors, antibiotics, and immune responses. This resilience makes biofilms particularly challenging in medical and industrial contexts, necessitating a deeper understanding of their composition and interactions to develop effective intervention strategies. This diversity contributes to biofilm resilience, as different species can metabolize a wide range of substrates and share resources. Additionally, heterogeneity within biofilms exists at the cellular level, with variations in gene expression, metabolic activity,

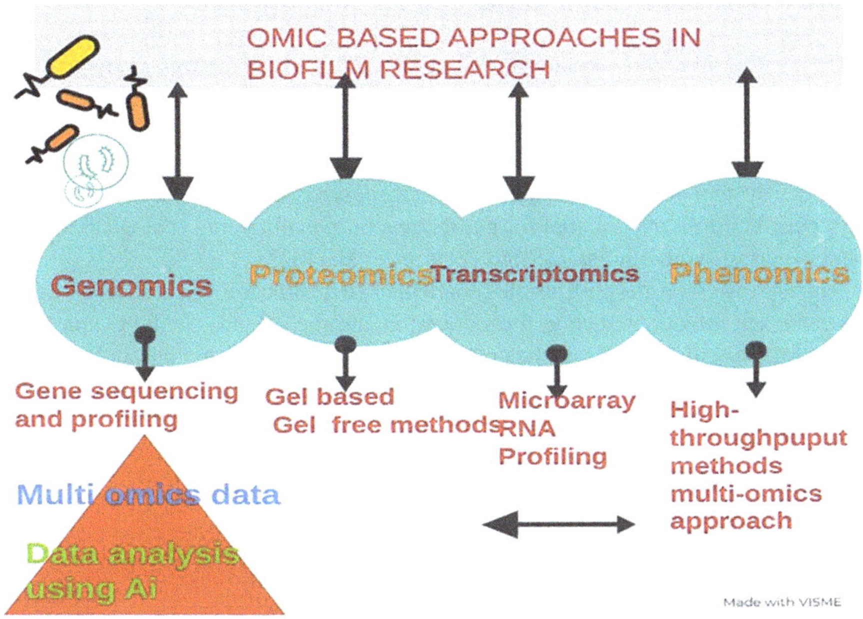

Fig. 10.1 OMIC-based approaches in Biofilm research

and spatial positioning, creating microenvironments that optimize survival. These structural and functional complexities underscore the challenges of eradicating biofilms and highlight their ecological and clinical significance (Flemming et al., 2016; Stewart & Franklin, 2008).

Biofilms exhibit unique physiological characteristics that confer remarkable metabolic versatility and stress tolerance, enabling microbial communities to thrive in diverse and often hostile environments. Within biofilms, cells can adopt distinct metabolic states, with some being highly active while others enter a dormant or slow-growing state, contributing to their resilience. The extracellular polymeric substances (EPS) matrix not only provides structural support but also acts as a barrier against desiccation, antibiotics, and immune responses. Biofilms can survive oxidative stress, nutrient limitation, and extreme pH or temperature through coordinated gene expression and metabolic adaptations. The remarkable resilience of biofilms plays a critical role in microbial persistence, especially in chronic infections and on medical devices, where standard antimicrobial therapies frequently fail. Biofilms serve as reservoirs for microbial cells, enabling recurrent infections and promoting horizontal gene transfer, which significantly contributes to the growing issue of antimicrobial resistance (Stewart & Franklin, 2008; Mah & O'Toole, 2001).

10.2.1 Microbial Communities and Biofilm Formation

Microbial biofilms are organized communities of microorganisms embedded in a self-produced extracellular matrix that allows them to adhere to surfaces across natural, industrial, and clinical environments. This matrix provides significant protection, enabling microbes to withstand environmental pressures, antibiotics, and immune system attacks, contributing to their remarkable resilience. In natural ecosystems, biofilms are essential for nutrient cycling, maintaining ecological balance and fostering symbiotic interactions (Sharma et al., 2023).

In contrast, biofilms pose serious challenges in healthcare, where they are linked to persistent infections such as those found in chronic wounds, dental plaques, and infections associated with medical devices. These infections are particularly difficult to treat due to the biofilm's protective nature. Gaining a deeper understanding of biofilm development, structure, and dynamics is crucial for mitigating their harmful effects while also harnessing their potential for beneficial uses, such as bioremediation and advancements in biotechnology (Flemming et al., 2016; Hall-Stoodley et al., 2004).

10.2.2 Multistage Process in Biofilm Formation

The first attachment of microbial cells to an opening surface, which is frequently facilitated by weak van der Waals forces or hydrophobic contacts, starts the multistage and dynamic process of biofilm formation. When cells create extracellular polymeric substances (EPS), which serve as the basis for the biofilm matrix, this connection develops into a more stable one. As the biofilm matures, it forms intricate three-dimensional structures with water channels that help with waste and nutrient exchange. Gene expression for biofilm formation, EPS production, and community behavior coordination is regulated by communication through quorum sensing, a density-dependent signaling mechanism. Cells either actively or passively separate from the biofilm during the last dispersion stage, which facilitates surface colonization and encourages the spread of microorganisms. Targeting biofilm-associated infections and stopping their development require an understanding of these stages (Hall-Stoodley et al., 2004; Flemming et al., 2016).

10.3 Genomic Approaches for Biofilm Research

The genetic foundations of biofilm development, growth, and persistence can now be examined by biofilm researchers using genomic approaches. Whole-genome sequencing (WGS) of biofilm-forming microbes has provided valuable new insights into the genetic repertoire that enables bacteria and other microorganisms to create biofilms O'Toole and Kolter (1998). This approach identifies genes that are critical for biofilm development, including those that encode surface adhesion molecules, biofilm matrix components (such as extracellular polymeric substances, or EPS), and regulatory proteins responsible for quorum sensing.

Additionally, genetic components supporting biofilm-specific characteristics like increased host immunological responses and antibiotic resistance have been identified by a comparison of the genomes of planktonic and biofilm-forming bacteria. Scientists can discover crucial genetic loci that contribute to the formation of biofilms by utilizing WGS in combination with bioinformatics tools. This opens the door to new treatment strategies intended to avoid diseases associated with biofilms.

10.3.1 Functional Genomics in Biofilm Research

Genomics provides functional gene expression profiles of bacteria. Because functional genomics makes it possible to analyze gene expression profiles in bacteria that create biofilms, biofilm research has been substantially enhanced. By measuring gene expression under biofilm settings, transcriptomic techniques—such as RNA sequencing—offer insights into the regulatory networks controlling the production of biofilms. A complex network of genes involved in signal transduction, stress response, metabolism, and virulence has been shown to closely govern the production of biofilms. Quorum sensing systems' significance in coordinating the production of biofilms is one significant finding from functional genomics. Researchers can examine the impact of gene knockouts on biofilm growth and discover important regulators of biofilm formation by modifying gene expression and employing reporter constructs. Furthermore, the targeted disruption of particular genes is made possible by the use of various tools such as transcription activator-like effector nucleases (TALENs), Retron Library Recombineering (RLR) Platform, and FANA Antisense Oligonucleotide (FANA ASO) technology. CRISPR-Cas9 and other genetic editing tools make it possible to identify the genes that are crucial for antibiotic resistance, biofilm stability, and persistence (Zhang et al., 2024).

10.3.2 Functional Genomics in Biofilm Phenotyping

Functional genomics helps to study gene functions by analyzing genome-wide data, by using techniques like transcriptomics, proteomics, and metabolomics. It aims to understand how genes and their products contribute to biological processes, phenotypes, and diseases. By connecting genetic information to phenotypic features, functional genomics is essential for improving our understanding of biofilm phenotyping. Structured microbial colonies called biofilms, which are anchored in an extracellular matrix that they manufacture on their own, have special traits that include heightened resilience to environmental stressors and antibiotics. Functional genomics makes it possible to identify the genes and processes necessary for biofilm development, maintenance, and dissemination using methods including transcriptomics, proteomics, and metabolomics. This method also makes it easier to identify the adaptive mechanisms and regulatory networks that support different biofilm morphologies. Functional genomics offers a thorough framework for investigating biofilm heterogeneity by fusing high-throughput data with computational modelling. This opens the door for new treatment approaches to address biofilm-associated illnesses and industrial fouling. Our ability to examine the genetic foundations of traits associated with biofilms is enhanced when functional genomics and phenomics are combined.

Our ability to examine the genetic foundations of traits associated with biofilms is enhanced when functional genomics and phenomics are combined. Doudna and Barrangou (2016) reported that Clustered Regularly Interspaced Short Palindromic Repeats. CRISPR-based tools, such as CRISPR-associated protein 9 (CRISPR-Cas9) and CRISPRi (CRISPR interference), are used to edit or regulate genes in microbial biofilms, enabling the identification of genes critical for biofilm development, maintenance, and dispersal. These technologies facilitate high-throughput functional genomics studies, allowing researchers to systematically disrupt or modulate gene expression to uncover phenotypic changes in biofilms under various conditions (van Opijnen & Camilli, 2013). These genomic techniques, when combined with phenotypic screening, enable scientists to map the functional landscape of genes associated with biofilms and discover novel targets for intervention. These techniques are particularly helpful for understanding the genetic components of antibiotic resistance and for developing strategies to arrest biofilm synthesis and reduce its pathogenic impact (O'Toole et al., 2000; Stewart & Franklin, 2008).

10.4 Phenomics and Microbial Biofilms

Phenomics is the comprehensive study of phenotypes based on its genetic, environmental, and developmental factors. It involves high-throughput methods to systematically measure and analyze traits across biological scales, including cellular, organismal, and ecological levels. The scope of phenomics spans diverse fields, such as linking genotype to phenotype, understanding organismal responses to environmental changes, advancing precision agriculture, and supporting personalized medicine. Multiomics methods helps researchers to gain insights on the complexity of biofilm formation using an interdisciplinary approach. As a transformative tool, phenomics accelerates discoveries by bridging the gap between molecular data and observable outcomes, enabling advancements in both fundamental and applied sciences (Houle et al., 2010; Furbank & Tester, 2011).

10.4.1 Phenomics Approaches in Microbial Biofilms: Applications and Insights

The application of phenomics in microbial biofilms extends to various fields. In healthcare, phenomic data can aid in the development of targeted antimicrobial therapies to combat biofilm-associated infections, such as those in chronic wounds or medical implants. In industrial settings, phenomics can help mitigate biofilm-related issues like biofouling and biocorrosion, by identifying conditions that suppress biofilm formation. Environmental applications include leveraging phenomics to engineer biofilms for bioremediation or wastewater treatment. Moreover, the integration of phenomics with computational tools, such as machine learning, enhances predictive modelling of biofilm behavior, accelerating the discovery of innovative solutions. As the field evolves, phenomics holds the potential to revolutionize our understanding and management of microbial biofilms, addressing critical challenges across diverse domains (Ali et al., 2023).

Unlike genomics, transcriptomics, proteomics, and metabolomics, which focus on molecular-level data such as genes, RNA, proteins, and metabolites, respectively, phenomics examines how these molecular components manifest as traits across biological scales. This field bridges the gap between molecular data and organismal characteristics, providing essential context for understanding genotype-phenotype relationships. By integrating high-throughput imaging, automated data collection, and computational analysis, phenomics enables the comprehensive mapping of traits under various conditions, aiding in the discovery of gene functions, developmental pathways, and environmental responses. Its scope extends across medicine, agriculture, and ecology, making it indispensable for understanding complex biological systems and addressing challenges like disease prediction, crop improvement, and environmental resilience (Houle et al., 2010; Furbank & Tester, 2011).

10.4.2 Phenomics Approaches for Studying Biofilm Formation

Phenomics enables the exploration of biofilm formation using high-throughput assays that systematically identify genes critical to this process. Techniques like transposon mutagenesis and RNA interference allow researchers to disrupt gene function and observe resulting phenotypic changes, uncovering the genetic networks and pathways involved in biofilm development. Phenotypic screening platforms are also employed to test the efficacy of potential biofilm inhibitors or disruptors, such as antimicrobial agents or environmental stressors. Advanced imaging technologies, including 3D confocal microscopy, live-cell microscopy, etc. can provide detailed insights into biofilm architecture and dynamics. These imaging tools allow real-time tracking of biofilm formation, maturation, and dispersion, offering valuable data on structural and functional phenotypes. Together, these phenomic approaches expand our understanding of biofilm biology and provide avenues for developing targeted anti-biofilm strategies (Yeung et al., 2022).

10.4.3 Application of Phenomics in Characterizing Biofilm Heterogeneity

The heterogeneity within biofilms, both at the cellular and community levels, is a critical factor contributing to their resilience and adaptability. Phenomics facilitates the analysis of single-cell phenotypic variation within biofilms, using techniques such as single-cell RNA sequencing, flow cytometry, and high-content microscopy. These methods reveal metabolic, physiological, and functional differences among individual cells, driven by spatial gradients of nutrients, oxygen, and signaling molecules within the biofilm matrix. This variation often results in the emergence of specialized subpopulations with distinct roles, such as persisters, which contribute to antimicrobial resistance, or dispersing cells, which aid in colonizing new environments. Phenomics-based approaches are instrumental in identifying and characterizing these subpopulations, providing insights into their contributions to biofilm stability, pathogenicity, and overall ecological function (Sadiq et al., 2017; Simkovsky et al., 2022).

10.4.4 Phenomics in Health and Disease

Biofilms play a critical role in human health, particularly in chronic infections where they contribute to persistent and recalcitrant conditions. Infections, such as lung disease, chronic UTI, and infective endocarditis, are often associated with biofilms, which provide a protective niche for pathogens against host immune responses and antibiotics. The mechanisms of biofilm-mediated antibiotic resistance include limited penetration of antimicrobial agents through the extracellular polymeric substances (EPS) matrix, altered metabolic states of cells that can withstand treatment. Medical implants, catheters, and prosthetics further complicate healthcare settings, often leading to device failure due to biofilm-associated systemic infections. Strategies to mitigate device-associated biofilm infections include surface modifications, incorporation of antimicrobial agents, and anti-biofilm forming agents. Emerging antibiofilm treatments viz., quorum sensing inhibitors (QSi) and phage therapy hold promise in targeting biofilm resilience and preventing chronic infections (Hall-Stoodley et al., 2004; Stewart, 2015).

10.4.5 Phenomics in Biofilm Pathogenicity

Phenomics has advanced the understanding of biofilm pathogenicity by identifying phenotypic traits that contribute to microbial virulence. Traits, such as EPS composition, quorum sensing activity, and metabolic adaptations, are critical for biofilm formation and persistence in host environments. By analyzing the relationship between biofilm phenotype and virulence, phenomics helps elucidate how specific traits enable pathogens to evade immune responses and establish chronic infections. High-throughput phenotyping platforms combined with genomic tools provide an understanding about biofilm-formers and their surfaces aiding in the identification of novel targets for therapeutic intervention. These approaches are crucial for addressing the growing threat of biofilm-associated infections and antimicrobial resistance (Flemming et al., 2016; Bjarnsholt, 2013).

10.4.6 Phenomics in Environmental and Agricultural Systems

Biofilms play essential roles in both environmental and agricultural systems, contributing to natural processes, such as wastewater treatment, bioremediation, and nutrient cycling. In wastewater treatment, biofilms on surfaces in bioreactors facilitate the degradation of organic pollutants, improving water quality. Similarly, biofilms contribute to bioremediation by breaking down hazardous substances in contaminated environments, such as heavy metals or petroleum hydrocarbons. In natural habitats, biofilms are found in riverbeds, soil, and on plant surfaces, where they interact with the surrounding ecosystem to enhance nutrient availability and protect against pathogens. In agriculture, plant-associated microbial biofilms are crucial for plant health, promoting growth through nutrient fixation, enhancing resistance to diseases, and improving stress tolerance. Additionally, biofilms provide sustainable alternatives to chemical agents in agriculture inputs. Phenomics provides a knowledge on various biofilm-related interactions in plant development, enabling the identification of better microbial communities that can be harnessed for agricultural innovation and environmental protection (Flemming et al., 2016; O'Toole et al., 2000).

10.4.7 Phenomics of Biofilms in Industrial Applications

Biofilms are also of significant importance in industrial applications, including food processing and fermentation, where they can both benefit and pose challenges. In food processing, biofilms on surfaces can enhance fermentation by providing a stable microbial community for consistent product quality, such as in dairy or brewing industries. However, unwanted biofilms can cause contamination and spoilage, leading to foodborne illnesses and reduced product quality. In industrial settings like water treatment plants or heat exchangers, biofilms may contribute to equipment fouling, corrosion, and reduced efficiency. To control biofilm growth in these settings, various strategies are employed, such as surface modifications, antimicrobial coatings, and the use of biofilm-disrupting agents. The continued development of biofilm management technologies is essential for improving industrial processes and ensuring hygiene and safety in food production (Stewart & Franklin, 2008; Venkatesan & Kim, 2015).

10.5 Current Challenges in Phenomics and Biofilm Research

Studying biofilms using phenomics helps to provide as an insight about the complexity of biofilm structure and its function. Biofilms form in dynamic, heterogeneous environments where microbial communities interact with surfaces, nutrients, and host organisms. Simulating these intricate conditions in vitro can be difficult, as standard laboratory models often fail to capture the full diversity of microbial interactions and environmental gradients that influence biofilm development.

Additionally, biofilms are spatially structured, with gradients of oxygen, nutrients, and waste products, making it challenging to study them in a way that fully reflects their physiological state in natural habitats. This complexity not only complicates the experimental design but also limits the ability to predict biofilm behavior in real-world settings, such as in clinical or industrial environments (Dzianach et al., 2019).

10.6 Exciting New Technologies in Biofilm Phenomics

Despite these challenges, advancements in technology are transforming biofilm research and expanding the capabilities of phenomics in this field. One exciting development is the use of various sequencing technologies, such as pyro sequencing, microarray technology, and single-cell RNA sequencing, which helps to understand the gene expression pattern of bacterial species embedded within each biofilm. This approach reveals the heterogeneity of biofilm communities, uncovering variations in metabolic states, stress responses, and population studies. Advancement of artificial intelligence-based techniques (AI techniques) and AI-driven data analysis enables researchers to process large volumes of high-dimensional data, such as those generated by imaging techniques or multiomics approaches, and identify patterns or phenotypic traits associated with biofilm formation, antimicrobial resistance, and pathogenicity. These technologies provide deeper insights into the dynamics of biofilm growth and behavior, facilitating the identification of key factors that contribute to their persistence and resistance Shapiro (2017).

10.7 Multiomics Approaches for Comprehensive Biofilm Phenotyping

Integration of various data from diverse layers of biological information provides an idea about biofilm formation, function, and how it adapts to changing environments. For example, by combining transcriptomic data with proteomic and metabolomic profiles, it becomes possible to link gene expression to protein production and metabolic activity, providing a comprehensive picture of biofilm physiology. These multiomics strategies, when coupled with advanced imaging and single-cell technologies, allow for the identification of novel biomarkers for biofilm-related diseases, as well as new therapeutic targets. As the resolution and integration of these technologies improve, the study of biofilms will move closer to a systems biology approach, enabling a more accurate understanding of biofilm-related phenomena and targeted treatments against biofilm formation in both medical and industrial settings (O'Donnell et al., 2020).

10.8 Conclusion

In conclusion, phenomics with the multiomics approach can provide significant insight into the roles of various molecules in biofilm formation and in pathogenesis. These high-throughput technologies have revealed the genetic and biochemical complexity of biofilms, shedding light on the molecular pathways that govern biofilm development, microbial interactions, and resistance mechanisms. By identifying key genes, metabolic shifts, and regulatory networks, omics research has provided valuable insights into how biofilms contribute to chronic infections, environmental processes, and industrial applications. The application of phenomics, in particular, has the potential to revolutionize the study of biofilms by enabling the precise characterization of phenotypic traits at the cellular and community levels, bridging the gap between genotype and phenotype and offering new avenues for therapeutic and technological interventions (Stewart & Franklin, 2008; Hemdan et al., 2019; Dutta et al., 2024).

Looking ahead, future biofilm research will benefit from further advancements in omics technologies. The development of single-cell omics and AI-driven data analysis will enable a deeper biofilm heterogeneity, microbial population their functional role. Additionally, the application of phenomics and gene editing tools, such as CRISPR, to study biofilm phenotypes will open new opportunities for identifying essential genes involved in biofilm persistence, virulence, and resistance and targeted drug therapeutic regimes (Stewart & Franklin, 2008).

10.9 Summary

In summary, the integration of multiomics with phenomics approaches has provided understanding of the molecular and biochemical mechanisms undergoing in biofilm development, including key genes involved in matrix production, surface adhesion, and quorum sensing regulation. OMICS has allowed for the identification of metabolic shifts, biofilm-specific metabolites, and microbial interactions within biofilms, which provides an understanding of bacterial communities embedded in biofilm and their multiresistant factors and their role in virulence and pathogenicity. As a result, phenomics along with other OMICS approaches has become an essential tool in biofilm research, offering a more comprehensive picture of microbial communities than traditional methods alone.

References

Ali, A., Zahra, A., Kamthan, M., Husain, F. M., Albalawi, T., Zubair, M., Alatawy, R., Abid, M., & Noorani, M. S. (2023). Microbial biofilms: Applications, clinical consequences, and alternative therapies. *Microorganisms, 11*(8), 1934. https://doi.org/10.3390/microorganisms11081934. PMID: 37630494; PMCID: PMC10459820.

Assisi, C., Forauer, E., Oliver, H. F., & Etter, A. J. (2021). Genomic and transcriptomic analysis of biofilm formation in persistent and transient *Listeria monocytogenes* isolates from the retail deli environment does not yield insight into persistence mechanisms. *Foodborne Pathogens and Disease, 18*(3), 179–188. https://doi.org/10.1089/fpd.2020.2817. Epub 23.

Barrangou, R., & Doudna, J. A. (2016). Applications of CRISPR technologies in research and beyond. *Nature Biotechnology, 34*(9), 933–941. https://doi.org/10.1038/nbt.3659

Bjarnsholt, T. (2013). The role of bacterial biofilms in chronic infections. *APMIS Supplementum, 121*(Suppl 136), 1–58. https://doi.org/10.1111/apm.12099

Costerton, J. W., Stewart, P. S., & Greenberg, E. P. (1999). Bacterial biofilms: A common cause of persistent infections. *Science, 284*(5418), 1318–1322. https://doi.org/10.1126/science.284.5418.1318

Davies, D. G., & Geesey, G. G. (1995). Bacterial biofilms and their role in corrosion. *Journal of Industrial Microbiology, 14*(4), 253–267. https://doi.org/10.1007/BF01569835

Donlan, R. M., & Costerton, J. W. (2002). Biofilms: Survival mechanisms of clinically relevant microorganisms. *Clinical Microbiology Reviews, 15*(2), 167–193. https://doi.org/10.1128/CMR.15.2.167-193.2002

Dutta, B., Chatterjee, D., Sarkar, N., Lahiri, D., Nag, M., & Ray, R. R. (2024). Multi-omics technology in detection of multispecies biofilm. *The Microbe, 4*, 100128, ISSN 2950-1946. https://doi.org/10.1016/j.microb..100128

Dzianach, P. A., Dykes, G. A., Strachan, N. J. C., Forbes, K. J., & Pérez-Reche, F. J. (2019). Challenges of biofilm control and utilization: Lessons from mathematical modelling. *Journal of the Royal Society Interface, 16*(155), 20190042. https://doi.org/10.1098/rsif.2019.0042. Epub 2019 June 12. PMID: 31185817; PMCID: PMC6597778.

Flemming, H.-W., & Wingender, J. (2010). The biofilm matrix. *Nature Reviews Microbiology, 8*(9), 623–633. https://doi.org/10.1038/nrmicro2415

Flemming, H. C., Wingender, J., Szewzyk, U., Steinberg, P., Rice, S. A., & Kjelleberg, S. (2016). Biofilms: An emergent form of bacterial life. *Nature Reviews Microbiology, 14*(9), 563–575. https://doi.org/10.1038/nrmicro.2016.94

Furbank, R. T., & Tester, M. (2011). Phenomics–technologies to relieve the phenotyping bottleneck. *Trends in Plant Science, 16*(12), 635–644. https://doi.org/10.1016/j.tplants.2011.09.005

Hall-Stoodley, L., & Stoodley, P. (2009). Evolving concepts in biofilm infections. *Cell Host & Microbe, 5*(3), 295–300. https://doi.org/10.1016/j.chom.2009.03.010

Hall-Stoodley, L., Costerton, J. W., & Stoodley, P. (2004). Bacterial biofilms: From the natural environment to infectious diseases. *Nature Reviews Microbiology, 2*(2), 95–108. https://doi.org/10.1038/nrmicro821

Hemdan, B. A., El-Liethy, M. A., ElMahdy, M. E. I., & El-Taweel, G. E. (2019). Metagenomics analysis of bacterial structure communities within natural biofilm. *Heliyon, 5*(8), e02271. https://doi.org/10.1016/j.heliyon.2019.e02271. PMID: 31485510; PMCID: PMC6716113.

Houle, D., Govindaraju, D. R., & Omholt, S. (2010). Phenomics: The next challenge. *Nature Reviews Genetics, 11*(12), 855–866. https://doi.org/10.1038/nrg2897

Mah, T. F., & O'Toole, G. A. (2001). Mechanisms of biofilm resistance to antimicrobial agents. *Trends in Microbiology, 9*(1), 34–39. https://doi.org/10.1016/S0966-842X(00)01913-2

Noto, M. J., & Miller, L. M. (2017). Omics technologies in biofilm research. *Biofilm, 6*, 18–29. https://doi.org/10.1016/j.bioflm.2017.07.003

O'Donnell, S. T., Ross, R. P., & Stanton, C. (2020). The progress of multi-omics technologies: Determining function in lactic acid bacteria using a systems level approach. *Frontiers in Microbiology, 10*, 3084. https://doi.org/10.3389/fmicb.2019.03084. PMID: 32047482; PMCID: PMC6997344.

O'Toole, G. A., & Kolter, R. (1998). Initiation of biofilm formation in Pseudomonas fluorescens WCS365 proceeds via multiple, convergent signaling pathways: A genetic analysis. *Journal of Bacteriology, 180*(24), 5635–5644. https://doi.org/10.1128/jb.180.24.5635-5644.1998

O'Toole, G., Kaplan, H. B., & Kolter, R. (2000). Biofilm formation as microbial development. *Annual Review of Microbiology, 54*, 49–79. https://doi.org/10.1146/annurev.micro.54.1.49

Sadiq, F. A., Flint, S., Li, Y., Ou, K., Yuan, L., & He, G. Q. (2017). Phenotypic and genetic heterogeneity within biofilms with particular emphasis on persistence and antimicrobial tolerance. *Future Microbiology, 12*, 1087–1107. https://doi.org/10.2217/fmb-2017-0042. Epub 2017 Aug 7.

Shapiro, J. A. (2017). Bacterial biofilms: An overview. *Journal of Bacteriology, 199*(1), 1–11. https://doi.org/10.1128/JB.00477-16

Sharma, S., Mohler, J., Mahajan, S. D., Schwartz, S. A., Bruggemann, L., & Aalinkeel, R. (2023). Microbial biofilm: A review on formation, infection, antibiotic resistance, control measures, and innovative treatment. *Microorganisms, 11*, 1614. https://doi.org/10.3390/microorganisms11061614

Simkovsky, R., Parnasa, R., Wang, J., Nagar, E., Zecharia, E., Suban, S., Yegorov, Y., Veltman, B., Sendersky, E., Schwarz, R., & Golden, S. S. (2022). Transcriptomic and phenomic investigations reveal elements in biofilm repression and formation in the cyanobacterium *Synechococcus elongatus* PCC 7942. *Frontiers in Microbiology, 13*, 899150. https://doi.org/10.3389/fmicb.2022.899150. PMID: 35814646; PMCID: PMC9260433.

Stewart, P. S. (2015). Antimicrobial tolerance in biofilms. *Microbiology Spectrum, 3*(3), MB-0010-2014. https://doi.org/10.1128/microbiolspec.MB-0010-2014

Stewart, P. S., & Franklin, M. J. (2008). Physiological heterogeneity in biofilms. *Nature Reviews Microbiology, 6*(3), 199–210. https://doi.org/10.1038/nrmicro1838

Van Opijnen, T., & Camilli, A. (2013). Genome-wide fitness and genetic interactions determined by Tn-seq, a high-throughput sequencing approach for microorganisms. *Methods in Molecular Biology, 966*, 301–316. https://doi.org/10.1007/978-1-62703-245-2_19

Venkatesan, J., & Kim, S. K. (2015). Industrial applications of microbial biofilms. *Microorganisms, 3*(4), 1050–1064. https://doi.org/10.3390/microorganisms3041050

Watnick, P. I., & Kolter, R. (2000). Biofilm formation by Gram-negative bacteria. *Annual Review of Microbiology, 54*, 731–758. https://doi.org/10.1146/annurev.micro.54.1.731

Yeung, C. H. L., Sahin, N., & Andrews, B. (2022). Phenomics approaches to understand genetic networks and gene function in yeast. *Biochemical Society Transactions, 50*(2), 713–721. https://doi.org/10.1042/BST20210285. PMID: 35285506; PMCID: PMC9162466.

Zhang, Z., Chen, G., Hussain, W., Pan, Y., Zhu, Y., Liu, Y., & Li, E. (2024). Machine learning and network analysis with focus on the biofilm in Staphylococcus aureus. *Computational and Structural Biotechnology Journal, 23*, 4148–4160, ISSN 2001-0370. https://doi.org/10.1016/j.csbj.2024.11.011

Comparative Genomics in Understanding Microbial Biofilms

11

Pooja Ramesh Swamy and Priyanka Sharma

11.1 Introduction

Over the past four decades, microbiology experts have classified microbes into two main environmental groups: (i) Single, autonomous, planktonic cells that float freely and (ii) Microbial aggregates or biofilms, which are clusters of bacteria. Initially, the term 'biofilm' referred exclusively to bacteria attached to surfaces. Recently, researchers have recognised that bacteria that form aggregates without attaching to surfaces can also be classified as biofilms (Costerton et al., 1978; McCoy et al., 1981; Sauer et al., 2022). Persistent biofilm infections are classified as either surface- or non-surface-associated. Infections linked to surfaces are commonly seen in individuals using medical devices or implants. Non-surface-related infections, such as those affecting soft tissue persistently, and chronic respiratory tract infections, are often linked to conditions, such as diabetes or compromised mucociliary clearance (Lebeaux et al., 2013). Understanding these classifications is essential to address the diverse challenges posed by biofilms in clinical settings.

Rapid technological advancements in biological research have enabled significant discoveries in the physiology, ecology and genetics of bacteria developing into biofilms. These include widespread utilisation of molecular, analytical and imaging techniques. For example, omics technologies have provided powerful tools for biofilm research. High-throughput DNA sequencing approaches, such as next-generation sequencing, have revealed the genetic coding potential of biofilm-forming organisms and the morphology of biofilm communities (Grzymski et al., 2008; Franklin et al., 2015). Transcriptomic techniques, including microarray, RNA-seq and RT-qPCR, have enhanced our understanding of global and localised gene

P. R. Swamy
The Scientific Way, Electronic City, Bengaluru, India

P. Sharma (✉)
Department of Tropical Disease Biology, Liverpool School of Tropical Medicine, Pembroke Place, Liverpool, UK

S. Busi et al. (eds.), *Omics Approaches in Biofilm Research*,
https://doi.org/10.1007/978-3-031-91863-6_11

expression in biofilms (Ishii et al., 2013). Additionally, mass spectrometry has advanced metabolomic and proteomic research, providing insight into biofilm functionality (Beale et al., 2013).

Both genomics and comparative genomics have emerged as transformative tools in this field. Genomics offers a comprehensive view of the genetic material of biofilm-forming organisms and highlights the genetic elements responsible for the development of biofilms, their durability and the capacity to withstand antibiotics. Comparative genomics extends this by identifying conserved pathways and unique adaptations across strains, species and environmental conditions (Shree et al., 2023). Studies of microorganisms, such as *Pseudomonas aeruginosa* and *Staphylococcus aureus*, have demonstrated how comparative genomics can elucidate the mechanisms underlying biofilm-associated infections. These insights have been instrumental in the development of biofilm-specific antibiotics and anti-adhesion strategies (Arciola et al., 2015; Hauser et al., 2016).

The integration of genomics with multi-omics approaches, machine learning and synthetic biology promises to revolutionise our ability to decode biofilm complexity and address biofilm-related challenges. This chapter explores these advancements and emphasises the role of comparative genomics in understanding and managing microbial biofilms in clinical and environmental contexts.

11.2 Biofilm

A biofilm is an organised community of microorganisms that cling together and is surrounded by a self-produced extracellular polymeric substance that adheres to either a living or a non-living surface (Wu et al., 2015). Biofilms, which are communities of microorganisms attached to surfaces, consist of multiple layers of cells enveloped in moist polymers (Kierek-Pearson & Karatan, 2005). Bacteria benefit from living in biofilms. The biofilm matrix comprises various components, including peptides like fibrin, saccharides such as alginate and external DNA. According to Potera (1996), biofilms are capable of forming on various surfaces, including biotic and abiotic ones, as well as within microaggregates. Biofilms were first studied in a patient suffering from recurrent bacteraemia after surface-attached bacteria were attached to a pacemaker lead (Thaarup & Bjarnsholt, 2021). An early mention of 'biofilm-forming bacteria' in medical literature was made by bacteria that adhered to the pacemaker lead. This has led to an increase in the interest in biofilm infections. Dr. R.J. Gibbons published his findings about the function of polysaccharide glycocalyx development in *Streptococcus mutans* on teeth almost 40 years ago, marking the earliest reports of the medical significance of bacterial communities now known as biofilms (Costerton et al., 1978). Biofilms are common in nature, as they represent a survival strategy for bacteria. In 1683, Antoni van Leeuwenhoek used a crude microscope to examine material from his own teeth to make observations and descriptions of biofilms. However, it was not until Nils Høiby noticed a connection between the cause of chronic infection and bacterial aggregation in patients with cystic fibrosis in the early 1970s that medical microbiologists became interested in the biofilm of microorganisms (Høiby, 2017; Vestby et al., 2020). Since then, numerous clinical illnesses have been linked to biofilm formation (Katharine & Karatan, 2005). The establishment of biofilms on the roots of plants and fungi may enhance bacterial acquisition and distribution of nutrients. Furthermore, it has been hypothesised that confinement of enzymes within the biofilm matrix could potentially expand the scope and improve the efficacy of organic matter degradation (Jass et al., 2002). Biofilms have been linked to microbially induced oilfield pipeline degradation, pipe blockages, ship hull fouling that increases drag and fuel expenses, cooling tower heat transfer declines and manufacturing line fouling that contaminates products (Vishwakarma, 2020).

The biofilm matrix creates an ideal environment for metabolic exchange by slowing the transport of tiny molecules into and out of the biofilm's structure. Consequently, several species or strains with different nutritional needs can live together in biofilms. Additionally, it has been proposed that the interaction of bacteria in biofilms promotes lateral transfer of genetic material. Component bacteria may benefit from this transfer to address novel environmental problems. Gene transfer has been observed in biofilms (Molin & Tolker-Nielsen, 2003). The biofilm matrix acts as a barrier against chemical attacks, such as chlorine (Norwood & Gilmour, 2000), hydrogen peroxide and physical forces (Elkins et al., 1999). Most drugs cannot kill biofilm bacteria and evade detection by avoiding phagocytosis (Mah et al., 2003; Stroh et al., 2011). Beyond the safeguarding effects of the matrix, bacteria within microbial films employ an array of strategies to counteract the host's defensive mechanism. They may inflict local tissue damage and create an acute infection by remaining dormant and hiding from the immune system. A reduced metabolic rate and slower rate of cell division may result from changes in bacterial metabolism, gene expression and protein synthesis. This change is due to their adaptation to ambient anoxic conditions and nutritional limitations within biofilms (Stoodley & Hall-Stoodley, 2009; Donlan & Costerton, 2002; Vestby et al., 2020).

The existence of a dormant subpopulation known as persisters is another significant biofilm feature linked to drug resistance (Jayaraman, 2008). *Staphylococcus* species were the first to be found to have persister cells, followed by *Escherichia coli*, *P. aeruginosa*, *Lactobacillus acidophilus* and *Candida* species (Mulcahy et al., 2010; Wuyts et al., 2018). Metabolically inert persister cells can withstand antibiotic treatment, because antibiotics are often effective against metabolically active cells (Miyaue et al., 2018).

11.3 Biofilm Formation Stages

Antibiotic resistance is supported by the distinct characteristics of biofilm cells that encourage infection. The formation and architecture of biofilms are influenced by the specific bacterial species and strains involved as well as the environmental conditions in which they develop (Costerton et al., 1978). A notable illustration of this phenomenon was observed in *P. aeruginosa* biofilms grown in the flow chambers. When glucose was used as the growth medium, these biofilms formed microcolonies with distinctive mushroom-like shapes. Conversely, when citrate serves as the carbon source, the resulting biofilms exhibit a 'flat' configuration (Klausen et al., 2003). Depending on the conditions, *S. aureus* can use different strategies to form biofilms.

The archetypes of biofilms consist of the following:

(i) Polysaccharide biofilm: Expression of polysaccharide intercellular adhesin and poly-N-acetylglucosamine
(ii) Protein/extracellular biofilm: Integrates extracellular DNA from lysed cells into the biofilm matrix and mediates cell-to-cell interactions via surface proteins.
(iii) Fibrin biofilm: Biofilm supported by a fibrin scaffold generated through coagulase-mediated plasminogen activation.
(iv) Amyloid biofilm: Phenol-soluble compounds are used to encourage biofilm formation and dissemination (Zapotoczna et al., 2016).

During infection, interactions between the bacteria and the host have a significant impact on biofilm production. In contrast to biofilms produced on polystyrene in bacteriological medium, *S. aureus* biofilms developed for less than 24 h on human plasma-conditioned surfaces and exposed to shear flow in chemically defined media to simulate human infection were noticeably more susceptible to rifampicin and vancomycin (Zapotoczna et al., 2015). Generally, there are a few significant events in the complicated processes of biofilm development.

There are five basic stages that make up the lifecycle of biofilms: (i) primary adhesion, (ii) permanent bonding, (iii) formation of small colonies, (iv) development of mature biofilm and (v) detachment (Sauer et al., 2022; Liu et al., 2024). When unattached pelagic cells adhere to a life-associated or inert surface, biofilms are formed. Water pipes; indwelling medical devices, such as catheters; and live tissues, such as gut and urinary tract epithelial cells, are just a few of the many surfaces that cells can adhere to (Donlan, 2002). Conventional biofilm formation models explain how single cells first adhere to a surface, forming a reversible attachment before adhering to an irreversible attachment. In response to signals, such as a spike in intracellular cyclic diguanylate monophosphate (c-di-GMP), a secondary messenger that is crucial for regulating biofilm formation, the biofilm starts to form microcolonies and moves into the maturation step after cell adhesion has been established (Berne et al., 2018). Cells divide to create microcolonies and secrete the extracellular matrix as the process progresses. The matrix, which is made up of a collection of different biopolymers known as exopolysaccharides, can comprise more than 90% of the biofilm volume (Flemming & Wingender, 2010). By providing mechanical stability, extracellular matrix protects against desiccation, UV light, biocide antibiotics and host immunological responses, all of which increase resistance (Koo et al., 2017; Fulaz et al., 2019). Cells can separate from the surface of the biofilm and migrate to colonise new substrates once they reach maturity (Rumbaugh & Sauer, 2020). Planting, weathering and shedding are considered processes pertaining to the release of cells in biofilms, but these processes are complicated. Large numbers of cells or microcolonies are quickly released from the biofilm during the active cell dispersal process known as 'seeding', which causes the biofilm to create hollow cavities (Kaplan, 2010; Rumbaugh & Sauer, 2020). Cells within the biofilm frequently initiate this response to environmental changes, such as stress from food shortages or the presence of antibiotics (Muhammad et al., 2020). While erosion occurs passively as a result of external pressures, such as mechanical processes, such as tooth brushing and shear flow, sloughing occurs when large portions of the biofilm unexpectedly separate from it over time (Kaplan, 2010; Fleming & Rumbaugh, 2017; Rumbaugh & Sauer, 2020).

The variables affecting cell attachment and biofilm formation were significant for biofilm formation. Attachment occurs more readily on surfaces that are textured, water-repellent and coated with surface conditioning films (Donlan, 2002). The chemistry of the surface is changed by conditioning films, which are made of organic compounds and encourage colonisation (Loeb & Neihof, 1975). Laminar flow velocity promotes biofilm growth, but hydrodynamic shear forces control early attachment. Interaction with surfaces is influenced by microbial properties, such as hydrophobicity, cell surface charge and structural appendages (such as flagella and fimbriae). Finally, environmental factors that affect adhesion and biofilm stability include pH, temperature and nutrient availability (Costerton et al., 1978; Fletcher & Loeb, 1979).

11.4 Biofilms-Associated Diseases

The pathophysiology of many chronic illnesses in different organ systems is greatly influenced by bacterial biofilms (Fig. 11.1). Biofilms have been linked to disorders, such as cholesteatoma and otitis media in the auditory system, which result in chronic inflammation and recurrent infections. Biofilm deposition on heart valves or arterial plaques is a feature of cardiovascular illnesses, such as atherosclerosis and infective endocarditis, which makes therapy more difficult and exacerbates the disease. Biofilms in the digestive tract have been linked to inflammatory bowel illnesses and chronic typhoid fever linked to gallstones, which may lead to colorectal cancer. Biofilm-related wound infections affect the integumentary system, causing chronic inflammation and hindering wound healing. Polymicrobial biofilms are linked to bacterial vaginosis and chronic endometritis in the reproductive tract, which can lead to recurring therapy-resistant infections. Biofilms in the respiratory system worsen persistent rhinosinusitis and play a role in cystic fibrosis and lung infections. Dental biofilms, which form in the mouth, are recognised as a cause of oral conditions such as gingivitis, periodontitis and dental caries (Huang et al., 2011; Vestby et al., 2020).

11.5 Key Genes in Biofilm Formation

Biofilm production is an intricate process that is governed by numerous genes across different bacterial species. This process encompasses various functions, including cellular cohesion, microbial group behaviour signalling, construction of

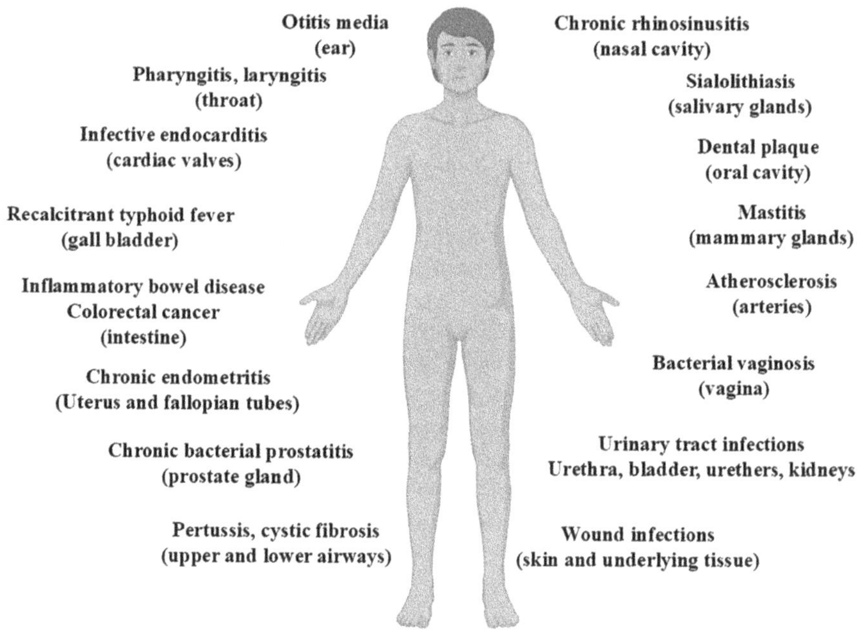

Fig. 11.1 Diseases associated with bacterial biofilms

cell membranes, physiological functions and mechanisms for coping with environmental challenges (Table 11.1). For example, the *icaADBC* operon in *S. aureus* and *S. epidermidis* serves as a crucial component in generating intercellular adhesins that stabilise the biofilm matrix. Similarly, *gbpA* in *S. mutans* contributes to structural integrity by producing polysaccharides essential for the biofilm matrix. In quorum sensing, genes such as *lasI* in *P. aeruginosa* and *luxS* in *S. mutans* facilitate bacterial communication and coordinate biofilm development through the synthesis of signalling molecules (Zuberi et al., 2017).

Cell wall synthesis is another vital aspect supported by genes, such as *PBP2B* and *glmM* in *S. gordonii*, which ensures bacterial structural stability under environmental stress. Metabolic regulation through genes, such as *ccpA*, in *S. mutans* links carbon metabolism to biofilm growth, emphasising the interconnection between metabolism and biofilm development (Zhang et al., 2021). Stress response genes, such as σB in *S. aureus* and *dgk* in *S. mutans*, play crucial roles in bacterial survival under adverse conditions by enhancing resilience to environmental changes (Ranganathan et al., 2020). This integrated genetic network of adhesion, quorum sensing, cell wall synthesis and stress responses underscores the adaptive capabilities of biofilms in diverse environments. Understanding these genetic factors is essential for the development of targeted strategies to combat biofilm-associated infections (Sahoo & Meshram, 2024).

Figure 11.2 illustrates the application of comparative genomics to identify key shared gene categories involved in biofilm formation for *P. aeruginosa and S. aureus.*

Table 11.1 Functions and key genes in biofilm formation across species

Gene	Primary function	Species	Category	Notes
abpA	Promotes adhesion by binding amylase	*Streptococcus gordonii*	Adhesion	Facilitates attachment to salivary surfaces (Zuberi et al., 2017)
sspA/B	Interacts with salivary proteins	*Streptococcus gordonii*	Adhesion	Contributes to oral biofilm formation (Rath et al., 2021)
icaADBC	Synthesises intercellular adhesins	*S. aureus, S. epidermidis*	Adhesion	Crucial for biofilm matrix stability (Hobley et al., 2015)
gbpA	Produces biofilm matrix polysaccharides	*Streptococcus mutans*	Adhesion	Enhances structural integrity (Limoli et al., 2015)
lasI	Regulates quorum-sensing molecules	*P. aeruginosa*	Quorum Sensing	Facilitates cell communication (Zhang et al., 2021)
luxS	Influences signalling pathways	*S. mutans, E. coli*	Quorum Sensing	Essential for autoinducer-2 synthesis (Zuberi et al., 2017)
PBP2B	Supports peptidoglycan synthesis	*Streptococcus gordonii*	Cell Wall	Ensures cell wall integrity (Huang et al., 2024; Park et al., 2020)
glmM	Aids in cell wall precursor synthesis	*Streptococcus gordonii*	Cell Wall	Maintains bacterial structural stability (Luo et al., 2023)
ccpA	Links carbon metabolism to biofilm growth	*Streptococcus mutans*	Metabolism	Regulates biofilm-related metabolic pathways (Zhang et al., 2021)
σB	Manages stress responses	*S. aureus, S. epidermidis*	Stress Response	Enhances survival under adverse conditions (Kazmierczak et al., 2003; Valle et al., 2019)
Dgk	Regulates bacterial stress response	*Streptococcus mutans*	Stress Response	Helps bacteria adapt to environmental changes (Lis & Kuramitsu, 2003; Shibata et al., 2009)

Comparative genomic analysis has revealed two main categories of genes that play crucial roles in both species: genes for adhesion and genes for extracellular polymeric substances (EPS) production. Adhesion genes facilitate the initial attachment of bacteria to surfaces, while EPS production genes are crucial for forming the biofilm matrix. This comparative approach provides insight into the core genetic elements of biofilm formation across these clinically relevant pathogens, highlighting the power of comparative genomics in elucidating common mechanisms of biofilm development.

By identifying these shared genetic elements, comparative genomics offers potential targets for developing strategies to combat biofilm-associated infections in both species.

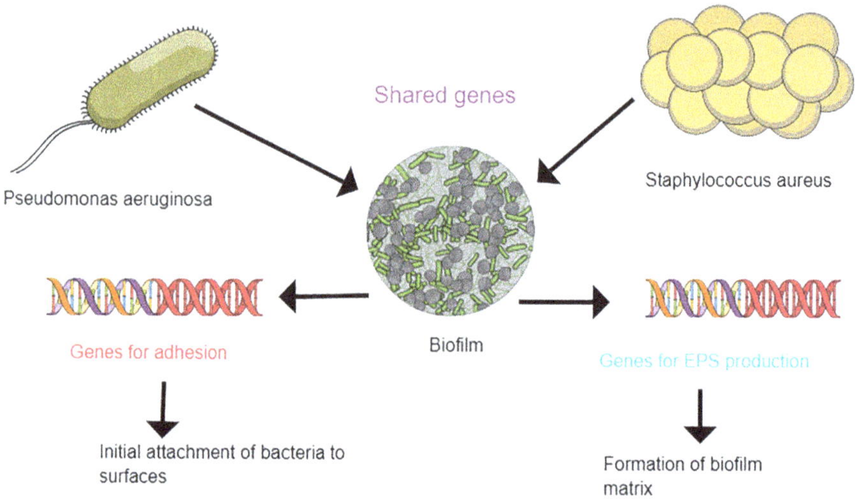

Fig. 11.2 Comparative genomics of shared gene categories in biofilm formation: *P. aeruginosa* and *S. aureus*

11.6 Approaches of Genomics and Metagenomics for Biofilm Studies

Researchers may now access genome-scale data on biofilms because of the development of high-throughput sequencing technologies such as Illumina, Pacific Biosciences [PacBio], SOLiD and Roche 454 (Metzker, 2010). Additionally, many of these technologies offer insight into epigenetics, which could be one of the mechanisms influencing the physiology and gene expression of bacterial biofilms (Korlach & Turner, 2010). Whole-genome sequences of thousands of different microorganisms have been obtained using these novel sequencing technologies. As the cost of acquiring genome sequences continues to decrease, it is now possible for scientists to gather the complete genome sequences of many more strains. These advanced tools have illuminated the community structure of natural biofilms, the roles of key genes in biofilm formation and the cellular activities within these structures. These findings set the stage for a deeper analysis through comparative genomics, which explores conserved and species-specific genetic elements (Franklin et al., 2015).

11.7 Comparative Genomics in Biofilm Research

Building on these genomic insights, comparative genomics delves into the genetic similarities and differences among organisms to uncover functional elements, evolutionary processes and adaptive mechanisms. In biofilm research, this approach identifies conserved genes essential for biofilm development, and species-specific adaptations influence resilience and virulence (Alföldi & Lindblad-Toh, 2013; Saini et al., 2024). Comparative genomics in biofilm research has been extended beyond the identification of conserved genes and species-specific adaptations. It encompasses analysis of entire genomes, including coding and non-coding regions, to uncover the genetic basis of biofilm formation, development and persistence. This approach allows researchers to trace the evolutionary history of biofilm-related genes, thereby revealing how different organisms have adapted to form and maintain complex microbial communities in diverse environments (Yang et al., 2023; Wang et al., 2023a, b).

The application of comparative genomics to biofilm studies has provided significant insights into the molecular pathway underlying biofilm production. By comparing the genomes of strains capable of biofilm production versus those lacking this ability within a species or across different species, researchers can identify the key genetic elements responsible for adhesion, extracellular matrix production and quorum sensing (Mirghani et al., 2022; Damyanova & Paunova-Krasteva, 2025). Furthermore, this approach revealed the importance of horizontal gene transfer in the acquisition of biofilm-related genes, thereby highlighting the role of mobile DNA sequences in the spread of biofilm-forming capabilities among bacterial populations. These findings offer significant implications for understanding the evolution of biofilm-forming traits and developing strategies to combat biofilm-associated infections and biofouling in industrial settings. Comparative genomics is a powerful approach for understanding the genetic evolution across species by examining and comparing genomic information (Bornstein et al., 2023; Ellegren, 2008). This field has significantly advanced our understanding of genome structure, function and evolutionary processes, providing insight into how natural selection shapes the evolution of DNA sequences (Ellegren, 2008). Notably, comparative genomics studies have shown that the functional divergence of protein-coding genes varies among different lineages and appears to be related to population size. For example, the proportion of non-synonymous to synonymous substitutions is greater in apes than in rodents, lending support to Ohta's nearly neutral theory of molecular evolution (Ellegren, 2008). This observation implies that mildly detrimental mutations play a role in protein evolution, to a degree that is inversely related to the effective size of the population.

In conclusion, comparative genomics has become an essential tool for exploring genetic evolution, uncovering the role of favourable Darwinian selection occurring in 10–40% of genes across various evolutionary lineages (Ellegren, 2008). It has also highlighted the importance of regulatory elements in trait evolution and provided valuable information on population history and demography through polymorphism data (Ellegren, 2008). As the field continues to evolve, integrating comparative genomics with other disciplines, such as ecology, archaeology and conservation biology, is crucial for a comprehensive understanding of genetic evolution and its implications for biodiversity and human health (Cassini, 2013; Lamichhaney et al., 2019).

11.8 Application of Comparative Genomics to Biofilm Studies

Comparative genomics has been the cornerstone for identifying the genetic determinants of biofilm formation (Nobrega & Pennacchio, 2004; Bevers et al., 2022). For example, studies on *P. aeruginosa* and *S. aureus* have demonstrated that comparative analysis can reveal the genetic adaptations that contribute to their survival in biofilm consortia. Comparative genomics is a powerful approach that involves analysing and comparing the genetic information of different organisms to understand their biology, evolution and functional characteristics (Bornstein et al., 2023; Sivashankari & Shanmughavel, 2007; Ellegren, 2008; Enard, 2014). It encompasses the study of genome structure, coding regions and non-coding regions to gain insight into the similarities and differences between genomes (Locey & White, 2011; Gorban et al., 2020). This field has become increasingly important for microbial classification, systematics and evolution, particularly with the exponential growth of fungi (Ivanova et al., 2009; Konkel & Slot, 2023). Comparative genomics enables researchers to investigate the genetic basis of organismal biology and ecology, identify pathogenicity-related chromosomes and to understand the evolution of toxin gene clusters (Konkel & Slot, 2023; Rep & Kistler, 2010). They also play a crucial role in detecting horizontally acquired genes, which are important in bacterial genomics (Bolotin & Hershberg, 2017; Groisman & Choi, 2023).

Comparative genomics has wide-ranging applications, including the identification of bacterial pathogens at the molecular level, genetic typing, identification of virulence-associated genes and elucidation of the underlying mechanisms of physiological adaptations to environmental fluctuations (Ivanova et al., 2009; Whitehead et al., 2012). Tools, such as DiGAlign and Genomicus, have been developed to facilitate comparative genomic analyses, providing versatile functionality and intuitive interfaces for researchers (Nguyen et al., 2021; Nishimura et al., 2024). As the field continues to advance, it holds promise for impacting various areas of research, such as zoonotic disease research, therapeutic development and climate change studies (Bornstein et al., 2023; Whitehead et al., 2012). Comparative genomics has emerged as a powerful tool for unravelling the genetic underpinnings of biofilm formation in various bacterial species. By comparing the genomes of biofilm- and non-biofilm-forming strains, researchers have identified specific genes and regulatory elements that play crucial roles for the formation and sustenance of biofilms. This approach has been particularly useful for studying clinically relevant pathogens, such as *P. aeruginosa* and *S. aureus*, which are notorious for their ability to form robust biofilms in medical settings (Müsken et al., 2009; Beaudoin et al., 2017). Applying comparative genomics to microbial film studies has yielded significant insights into adaptive strategies employed by bacteria within biofilm communities. For example, analyses have revealed genetic modifications that enhance surface adhesion, promote the production of extracellular polymeric substances and facilitate cell-to-cell communication within the biofilm matrix. These studies have also uncovered novel targets for potential therapeutic interventions aimed at disrupting biofilm formation or enhancing susceptibility to antimicrobial agents. As genomic technologies continue to advance, the amalgamation of comparative genomic analyses with alternative omics methodologies, namely transcriptomics and proteomics, promises to provide a more comprehensive understanding of the complex biological processes underlying biofilm formation and persistence (Müsken et al., 2009; Beaudoin et al., 2017).

Comparative genomics has been instrumental in identifying the genetic determinants of biofilm produced by *P. aeruginosa* and *S. aureus*. Comprehensive screening of the *P. aeruginosa* PA14 mutant library revealed 394 genetic determinants for reduced biofilm production and 285 determinants for enhanced biofilm production (Müsken et al., 2009). This study highlighted the complexity of biofilm formation. The study found that 67% of the detected mutants impacted genes coding putative proteins, hinting at the existence of yet-to-be-discovered developmental pathways.

Interestingly, the interaction between *P. aeruginosa* and *S. aureus* during biofilm formation is complex. Although *P. aeruginosa* generally inhibits *S. aureus* biofilm formation, some studies have shown that certain isolates of *P. aeruginosa* can form different biofilm architectures in the presence of *S. aureus* exoproducts (Beaudoin et al., 2017). This interaction relies on Psl generated by *P. aeruginosa* and the staphylococcal protein A from *S. aureus*, highlighting the importance of specific genetic elements in interspecies biofilm formation. In conclusion, comparative genomics revealed that biofilm synthesis by *P. aeruginosa* and *S. aureus* involves intricate interplay between various genetic determinants. These encompass genes associated with survival in low-oxygen environments, arginine metabolism, signalling via alkyl-quinolones, maintenance of pH balance and mechanisms for repairing DNA (Müsken et al., 2009). Furthermore, the genetic basis of the interspecies interactions in biofilm formation adds another layer of complexity, emphasising the need for further genomic studies to fully understand these processes.

11.9 Role of Genomic Diversity in Biofilm Resilience

Genomic diversity, fuelled by horizontal gene transfer and mutations, plays a pivotal role in biofilm resilience. For instance, small colony variants of *S. aureus* exhibit increased survival in biofilm consortia with *P. aeruginosa*, a phenomenon linked to altered gene expression profiles, such as higher *sigB* and *sarA* activities (Mitchell et al., 2010; Mirani et al., 2018). Genomic diversity within biofilm communities significantly contributes to their resilience and adaptability to various environments. The observed genetic variation stems from several mechanisms of horizontal genetic exchange, such as conjugation, conversion and transduction, as well as spontaneous mutations in genetic material (Bonham et al., 2017; Fall et al., 2007). These processes allow the rapid exchange and acquisition of genetic material, enabling biofilm-forming bacteria to adapt to changing conditions and resist antimicrobial treatments. Small colony variants (SCVs) of *S. aureus* exemplify this phenomenon as these variants demonstrate enhanced survival capabilities when coexisting with *P. aeruginosa* in mixed-species biofilms. The increased resilience of *S. aureus* SCVs in biofilm consortia with *P. aeruginosa* was attributed to specific alterations in the gene expression patterns. Notably, these variants exhibit upregulated *sigB* and *sarA* gene activities, which are known to play crucial roles in the stress response and virulence regulation (Riordan et al., 2006; Jenul & Horswill, 2019). This genetic adaptation allows *S. aureus* to withstand the competitive pressures exerted by *P. aeruginosa* and persist within the biofilm structure. Such genomic diversity not only enhances the survival of individual species but also contributes to the overall stability and functional diversity of the biofilm community, making it more resistant to external stressors and antimicrobial interventions (Mirghani et al., 2022; Agarwal et al., 2024).

Horizontal gene transfer (HGT) and mutations are significant contributors to genomic diversity and biofilm resilience. HGT has been recognised as a major force in bacterial genome evolution, enabling microorganisms to swiftly acquire novel abilities and to adjust to fluctuating environments (Fall et al., 2007; Bonham et al., 2017). In biofilm communities, HGT has been shown to occur frequently, with transferred genes often involved in beneficial functions for bacteria living in specific niches, such as nutrient transport, sugar degradation and stress response (Song et al., 2021). Interestingly, the outcome of HGT about biofilm resilience is by no means solely reliant on the acquisition of new genes. The transfer of exogenous genes can extensively modify the entire proteome, resulting in substantial modifications to existing genetic and phenotypic characteristics (Bedhomme et al., 2019). This allows bacteria that receive genetic material can investigate novel modes of operation, extending far beyond the simple acquisition of a new capability. Mutations play a crucial role in the resilience of biofilms. For instance, experimental evolutionary studies have shown that mutations can occur in antibiotic off-target genes and genes responsive to translational stress, depending on the growth conditions and transferred gene version (Bedhomme et al., 2019). In conclusion, the interplay between HGT and mutations contributes significantly to genomic diversity and biofilm resilience. Although HGT allows the rapid acquisition of potentially beneficial genes, mutations fine-tune the integration and expression of these genes in the new genomic context. This dynamic process enables biofilm communities to adapt quickly to environmental challenges, enhancing their resilience and survival in various ecological niches (Liang et al., 2020; Song et al., 2021).

11.10 Enhancing Understanding of Biofilm-Forming Organisms

Biofilm-forming bacteria display phenotypic traits, such as increased antibiotic resistance and stress tolerance. These phenotypes are linked to specific genetic markers, such as those associated with efflux pumps and biofilm matrix production (Feldbauer et al., 2015; Ciofu et al., 2017). Biofilm-forming bacteria exhibit phenotypic traits that contribute to their survival and persistence in various environments. One of the most significant traits is their increased resistance to antibiotics, which poses a major challenge in the clinical setting. This enhanced resistance is often attributed to the physical barrier provided by the extracellular matrix of the biofilm as well as the altered metabolic state of bacteria within the biofilm structure (Anderson & O'Toole, 2008; Høiby et al., 2010). Additionally, biofilm-forming bacteria demonstrate an increased tolerance to various environmental stresses, including pH fluctuations, temperature changes and oxidative stress. These traits enable them to thrive under hostile conditions, where planktonic bacteria may struggle to survive (Olsen, 2015). Phenotypic characteristics of biofilm-forming bacteria are closely linked to specific genetic markers. For instance, genes associated with efflux pump systems play a crucial role in antibiotic resistance by actively expelling antimicrobial compounds from the bacterial cells. Similarly, genes responsible for the production of extracellular polymeric substances (EPS) are essential for biofilm matrix formation, which provides structural integrity and protection to the bacterial community (Vu et al., 2009; Flemming et al., 2016). Other genetic markers may be involved in quorum sensing, a communication system that allows bacteria to coordinate their behaviour within the biofilm. Understanding these genetic underpinnings is vital for developing targeted strategies to combat biofilm-associated infections and mitigate their impact in various sectors, including healthcare and industrial settings (Gunathilaka et al., 2024).

Biofilm-forming bacteria exhibit increased antibiotic resistance and stress tolerance compared with their planktonic counterparts. These phenotypic traits are associated with various genetic markers and mechanisms that are specific to biofilm lifestyle (Ciofu et al., 2017; Olsen, 2015). Biofilms demonstrate enhanced antibiotic resistance through multiple mechanisms, including decreased diffusion of antibiotics through the matrix, altered metabolic activity due to nutrient and oxygen gradients, and formation of persister cells (Anderson & O'Toole, 2008; Høiby et al., 2010). Genetic factors contributing to biofilm-specific resistance include mutations in the regulators of efflux pumps, as observed in *Acinetobacter baumannii* biofilms exposed to ciprofloxacin (Haddadi et al., 2024). Additionally, biofilm formation is positively correlated with resistance to certain antibiotics, such as ciprofloxacin and trimethoprim-sulfamethoxazole, along with the presence of quinolone resistance genes carried on plasmids, such as *qnrS* (Gunathilaka et al., 2024). Interestingly, biofilm-adapted populations may exhibit evolutionary dynamics different from those of planktonic populations. Although planktonic bacteria tend to acquire mutations in primary drug targets, biofilm-adapted bacteria may develop mutations in regulatory genes, leading to a trade-off between resistance and overall fitness (Ahmed et al., 2018). This highlights the complex interplay between genetic markers, phenotypic traits and biofilm lifestyle in the development of antibiotic resistance and stress tolerance.

11.11 Case Studies: *S. aureus* and *P. aeruginosa*

Comparative studies on *S. aureus* and *P. aeruginosa* have revealed intricate interspecies interactions and biofilm dynamics in polymicrobial environments. As shown in Table 11.2, *P. aeruginosa* typically dominates in co-culture conditions by producing factors, such as cis-2-decanoic acid, which inhibits *S. aureus* biofilm formation

Table 11.2 Analysing biofilm formation and species interactions: A comparative examination of *S. aureus* and *P. aeruginosa*

Study focus	Key findings	Regulatory factors/ Interactions	References
Biofilm Dynamics in polymicrobial environments	In mixed biofilms, *P. aeruginosa* predominantly outcompetes *S. aureus*, which survives as small colony variants	Environmental influences, such as host proteins like calprotectin, regulate competition and growth	Miller and Bassler (2016)
Competitive Interactions in biofilm formation	*P. aeruginosa* secretes cis-2-decanoic acid, which inhibits *S. aureus* biofilm formation and reduces its growth	Exoproducts from *P. aeruginosa* disrupt *S. aureus* biofilm development in vitro	Mirani et al. (2018)
Contradictory results in co-culture	Certain strains of *S. aureus* enhance *P. aeruginosa* motility without negatively affecting biofilm formation	Interaction triggers exploratory motility in *P. aeruginosa* and invasion of *S. aureus* colonies via type IV pili	Woods et al. (2019), Limoli et al. (2019)
Modulation of interspecies interactions	Interaction outcomes depend on the specific strains and environmental factors influencing their biofilms	Traits such as colony size, hydrophobicity and growth rate impact the competitive dynamics	Wu et al. (2019)

and reduces its growth (Mirani et al., 2018; Lopes et al., 2023; Najeeb et al., 2024). Despite this competitive advantage, *S. aureus* adapts by forming small colony variants, enabling its persistence in hostile biofilm settings (Miller & Bassler, 2016). Environmental factors such as the host-derived protein calprotectin play a significant role in modulating these interactions. Calprotectin sequesters essential metals like zinc and manganese, altering the competitive balance between *S. aureus* and *P. aeruginosa* based on their respective metal acquisition strategies (Miller & Bassler, 2016). Some studies have reported contradictory results. As shown in Table 11.2, specific *S. aureus* strains enhanced *P. aeruginosa* motility without impairing biofilm formation. This interaction triggers exploratory motility in *P. aeruginosa*, allowing it to invade *S. aureus* colonies via type IV pili (Limoli et al., 2019). Overall, the interactions between these species are strain-specific and heavily influenced by environmental conditions. Factors, such as growth rate, cell surface hydrophobicity and the presence of small-colony variants significantly affect biofilm consortia dynamics (Proctor et al., 2006). Understanding these complex interspecies interactions is essential to develop effective antimicrobial strategies against polymicrobial infections.

11.12 Advanced Techniques in Biofilm Genomics

11.12.1 Single-Cell Genomics

Single-cell genomics has been instrumental in studying biofilm heterogeneity. This technique has revealed how genetic variation within individual cells contributes to the overall functionality and resilience of biofilms (Pal, 2021; Yan et al., 2024). Single-cell genomics has revolutionised our understanding of biofilm heterogeneity by providing unprecedented insights into the genetic makeup and functional diversity of individual cells within complex microbial communities. This technique allows researchers to examine genetic variations, gene expression patterns and metabolic activities of single cells, thereby revealing the intricate cellular heterogeneity that exists within biofilms (Yan et al., 2024; Monteith et al., 2023). By analysing individual cells, scientists can identify subpopulations with distinct genetic profiles, metabolic capabilities and functional roles, shedding light on how these diverse cellular components contribute to the overall structure, stability and resilience of biofilms (Blainey, 2013).

The application of single-cell genomics in biofilm research has led to several important discoveries. Genetic diversity within biofilms is often much higher than previously thought, with individual cells exhibiting unique genetic adaptations that contribute to the overall fitness of the community. This technique has also helped elucidate the mechanisms underlying antibiotic resistance in biofilms, showing how certain cell subpopulations can survive antibiotic treatment and potentially serve as reservoirs for recolonisation (Paolillo et al., 2019). Furthermore, single-cell genomics has provided insights into the spatial organisation of different cell types within biofilms, helping researchers understand how this organisation influences biofilm formation, maintenance and dispersal processes (Mclean et al., 2013; Wang et al., 2023a, b). Single-cell genomics has emerged as a powerful tool to investigate the heterogeneity and resilience of biofilms. This method enables the identification of cell diversity inside a sample without losing the information that occurs in the bulk tissue (Paolillo et al., 2019). By enabling the study of uncultivated microorganisms and recognising the heterogeneity among seemingly similar cells, single-cell genomics has significantly advanced our understanding of microbial communities and population biology (Blainey, 2013). Interestingly, single-cell genome sequencing has revealed surprising diversity within biofilms in critical areas, such as healthcare facilities. For instance, a study of a hospital bathroom sink drain biofilm identified 25 different genera, including environmental species, human commensals and opportunistic pathogens (Mclean et al., 2013). This highlights the potential of single-cell genomics to capture novel genomes and enable the comparative genomic analysis of strain variations in complex biofilm samples. In conclusion, single-cell genomics has revolutionised our understanding of biofilm heterogeneity and resilience. The development of optimised methods, such as BaSSSh-seq, has allowed the study of transcriptional heterogeneity and regulatory networks within biofilm subpopulations (Homberger et al., 2023). These advancements have not only improved our knowledge of biofilm formation and persistence but have also opened new avenues for developing targeted strategies to combat biofilm-related infections and antibiotic resistance (Yan & Bassler, 2019). As this field continues to evolve, single-cell genomics promises to transform cancer research and precision medicine further by enhancing our understanding of gene regulation and disease mechanisms (Lim et al., 2020; Nomura, 2020).

11.12.2 Metagenomics

Metagenomic approaches have enabled researchers to study the genetic composition of biofilm communities. They have been used to elucidate the interactions and cooperative behaviours among microbial species in mixed-species biofilms (Van Merode et al., 2006). Metagenomic approaches have revolutionised our understanding of biofilm communities by providing a comprehensive view of their genetic composition. This technique allows researchers to analyse the collective genomes of all microorganisms present in a biofilm, including those that are difficult or impossible to culture under laboratory conditions. Metagenomics offers insights into the functional potential and diversity of biofilm communities by examining genetic material extracted directly from environmental samples (Mclean & Kakirde, 2013). The application of metagenomics to mixed-species biofilms has been particularly valuable for elucidating the complex interactions and cooperative behaviours among different microbial species. These studies have revealed intricate networks of metabolic dependencies, communication systems and synergistic relationships that contribute to the overall stability and resilience of biofilms (Majumdar & Roy, 2020). Furthermore, metagenomic analyses have helped to identify key genes and pathways involved in biofilm formation, maintenance and dispersal, providing potential targets for biofilm control strategies in various fields, including medicine, industry and environmental management (Rosario & Breitbart, 2011).

Metagenomic approaches have revolutionised our understanding of biofilm communities by providing a comprehensive view of their genetic composition and functional potential (Dopheide et al., 2015). These techniques allow the exploration of microbial communities in various environments without the need for cultivation, thereby revealing the full extent of microbial gene diversity (Gibbons & Gilbert, 2015). In particular, shotgun metagenomic sequencing has enabled researchers to detect and characterise the diversity and function of complex microbial communities with greater resolution, identifying species and strains across phyla as well as their functional content (Doonan et al., 2016). This approach has demonstrated that dominant viruses in environmental communities are often not well represented by cultured viruses in existing sequence databases, highlighting the importance of culture-independent methods (Rosario & Breitbart, 2011). Interestingly, metagenomic studies have also revealed unexpected virus-host interactions, revealing genes that may allow viruses to manipulate their hosts in previously unknown ways (Rosario & Breitbart, 2011). This finding challenges the traditional view of bacteriophages as mere predators, and suggests a more complex relationship between viruses and their bacterial hosts (Fernández et al., 2018). However, challenges remain in annotating the functions because of the limited functional databases for environmental bacteria compared to model organisms (Coelho & Cock, 2020). Despite these limitations, metagenomic approaches continue to offer valuable insights into the structure, heterogeneity and ecology of microbial consortia, including biofilms, and hold great potential for advancing fields, such as evolutionary biology, pathogen surveillance and biotechnology (Sabra et al., 2010; Rosario & Breitbart, 2011; Alves et al., 2018).

11.13 Potential Therapeutic Targets

11.13.1 Identification of Targets Through Comparative Genomic Analysis

Targeting biofilm-specific pathways, such as quorum sensing and exopolysaccharide production, offers promising therapeutic avenues. For instance, quorum-sensing inhibitors have been shown to be effective in disrupting biofilm formation by *P. aeruginosa* (Zeng et al., 2008; Li et al., 2022).

Comparative genomic analysis has emerged as a powerful tool for identifying potential targets in biofilm-specific pathways, particularly focusing on quorum sensing and exopolysaccharide production mechanisms. This approach allows researchers to identify conserved genes and regulatory elements across different bacterial species that are crucial for biofilm formation and maintenance (Sauer, 2003). By comparing the genomes of biofilm and non-biofilm-forming strains, scientists can identify unique genetic signatures associated with biofilm development, thereby providing valuable insights into potential therapeutic targets. The efficacy of targeting quorum sensing pathways, as demonstrated in *P. aeruginosa*, highlights the potential of this approach in combating biofilm-related infections. Quorum sensing inhibitors interfere with bacterial communication systems that coordinate biofilm formation and effectively disrupt the process (Majumdar, 2020). Similarly, targeting exopolysaccharide production pathways can weaken the structural integrity of biofilms, making them more susceptible to antimicrobial treatments. These strategies, informed by comparative genomic analysis, offer promising avenues for developing novel antibiofilm therapies that could potentially overcome the limitations of traditional antibiotic treatments in addressing biofilm-associated infections.

Targeting quorum sensing (QS) has emerged as a promising approach to disrupt biofilm formation and combat bacterial infection. QS inhibitors (QSIs) and quorum quenching enzymes (QQEs) have shown significant potential to interfere with bacterial communication and reduce virulence factors. By targeting QS-regulated virulence, pathogens can be disarmed without exerting pressure on their growth, thereby potentially preventing the emergence of resistance (Ahmed et al., 2019). Interestingly, the combination of QSIs and QQEs demonstrated enhanced efficacy in blocking multiple QS systems in individual bacterial species. For instance, a study using AiiA as a QQE and G1 as a QSI almost completely blocked *P. aeruginosa* Las and Rhl QS systems (Fong et al., 2018). This synergistic approach addresses the challenges posed by the presence of multiple QS systems in the bacteria. In conclusion, targeting QS aids in biofilm disruption by interfering with bacterial communication, reducing virulence factor production and increasing biofilm sensitivity to antibiotics (Fekrirad et al., 2019). This strategy has shown promise in combating various pathogens including *P. aeruginosa* and *A. baumannii* (Daoud et al., 2013). The development of novel QS inhibitors, such as 1,3,4-oxadiazoles (Belen'Kii et al., 1998), and the combination of QSIs with other antimicrobial strategies, such as therapy (Sun et al., 2019), offer exciting possibilities for future anti-virulence treatments.

11.13.2 Implications for Biofilm Control and Treatment

Comparative genomics provides a framework for developing combination therapies that target biofilm-specific genes, while enhancing the efficacy of traditional antibiotics. This approach can mitigate infection-linked biofilms, particularly in the clinical and industrial settings (Bornstein et al., 2023). Comparative genomics offers valuable insights into the genetic basis of biofilm formation and persistence, enabling the development of targeted therapeutic strategies. By identifying biofilm-specific genes and their regulatory networks, researchers can design combination therapies that can simultaneously disrupt biofilm formation and enhance the efficacy of conventional antibiotics. This approach holds promise for addressing the challenges posed by biofilm-associated infections in both clinical and industrial settings where traditional antimicrobial treatments are often inadequate (Varadarajan et al., 2020). The implications of this approach extend beyond mere biofilm control, potentially revolutionising treatment protocols for chronic infections and industrial contamination. By targeting biofilm-specific genes, it may be possible to prevent initial biofilm formation or destabilise existing biofilms, making them more susceptible to antibiotics and host immune responses. Additionally, this approach could result in the creation of novel antibiofilm agents that work synergistically with existing antimicrobial compounds, reducing the likelihood of resistance development and improving overall treatment outcomes (Rima et al., 2022). Research in this field is likely to yield new tools and techniques for managing biofilms in diverse environments, from medical devices and implants to industrial pipelines and water-treatment systems.

Comparative genomics has emerged as a powerful tool for advancing our understanding of biofilm formation and developing targeted therapies against biofilm associated infections. By comparing genetic information across organisms, researchers can identify key genes and pathways involved in biofilm development, antimicrobial resistance and virulence factors (Bornstein et al., 2023). This approach has been particularly useful in studying important clinical pathogens such as *P. aeruginosa*, where comparative genomics has revealed unique and conserved genes involved in biofilm growth and antibiotic resistance (Varadarajan et al., 2020). Interestingly, comparative genomics has also revealed novel essential genes that have been previously overlooked when mapping data against reference strains. For instance, in *P. aeruginosa* MPAO1, six general essential genes, including an antitoxin, were identified that had been missed when using the PAO1 reference strain (Varadarajan et al., 2020). This highlights the importance of using strain-specific genomic data in biofilm research. In conclusion, comparative genomics is driving the development of innovative multi-target approaches to combat biofilms. These comprise botanical extracts, phages, antimicrobial photodynamic therapy, CRISPR-Cas gene editing and nanoscale methods (Zhang et al., 2022). By integrating genomic data with phenotypic and proteomic information, researchers can gain a more comprehensive understanding of biofilm development, antibiotic resistance and the evolution of resistance, leading to improved therapeutic strategies against biofilm-associated infections (Varadarajan et al., 2020).

11.14 Future Directions in Comparative Genomics and Biofilm Research

Future studies should focus on integrating omics approaches to identify novel therapeutic targets. Leveraging advanced genomic techniques, such as single-cell genomics and metagenomics, will further refine the strategies for biofilm prevention and control. Future comparative genomics and biofilm studies should prioritise the integration of various omics approaches to uncover novel therapeutic targets (Yang et al., 2022; Imchen et al., 2022). Researchers can obtain more holistic insights into the intricate interactions within biofilms and uncover their potential weaknesses through the synthesis of data derived from genomic, transcriptomic, proteomic and metabolomic analyses. This comprehensive approach utilising multiple omics technologies will facilitate the creation of precise and efficient methods for preventing and eliminating biofilms. Leveraging advanced genomic techniques, such as single-cell genomics and metagenomics, will further refine the strategies for biofilm management (Almatroudi, 2024). Diversity within biofilm communities can be elucidated through the application of single-cell genomic techniques, revealing subpopulations that may be more resistant to treatment or responsible for biofilm initiation. Metagenomics, however, can help elucidate the interactions between different microbial species within biofilms, potentially uncovering the synergistic relationships that contribute to biofilm formation and persistence (Imchen et al., 2022). These approaches, combined with functional genomics and comparative analyses across multiple species, will lead to a more nuanced understanding of biofilm dynamics and inform the development of novel antibiofilm therapies.

Omics technologies have significantly advanced our understanding of microbial biofilms, thereby enabling the design of effective management strategies in various industries. The integration of different omics tools has provided in-depth knowledge on biofilm formation mechanisms, specific molecular phenotypes and potential control methods (Ban et al., 2012; Buttner et al., 2015). Transcriptomic, proteomic, metabolomic and multi-omics approaches are particularly valuable for studying microbial biofilms, offering a holistic view of biofilm biology.

11.15 Conclusion

Comparative genomics has been used to elucidate the genetic and phenotypic factors that drive biofilm formation and resilience. Insights from case studies and advanced genomic techniques have significantly enhanced our understanding of biofilm biology. Comparative genomics has substantially contributed to our understanding of biofilm formation and resilience. By analysing and comparing the genomes of various microorganisms, researchers have identified the key genetic elements and regulatory pathways that play crucial roles in biofilm development and maintenance (Sharma et al., 2023). These studies have revealed a complex interplay between different genes and their products, shedding light on the molecular mechanisms underlying the transition from planktonic to sessile lifestyles.

Case studies focusing on specific microbial species have provided valuable insights into the diversity of biofilm-formation strategies across different organisms. Advanced genomic techniques such as high-throughput sequencing, transcriptomics and proteomics have enabled researchers to examine biofilm formation in unprecedented detail (Franklin et al., 2015; Vohra et al., 2024). These approaches have uncovered novel gene-regulatory networks involved in biofilm development as well as species-specific adaptations that contribute to the resilience of biofilm communities in diverse environments. The integration of these findings has led to a more comprehensive understanding of biofilm biology, paving the path to innovative actions to control along with manipulate biofilms within various contexts, from medical applications to industrial settings. Comparative genomics has illuminated the genetic and phenotypic factors that drive biofilm formation and resilience in various microbial species. By comparing the genomes of biofilm-forming strains, researchers have identified key genes and molecular mechanisms associated with this complex phenotype (Sharma et al., 2023).

Studies have revealed that biofilm formation is influenced by multiple genetic factors, including genes involved in cellular adherence, motility, glycosylation, protective layer production and the oxidative strain response (Pascoe et al., 2015). For instance, genome-wide association studies of *Campylobacter jejuni* have identified 13 genes that are robustly associated with biofilm formation, with different genetic determinants in host generalist clonal complexes (Pascoe et al., 2015). This suggests that enhanced biofilm formation can evolve from different genetic backgrounds, potentially contributing to colonisation of multiple hosts and transmission to humans. Interestingly, the genetic basis of biofilm formation varies significantly between and within species. In *Bacillus cereus*, the ability to form biofilms was found to correlate with the synthesis levels of genes involved in cell specialisation and mobility, as well as phenotypic characteristics such as hydrophobicity (Majed et al., 2016). In *Candida tropicalis*, biofilm formation is regulated by a complex network of genes and signalling pathways, including environmental factors, morphological alterations and signal molecules for population density also playing crucial roles (De Souza et al., 2023). Advanced genomic techniques combined with traditional microbiological methods have greatly enhanced our understanding of biofilm formation and resilience. The integration of comparative genomics with high-throughput phenotyping and machine learning approaches offers promising avenues for further elucidation of genotype–phenotype relationships in biofilm formation (Khordad & Mercer, 2017; Elison & Acar, 2018; Bornstein et al., 2023). This integrated approach can provide a more comprehensive understanding of the genetic and molecular pathways underlying biofilm genesis, potentially leading to new strategies for combating biofilm-associated infections and addressing the challenges posed by antimicrobial resistance in clinical settings (Sun et al., 2013; Pinu et al., 2019; Sanches et al., 2024).

References

Agarwal, H., Gurnani, B., Pippal, B., & Jain, N. (2024). Capturing the micro-communities: Insights into biogenesis and architecture of bacterial biofilms. *BBA Advances, 7*, 100133.

Ahmed, M. N., Porse, A., Sommer, M. O. A., Høiby, N., & Ciofu, O. (2018). Evolution of antibiotic resistance in biofilm and planktonic Pseudomonas aeruginosa populations exposed to subinhibitory levels of ciprofloxacin. *Antimicrobial Agents and Chemotherapy, 62*(8), e00320-18.

Ahmed, S. A. K. S., Rudden, M., Dooley, J. S. G., Banat, I. M., Marchant, R., & Smyth, T. J. (2019). Natural quorum sensing inhibitors effectively downregulate gene expression of Pseudomonas aeruginosa virulence factors. *Applied Microbiology and Biotechnology, 103*(8), 3521–3535.

Alföldi, J., & Lindblad-Toh, K. (2013). Comparative genomics as a tool to understand evolution and disease. *Genome Research, 23*, 1063–1068.

Almatroudi, A. (2024). Investigating biofilms: Advanced methods for comprehending microbial behavior and antibiotic resistance. *Frontiers in Bioscience (Landmark Edition), 29*, 133.

Alves, L. F., Westmann, C. A., Lovate, G. L., de Siqueira, G. M. V., Borelli, T. C., & Guazzaroni, M. E. (2018). Metagenomic approaches for understanding new concepts in microbial science. *International Journal of Genomics, 2018*, 2312987.

Anderson, G. G., & O'Toole, G. A. (2008). Innate and induced resistance mechanisms of bacterial biofilms. *Current Topics in Microbiology and Immunology, 322*, 85–105.

Arciola, C. R., Campoccia, D., & Montanaro, L. (2015). Implant infections: Adhesion, biofilm formation, and antibiotic resistance. *Nature Reviews Microbiology, 13*, 806–822.

Ban, H., Meeson, A., Preston, T. J., & Welchman, A. E. (2012). The integration of motion and disparity cues to depth in dorsal visual cortex. *Nature Neuroscience, 15*(4), 636–643.

Beale, D. J., Barratt, R., Marlow, D. R., Dunn, M. S., Palombo, E. A., Morrison, P. D., & Key, C. (2013). Application of metabolomics to understanding biofilms in water distribution systems: A pilot study. *Biofouling, 29*, 283–294.

Beaudoin, T., Conway, T., Brouillette, E., et al. (2017). Interactions between Pseudomonas aeruginosa and Staphylococcus aureus within mixed-species biofilms: Impact of biofilm matrix structure. *Molecular Microbiology, 104*, 364–376.

Bedhomme, S., Amorós-Moya, D., Valverde, C., et al. (2019). Integration of horizontally transferred genes into bacterial genomes: The impact on biofilm resilience. *Nature Reviews Microbiology, 17*, 116–126.

Belen'Kii, L. I., Krayushkin, M. M., Poddubnyi, I. S., & Luiksaar, S. I. (1998). Novel syntheses of symmetrical 2,5-diaryl-1,3,4-oxadiazoles and 1,4-phenylenebis-1,3,4-oxadiazoles. *Russian Chemical Bulletin, 47*(11), 2238–2245.

Berne, C., Ellison, C. K., Ducret, A., & Brun, Y. V. (2018). Bacterial adhesion at the single-cell level. *Nature Reviews Microbiology, 16*, 616–627.

Bevers, R., Arts, J., Loo, I., Geurts, J., & Voort, M. (2022). The role of material technologies targeting P. Aeruginosa and S. Aureus quorum sensing in biofilm formation. *Medical Research Archives, 10*(10).

Blainey, P. C. (2013). The future is now: Single-cell genomics of bacteria and archaea. *FEMS Microbiology Reviews, 37*, 407–427.

Bolotin, E., & Hershberg, R. (2017). Horizontally acquired genes are often shared between closely related bacterial species. *Frontiers in Microbiology, 8*, 1536.

Bonham, K. S., Wolfe, B. E., & Dutton, R. J. (2017). Extensive horizontal gene transfer in biofilm-communities. *The ISME Journal, 11*, 923–934.

Bornstein, J. C., Mayrhofer, G., McKeown, S. J., Nguyen, T. V., & Young, H. M. (2023). Comparative genomics sheds light on the evolutionary basis of neural regulation in the gastrointestinal tract. *PLoS Biology, 21*, e3002186.

Buttner, H., Mack, D., & Rohde, H. (2015). Structural basis of Staphylococcus epidermidis biofilm formation: Mechanisms and molecular interactions. *Frontiers in Cellular and Infection Microbiology, 5*(38), 14.

Cassini, M. H. (2013). *Distribution ecology in conservation biology* (pp. 145–162). Springer New York.

Ciofu, O., Moser, C., Jensen, P. Ø., et al. (2017). Tolerance and resistance of microbial biofilms. *Nature Reviews Microbiology, 15*, 71–81.

Coelho, S. M., & Cock, J. M. (2020). Brown algal model organisms. *Annual Review of Genetics, 54*(1), 71–92.

Costerton, J. W., Geesey, G. G., & Cheng, K. J. (1978). How bacteria stick. *Scientific American, 238*, 86–95.

Damyanova, T., & Paunova-Krasteva, T. (2025). What we still don't know about biofilms-Current overview and key research information. *Microbiology Research, 16*, 46.

Daoud, Z., Mansour, N., & Masri, K. (2013). Synergistic combination of Carbapenems and Colistin against P. aeruginosa and A. baumannii. *Open Journal of Medical Microbiology, 03*(04), 253–258.

De Souza, C. M., Furlaneto-Maia, L., Furlaneto, M. C., & Dos Santos, M. M. (2023). Adhesion and biofilm formation by the opportunistic pathogen Candida tropicalis: What do we know? *Canadian Journal of Microbiology, 69*, 207–218.

Donlan, R. M. (2002). Biofilms: Microbial life on surfaces. *Emerging Infectious Diseases, 8*, 881–890.

Donlan, R. M., & Costerton, J. W. (2002). Biofilms: Survival mechanisms of clinically relevant microorganisms. *Clinical Microbiology Reviews, 15*, 167–193.

Doonan, J., Mcdonald, J. E., Denman, S., & Golyshin, P. N. (2016). *Shotgun metagenomic sequencing analysis of soft-rot Enterobacteriaceae in polymicrobial communities* (Vol. 1539, pp. 85–97). Springer.

Dopheide, A., He, Z., Lewis, G. D., Zhou, J., & Lear, G. (2015). Functional gene composition, diversity and redundancy in microbial stream biofilm communities. *PLoS One, 10*(4), e0123179.

Elison, G. L., & Acar, M. (2018). Scarless genome editing: Progress towards understanding genotype-phenotype relationships. *Current Genetics, 64*(6), 1229–1238.

Elkins, J. G., Hassett, D. J., Stewart, P. S., Schweizer, H. P., & McDermott, T. R. (1999). Protective role of catalase in Pseudomonas aeruginosa biofilm resistance to hydrogen peroxide. *Applied and Environmental Microbiology, 65*, 4594–4600.

Ellegren, H. (2008). Comparative genomics and the study of evolution by natural selection. *Molecular Ecology, 17*(21), 4586–4596.

Enard, W. (2014). Comparative genomics of brain size evolution. *Frontiers in Human Neuroscience, 8*, 345.

Fall, S., Mercier, A., Bertolla, F., et al. (2007). Horizontal gene transfer in microbial biofilms: A case study of Pseudomonas putida. *Microbial Ecology, 53*, 366–374.

Fernández, L,. Rodríguez, A., & García, P. (2018). Phage or foe: An insight into the impact of viral predation on microbial communities. *The ISME Journal, 12*, 1171–1179.

Fekrirad, Z., Kashef, N., & Arefian, E. (2019). Photodynamic inactivation diminishes quorum sensing-mediated virulence factor production and biofilm formation of Serratia marcescens. *World Journal of Microbiology and Biotechnology, 35*(12), 191.

Feldbauer, K., Brugirard-Ricaud, K., Juge, N., et al. (2015). Decoding biofilm resistance: Genetic determinants and mechanisms. *Frontiers in Microbiology, 6*, 473.

Fleming, D., & Rumbaugh, K. P. (2017). Approaches to dispersing medical biofilms. *Microorganisms, 5*, 15.

Flemming, H. C., & Wingender, J. (2010). The biofilm matrix. *Nature Reviews Microbiology, 8*, 623–633.

Flemming, H.-C., Wingender, J., & Neu, T. R. (2016). The perfect slime: Microbial extracellular polymeric substances (EPS). *Water Intelligence Online, 15*(0), 9781780407425.

Fletcher, M., & Loeb, G. I. (1979). Influence of substratum characteristics on the attachment of a marine pseudomonad to solid surfaces. *Applied and Environmental Microbiology, 37*, 67–72.

Fong, J., Givskov, M., Liu, X.-W., Zhang, C., Su, H., Bin, W., Yang, R., Nielsen, T. E., Yang, L., Tan, S. K., & Boo, Z. Z. (2018). Combination therapy strategy of quorum quenching enzyme and quorum sensing inhibitor in suppressing multiple quorum sensing pathways of P. aeruginosa. *Scientific Reports, 8*(1), 1155.

Franklin, M. J., Chang, C., Akiyama, T., & Bothner, B. (2015). New technologies for studying biofilms. *Microbiology Spectrum, 3*(4), MB0016-2014.

Fulaz, S., Vitale, S., Quinn, L., & Casey, E. (2019). Nanoparticle–biofilm interactions: The role of the EPS matrix. *Trends in Microbiology, 27*, 915.

Gibbons, S. M., & Gilbert, J. A. (2015). Microbial diversity – exploration of natural ecosystems and microbiomes. *Current Opinion in Genetics & Development, 35*, 66–72.

Gorban, I. K., Senashova, M. Y., & Sadovsky, M. G. (2020). Towards the structuredness of non-coding regions of chloroplast genomes. *Journal of Physics: Conference Series, 1679*(4), 042004.

Groisman, E. A., & Choi, J. (2023). Advancing evolution: Bacteria break down gene silencer to express horizontally acquired genes. *BioEssays: News and Reviews in Molecular, Cellular and Developmental Biology, 45*(10), e2300062.

Grzymski, J. J., Murray, A. E., Campbell, B. J., Kaplarevic, M., Gao, G. R., Lee, C., Daniel, R., Ghadiri, A., Feldman, R. A., & Cary, S. C. (2008). Metagenome analysis of an extreme microbial symbiosis reveals eurythermal adaptation and metabolic flexibility. *Proceedings of the National Academy of Sciences USA, 105*, 17516–17521.

Gunathilaka, G. U., Fernando, D. M., Mulvey, M. R., et al. (2024). Antibiotic resistance and biofilm formation in clinical isolates of Acinetobacter baumannii. *Journal of Medical Microbiology, 73*, 455–466.

Haddadi, A., Angaji, S. A., Babapour, E., Mirnejad, R., & Amirmozafari, N. (2024). Association between drug efflux pumps and resistance to ciprofloxacin in clinical isolates of Acinetobacter baumannii. *Iranian Journal of Medical Microbiology, 18*(2), 113–122.

Hauser, A. R., Jain, M., Bar-Meir, M., & McColley, S. A. (2016). Clinical significance of microbial biofilms in Pseudomonas aeruginosa infections. *Clinical Microbiology Reviews, 29*, 707–742.

Hobley, L., Harkins, C., MacPhee, C. E., & Stanley-Wall, N. R. (2015). Giving structure to the biofilm matrix: An overview of individual strategies and emerging common themes. *FEMS Microbiology Reviews, 39*(5), 649–669.

Høiby, N. (2017). A short history of microbial biofilms and biofilm infections. *APMIS, 125*, 272–275.

Høiby, N., Bjarnsholt, T., Givskov, M., et al. (2010). Antibiotic resistance of bacterial biofilms. *International Journal of Antimicrobial Agents, 35*, 322–332.

Homberger, C., Hayward, R. J., Barquist, L., & Vogel, J. (2023). Improved bacterial single-cell RNA-Seq through automated MATQ-Seq and Cas9-based removal of rRNA reads. *mBio, 14*(2), e0355722.

Huang, R., Li, M., & Gregory, R. L. (2011). Bacterial interactions in dental biofilm. *Virulence, 2*, 435–444.

Huang, L. D., Gou, X. Y., Yang, M. J., Li, M. J., Chen, S. N., Yan, J., Liu, X. X., & Sun, A. H. (2024). Peptidoglycan biosynthesis-associated enzymatic kinetic characteristics and β-lactam antibiotic inhibitory effects of different Streptococcus pneumoniae penicillin-binding proteins. *International Journal of Biological Macromolecules, 254*, 127784.

Imchen, M., Anju, V. T., Busi, S., Mohan, M. S., Subhaswaraj, P., Dyavaiah, M., & Kumavath, R. (2022). Metagenomic insights into taxonomic, functional diversity and inhibitors of microbial biofilms. *Microbiological Research, 265*, 127207.

Ishii, S., Suzuki, S., Norden-Krichmar, T. M., Tenney, A., Chain, P. S., Scholz, M. B., Nealson, K. H., & Bretschger, O. (2013). A novel metatranscriptomic approach to identify geneexpression dynamics during extracellular electron transfer. *Nature Communications, 4*, 1601.

Ivanova, N., Gronow, S., Lapidus, A., et al. (2009). Genome sequence of the soil bacterium Streptomyces scabiei and its comparison to other actinobacteria. *Journal of Bacteriology, 191*, 2404–2417.

Jass, J., Roberts, S. K., & Lappin-Scott, H. M. (2002). *Enzymes in the environment. Activity, ecology, and applications* (pp. 307–326). Marcel Dekker.

Jayaraman, R. (2008). Bacterial persistence: Some new insights into an old phenomenon. *Journal of Biosciences, 33*, 795–805.

Jenul, C., & Horswill, A. R. (2019). Regulation of Staphylococcus aureus Virulence. *Microbiology Spectrum, 7*(2). https://doi.org/10.1128/microbiolspec.gpp3-0031-2018

Kaplan, J. B. (2010). Biofilm dispersal: Mechanisms, clinical implications, and potential therapeutic uses. *Journal of Dental Research, 89*, 205–218.

Katharine, K., & Karatan, E. (2005). Biofilm development in bacteria. *Advances in Applied Microbiology, 57*, 79–111.

Kazmierczak, M. J., Mithoe, S. C., Boor, K. J., & Wiedmann, M. (2003). Listeria monocytogenes sigma B regulates stress response and virulence functions. *Journal of Bacteriology, 185*, 5722–5734.

Khordad, M., & Mercer, R. E. (2017). Identifying genotype-phenotype relationships in biomedical text. *Journal of Biomedical Semantics, 8*(1).

Kierek-Pearson, K., & Karatan, E. (2005). Biofilm development in bacteria. *Advances in Applied Microbiology, 57*, 79–111.

Klausen, M., Heydorn, A., Ragas, P., Lambertsen, L., Aaes-Jørgensen, A., Molin, S., & Tolker-Nielsen, T. (2003). Biofilm formation by Pseudomonas aeruginosa wild type, flagella and type IV pili mutants. *Molecular Microbiology, 48*, 1511–1524.

Konkel, Z., & Slot, J. C. (2023). Evolutionary insights from comparative genomics of fungal secondary metabolites. *Fungal Genetics and Biology, 168*, 103634.

Koo, H., Allan, R. N., Howlin, R. P., Stoodley, P., & Hall-Stoodley, L. (2017). Targeting microbial biofilms: Current and prospective therapeutic strategies. *Nature Reviews Microbiology, 15*(12), 740.

Korlach, J., & Turner, S. W. (2010). Direct detection of DNA methylation during single-molecule, real-time sequencing. *Nature Methods, 7*, 461–465.

Lamichhaney, S., Card, D. C., Grayson, P., Tonini, J. F. R., Bravo, G. A., Näpflin, K., Termignoni-Garcia, F., Torres, C., Burbrink, F., Clarke, J. A., Sackton, T. B., & Edwards, S. V. (2019). Integrating natural history collections and comparative genomics to study the genetic architecture of convergent evolution. *Philosophical Transactions of the Royal Society of London. Series B, Biological Sciences, 374*(1777), 20180248.

Lebeaux, D., Chauhan, A., Rendueles, O., & Beloin, C. (2013). From in vitro to in vivo models of bacterial biofilm-related infections. *Pathogens, 2013*, 288–356.

Li, Q., Mao, S., Ye, X., & Wang, H. (2022). The molecular architecture of Pseudomonas aeruginosa quorum-sensing inhibitors. *Marine Drugs, 20*(8), 488.

Liang, J., Haney, C. H., Rees, S., et al. (2020). Regulation of biofilm-forming pathogens: Opportunities and challenges. *Cell Biology International, 44*, 585–603.

Lim, B., Lin, Y., & Navin, N. (2020). Advancing cancer research and medicine with single-cell genomics. *Cancer Cell, 37*(4), 456–470.

Limoli, D. H., Jones, C. J., & Wozniak, D. J. (2015). Bacterial extracellular polysaccharides in biofilm formation and function. *Microbiology Spectrum, 3*(3). https://doi.org/10.1128/microbiolspec.MB-0011-2014

Limoli, D. H., Warren, E. A., O'Toole, G. A., Cheung, A. L., Yarrington, K. D., & Donegan, N. P. (2019). Interspecies interactions induce exp loratory motility in Pseudomonas aeruginosa. *elife, 8*, e47365.

Lis, M., & Kuramitsu, H. K. (2003). The stress-responsive dgk gene from Streptococcus mutans encodes a putative undecaprenol kinase activity. *Infection and Immunity, 71*(4), 1938–1943.

Liu, H. Y., Prentice, E. L., & Webber, M. A. (2024). Mechanisms of antimicrobial resistance in biofilms. *npj Antimicrobial Resistance, 2*, 27.

Locey, K. J., & White, E. P. (2011). Simple structural differences between coding and noncoding DNA. *PLoS One, 6*(2), e14651.

Loeb, G. I., & Neihof, R. A. (1975). Marine conditioning films. *Advances in Chemistry, 145*, 319–335.

Lopes, L. Q. S., Marquezan, F. K., Christ Vianna Santos, R., Henrique Fortes Guerim, P., & Marquezan, P. K. (2023). The influence of different culture media on Pseudomonas aeruginosa, Escherichia coli, and Staphylococcus aureus biofilm formation. *Bioscience Journal, 39*, e39096.

Luo, Z. X., Li, Y., Liu, M. F., & Zhao R. (2023). Ciprofloxacin enhances the biofilm formation of Staphylococcus aureus via an agrC-dependent mechanism. *Frontiers in Microbiology, 21*, 14:1328947.

Mah, T. F., Pitts, B., Pellock, B., Walker, G. C., Stewart, P. S., & O'Toole, G. A. (2003). A genetic basis for Pseudomonas aeruginosa biofilm antibiotic resistance. *Nature, 426*, 306–310.

Majed, R., Faille, C., Gohar, M., & Kallassy, M. (2016). Bacillus cereus biofilms-same, only different. *Frontiers in Microbiology, 7*(745).

Majumdar, S. (2020). *Anti-quorum sensing systems and biofilm formation* (pp. 293–303). Springer.

Majumdar, S., & Roy, S. (2020). *Electrical communication systems in bacterial biofilms and ion-channels* (pp. 127–135). Springer Singapore.

McCoy, W. F., Bryers, J. D., Robbins, J., & Costerton, J. W. (1981). Observations of fouling biofilm formation. *Canadian Journal of Microbiology, 27*, 910–917.

Mclean, R., & Kakirde, K. (2013). Enhancing metagenomics investigations of microbial interactions with biofilm technology. *International Journal of Molecular Sciences, 14*(11), 22246–22257.

Mclean, J. S., Brami, D., Badger, J. H., Lasken, R. S., Ziegler, M. G., Nurk, S., Durkin, S., Hall, A. P., Nealson, K. H., Edlund, A., Yee-Greenbaum, J., Tesler, G., Venter, J. C., Pevzner, P. A., Lombardo, M., Allen, L., Torriani, F., Reed, S., Lesin, V., et al. (2013). Genome of the pathogen Porphyromonas gingivalis recovered from a biofilm in a hospital sink using a high-throughput single-cell genomics platform. *Genome Research, 23*(5), 867–877.

Metzker, M. L. (2010). Sequencing technologies: The next generation. *Nature Reviews Genetics, 11*, 31–46.

Miller, M. B., & Bassler, B. L. (2016). Quorum sensing in bacteria. *Annual Review of Microbiology, 55*, 165–199.

Mirani, Z. A., Urooj, S., Fatima, A., Aziz, M., Khan, M., & Abbas, T. (2018). Relationship of cell surface hydrophobicity with biofilm formation and growth rate: A study on Pseudomonas aeruginosa, Staphylococcus aureus, and Escherichia coli. *Iranian Journal of Basic Medical Sciences, 21*(7), 760–769.

Mirghani, R., Saba, T., Khaliq, H., Mitchell, J., Do, L., Chambi, L., et al. (2022). Biofilms: Formation, drug resistance and alternatives to conventional approaches. *AIMS Microbiology, 8*, 239–277.

Mitchell, G., Frost, E. H., Déziel, E., Malouin, F., Michaud, S., Asselin, A.-E., Séguin, D. L., & Cantin, A. M. (2010). Staphylococcus aureus sigma B-dependent emergence of small-colony variants and biofilm production following exposure to Pseudomonas aeruginosa 4-hydroxy-2-heptylquinoline-N- oxide. *BMC Microbiology, 10*(1), 33.

Miyaue, S., Suzuki, E., Komiyama, Y., Kondo, Y., Morikawa, M., & Maeda, S. (2018). Bacterial memory of persisters: Bacterial persister cells can retain their phenotype for days or weeks after withdrawal from colony–biofilm culture. *Frontiers in Microbiology, 9*, 1396.

Molin, S., & Tolker-Nielsen, T. (2003). Gene transfer occurs with enhanced efficiency in biofilms and induces enhanced stabilisation of the biofilm structure. *Current Opinion in Biotechnology, 14*, 255–261.

Monteith, W., Pascoe, B., Mourkas, E., Clark, J., Hakim, M., Hitchings, M. D., et al. (2023). Contrasting genes conferring short- and long-term biofilm adaptation in Listeria. *Microbial Genomics, 9*, 001114.

Muhammad, M. H., Idris, A. L., Fan, X., Guo, Y., Yu, Y., Jin, X., Qiu, J., Guan, X., & Huang, T. (2020). Beyond Risk: Bacterial biofilms and their regulating approaches. *Frontiers in Microbiology, 11*, 928.

Mulcahy, L. R., Burns, J. L., Lory, S., & Lewis, K. (2010). Emergence of Pseudomonas aeruginosa strains producing high levels of persister cells in patients with cystic fibrosis. *Journal of Bacteriology, 192*, 6191–6199.

Müsken, M., Klimmek, K., Sauer-Heilborn, A., Donnert, M., Sedlacek, L., Suerbaum, S., & Häussler, S. (2009). Towards individualized diagnostics of biofilm-associated infections: A case study. *npj Biofilms and Microbiomes, 5*, 57.

Najeeb, A. S., Kamil, I. S., & Khalaf, N. M. (2024). Interference between S. Aureus and P. Aeruginosa clinical isolates. *Academia Open, 9*(2).

Nguyen, N. T. T., Dufayard, J. F., Louis, A., Vincens, P., & Roest Crollius, H. (2021). Genomicus in 2022: Comparative tools for thousands of genomes and reconstructed ancestors. *Nucleic Acids Research, 50*(D1), D1025–D1031.

Nishimura, Y., Yamada, K., Okazaki, Y., & Ogata, H. (2024). DiGAlign: Versatile and interactive visualization of sequence alignment for comparative genomics. *Microbes and Environments, 39*(1), n/a.

Nobrega, M. A., & Pennacchio, L. A. (2004). Comparative genomic analysis as a tool for biological discovery. *The Journal of physiology, 554*(Pt 1), 31–39.

Nomura, S. (2020). Single-cell genomics to understand disease pathogenesis. *Journal of Human Genetics, 66*(1), 75–84.

Norwood, D. E., & Gilmour, A. (2000). The growth and resistance to sodium hypochlorite of Listeria monocytogenes in a steady-state multispecies biofilm. *Journal of Applied Microbiology, 88*, 512–520.

Olsen, I. (2015). Biofilm-specific antibiotic tolerance and resistance. *European Journal of Clinical Microbiology & Infectious Diseases, 34*, 877–886.

Pal, A. (2021). *Single-cell genomics* (pp. 683–695). Springer US.

Paolillo, C., Londin, E., & Fortina, P. (2019). Single-cell genomics. *Clinical Chemistry, 65*, 972–985.

Park, O.-J., Jeong, S., Kwon, Y., Han, S. H., Yun, C.-H., Park, T. H., et al. (2020). Streptococcus gordonii: Pathogenesis and host response to its cell wall components. *Microorganisms, 8*(12), 1852.

Pascoe, B., Méric, G., Murray, S., Yahara, K., Mageiros, L., Bowen, R., et al. (2015). Enhanced biofilm formation and multi-host transmission evolve from divergent genetic backgrounds in Campylobacter jejuni. *Environmental Microbiology, 17*, 4779–4789.

Pinu, F. R., Beale, D. J., Paten, A. M., Kouremenos, K., Swarup, S., Schirra, H. J., & Wishart, D. (2019). Systems biology and multi-omics integration: Viewpoints from the metabolomics research community. *Metabolites, 9*(4), 76.

Potera, C. (1996). Biofilms invade microbiology. *Science, 273*, 1795–1797.

Proctor, R. A., Kahl, B. C., Becker, K., Mcnamara, P., Peters, G., Herrmann, M., & Von Eiff, C. (2006). Small colony variants: A pathogenic form of bacteria that facilitates persistent and recurrent infections. *Nature Reviews Microbiology, 4*(4), 295–305.

Ranganathan, N., Johnson, R., & Edwards, A. M. (2020). The general stress response of Staphylococcus aureus promotes tolerance of antibiotics and survival in whole human blood. *Microbiology (Reading, England), 166*(11), 1088–1094.

Rath, S., Bal, S. C. B., & Dubey, D. (2021). Oral biofilm: Development mechanism, multidrug resistance, and their effective management with novel techniques. *Rambam Maimonides Medical Journal, 12*(1), e0004.

Rep, M., & Kistler, H. C. (2010). The genomic organization of plant pathogenicity in Fusarium species. *Current Opinion in Plant Biology, 13*, 420–426.

Rima, M., El Garah, F., Roques, C., & Chbani, A. (2022). Seaweed extracts as an effective gateway in the search for novel antibiofilm agents against Staphylococcus aureus. *Plants, 11*(17), 2285.

Riordan, J. T., O'Leary, J. O., & Gustafson, J. E. (2006). Contributions of sigB and sarA to distinct multiple antimicrobial resistance mechanisms of Staphylococcus aureus. *International Journal of Antimicrobial Agents, 28*(1), 54–61.

Rosario, K., & Breitbart, M. (2011). Exploring the viral world through metagenomics. *Current Opinion in Virology, 1*, 289–297.

Rumbaugh, K. P., & Sauer, K. (2020). Biofilm dispersion. *Nature Reviews Microbiology, 18*, 571–586.

Sabra, W., Zeng, A., Tjahjasari, D., & Dietz, D. (2010). Biosystems analysis and engineering of microbial consortia for industrial biotechnology. *Engineering in Life Sciences, 10*(5), 407–421.

Sahoo, K., & Meshram, S. (2024). Biofilm formation in chronic infections: A comprehensive review of pathogenesis, clinical implications, and novel therapeutic approaches. *Cureus, 16*, e70629.

Saini, P., Bandsode, V., Singh, A., Mendem, S. K., Semmler, T., Alam, M., et al. (2024). Genomic insights into virulence, antimicrobial resistance, and adaptation acumen of Escherichia coli isolated from an urban environment. *mBio, 15*, e0354523.

Sanches, P. H. G., de Melo, N. C., Porcari, A. M., & de Carvalho, L. M. (2024). Integrating molecular perspectives: Strategies for comprehensive multi-omics integrative data analysis and machine learning applications in transcriptomics, proteomics, and metabolomics. *Biology, 13*(11), 848.

Sauer, K. (2003). The genomics and proteomics of biofilm formation. *Genome Biology, 4*, 219.

Sauer, K., Stoodley, P., Goeres, D. M., Hall-Stoodley, L., Burmølle, M., Stewart, P. S., & Bjarnsholt, T. (2022). The biofilm life cycle: Expanding the conceptual model of biofilm formation. *Nature Reviews Microbiology, 20*, 608–620.

Sharma, S., Mohler, J., Mahajan, S. D., Schwartz, S. A., Bruggemann, L., & Aalinkeel, R. (2023). Microbial biofilm: A review on formation, infection, antibiotic resistance, control measures, and innovative treatment. *Microorganisms, 11*, 1614.

Shibata, Y., Van Der Ploeg, J. R., Takeshita, T., Kozuki, T., Yamashita, Y., Saito, N., et al. (2009). Kinase activity of the dgk gene product is involved in the virulence of Streptococcus mutans. *Microbiology, 155*(2), 557–565.

Shree, P., Singh, C. K., Sodhi, K. K., Surya, J. N., & Singh, D. K. (2023). Biofilms: Understanding the structure and contribution towards bacterial resistance in antibiotics. *Medicine in Microecology, 16*, 100084.

Sivashankari, S., & Shanmughavel, P. (2007). Comparative genomics – a perspective. *Bioinformation, 1*(9), 376–378.

Song, F., Koo, H., & Ren, D. (2021). Effects of material properties on bacterial adhesion and biofilm formation. *Journal of Dental Research, 100*, 1027–1034.

Stoodley, P., & Hall-Stoodley, L. (2009). Evolving concepts in biofilm infections. *Cellular Microbiology, 11*, 1034–1043.

Stroh, P., Günther, F., Meyle, E., Prior, B., Wagner, C., & Hänsch, G. M. (2011). Host defence against Staphylococcus aureus biofilms by polymorphonuclear neutrophils: Oxygen radical production but not phagocytosis depends on opsonisation with immunoglobulin G. *Immunobiology, 216*(3), 351–357.

Sun, F., Ling, Y., Mao, P., Chen, H., Zhou, D., Xia, P., & Qu, F. (2013). Biofilm-associated infections: Antibiotic resistance and novel therapeutic strategies. *Future Microbiology, 8*(7), 877–886.

Sun, Y., Dowd, S. E., Smith, E., Rhoads, D. D., & Wolcott, R. D. (2019). In vitro multispecies Lubbock chronic wound biofilm model. *Wound Repair and Regeneration, 16*, 805–813.

Thaarup, I. C., & Bjarnsholt, T. (2021). Current in vitro biofilm infected chronic wound models for developing new treatment possibilities. *Advances in Wound Care, 10*, 91–102.

Valle, J., Lasa, I., & Echeverz, M. (2019). σB inhibits poly-N-acetylglucosamine exopolysaccharide synthesis and biofilm formation in Staphylococcus aureus. *Journal of Bacteriology, 201*(11), e00098-19.

Van Merode, A. E. J., Pothoven, D. C., Krom, B. P., Van Der Mei, H. C., & Busscher, H. J. (2006). Surface charge influences enterococcal prevalence in mixed-species biofilms. *Journal of Applied Microbiology, 102*(5), 1254–1260.

Varadarajan, A. R., Gavioli, E. M., Srivatsan, A., Chandler-Brown, D., Song, W., Dewar, R. S., et al. (2020). An integrated model system to gain mechanistic insights into biofilm-associated antimicrobial resistance in Pseudomonas aeruginosa MPAO1. *NPJ Biofilms and Microbiomes, 6*, 46.

Vestby, L. K., Grønseth, T., Simm, R., & Nesse, L. L. (2020). Bacterial biofilm and its role in the pathogenesis of disease. *Antibiotics, 9*, 59.

Vishwakarma, V. (2020). Impact of environmental biofilms: Industrial components and its remediation. *Journal of Basic Microbiology, 60*, 198–206.

Vohra, M., Kour, A., Kalia, N. P., Kumar, M., Sharma, S., Jaglan, S., et al. (2024). A comprehensive review of genomics, transcriptomics, proteomics, and metabolomic insights into the differentiation of Pseudomonas aeruginosa from the planktonic to biofilm state: A multi-omics approach. *International Journal of Biological Macromolecules, 257*, 128563.

Vu, B., Chen, M., Crawford, R. J., & Ivanova, E. P. (2009). Bacterial extracellular polysaccharides involved in biofilm formation. *Molecules (Basel, Switzerland), 14*(7), 2535–2554.

Wang, D., Fletcher, G. C., Gagic, D., On, S. L. W., Palmer, J. S., & Flint, S. H. (2023a). Comparative genome identification of accessory genes associated with strong biofilm formation in Vibrio parahaemolyticus. *Food Research International, 166*, 112605.

Wang, X., Liu, M., Yu, C., Li, J., & Zhou, X. (2023b). Biofilm formation: Mechanistic insights and therapeutic targets. *Molecular Biomedicine, 4*(1), 49.

Whitehead, N. A., Barnard, A. M., Slater, H., Simpson, N. J., & Salmond, G. P. (2012). Quorum-sensing in Gramnegative bacteria. *FEMS Microbiology Reviews, 25*, 365–404.

Woods, P. W., Haynes, Z. M., Mina, E. G., & Marques, C. N. H. (2019). Maintenance of S. aureus in co-culture with P. aeruginosa while growing as biofilms. *Frontiers in Microbiology, 9*, 3291.

Wu, H., Moser, C., Wang, H. Z., Høiby, N., & Song, Z. J. (2015). Strategies for combating bacterial biofilm infections. *International Journal of Oral Science, 7*, 1–7.

Wu, Q., Zhang, X., Jia, S., Li, J., & Li, P. (2019). Effects of the cultivable bacteria attached to Microcystis colonies on the colony size and growth of Microcystis. *Journal of Freshwater Ecology, 34*(1), 663–673.

Wuyts, J., Van Dijck, P., & Holtappels, M. (2018). Fungal persister cells: The basis for recalcitrant infections? *PLoS Pathogens, 14*, e1007017.

Yan, J., & Bassler, B. L. (2019). Surviving as a community: Antibiotic tolerance and persistence in bacterial biofilms. *Cell Host Microbe, 26*, 15–21.

Yan, X., Liao, H., Wang, C., Huang, C., Zhang, W., Guo, C., & Pu, Y. (2024). An improved bacterial single-cell RNA-seq reveals biofilm heterogeneity. *elife, 13*, RP97543.

Yang, Y., Fang, A., Feng, K., Zhang, D., Zhou, H., & Xing, D. (2022). Single-cell metagenomics and metagenomics approaches reveal extracellular electron transfer of psychrophilic electroactive biofilms. *Science of The Total Environment, 836*, 155606.

Yang, X., Tran, F., & Zhang, P. (2023). Comparative genomic analyses of Escherichia coli from a meat processing environment in relation to their biofilm formation and persistence. *Microbiology Spectrum, 11*, e0018323.

Zapotoczna, M., McCarthy, H., Rudkin, J. K., O'Gara, J. P., & O'Neill, E. (2015). An essential role for coagulase in Staphylococcus aureus biofilm development reveals new therapeutic possibilities for device-related infections. *Journal of Infectious Diseases, 212*, 1883–1893.

Zapotoczna, M., O'Neill, E., & O'Gara, J. P. (2016). Untangling the diverse and redundant mechanisms of Staphylococcus aureus biofilm formation. *PLoS Pathogens, 12*, e1005671.

Zeng, Z., Xue, X., Cao, L., Qian, L., Huang, Y., Shen, Y., Tan, H., & Zhou, S. (2008). Virtual screening for novel quorum sensing inhibitors to eradicate biofilm formation of Pseudomonas aeruginosa. *Applied Microbiology and Biotechnology, 79*(1), 119–126.

Zhang, T., Zhang, Y., Chen, Z., et al. (2021). The regulatory mechanisms of bacterial biofilm formation. *Frontiers in Microbiology, 12*, 728567.

Zhang, D., Siegwart, D. J., Xu, J., Taylor, A. M., Wang, G., Zhu, H., Yu, X., Wei, T., Hong, Y., Farbiak, L., & Johnson, L. T. (2022). Enhancing CRISPR/Cas gene editing through modulating cellular mechanical properties for cancer therapy. *Nature Nanotechnology, 17*(7), 777–787.

Zuberi, A., Misba, L., & Khan, A. U. (2017). CRISPR interference (CRISPRi) inhibition of luxS gene expression in E. coli: An approach to inhibit biofilm. *Frontiers in Cellular and Infection Microbiology, 7*, 214.

Lipidomics in Understanding Microbial Biofilms 12

Himani Meena, Sinjini Nandy, Vishal S. Patil, Rahul Sharma, and Smita Kumari

12.1 Introduction

12.1.1 Microbial Biofilm

Microbial colonization during biofilm formation exhibits a unique growth pattern over single cells with specific surface properties in a complex environmental condition (Kadirvelu et al., 2024). Microbial cell membranes are composed of higher molecular weight lipid molecules playing important roles in signal transduction, preventing antibiotic penetration, and also acting as an energy storage unit (Ali et al., 2023). The lipid membrane can highly influence the bacterial secretion system through changes in the membrane fluidity as well as the membrane selectivity based on the external environmental factors. The estimation of lipid profile using a lipidomic approach provides insights into cell viability apart from the non-viable cells through phospholipid fatty acid quantification (Sharma et al., 2024).

H. Meena (✉)
Department of Pathology and Genomic Medicine, Houston Methodist Research Institute, Houston, TX, USA

S. Nandy
University of Texas at Dallas, Richardson, TX, USA

V. S. Patil
R.C. Patel Institute of Pharmaceutical Education and Research, Shirpur, Maharashtra, India

R. Sharma
Environmental Biotechnology, Genetics and Molecular Biology Division, ICMR – National Institute for Research in Environmental Health, Bhopal, Madhya Pradesh, India

S. Kumari
College of Pharmacy, Ohio State University, Columbus, OH, USA

S. Busi et al. (eds.), *Omics Approaches in Biofilm Research*,
https://doi.org/10.1007/978-3-031-91863-6_12

12.1.2 Understanding the Microbial Colonization

Microbial communities are highly compact, which could withstand antimicrobials, acculturate fluctuating environmental conditions, and be able to defy innate immunity response in the host in order to thrive in the environment (Li et al., 2025). However, it is unknown how these microbial communities are developing specific phenomena while lasting forever on the surfaces with active metabolism. Several studies have been conducted and concluded that metabolite exchange during microbial growth is as vital as the physical barrier. Spatial microbial arrangement and survival could be enhanced via metabolite cross-feeding, local supply chains, and feedback signaling via creating nutrient-deprived interactions (Zhao et al., 2024). Therefore, the ability of microbial cells to exhibit such adaptability and pliancy resulting in intricate metabolic exchanges still remains unknown and requires proper scientific investigation (Li et al., 2024).

12.1.3 Role of Lipids in Microbial Membrane

Microbial membrane lipids play an essential role in pathogenicity and are significantly involved with microbial adhesion, signal transduction pathways, conformational changes or adaptations, and immune response modulation. Several scientific reports have defined the presence of different lipid molecules and their role in the microbial biofilm as a biomarker. Several biomarkers have been reported as associated with algal, fungal, and bacterial infections, some of which were species-specific and depended on planktonic and biofilm-associated (Wood, 2024). Wood reported the presence of fungal lipids, that is, ergosterol, sophorolipids, and glycosyl inositol-phosphoryl ceramides, involved with host pathogenicity. Many mycobacterial biomarkers are reported mycolic acids, trehalose, and glycopeptido-lipids, which mainly participate in bacterial motility and invasive biofilm development.

Bacterial membranes are divided into two groups based on their cell wall structure: Gram-positive and Gram-negative bacteria. Gram-positive bacteria are highly susceptible to antibiotics, as they directly interact with a compact peptidoglycan layer and allow antibiotic penetration. The peptidoglycan layer forms a dense layer between the cytoplasmic membrane and cell wall with anchored teichoic acid and lipoteichoic acid, which creates a negatively charged surface (Balleza et al., 2019; Sreedharan et al., 2023), as shown in Fig. 12.1.

However, reports suggested using precursor molecules as biomarkers to avoid the hassle of the extraction process. These biomolecules are identified as dihexosyl diacylglycerols, and glycerol phosphate forms a bond at a 6-hydroxy group of terminal hexose referred to as a lipoteichoic acid primer (LTA). Lipopeptides are amphiphilic cyclic and linear compounds that are biosynthesized by non-ribosomal peptide synthases. Tripeptides and pentapeptides contribute to the cell envelope function and possess antimicrobial properties. Abdelli et al. (2019) reported a few surfactin molecules from lipolytic *Bacillus safensis* that possess antibiofilm and antitumor properties.

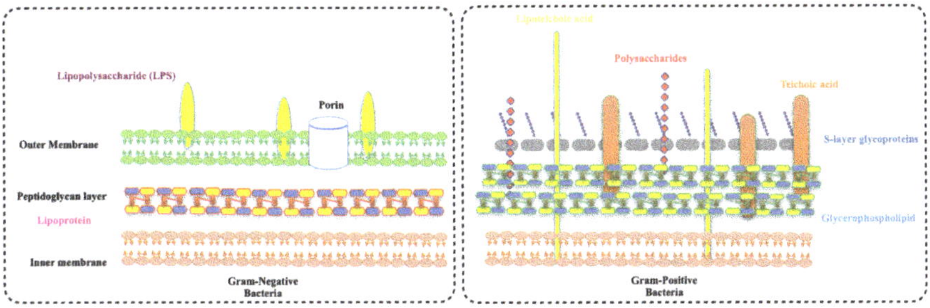

Fig. 12.1 The figure illustrates a symmetrical representation of the bacterial cell membrane along with the lipid species that are anchored

Bonnichsen et al. (2015) documented a study showing the production of viscosin lipopeptide by *Pseudomonas fluorescens* that could be used as a potent biofilm dispersal agent. Similarly, Das et al. (2008) documented the secretion of antimicrobial lipopeptides, surfactin, and lichenysin from *Bacillus circulans* against *Staphylococcus aureus*. Another excellent example of bacterial membrane lipid is cardiolipin, produced by *S. aureus*, a membrane stabilizer incorporated as phospholipid. Cardiolipin interacts with the bacterial two-component systems (SaeRS) and regulates the production of virulence factors (Liu et al., 2016; Yeo et al., 2023). *Streptomyces* also produces cardiolipin which plays a crucial role in morphogenesis, and controls hyphae formation and spore production (Jyothikumar et al., 2012). The QS system is a well-known signaling system in bacterial species. It is known for its involvement with cell-cell communication, leading to virulence factor production and biofilm formation via small signaling peptides. Bacterial infection is spread via excess production of these oligopeptides, which bind to membrane-anchored receptor proteins, auto-phosphorylate, and trigger intracellular pathways for enhanced pathogenicity (Kumar et al., 2023; Monnet & Gardan, 2015; Slamti & Lereclus, 2019).

However, Gram-negative bacteria do not have a cell wall; instead, they possess a thin peptidoglycan layer. Lipid A is an essential cell membrane component spanning the inner and outer membranes. A diversified Lipid A precursor is identified based on bacterial species, aminoacyl fatty acyls of hydroxy fatty acids (FAHFAs) consisting of glycine-serine-FAHFAs, glycine-serine-HFAs, and glycine-serine-ornitho-FAHFAs created via acylation of short-chain 3-hydroxy fatty acids with added ornithine to the free carboxy group of the hydroxy fatty acid. Glucosaminyl phosphatidyl glycerol (GlcN-PG) is known for protein translocation across the membranes and is part of the transportation system in microbial systems. Bacterial cells convert phosphatidyl glycerol into GlcN-PG, mainly involved with biofilm formation during pseudomonal infection (Khan et al., 2024). Glycerol phosphoryl ceramides and ethanolamines are involved microbial growth, pathogenicity, and trigger immune responses during chronic infection (Garsin, 2010). Another excellent example of membrane lipid is N-acetyl homo-serine lactone (AHL) signaling molecules involved with quorum-sensing mechanism in Gram-negative bacteria, associated with protease, rhamnolipid, elastase, antibiotic resistance, and biofilm formation. The mechanism involves binding signal molecules, C_3/C_4-oxo-C_{12}-HSL to its cognate receptors, LasR/RhlR in high cell density and initiating virulence factor production (Zhao et al., 2020; Jin et al., 2021).

A diversified lipid profile is essential for microbial strains, and a slight alteration can affect the physical properties of the membrane, such as membrane fluidity, selectivity, and protein or ion channels. In the chapter, we provide detailed information about sample preparation, techniques used for lipid profiling in microbial biofilm analysis, and the future applications in disease diagnosis.

12.2 Lipidomic Techniques and Methodologies

Lipidomic is used for lipid profiling in microbes to analyze changes in lipid shape, size, and molecular weight in normal and diseased conditions. The comprehensive analysis requires essential tools, such as ion mobility (IM-MS), electrospray ionization (ESI-MS), matrix-assisted laser desorption and ionization (MALDI-MS), gas chromatography (GC-MS), liquid chromatography-MS, used for quantifying preprocessed lipid species, the shotgun technique for lipidome profiling by mass spectrometry, LC-OzID-IMS-MS analysis for determining lipid carbon-carbon double bonds.

12.2.1 Sample Preparation and Extraction

Sample preparation and lipid extraction from microbial membranes is an essential procedure for lipidomics to prevent contamination and provide accurate results with efficacy. Finding appropriate solvents for lipid extraction based on their nature could save significant time and resources. A polar solvent helps extract lipid species from the attached protein complex; however, a non-polar solvent removes neutral lipids. Based on the rigidity of the membrane, microbial cells require various methodologies, such as chemical, mechanical, and enzymatic treatment (Saini et al., 2021; Zainuddin et al., 2021).

This section provides detailed information regarding lipid extraction using Bligh and Dyer method, Folch Method, Methyl tert-butyl ether (MTBE) method, and cell disruption methods (bead-beating, sonication, detergents or alkali treatment, and desalting and drying.

12.2.1.1 Bligh and Dyer Method & Folch Method

The Bligh and Dyer method utilizes a combination of solvents, chloroform:methanol:water (1:2:0.25; v/v/v), which could be modified based on the lipid species. The reaction occurs in methanol, a monophasic ternary solvent that transforms into a biphasic solvent by adding chloroform and water. The solvent mixture separates and yields two phases; removing the upper layer would result in concentrated lipid molecules in the lower phase. Many groups reported the modified

Bligh and Dyer method, using 1M NaCl to separate acidic lipids from the remaining lipid content. Adding 5% trichloroacetic acid can enhance lipid content sixfold in addition to the unmodified method (Mazzella et al., 2004; Ulmer et al., 2018). The Folch method is a well-known method for lipid extraction, in some ways similar to the Bligh and Dyer method, but it performs extraction with chloroform::methanol ratio (2:1) to maximize the lipid yield. The phase separation method employs centrifugal force to limit non-lipid molecules from contaminating lipid content. The lower layer comprises lipid content and the upper layer facilitates polar metabolite and protein molecules. The method is highly recommended for the extraction and separation of membrane lipids identifies as phospholipids, triglycerides, and steroids molecules. The Bligh and Dyer & Folch methods are experimentally approved for chromatographic analytical techniques (Breil et al., 2017) as demonstrated in Fig. 12.2.

12.2.1.2 MTBE (Methyl tert-butyl ether) Method

Another excellent method for lipid extraction is the Methyl tert-butyl ether or MTBE technique, which employs ether for extracting glycerophospholipids and sphingolipids from the membrane. The two-phase facilitate the separation of non-lipid species in the aqueous phase from the ether phase containing the essential lipid species (Kehelpannala et al., 2020).

12.2.1.3 Cell Disruption

Cell disruption is an excellent way to break the rigid cell walls and release the anchored biomolecules for identification and quantification. Researchers have developed many approaches in breaking cell walls, including mechanical, chemical, heat, or electric currents (sonication) that require low-energy consumption and provide efficient cell breakage (Nikolic et al., 2022; Rebollo-Ramirez et al., 2018).

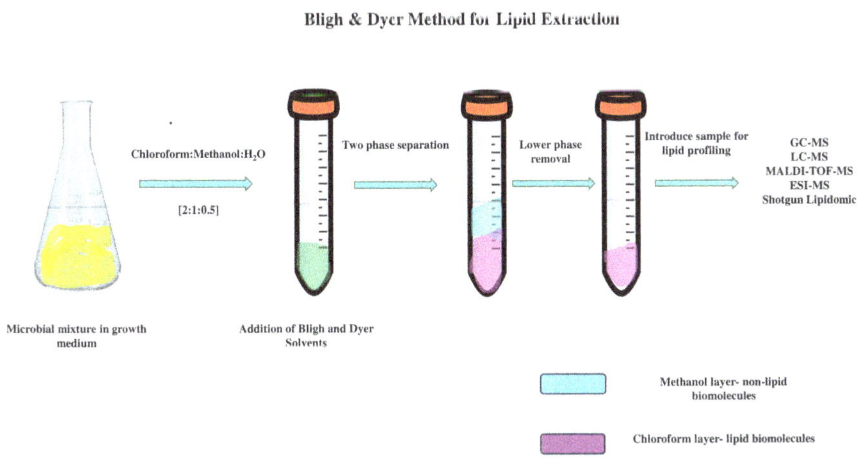

Fig. 12.2 Bligh and dyer method used for lipid extraction from microbial membrane

12.2.2 Analytical Techniques

12.2.2.1 Chromatography Techniques

Gas chromatography is an influential analytical tool to determine and characterize volatile compounds and metabolites with lower molecular weight. The techniques rely on the efficacy of the stationary phase to bind with lipid species via interactions and volatile properties along with an inert gas used as the mobile gas phase (Crick & Guan, 2015; Sherman et al., 2022). The technique incorporates vaporization of the lipid-containing sample, which runs over the stationary column and moves along with the inert gas to separate lipid species at their particular retention time (Allwood et al., 2014; Basile et al., 1998; Mondello et al., 2010).

The individual lipids can further subject to be analyzed by flame ionization detectors and mass spectrometry. The GC is a high-resolution technique that can identify biological molecules due to efficient sensitivity and the ability to analyze complex biological mixtures (Yang et al., 2021). Ishida et al. (2006) analyzed lipid profile in *E. coli* using thermally assisted hydrolysis and methylation-GC and MALDI-MS combined. The results displayed mass spectra peaks at methyl esters, especially for phosphatidylethanolamines (PEs) and phosphatidylglycerols (PGs). Mező et al. (2022) documented unsaturated and saturated fatty acids in membrane lipids in *P. aeruginosa*.

Liquid chromatography (LC) is an important technique applied for identifying and characterizing lipid-specific molecules. Like GC, the method involves a stationary column and a mobile phase, mainly solvents that flow through the column during separation (Sartorio et al., 2022). The lipids gradually separate based on their polarity and hydrophobicity to specific retention times and can be detected by using mass spectrometry and UV detection.

The lipidomic analysis is mainly assisted by high-performance liquid chromatography (HPLC) and ultra-HPLC. The group investigated the involvement of membrane lipids in the outer membrane vesicle formation using LC-MS. The results predicted negative ion m/z values specifically belonged to the membrane lipids, glycerophospholipids, serine dipeptides, and sphingolipids in *Bacteroides thetaiotaomicron*. Most of the glycerophospholipids identified were phosphatidylethanolamine at m/z 620.44, 634.5, 648.5, 662.5, and 676.5, whereas phosphatidyl inositol was at m/z 767.5, 781.5, and 795.5 (Valguarnera et al., 2018). Lopes et al. (2018) investigate lipid profiles in *Bacillus licheniformis* and documented the presence of phosphatidylethanolamines, phosphatidylglycerols, lysyl-PG, cardiolipins, monoglycosyldiacylglycerols, diglycosyldiacylglycerols, lipoteichoic acid primer (LTAP), and mono-alanylated LTAP.

12.2.2.2 Mass Spectrometry (MS)

Matrix-Assisted Laser Desorption/Ionization (MALDI-MS)

Matrix-assisted laser desorption-ionization time of flight mass spectrometry (MALDI-TOF MS) is widely used to identify microorganisms in clinical settings and has transformed the identification method compared to the traditional methods due to the availability of a vast dataset of microorganisms (Liang et al., 2018). It identifies microbial strains based on their phenotypic, genotypic, and biochemical patterns among microbes. The accuracy of microbial identification depends on the experience and sample preparation (Furniss et al., 2020). The analytical technique applies particle ionization to separate the lipids according to their mass to the ratio that can be detected at the end of a time-of-flight route. The identified molecules or lipids will show the mass: charge values on the X-axis and intensity at the Y-axis compared to the present dataset (Rychert, 2019).

MALDI-TOF detects species-specific conserved biomarkers and provides accurate microbial identification results through database spectra. MALDI-TOF could identify antimicrobial resistance (AMR) that is predictable for the microbes and could be treatable in time without increasing the chances of false results. Sorensen et al. (2020) determined lipid A modification that was directly associated with AMR. The group investigated the colistin-resistance plasmid, which is conferred by the *mcr-1* gene that encodes for phosphoethanolamine transferase. The enzyme modifies the terminal phosphate of lipid A, further capping the negative charge at the bacterial surface and minimizing the membrane interactions with cationic drug molecules. Four microbial strains, *E. coli*, *Klebsiella pneumoniae*, *Acinetobacter baumannii*, and *Pseudomonas aeruginosa,* were subjected to the analysis. They identified modified lipid A molecules linked to colistin resistance in Gram-negative bacteria. The lipid A modification was identified as an increased in the mass spectra compared to wild bacterial cells that remained the same (Δm/z 123) (Furniss et al., 2019).

A group reported a change in mass spectra shift in the presence of polymyxin resistance in *A. baumannii.* The native lipid A molecules displayed mass spectra peaks at m/z 1728.1 and m/z 1910.3, each peak belonged to unmodified bis-phosphorylated hexa-acyl (C_{12}) and bis-phosphorylated hepta-acyl lipid A (C_{14}) in polymyxin-susceptible strains. The group observed lipid A modification in the bis-phosphorylated hepta-acyl group and a peak shift from m/z 1910.3 to m/z 1935.3 and m/z 2033.3, leading to colistin resistance (Dortet et al., 2018; Fondrie et al., 2018).

Electrospray Ionization Tandem Mass Spectrometry (ESI-MS)

ESI-MS is an excellent tool for identifying and quantifying microbial membrane lipids. ESI-MS approaches lipid extraction via two distinct methods. The first method applies LC, which helps to elute lipid molecules that are further subjected to online mass spectrometry for identification (Randolph et al., 2020). The second technique extracts lipids via high-performance liquid chromatography (HPLC) or reverse-phase HPLC, separating microbial lipid species. Although LC-MS and HPLC are the ideal lipid extraction and identification methods, the individual retention time could affect ionization efficiency and change mass spectrum peaks (Ståhlman et al., 2009; Narváez-Rivas et al., 2016). The analysis is highly dependent on structural differences of glycerophospholipids and their isomers based on double bond position, their fatty acid chain lengths, and substitution group position defines m/z values.

A research group established a new method, laser ablation electrospray ionization mass spectrometry (LAESI-MS), to analyze polymicrobial biofilm. They provided enough evidence to prove the co-colonization of *P. aeruginosa* and *S. aureus* without disrupting microbial harmony and a well-defined microbial biofilm. Dean et al. (2015) recorded mass spectrum peaks corresponding to individual microbial strains, the ions m/z 241.4 and 941.6 assigned to *S. aureus*, and ions m/z 716.6, 281.4, and 255.3 reported for *P. aeruginosa* in a polymicrobial community. A novel triacylglycerol termed mono-meromycolyl diacylglycerol (MMDAG) was isolated from *Mycobacterium smegmatis*. The isomers reported belonged to Δ918:1- and 16:0-fatty acyl substituents located at sn-1 and sn-2 using a multiple-stage linear ion-trap mass spectrometric approach with high-resolution mass spectrometry (Purdy et al., 2013). Gonzalez et al. (2023) performed growth analysis in *E. coli* cells and observed a peak in phosphatidylethanolamines and phosphatidylglycerols levels over time, substituting alkylene for cyclopropyl in fatty acid chains.

Shotgun Lipidomics

Shotgun lipidomics is a promising tool for lipid profiling in microbial biofilm to identify and quantify lipid species in bacterial cells. This technique could be helpful in the therapeutic area as it elevates the understanding level of cell membrane components and protein-lipid interaction to mitigate antibiotic resistance or provide insightful information regarding biofilm formation. Kleiboeker et al. (2023) identified novel lipid structures DHC-1-P and GS-PA in the membranes of two bacterial strains, *Bacteroides fragilis* and *Porphyromonas gingivalis*, which belong to the colon and oral cavity, respectively. However, the lipids structurally differed in each strain, as the GS-PA lipid displayed a mass spectra peak at m/z 1227.91, representing a 15:0/βh16:0-GS-14:0/15:0-PA in *B. fragilis*, whereas *P. gingivalis* showed the GS-PA at mass spectra peak at 15:0/βh17:0-GS-15:0/13:0-PA. The author proved that phosphatidylinositol (PI) and inositol phosphoryl ceramide are distinguished with GS-PA in *B. fragilis* but lack in *P. gingivalis*.

The study reports the connection between fatty acid biosynthesis and intact lipid biosynthesis. It provides enough data to prove the fatty acid distribution among lipid species, which is required for looking into a deeper understanding of the biosynthesis pathway of phospholipids and non-phospholipids. The data reveals fragment ions for respective carboxylate ions and displays fatty acyl substituent loss while observing a weak ion at m/z 153. These m/z values were equal to (M-H)- ions of PG associated with fatty acid species PG-32:1, PG-32:0, PG-34:2, PG-34:1, PG-35:2, PG-35:1, PG-36:2, PG-37:2, and PG-38:2 (Basconcillo et al., 2008).

12.2.2.3 Nuclear Magnetic Resonance (NMR) Spectroscopy

Nuclear magnetic resonance (NMR) is another recommended tool for studying the structural elucidation of lipids, providing insightful details about atomic arrangements in lipid molecules. It is important to understand the saturation level of lipids by analyzing chemical shifts and splitting patterns in NMR spectra and investigating saturated and unsaturated fatty acids (Medeiros-Silva et al., 2018). This technique also defines the available functional moieties within lipid molecules that could be specific to the lipid architecture. The study identified phosphatidylglycerol at 0.65 ppm, cardiolipin at 0.38 ppm, and phosphatidylethanolamine (PE) in *E. coli* using NMR to investigate lipid profiles at different growth stages (Laydevant et al., 2021). Willmann et al. (2010) utilized a combination of HPLC, mass spectrometry (MS), and nuclear magnetic resonance (NMR) to analyze membrane lipids in microbial cells.

Each technique may be combined or tailored depending on the specific goals of the microbial lipidomic study, such as lipid profiling, structure elucidation, or metabolic flux analysis.

12.3 Applications of Lipidomics in Studying Biofilms

12.3.1 Structural Integrity and Fluidity

The bacterial membrane is a lipid bilayer that is responsible for channeling metabolite transportation and nutrient uptake from the environment. A salient feature of the membrane is the co-localization of important lipid biomolecules, which could be modified under reduced temperature. The modified lipid membrane prevents gel phase formation in membranes without rupturing the membrane (Bramkamp, 2022; Lee et al., 2024).

MALDI-TOF analyzes the complexity of lipid remodeling during adaptive responses, focusing specifically on the acyl chain composition of membrane lipids. Gram-negative bacteria, *A. baumannii* is known for pathogenicity and survival strategies under harsh environmental conditions, leading to antibiotic resistance. Glycerophospholipid (GPL) is integral to the bacterial membrane and helps maintain membrane structural integrity and fluidity under colder conditions. Under colder conditions, the bacterial cell go physiological adaptations, an increased level of palmitoleic acid (C16:1) and oleic acid (C18:1). Further, LC-HRMS2 reported a spectrum shift into phosphatidylethanolamine (PE) and phosphatidylglycerol (PG) leading to membrane modification.

Dessenne et al. (2024) reported different morphological patterns in bacterial motility and biofilm formation at low temperatures known as homeoviscous adaptations (Herrera et al., 2021). Chwastek et al. (2020) used the shotgun lipidomic tool to reveal headgroup-specific effects of acyl chain variability on membrane properties. The study revealed lipid modification required for structural integrity in *Methylobacterium extorquens*. Zielińska et al. (2020) explained that peptidoglycan synthesis and MreB movement is indirectly regulated by flotillin as it plays crucial role in membrane fluidity.

During growth analysis, *Shewanella putrefaciens* reported increased membrane fluidity at lower temperatures in ultra-HPLC (UHPLC-ESI-MS/MS). The profile displayed elevated membrane lipid levels, glycerolipids, and fatty acids with saccharolipids. When exposed to colder conditions, the results indicated a significant change in unsaturated and saturated fatty acid in the membrane. The changes indirectly control membrane fluidity under harsh environmental conditions and enhance bacterial survival (Gao et al., 2019).

12.3.2 Cell Signaling and Communication

Membrane-anchored lipid molecules play a significant role in bacterial cell signaling and communication and provide structural and functional support to the membrane. AHL molecules or peptides act as signaling molecules, bind with respective cognate receptor proteins, and initiate intracellular signal transduction or enhance pathogenicity in high cell-density environments. The signaling pathway may control microbial physiological alteration or host-pathogen interactions in planktonic or biofilm state (Fyfe et al., 2023; Garnier et al., 2006). Gosset-Erard et al. (2024) studied AHL secreted by *Burkholderia* and *Paraburkholderia*, which also works as a signaling molecule for the quorum-sensing pathway that regulates pathogenicity and virulence factor production. The group explored different aspects of the structural diversity of AHL within the community and co-localized C=C bonds using LC–MS/MS analysis.

The research group investigated the production of signaling molecules, that is, 2-alkyl-4-quinolones (AHQs) and 2-heptyl-3-hydroxy-4(1H)-quinolone (PQS) (García-Reyes et al., 2019) in *P. aeruginosa* under hypochlorous acid stress (HOCI). The hypochlorous acid is a stressor molecule that neutrophils produce during bacterial infection as an immune response. Laser desorption rapid evaporative ionization mass spectrometry (LD-REIMS) analysis reported a lower pyocyanin and elastase concentration under HOCI stress, suggesting downregulation of virulence factor production (Bradley et al., 2024; Farrant et al., 2020; Groitl et al., 2017). Bardin et al. (2018) studied pathogenicity-associated bacterial components, phospholipids, rhamnolipids, and QS-regulated metabolites using rapid evaporative ionization mass spectrometry (REI-MS) in *P. aeruginosa* clinical strain isolated from a cystic fibrotic patient. Gao et al. (2023) studied quorum-sensing mechanisms in *Aspergillus ochraceus* and characterized signaling molecules as oxylipins, which modulate hyphae production, morphology, and secondary metabolite production and infect host cells. Also, hydroxyoctadecadienoic acid and prostaglandins are reported as quorum-sensing regulating signaling molecules in the study.

A research group explored the lipid profile of two cellular motility states, swarming and swimming, in *Paenibacillus polymyxa*. The lipidomic analysis indicated lipid modifications correlating with motility, as swimming motility occurs in liquid, while swarming motility is displayed on a semi-solid surface. The study explored the impact of external stimuli and lipid remodeling on bacterial motilities. An increased level of glycerophospholipid was observed in swimming and glycerol-3-phosphate in swarming cells. Also, they noticed high expression levels of polyketides, antibiotics, and surfactin, along with glycoside hydrolase production, contributing extensively to exopolysaccharide matrix degradation, promoting flagellar rotation and biofilm formation (Poudel et al., 2019).

12.3.3 Antimicrobial Resistance

Lipid profiling of microbial membranes is an important aspect of microbial resistance against antimicrobial agents or external environmental stimuli. Alteration in microbial lipid profiling could enhance membrane fluidity and permeability, leading to antibiotic resistance, as particular lipid modification could change bacterial surface charge and enhance hydrophobicity, making antibiotic penetration and elevated efflux pump expression impossible. Understanding the mechanism behind lipid profiling is important to crack the mystery of antibiotic resistance (Delcour, 2008; Dunnick & O'Leary, 1970; MacDermott-Opeskin et al., 2022). The membrane proteins and lipids are known for their involvement with bacterial membranes and regulate antibiotic resistance. The mechanism involves interaction between membrane carotenoid and flotillin protein, leading to the formation of a functional membrane microdomain and facilitating efficient oligomerization. PBP2a is an important factor associated with penicillin resistance, but the disturbance of the FMM complex could increase penicillin susceptibility (García-Fernández et al., 2017).

A bacterial membrane lipid, methyl-(Z)-11-tetradecenoate acid from *E. coli*, was reported to facilitate antibiotic resistance against cationic drugs (Doma et al., 2022). MALDI-TOF combined with Fourier transform ion cyclotron resonance-MS was used for lipid profiling in *E. coli*. Lipid profile alterations are investigated to play an important role in membrane fluidity and antibiotic resistance against norfloxacin drugs, mainly involving upregulated fatty acids while glycerophospholipids were downregulated (Schenk et al., 2015). Yehia et al. (2015) reported an increase in membrane lipid and noted antibiotic resistance against two drugs, tetracycline (TE) and ciprofloxacin (CIP), with an increasing level of phospholipid content in *P. aeruginosa*.

Another excellent example of membrane lipid-associated antibiotic resistance in rhodococci grown on n-alkaline-containing media was investigated. An elevated level of cardiolipin and phosphatidylglycerol was reported. At the same time, most of it was made up of straight-chain fatty acids (C16:0, C18:0, C21:0) that promote antibiotic resistance towards aminoglycosides, macrolides, β-lactams, and aromatic compounds (Kuyukina et al., 2000). Bisignano et al. (2019) analyzed lipid profiles in *S. aureus* and observed elevated levels of anteiso-pentadecanoic acid (anteiso-C15:0), and anteiso-heptadecanoic acid (anteiso-C17:0) and iso-pentadecanoic acid (iso-C15:0). Iso-tridecanoic acid (iso-C13:0) and anteiso-tridecanoic acid (anteiso-C13:0), iso-hexadecanoic acid (iso-C16:0), and anteiso-hexadecanoic acid (anteiso-C16:0) were observed at lower levels associated with bacterial site of infection, antibiotic resistance, and cell-surface hydrophobicity in *S. aureus*.

The group previously reported daptomycin and bacitracin resistance associated with amino acid substitution at position 22, replacing alanine with aspartate in *Enterococcus faecalis*. The group investigated MadRS system, known for providing resistance against bacitracin via overexpression of an ABC transporter, MadLM. The group hypothesized that the MadEFG might be associated with daptomycin resistance and showcased a multilayered defense system against the antibiotic (Miller et al., 2024). Smith et al. (2022) predicted colistin resistance in *E. faecalis* using

MALDI-TOF and recorded an increased level of phosphoethanolamine and 4-amino-4-deoxy-l-arabinose (Ara4N), further contributing to a decreased electronegativity, increased membrane fluidity, and enhanced antibiotic resistance.

Ma et al. (2024) demonstrated that EptA, an enzyme that acts as a transferase in bacterial membranes, and targets phosphoethanolamine (PE). The lipid profile alteration triggers structural conformational changes in the enzyme and activates catalytic site and promotes ligand binding (Anandan et al., 2021). The group illustrated that a significant decrease in PE level in the membranes might block EptA activation and enhance antimicrobial resistance in Gram-negative bacteria. Kahler et al. (2018) and Freire et al. (2021) reported similar EptA function in the *Neisseria* group (*N. gonorrhoeae* and *N. meningitidis*) and *P. aeruginosa*, respectively. The studies demonstrated a significant role of EptA in the modulation of antibiotic resistance via lipid remodeling. Li et al. (2024) investigated the role of outer membrane vesicles consisting of membrane lipid molecules involved with bioactive transportation to the bacterial surface, enhanced antibiotic resistance, and modulates host-immune response in *K. pneumoniae*. The OL-synthesizing operon (olsBA) is highly expressed under phosphate depletion and regulates the production of ornithine lipid, an essential bacterial membrane lipid associated with bacterial persistence within the microbial community (Kim et al., 2018).

12.4 Bacterial Extracellular Vesicles (BEVs)/Membrane Vesicles (MVs)

Bacterial extracellular vesicles (BEVs) are peculiar cargo cells containing bioactive molecules produced by Gram-positive and Gram-negative species. BEVs are nanoparticles sizing from 40-400 nm and specialized in transporting bioactive molecules, as explained in Fig. 12.3. BEV acts as a signaling mechanism to mediate communication within microbial communities, is involved with pathogenicity, horizontal gene transfer, cell detoxification, AMR, and regulates host-immune response. BEVs could be defined as a biomarker that helps to understand the disease condition involved in the infection (Gilmore et al., 2021; Li et al., 2020; Liu et al., 2024; McMillan & Kuehn, 2021; Schwechheimer & Kuehn, 2015; Toyofuku et al., 2023). BEV or membrane vesicles could be generated by two distinct procedures: B-type and E-type. The B-type membrane vesicles are produced by blebbing of the outer membrane. Meanwhile, the E-type arises from cell lysis, which enables bending and self-annealing of exposed membrane sediments. The explosive cell lysis occurs under extraneous stimuli and provokes gene expression associated with prophage-derived endolysins. The endolysin degrades the peptidoglycan layer with specific structures and lipid contents (Toyofuku et al., 2018). Turnbull et al. (2016) reported that an essential prophage endolysin relies on the R- and F- pyocin gene clusters required for explosive cell lysis. Toyofuku et al. (2013) demonstrated enhanced pyocin production under denitrifying conditions, activating outer membrane vesicle production in *P. aeruginosa*. The PQS system is activated by nitric oxide, triggering an SOS response associated with increased membrane vesicles. Mandal et al. (2021)

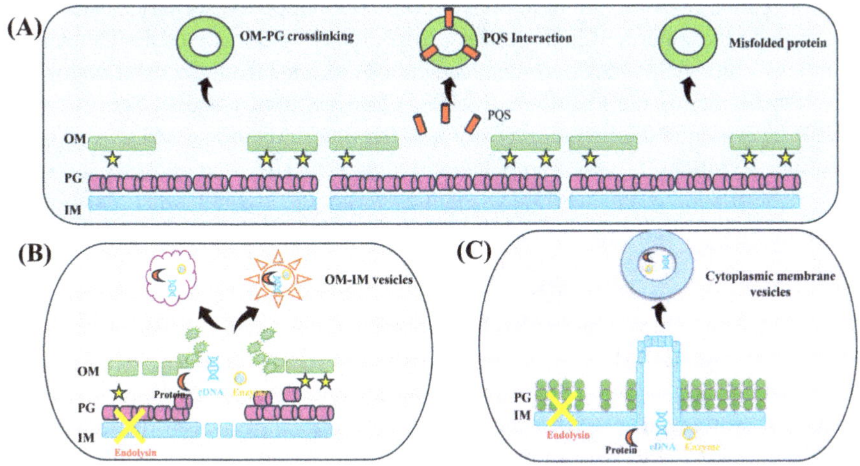

Fig. 12.3 The image demonstrates the processes involved in the formation of outer membrane vesicles (OMVs) and extracellular vesicles (EVs) through mechanisms such as membrane blebbing in Gram-negative (**a**) and the action of endolysin in Gram-negative (**b**) and Gram-positive (**c**)

studied *E. coli* for membrane vesicle production via explosive cell lysis eliciting via bacteriophage infection T4 and T7. Cooke et al. (2019) measured the MVs diameter and concentration using a single nanoparticle tracking technique, reporting various MVs based on the growth condition. The group investigated the physical size of OMVs associated with planktonic and biofilm of *P. aeruginosa* PA14. The study states that smaller-sized OMVs are more virulent than the larger ones secreted by planktonic and biofilm-embedded, respectively, indicating a negative reciprocity between vesicle size and exploitative nature. OMVs promote horizontal gene transfer in *P. aeruginosa* in planktonic and biofilm phases (Johnston et al., 2023). Meanwhile, several reports demonstrated a few alternative routes for membrane vesiculation, suggesting that negatively charged oligosaccharides influences divalent cations, thus destabilizing LPS and forming vesicles. Another suggestion is increased repulsion caused by the phosphate group enlarging the outer layer and initiating vesiculation. The OMVs biogenesis is induced by temperature variation, which influences membrane integrity and fluidity (Juodeikis & Carding, 2022).

12.4.1 BEVs/MVs Lipid Composition and Their Role in Biofilm Development

Lipidomics has been one of the approaches to studying and analyzing various lipids involved with BEV generation. The lipid analysis generally focuses on the structure of phospholipids, hydrophobic head groups, hydrophilic chains, chain length, and saturated fatty acid terminal moieties. The lipidomic analysis shows an abundance of phosphoglycerolipids, glycerolipids, and lipopolysaccharides in MVs. In

Gram-negative bacteria, the lipid bilayer consists of lipopolysaccharides and phospholipids at the outer and inner leaflets, respectively, and involves the OmpC-Mla complex, which helps transport misplaced phospholipids from the outer membrane to the inner membrane (Yeow et al., 2023). As mentioned above, Sartorio et al. (2022) investigated the lipid composition of the outer membrane and OMVs in the intestinal commensal, *Bacteroides thetaiotaomicron*. The lipid analysis showed a high level of sphingolipid in OMVs such as ceramide, ethanolamine phosphoceramide and inositol phosphoceramide compared to the outer membrane. The OMVs are recognized for their role in maintaining gut homeostasis. Valguarnera et al. (2018) performed the lipid analysis using mass spectrometry and reported the presence of a terminal moiety termed as lipoprotein export sequence (LES) at the end of lipoproteins within OMVs in *B. thetaiotaomicron*. The LES is a conserved motif that helps flip specific lipoproteins from the intracellular region and expose them on the OMVs surface. The studies suggest that the LES translocates SusG from the periplasmic side to the extracellular phase. The author also reported an elicited level of SusG in OMVs compared to the outer membrane, while SusC was highly present in the outer membrane.

Furthermore, Li et al. (2021) investigated the role of OMVs from *mcr-1 +/− E. coli in antibiotic resistance from the polymyxin drug colistin. The researcher indicated OMVs mediated colistin resistance in mcr-1* positive strains of *E.coli* via a higher abundance of modified lipid A moiety with phosphoethanolamine (pEtN), thus hindering lipid A and colistin interactions. They demonstrated that the two pEtN, Lipid-A-P and LipidA-2P, interact with the Lipid-A terminal and display spectrum at m/z 18.39.2 and m/z 1919.2, respectively. Another excellent example of membrane fluidity was reported in *Caulobacter crescentus*. The group investigated the role of a novel glycosphingolipid (GSL-2) in membrane integrity under phosphate depletion. The study supported the claim that ceramides support stalk filamentation, membrane blebbing, and increased colistin resistance (De Young et al., 2019; Stankeviciute et al., 2019). Zhao et al. (2021) reported structural changes in a mixed biofilm of *P. aeruginosa* and *S. aureus* in response to OMVs released by *P. aeruginosa* containing PQS molecules inhibiting *S. aureus* mediating interspecies competition. *Helicobacter pylori*, an opportunistic bacteria-causing agent of gastric ulcers, releases OMVs during infection. The group reported eDNA transportation and protection against degrading enzymes via OMVs in planktonic and biofilm phases, suggesting exposure of eDNA on the OMVs in microbial aggregation (Fernandez-Gonzalez & Backert, 2014; Grande et al., 2015). Gram-negative bacteria, *P. gingivalis*, prevents *S. aureus* colonization and eradicates it via OMVs secretion, enhancing bacterial aggregation and neutrophil killing. The OMVs facilitate bacterial transportation to the bloodstream, causing bacteremia (Du Teil Espina et al., 2022). Potapova et al. (2024) explored factors involved with biofilm matrix formation by *Vibrio cholerae*. The analysis revealed the involvement of outer membrane proteins RbmA, RbmC, Bap1, and eDNA with OMVs in biofilm assembly.

As discussed earlier, Gram-positive bacteria such as *S. aureus, Streptococcus* spp., *Mycobacterium, and Bacillus* spp. do not possess a thick layer of outer membrane; instead, they have a highly dense peptidoglycan layer, questioning the release of EVs through the thick layer (Briaud & Carroll, 2020). BEVs in Gram-positive are responsible for interspecies/intraspecies competition, lateral gene transfer, host invasion, and infection. Gram-positive EVs are mostly cytoplasmic due to the lack of an outer membrane and contain phosphatidylglycerol, myristic, and palmitic acid as the lipid content apart from the bacterial membrane (Bose et al., 2020). Mandelbaum et al. (2023) documented immuno-modulatory effect of EVs secreted by *Bifidobacterium longum.* The study showed induced production of IL-10 secretion from splenocytes and dendritic cells (DC)-CD4+ T cells through EVs. Chebotar et al. (2013) reported EVs interacting with *S. aureus* biofilm interacting with host-derived neutrophils, inducing EV secretion in the extracellular milieu. They treated secreted EVs with degrading enzymes such as phospholipase C, DNase, and proteinase K. Treatment of EVs with phospholipase C resulted in complete demolition, proteinase K partially impaired the vesicle structure, whereas DNase does not affect the vesicles. This study suggested the involvement of membrane lipids, phospholipids, and proteins in vesicle formation. Another report indicated the EV production in biofilm phase of *Streptococcus mutan* is higher than the planktonic cells and induced under harsh environmental conditions. Additionally, the EVs facilitate eDNA transportation during the early colonization, prompting in elicited eDNA level with the growth phase development (Leiva-Sabadini et al., 2024; Liao et al., 2014).

Listeria monocytogenes is known for secreting listeriolysin O and phosphatidylinositol phospholipase C via EVs during infection. The lipid analysis revealed an abundance of cardiolipin, saturated fatty acids, phosphatidyl glycerol, phosphatidyl ethanolamine, sphingolipid, and triglyceride in planktonic cells EVs compared to the cells from microbial community (Coelho et al., 2018).

Finally, BEVs are exclusive, specialized membrane vesicles derived from the bacterial outer or inner membrane. The membrane vesicles are enriched in lipid content from the parental membrane and inducer antibiotic treatment, temperature fluctuations and decreased nutrient availability. The lipid composition is highly bacterial species-specific, and are small treasures that facilitates bacterial surface studies and potential for biomedical applications.

12.5 Conclusions and Future Prospects

Lipidomic analysis is the biofilm research field that has opened multiple ways to analyze the complex microbial community that could provide significant insightful details on the complex structure formation and factors associated with stability. The methodologies, chromatography, mass spectrometry, and NMR, applied for the lipidomic analysis are of great interest, as each technique resolves unique properties of lipid profiles of individual microbial strains and understands their role in membrane integrity and fluidity, cell signaling and communication, and intercellular pathways associated with functional lipids. The lipidomic analysis may provide opportunities for further enhancing our knowledge regarding biofilm dynamics associated with antibiotic resistance and pathogenicity as displayed in Fig. 12.4. The researcher

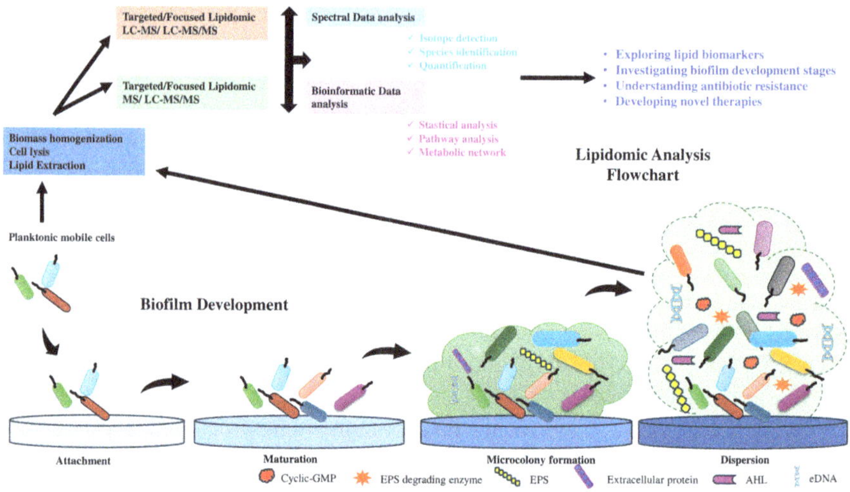

Fig. 12.4 The illustration shows a schematic overview of research focused on lipidomic analysis and their application in biofilm research

might want to dive deep into the untapped arena of biofilm formation and investigate how such a process occurs under harsh environmental conditions and protects persistent cells. However, the lipidomic methodologies need improvement while analyzing results using high-throughput screening and computational methods, which could extensively explore lipid species in bacterial membrane vesicles and microbial communities. Several interdisciplinary techniques are available, such as genomic, proteomic, transcriptomic, and metabolomic with lipidomic, which could be powerful tools to investigate microbial biofilms and uncover comprehensive details of membrane-lipid molecules in biofilm.

References

Abdelli, F., Jardak, M., Elloumi, J., Stien, D., Cherif, S., Mnif, S., & Aifa, S. (2019). Antibacterial, anti-adherent and cytotoxic activities of surfactin(s) from a lipolytic strain Bacillus safensis F4. *Biodegradation, 30*(4), 287–300. https://doi.org/10.1007/s10532-018-09865-4

Ali, A., Zahra, A., Kamthan, M., Husain, F. M., Albalawi, T., Zubair, M., Alatawy, R., Abid, M., & Noorani, M. S. (2023). Microbial biofilms: Applications, clinical consequences, and alternative therapies. *Microorganisms, 11*(8), 1934. https://doi.org/10.3390/microorganisms11081934

Allwood, J. W., AlRabiah, H., Correa, E., Vaughan, A., Xu, Y., Upton, M., & Goodacre, R. (2014). A workflow for bacterial metabolic fingerprinting and lipid profiling: Application to ciprofloxacin challenged Escherichia coli. *Metabolomics, 11*(2), 438–453. https://doi.org/10.1007/s11306-014-0674-6

Anandan, A., Dunstan, N. W., Ryan, T. M., Mertens, H. D. T., Lim, K. Y. L., Evans, G. L., Kahler, C. M., & Vrielink, A. (2021). Conformational flexibility of EptA driven by an interdomain helix provides insights for enzyme–substrate recognition. *IUCrJ, 8*(5), 732–746. https://doi.org/10.1107/s2052252521005613

Balleza, D., Alessandrini, A., & García, M. J. B. (2019). Role of lipid composition, physicochemical interactions, and membrane mechanics in the molecular actions of microbial cyclic lipopeptides. *The Journal of Membrane Biology, 252*(2–3), 131–157. https://doi.org/10.1007/s00232-019-00067-4

Bardin, E. E., Cameron, S. J. S., Perdones-Montero, A., Hardiman, K., Bolt, F., Alton, E. W. F. W., Bush, A., Davies, J. C., & Takáts, Z. (2018). Metabolic phenotyping and strain characterisation of Pseudomonas aeruginosa isolates from cystic fibrosis patients using rapid evaporative ionisation mass spectrometry. *Scientific Reports, 8*(1), 10952. https://doi.org/10.1038/s41598-018-28665-7

Basconcillo, L. S., Zaheer, R., Finan, T. M., & McCarry, B. E. (2008). A shotgun lipidomics approach in Sinorhizobium meliloti as a tool in functional genomics. *Journal of Lipid Research, 50*(6), 1120–1132. https://doi.org/10.1194/jlr.m800443-jlr200

Basile, F., Beverly, M. B., Voorhees, K. J., & Hadfield, T. L. (1998). Pathogenic bacteria: Their detection and differentiation by rapid lipid profiling with pyrolysis mass spectrometry. *TrAC Trends in Analytical Chemistry, 17*(2), 95–109. https://doi.org/10.1016/s0165-9936(97)00103-9

Bisignano, C., Ginestra, G., Smeriglio, A., La Camera, E., Crisafi, G., Franchina, F. A., Tranchida, P. Q., Alibrandi, A., Trombetta, D., Mondello, L., & Mandalari, G. (2019). Study of the lipid profile of ATCC and clinical strains of Staphylococcus aureus in relation to their antibiotic resistance. *Molecules, 24*(7), 1276. https://doi.org/10.3390/molecules24071276

Bonnichsen, L., Svenningsen, N. B., Rybtke, M., De Bruijn, I., Raaijmakers, J. M., Tolker-Nielsen, T., & Nybroe, O. (2015). Lipopeptide biosurfactant viscosin enhances dispersal of Pseudomonas fluorescens SBW25 biofilms. *Microbiology, 161*(12), 2289–2297. https://doi.org/10.1099/mic.0.000191

Bose, S., Aggarwal, S., Singh, D. V., & Acharya, N. (2020). Extracellular vesicles: An emerging platform in gram-positive bacteria. *Microbial Cell, 7*(12), 312–322. https://doi.org/10.15698/mic2020.12.737

Bradley, R., Simon, D., Spiga, L., Xiang, Y., Takats, Z., & Williams, H. (2024). Laser desorption rapid evaporative ionization mass spectrometry (LD-REIMS) demonstrates a direct impact of hypochlorous acid stress on PQS-mediated quorum sensing in Pseudomonas aeruginosa. *mSystems, 9*(4), e0116523. https://doi.org/10.1128/msystems.01165-23

Bramkamp, M. (2022). Fluidity is the way to life: Lipid phase separation in bacterial membranes. *The EMBO Journal, 41*(5). https://doi.org/10.15252/embj.2022110737

Breil, C., Vian, M. A., Zemb, T., Kunz, W., & Chemat, F. (2017). "Bligh and dyer" and Folch methods for solid–liquid–liquid extraction of lipids from microorganisms. Comprehension of Solvatation mechanisms and towards substitution with alternative solvents. *International Journal of Molecular Sciences, 18*(4), 708. https://doi.org/10.3390/ijms18040708

Briaud, P., & Carroll, R. K. (2020). Extracellular vesicle biogenesis and functions in Gram-positive bacteria. *Infection and Immunity, 88*(12). https://doi.org/10.1128/iai.00433-20

Chebotar, I. V., Konchakova, E. D., & Maianskii, A. N. (2013). Vesicle formation as a result of interaction between polymorphonuclear neutrophils and Staphylococcus aureus biofilm. *Journal of Medical Microbiology, 62*(8), 1153–1159. https://doi.org/10.1099/jmm.0.048967-0

Chwastek, G., Surma, M. A., Rizk, S., Grosser, D., Lavrynenko, O., Rucińska, M., Jambor, H., & Sáenz, J. (2020). Principles of membrane adaptation revealed through environmentally induced bacterial Lipidome remodeling. *Cell Reports, 32*(12), 108165. https://doi.org/10.1016/j.celrep.2020.108165

Coelho, C., Brown, L., Maryam, M., Vij, R., Smith, D. F., Burnet, M. C., Kyle, J. E., Heyman, H. M., Ramirez, J., Prados-Rosales, R., Lauvau, G., Nakayasu, E. S., Brady, N. R., Hamacher-Brady, A., Coppens, I., & Casadevall, A. (2018). Listeria monocytogenes virulence factors, including listeriolysin O, are secreted in biologically active extracellular vesicles. *Journal of Biological Chemistry, 294*(4), 1202–1217. https://doi.org/10.1074/jbc.ra118.006472

Cooke, A. C., Nello, A. V., Ernst, R. K., & Schertzer, J. W. (2019). Analysis of Pseudomonas aeruginosa biofilm membrane vesicles supports multiple mechanisms of biogenesis. *PLoS One, 14*(2), e0212275. https://doi.org/10.1371/journal.pone.0212275

Crick, P. J., & Guan, X. L. (2015). Lipid metabolism in mycobacteria—Insights using mass spectrometry-based lipidomics. *Biochimica Et Biophysica Acta (BBA) – Molecular and Cell Biology of Lipids, 1861*(1), 60–67. https://doi.org/10.1016/j.bbalip.2015.10.007

De Young, K. D., Stankeviciute, G., & Klein, E. A. (2019). Sugar-phosphate metabolism regulates stationary-phase entry and stalk elongation in Caulobacter crescentus. *Journal of Bacteriology, 202*(4). https://doi.org/10.1128/jb.00468-19

Dean, S. N., Walsh, C., Goodman, H., & Van Hoek, M. L. (2015). Analysis of mixed biofilm (Staphylococcus aureusandPseudomonas aeruginosa)by laser ablation electrospray ionization mass spectrometry. *Biofouling, 31*(2), 151–161. https://doi.org/10.1080/08927014.2015.1011067

Delcour, A. H. (2008). Outer membrane permeability and antibiotic resistance. *Biochimica Et Biophysica Acta (BBA) – Proteins and Proteomics, 1794*(5), 808–816. https://doi.org/10.1016/j.bbapap.2008.11.005

Dessenne, C., Ménart, B., Acket, S., Dewulf, G., Guerardel, Y., Vidal, O., & Rossez, Y. (2024). Lipidomic analyses reveal distinctive variations in homeoviscous adaptation among clinical strains of Acinetobacter baumannii, providing insights from an environmental adaptation perspective. *Microbiology Spectrum, 12*(10). https://doi.org/10.1128/spectrum.00757-24

Doma, A. O., Cristina, R. T., Muselin, F., Dumitrescu, E., Dégi, J., Imre, K., Boldea, M., Vlad, D. C., Popescu, R., Cimporescu, A., & Bratu, D. C. (2022). The role of methyl-(Z)-11-tetradecenoate acid from the bacterial membrane lipid composition in Escherichia coli antibiotic resistance. *BioMed Research International, 2022*, 1–10. https://doi.org/10.1155/2022/6028045

Dortet, L., Potron, A., Bonnin, R. A., Plesiat, P., Naas, T., Filloux, A., & Larrouy-Maumus, G. (2018). Rapid detection of colistin resistance in Acinetobacter baumannii using MALDI-TOF-based lipidomics on intact bacteria. *Scientific Reports, 8*(1), 16910. https://doi.org/10.1038/s41598-018-35041-y

Du Teil Espina, M., Rubio, A. H., Fu, Y., López-Álvarez, M., Gabarrini, G., & Van Dijl, J. M. (2022). Outer membrane vesicles of the oral pathogen Porphyromonas gingivalis promote aggregation and phagocytosis of Staphylococcus aureus. *Frontiers in Oral Health, 3*. https://doi.org/10.3389/froh.2022.948524

Dunnick, J. K., & O'Leary, W. M. (1970). Correlation of bacterial lipid composition with antibiotic resistance. *Journal of Bacteriology, 101*(3), 892–900. https://doi.org/10.1128/jb.101.3.892-900.1970

Farrant, K. V., Spiga, L., Davies, J. C., & Williams, H. D. (2020). Response of Pseudomonas aeruginosa to the innate immune system-derived oxidants Hypochlorous acid and Hypothiocyanous acid. *Journal of Bacteriology, 203*(2). https://doi.org/10.1128/jb.00300-20

Fernandez-Gonzalez, E., & Backert, S. (2014). DNA transfer in the gastric pathogen Helicobacter pylori. *Journal of Gastroenterology, 49*(4), 594–604. https://doi.org/10.1007/s00535-014-0938-y

Fondrie, W. E., Liang, T., Oyler, B. L., Leung, L. M., Ernst, R. K., Strickland, D. K., & Goodlett, D. R. (2018). Pathogen identification direct from polymicrobial specimens using membrane glycolipids. *Scientific Reports, 8*(1), 15857. https://doi.org/10.1038/s41598-018-33681-8

Freire, C. M. A. D. S., Taunay-Rodrigues, A., Gonzatti, M. B., Fonseca, F. M. P., & Da Cruz Freire, J. E. (2021). New insights about the EptA protein and its correlation with the pmrC gene in polymyxin resistance in Pseudomonas aeruginosa. *Current Research in Microbial Sciences, 2*, 100042. https://doi.org/10.1016/j.crmicr.2021.100042

Furniss, R. C. D., Dortet, L., Bolland, W., Drews, O., Sparbier, K., Bonnin, R. A., Filloux, A., Kostrzewa, M., Mavridou, D. A. I., & Larrouy-Maumus, G. (2019). Detection of Colistin resistance in Escherichia coli by use of the MALDI Biotyper Sirius mass spectrometry system. *Journal of Clinical Microbiology, 57*(12). https://doi.org/10.1128/jcm.01427-19

Furniss, R. C. D., Kostrzewa, M., Mavridou, D. A. I., & Larrouy-Maumus, G. (2020). The clue is in the lipid A: Rapid detection of colistin resistance. *PLoS Pathogens, 16*(4), e1008331. https://doi.org/10.1371/journal.ppat.1008331

Fyfe, J., Casari, I., Manfredi, M., & Falasca, M. (2023). Role of lipid signalling in extracellular vesicles-mediated cell-to-cell communication. *Cytokine & Growth Factor Reviews, 73*, 20–26. https://doi.org/10.1016/j.cytogfr.2023.08.006

Gao, X., Liu, W., Mei, J., & Xie, J. (2019). Quantitative analysis of cold stress inducing Lipidomic changes in Shewanella putrefaciens using UHPLC-ESI-MS/MS. *Molecules, 24*(24), 4609. https://doi.org/10.3390/molecules24244609

Gao, J., Liu, H., Zhang, Z., & Liang, Z. (2023). Quorum sensing-mediated lipid oxidation further regulating the environmental adaptability of aspergillus ochraceus. *Metabolites, 13*(4), 491. https://doi.org/10.3390/metabo13040491

García-Fernández, E., Koch, G., Wagner, R. M., Fekete, A., Stengel, S. T., Schneider, J., Mielich-Süss, B., Geibel, S., Markert, S. M., Stigloher, C., & Lopez, D. (2017). Membrane microdomain disassembly inhibits MRSA antibiotic resistance. *Cell, 171*(6), 1354–1367.e20. https://doi.org/10.1016/j.cell.2017.10.012

García-Reyes, S., Soberón-Chávez, G., & Cocotl-Yanez, M. (2019). The third quorum-sensing system of Pseudomonas aeruginosa: Pseudomonas quinolone signal and the enigmatic PqsE protein. *Journal of Medical Microbiology, 69*(1), 25–34. https://doi.org/10.1099/jmm.0.001116

Garnier, M., Dufourc, E. J., & Larijani, B. (2006). Characterisation of lipids in cell signalling and membrane dynamics by nuclear magnetic resonance spectroscopy and mass spectrometry. *Signal Transduction, 6*(2), 133–143. https://doi.org/10.1002/sita.200500077

Garsin, D. A. (2010). Ethanolamine utilization in bacterial pathogens: Roles and regulation. *Nature Reviews Microbiology, 8*(4), 290–295. https://doi.org/10.1038/nrmicro2334

Gilmore, W. J., Bitto, N. J., & Kaparakis-Liaskos, M. (2021). Pathogenesis mediated by bacterial membrane vesicles. In *Sub-cellular biochemistry/subcellular biochemistry* (pp. 101–150). https://doi.org/10.1007/978-3-030-67171-6_6

Gonzalez, L. E., Snyder, D. T., Casey, H., Hu, Y., Wang, D. M., Guetzloff, M., Huckaby, N., Dziekonski, E. T., Wells, J. M., & Cooks, R. G. (2023). Machine-learning classification of bacteria using two-dimensional tandem mass spectrometry. *Analytical Chemistry, 95*(46), 17082–17088. https://doi.org/10.1021/acs.analchem.3c04016

Gosset-Erard, C., Han, G., Kyrko, D., Hueber, A., Nay, B., Eparvier, V., & Touboul, D. (2024). Structural characterization of N-acyl-homoserine lactones from bacterial quorum sensing using LC–MS/MS analyses after Paternò-Büchi derivatization in solution. *Analytical and Bioanalytical Chemistry, 416*(25), 5431–5443. https://doi.org/10.1007/s00216-024-05355-0

Grande, R., Di Marcantonio, M. C., Robuffo, I., Pompilio, A., Celia, C., Di Marzio, L., Paolino, D., Codagnone, M., Muraro, R., Stoodley, P., Hall-Stoodley, L., & Mincione, G. (2015). Helicobacter pylori ATCC 43629/NCTC 11639 outer membrane vesicles (OMVs) from biofilm and planktonic phase associated with extracellular DNA (eDNA). *Frontiers In Microbiology, 6*. https://doi.org/10.3389/fmicb.2015.01369

Groitl, B., Dahl, J., Schroeder, J. W., & Jakob, U. (2017). Pseudomonas aeruginosa defense systems against microbicidal oxidants. *Molecular Microbiology, 106*(3), 335–350. https://doi.org/10.1111/mmi.13768

Herrera, C. M., Voss, B. J., & Trent, M. S. (2021). Homeoviscous adaptation of the Acinetobacter baumannii outer membrane: Alteration of Lipooligosaccharide structure during cold stress. *MBio, 12*(4), e0129521. https://doi.org/10.1128/mbio.01295-21

Ishida, Y., Kitagawa, K., Nakayama, A., & Ohtani, H. (2006). Complementary analysis of lipids in whole bacteria cells by thermally assisted hydrolysis and methylation-GC and MALDI-MS combined with on-probe sample pretreatment. *Journal of Analytical and Applied Pyrolysis, 77*(2), 116–120. https://doi.org/10.1016/j.jaap.2006.02.006

Jin, T., Patel, S. J., & Van Lehn, R. C. (2021). Molecular simulations of lipid membrane partitioning and translocation by bacterial quorum sensing modulators. *PLoS One, 16*(2), e0246187. https://doi.org/10.1371/journal.pone.0246187

Johnston, E. L., Zavan, L., Bitto, N. J., Petrovski, S., Hill, A. F., & Kaparakis-Liaskos, M. (2023). Planktonic and biofilm-derived Pseudomonas aeruginosa outer membrane vesicles facilitate horizontal gene transfer of plasmid DNA. *Microbiology Spectrum, 11*(2). https://doi.org/10.1128/spectrum.05179-22

Juodeikis, R., & Carding, S. R. (2022). Outer membrane vesicles: Biogenesis, functions, and issues. *Microbiology and Molecular Biology Reviews, 86*(4), e0003222. https://doi.org/10.1128/mmbr.00032-22

Jyothikumar, V., Klanbut, K., Tiong, J., Roxburgh, J. S., Hunter, I. S., Smith, T. K., & Herron, P. R. (2012). Cardiolipin synthase is required for Streptomyces coelicolor morphogenesis. *Molecular Microbiology, 84*(1), 181–197. https://doi.org/10.1111/j.1365-2958.2012.08018.x

Kadirvelu, L., Sivaramalingam, S. S., Jothivel, D., Chithiraiselvan, D. D., Govindarajan, D. K., & Kandaswamy, K. (2024). A review on antimicrobial strategies in mitigating biofilm-associated infections on medical implants. *Current Research in Microbial Sciences, 6*, 100231. https://doi.org/10.1016/j.crmicr.2024.100231

Kahler, C. M., Nawrocki, K. L., Anandan, A., Vrielink, A., & Shafer, W. M. (2018). Structure-function relationships of the neisserial EPTA enzyme responsible for phosphoethanolamine decoration of lipid A: Rationale for drug targeting. *Frontiers in Microbiology, 9*. https://doi.org/10.3389/fmicb.2018.01922

Kehelpannala, C., Rupasinghe, T. W. T., Hennessy, T., Bradley, D., Ebert, B., & Roessner, U. (2020). A comprehensive comparison of four methods for extracting lipids from Arabidopsis tissues. *Plant Methods, 16*(1), 155. https://doi.org/10.1186/s13007-020-00697-z

Khan, F. Z., Palmer, K. L., & Guan, Z. (2024). Biosynthesis of glucosaminyl phosphatidylglycerol inPseudomonas aeruginosa. *bioRxiv (Cold Spring Harbor Laboratory)*. https://doi.org/10.1101/2024.10.10.617631

Kim, S., Park, S., Li, X., Choi, Y., Im, D., & Lee, J. (2018). Bacterial ornithine lipid, a surrogate membrane lipid under phosphate-limiting conditions, plays important roles in bacterial persistence and interaction with host. *Environmental Microbiology, 20*(11), 3992–4008. https://doi.org/10.1111/1462-2920.14430

Kleiboeker, B. A., Frankfater, C., Davey, M. E., & Hsu, F. (2023). Lipidomic analysis of Porphyromonas gingivalis reveals novel glycerol bisphosphoceramide, phosphatidyl-, and phosphoglycerol dipeptide lipid families. *Journal of Lipid Research, 64*(12), 100470. https://doi.org/10.1016/j.jlr.2023.100470

Kumar, S., Balaya, R. D. A., Kanekar, S., Raju, R., Prasad, T. S. K., & Kandasamy, R. K. (2023). Computational tools for exploring peptide-membrane interactions in gram-positive bacteria. *Computational and Structural Biotechnology Journal, 21*, 1995–2008. https://doi.org/10.1016/j.csbj.2023.02.051

Kuyukina, M. S., Ivshina, I. B., Rychkova, M. I., & Chumakov, O. B. (2000). Effect of cell lipid composition on the formation of nonspecific antibiotic resistance in alkanotrophic rhodococci. *Microbiology, 69*(1), 51–57. https://doi.org/10.1007/bf02757257

Laydevant, F., Mahabadi, M., Llido, P., Bourgouin, J., Caron, L., Arnold, A. A., Marcotte, I., & Warschawski, D. E. (2021). Growth-phase dependence of bacterial membrane lipid profile and labeling for in-cell solid-state NMR applications. *Biochimica Et Biophysica Acta (BBA) - Biomembranes, 1864*(2), 183819. https://doi.org/10.1016/j.bbamem.2021.183819

Lee, T., Charchar, P., Separovic, F., Reid, G. E., Yarovsky, I., & Aguilar, M. (2024). The intricate link between membrane lipid structure and composition and membrane structural properties in bacterial membranes. *Chemical Science, 15*(10), 3408–3427. https://doi.org/10.1039/d3sc04523d

Leiva-Sabadini, C., Saavedra, P., Inostroza, C., & Aguayo, S. (2024). Extracellular vesicle production by oral bacteria related to dental caries and periodontal disease: Role in microbe-host and interspecies interactions. *Critical Reviews in Microbiology*, 1–18. https://doi.org/10.1080/1040841x.2024.2427656

Li, M., Zhou, H., Yang, C., Wu, Y., Zhou, X., Liu, H., & Wang, Y. (2020). Bacterial outer membrane vesicles as a platform for biomedical applications: An update. *Journal of Controlled Release, 323*, 253–268. https://doi.org/10.1016/j.jconrel.2020.04.031

Li, X., Sun, L., Li, C., Yang, X., Wang, X., Hu, X., Nie, T., Zhang, Y., & You, X. (2021). The attenuated protective effect of outer membrane vesicles produced by a mcr-1 positive strain on Colistin sensitive Escherichia coli. *Frontiers in Cellular and Infection Microbiology, 11*. https://doi.org/10.3389/fcimb.2021.701625

Li, L., Gao, X., Li, M., Liu, Y., Ma, J., Wang, X., Yu, Z., Cheng, W., Zhang, W., Sun, H., Song, X., & Wang, Z. (2024). Relationship between biofilm formation and antibiotic resistance of Klebsiella pneumoniae and updates on antibiofilm therapeutic strategies. *Frontiers in Cellular and Infection Microbiology, 14*. https://doi.org/10.3389/fcimb.2024.1324895

Li, L., Xu, X., Cheng, P., Yu, Z., Li, M., Yu, Z., Cheng, W., Zhang, W., Sun, H., & Song, X. (2025). Klebsiella pneumoniae derived outer membrane vesicles mediated bacterial virulence, antibiotic resistance, host immune responses and clinical applications. *Virulence, 16*(1). https://doi.org/10.1080/21505594.2025.2449722

Liang, T., Leung, L. M., Opene, B., Fondrie, W. E., Lee, Y. I., Chandler, C. E., Yoon, S. H., Doi, Y., Ernst, R. K., & Goodlett, D. R. (2018). Rapid microbial identification and antibiotic resistance detection by mass spectrometric analysis of membrane lipids. *Analytical Chemistry, 91*(2), 1286–1294. https://doi.org/10.1021/acs.analchem.8b02611

Liao, S., Klein, M. I., Heim, K. P., Fan, Y., Bitoun, J. P., Ahn, S., Burne, R. A., Koo, H., Brady, L. J., & Wen, Z. T. (2014). Streptococcus mutans extracellular DNA is upregulated during growth in biofilms, actively released via membrane vesicles, and influenced by components of the protein secretion machinery. *Journal of Bacteriology, 196*(13), 2355–2366. https://doi.org/10.1128/jb.01493-14

Liu, Q., Yeo, W., & Bae, T. (2016). The SaeRS two-component system of Staphylococcus aureus. *Genes, 7*(10), 81. https://doi.org/10.3390/genes7100081

Liu, C., Yazdani, N., Moran, C. S., Salomon, C., Seneviratne, C. J., Ivanovski, S., & Han, P. (2024). Unveiling clinical applications of bacterial extracellular vesicles as natural nanomaterials in disease diagnosis and therapeutics. *Acta Biomaterialia, 180*, 18–45. https://doi.org/10.1016/j.actbio.2024.04.022

Lopes, C., Barbosa, J., Maciel, E., Da Costa, E., Alves, E., Domingues, P., Mendo, S., & Domingues, M. R. M. (2018). Lipidomic signature of Bacillus licheniformis I89 during the different growth phases unravelled by high-resolution liquid chromatography-mass spectrometry. *Archives of Biochemistry and Biophysics, 663*, 83–94. https://doi.org/10.1016/j.abb.2018.12.024

Ma, Z., Nang, S. C., Liu, Z., Zhu, J., Mu, K., Xu, L., Xiao, M., Wang, L., Li, J., & Jiang, X. (2024). Membrane lipid homeostasis dually regulates conformational transition of phosphoethanolamine transferase EptA. *Nature Communications, 15*(1). https://doi.org/10.1038/s41467-024-54607-1

MacDermott-Opeskin, H. I., Gupta, V., & O'Mara, M. L. (2022). Lipid-mediated antimicrobial resistance: A phantom menace or a new hope? *Biophysical Reviews, 14*(1), 145–162. https://doi.org/10.1007/s12551-021-00912-8

Mandal, P. K., Ballerin, G., Nolan, L. M., Petty, N. K., & Whitchurch, C. B. (2021). Bacteriophage infection of Escherichia coli leads to the formation of membrane vesicles via both explosive cell lysis and membrane blebbing. *Microbiology, 167*(4). https://doi.org/10.1099/mic.0.001021

Mandelbaum, N., Zhang, L., Carasso, S., Ziv, T., Lifshiz-Simon, S., Davidovich, I., Luz, I., Berinstein, E., Gefen, T., Cooks, T., Talmon, Y., Balskus, E. P., & Geva-Zatorsky, N. (2023). Extracellular vesicles of the Gram-positive gut symbiont Bifidobacterium longum induce immune-modulatory, anti-inflammatory effects. *npj Biofilms and Microbiomes, 9*(1), 30. https://doi.org/10.1038/s41522-023-00400-9

Mazzella, N., Molinet, J., Syakti, A. D., Dodi, A., Doumenq, P., Artaud, J., & Bertrand, J. (2004). Bacterial phospholipid molecular species analysis by ion-pair reversed-phase HPLC/ESI/MS. *Journal of Lipid Research, 45*(7), 1355–1363. https://doi.org/10.1194/jlr.d300040-jlr200

McMillan, H. M., & Kuehn, M. J. (2021). The extracellular vesicle generation paradox: A bacterial point of view. *The EMBO Journal, 40*(21). https://doi.org/10.15252/embj.2021108174

Medeiros-Silva, J., Jekhmane, S., Paioni, A. L., Gawarecka, K., Baldus, M., Swiezewska, E., Breukink, E., & Weingarth, M. (2018). High-resolution NMR studies of antibiotics in cellular membranes. *Nature Communications, 9*(1). https://doi.org/10.1038/s41467-018-06314-x

Mező, E., Hartmann-Balogh, F., Horváth, I. M., Bufa, A., Marosvölgyi, T., Kocsis, B., & Makszin, L. (2022). Effect of culture conditions on fatty acid profiles of bacteria and lipopolysaccharides of the genus pseudomonas—GC-MS analysis on ionic liquid-based column. *Molecules, 27*(20), 6930. https://doi.org/10.3390/molecules27206930

Miller, W. R., Nguyen, A., Singh, K. V., Rizvi, S., Khan, A., Erickson, S. G., Egge, S. L., Cruz, M., Dinh, A. Q., Diaz, L., Thornton, P. C., Zhang, R., Xu, L., Garsin, D. A., Shamoo, Y., & Arias, C. A. (2024). Membrane lipids augment cell envelope stress signaling via the MadRS system to defend against antimicrobial peptides and antibiotics in enterococcus faecalis. *The Journal of Infectious Diseases., 231*, 307. https://doi.org/10.1093/infdis/jiae173

Mondello, L., Tranchida, P. Q., Purcaro, G., Fanali, C., Dugo, P., La Camera, E., & Bisignano, C. (2010). *Identification of the bacterial cellular lipid fraction by using Fast GC × GC-MS and innovative MS libraries* (NATO science for peace and security series. A, Chemistry and biology) (pp. 231–244). https://doi.org/10.1007/978-90-481-9815-3_15

Monnet, V., & Gardan, R. (2015). Quorum-sensing regulators in Gram-positive bacteria: 'cherchez le peptide. *Molecular Microbiology, 97*(2), 181–184. https://doi.org/10.1111/mmi.13060

Narváez-Rivas, M., Vu, N., Chen, G., & Zhang, Q. (2016). Off-line mixed-mode liquid chromatography coupled with reversed phase high performance liquid chromatography-high resolution mass spectrometry to improve coverage in lipidomics analysis. *Analytica Chimica Acta, 954*, 140–150. https://doi.org/10.1016/j.aca.2016.12.003

Nikolic, P., Mudgil, P., Harman, D. G., & Whitehall, J. (2022). Untargeted lipidomic differences between clinical strains of methicillin-sensitive and methicillin-resistant Staphylococcus aureus. *Infectious Diseases, 54*(7), 497–507. https://doi.org/10.1080/23744235.2022.2049863

Potapova, A., Garvey, W., Dahl, P., Guo, S., Chang, Y., Schwechheimer, C., Trebino, M. A., Floyd, K. A., Phinney, B. S., Liu, J., Malvankar, N. S., & Yildiz, F. H. (2024). Outer membrane vesicles and the outer membrane protein OmpU govern Vibrio cholerae biofilm matrix assembly. *MBio, 15*(2), e0330423. https://doi.org/10.1128/mbio.03304-23

Poudel, S., Giannone, R. J., Farmer, A. T., Campagna, S. R., Bible, A. N., Morrell-Falvey, J. L., Elkins, J. G., & Hettich, R. L. (2019). Integrated proteomics and Lipidomics reveal that the swarming motility of Paenibacillus polymyxa is characterized by phospholipid modification, surfactant deployment, and Flagellar specialization relative to swimming motility. *Frontiers in Microbiology, 10*. https://doi.org/10.3389/fmicb.2019.02594

Purdy, G. E., Pacheco, S., Turk, J., & Hsu, F. (2013). Characterization of mycobacterial triacylglycerols and monomeromycolyl diacylglycerols from mycobacterium smegmatis biofilm by electrospray ionization multiple-stage and high-resolution mass spectrometry. *Analytical and Bioanalytical Chemistry, 405*(23), 7415–7426. https://doi.org/10.1007/s00216-013-7179-4

Randolph, C. E., Blanksby, S. J., & McLuckey, S. A. (2020). Enhancing detection and characterization of lipids using charge manipulation in electrospray ionization-tandem mass spectrometry. *Chemistry and Physics of Lipids, 232*, 104970. https://doi.org/10.1016/j.chemphyslip.2020.104970

Rebollo-Ramirez, S., Krokowski, S., Lobato-Márquez, D., Thomson, M., Pennisi, I., Mostowy, S., & Larrouy-Maumus, G. (2018). Intact cell lipidomics reveal changes to the ratio of cardiolipins to phosphatidylinositols in response to kanamycin in HELA and primary cells. *Chemical Research in Toxicology, 31*(8), 688–696. https://doi.org/10.1021/acs.chemrestox.8b00038

Rychert, J. (2019). Benefits and limitations of MALDI-TOF mass spectrometry for the identification of microorganisms. *Journal of Infectiology, 2*(4), 1–5. https://doi.org/10.29245/2689-9981/2019/4.1142

Saini, R. K., Prasad, P., Shang, X., & Keum, Y. (2021). Advances in lipid extraction methods—A review. *International Journal of Molecular Sciences, 22*(24), 13643. https://doi.org/10.3390/ijms222413643

Sartorio, M. G., Valguarnera, E., Hsu, F., & Feldman, M. F. (2022). Lipidomics analysis of outer membrane vesicles and elucidation of the inositol Phosphoceramide biosynthetic pathway in Bacteroides thetaiotaomicron. *Microbiology Spectrum, 10*(1). https://doi.org/10.1128/spectrum.00634-21

Schenk, E. R., Nau, F., Thompson, C. J., Tse-Dinh, Y., & Fernandez-Lima, F. (2015). Changes in lipid distribution inE. colistrains in response to norfloxacin. *Journal of Mass Spectrometry, 50*(1), 88–94. https://doi.org/10.1002/jms.3500

Schwechheimer, C., & Kuehn, M. J. (2015). Outer-membrane vesicles from Gram-negative bacteria: Biogenesis and functions. *Nature Reviews Microbiology, 13*(10), 605–619. https://doi.org/10.1038/nrmicro3525

Sharma, S., Mohler, J., Mahajan, S. D., Schwartz, S. A., Bruggemann, L., & Aalinkeel, R. (2024). Correction: Sharma et al. Microbial biofilm: A review on formation, infection, antibiotic resistance, control measures, and innovative treatment. Microorganisms 2023, 11, 1614. *Microorganisms, 12*(10), 1961. https://doi.org/10.3390/microorganisms12101961

Sherman, M. E., Smith, R. D., Gardner, F. M., Goodlett, D. R., & Ernst, R. K. (2022). A sensitive GC-MS method for quantitation of lipid A backbone components and terminal phosphate modifications. *Journal of the American Society for Mass Spectrometry, 33*(12), 2301–2309. https://doi.org/10.1021/jasms.2c00266

Slamti, L., & Lereclus, D. (2019). The oligopeptide ABC-importers are essential communication channels in Gram-positive bacteria. *Research in Microbiology, 170*(8), 338–344. https://doi.org/10.1016/j.resmic.2019.07.004

Smith, R. D., McElheny, C. L., Izac, J. R., Gardner, F. M., Chandler, C. E., Goodlett, D. R., Doi, Y., Johnson, J. K., & Ernst, R. K. (2022). A novel Lipid-Based MALDI-TOF assay for the rapid detection of Colistin-Resistant enterobacter species. *Microbiology Spectrum, 10*(1). https://doi.org/10.1128/spectrum.01445-21

Sorensen, M., Chandler, C. E., Gardner, F. M., Ramadan, S., Khot, P. D., Leung, L. M., Farrance, C. E., Goodlett, D. R., Ernst, R. K., & Nilsson, E. (2020). Rapid microbial identification and colistin resistance detection via MALDI-TOF MS using a novel on-target extraction of membrane lipids. *Scientific Reports, 10*(1), 21536. https://doi.org/10.1038/s41598-020-78401-3

Sreedharan, S. M., Rishi, N., & Singh, R. (2023). Microbial lipopeptides: Properties, mechanics and engineering for novel lipopeptides. *Microbiological Research, 271*, 127363. https://doi.org/10.1016/j.micres.2023.127363

Stankeviciute, G., Guan, Z., Goldfine, H., & Klein, E. A. (2019). Caulobacter crescentus adapts to phosphate starvation by synthesizing anionic Glycoglycerolipids and a novel glycosphingolipid. *MBio, 10*(2). https://doi.org/10.1128/mbio.00107-19

Toyofuku, M., Zhou, S., Sawada, I., Takaya, N., Uchiyama, H., & Nomura, N. (2013). Membrane vesicle formation is associated with pyocin production under denitrifying conditions in Pseudomonas aeruginosa PAO1. *Environmental Microbiology, 16*(9), 2927–2938. https://doi.org/10.1111/1462-2920.12260

Toyofuku, M., Nomura, N., & Eberl, L. (2018). Types and origins of bacterial membrane vesicles. *Nature Reviews Microbiology, 17*(1), 13–24. https://doi.org/10.1038/s41579-018-0112-2

Toyofuku, M., Schild, S., Kaparakis-Liaskos, M., & Eberl, L. (2023). Composition and functions of bacterial membrane vesicles. *Nature Reviews Microbiology, 21*(7), 415–430. https://doi.org/10.1038/s41579-023-00875-5

Turnbull, L., Toyofuku, M., Hynen, A. L., Kurosawa, M., Pessi, G., Petty, N. K., Osvath, S. R., Cárcamo-Oyarce, G., Gloag, E. S., Shimoni, R., Omasits, U., Ito, S., Yap, X., Monahan, L. G., Cavaliere, R., Ahrens, C. H., Charles, I. G., Nomura, N., Eberl, L., & Whitchurch, C. B. (2016). Explosive cell lysis as a mechanism for the biogenesis of bacterial membrane vesicles and biofilms. *Nature Communications, 7*(1). https://doi.org/10.1038/ncomms11220

Ulmer, C. Z., Jones, C. M., Yost, R. A., Garrett, T. J., & Bowden, J. A. (2018). Optimization of Folch, Bligh-Dyer, and Matyash sample-to-extraction solvent ratios for human plasma-based lipidomics studies. *Analytica Chimica Acta, 1037*, 351–357. https://doi.org/10.1016/j.aca.2018.08.004

Valguarnera, E., Scott, N. E., Azimzadeh, P., & Feldman, M. F. (2018). Surface exposure and packing of lipoproteins into outer membrane vesicles are coupled processes in Bacteroides. *mSphere, 3*(6). https://doi.org/10.1128/msphere.00559-18

Willmann, J., Thiele, H., & Leibfritz, D. (2010). Combined reversed phase HPLC, mass spectrometry, and NMR spectroscopy for a fast separation and efficient identification of phosphatidylcholines. *BioMed Research International, 2011*(1). https://doi.org/10.1155/2011/385786

Wood, P. L. (2024). Metabolic and lipid biomarkers for pathogenic algae, fungi, cyanobacteria, mycobacteria, Gram-Positive bacteria, and Gram-Negative bacteria. *Metabolites, 14*(7), 378. https://doi.org/10.3390/metabo14070378

Yang, C., Lu, L., Liao, L., Zhang, B., Zeng, M., Zou, K., Liu, X., & Zhang, M. (2021). Establishment of GC–MS method for the determination of Pseudomonas aeruginosa biofilm and its application in metabolite enrichment analysis. *Journal of Chromatography B, 1179*, 122839. https://doi.org/10.1016/j.jchromb.2021.122839

Yehia, H. M., Hassanein, W. A., & Ibraheim, S. M. (2015). Studies on molecular characterizations of the outer membrane proteins, lipids profile, and exopolysaccharides of antibiotic resistant StrainPseudomonas aeruginosa. *BioMed Research International, 2015*, 1–7. https://doi.org/10.1155/2015/651464

Yeo, W., Dyzenhaus, S., Torres, V. J., Brinsmade, S. R., & Bae, T. (2023). Regulation of bacterial two-component systems by Cardiolipin. *Infection and Immunity, 91*(4), e0004623. https://doi.org/10.1128/iai.00046-23

Yeow, J., Luo, M., & Chng, S. (2023). Molecular mechanism of phospholipid transport at the bacterial outer membrane interface. *Nature Communications, 14*(1). https://doi.org/10.1038/s41467-023-44144-8

Zainuddin, M. F., Fai, C. K., Ariff, A. B., Rios-Solis, L., & Halim, M. (2021). Current pretreatment/cell disruption and extraction methods used to improve intracellular lipid recovery from oleaginous yeasts. *Microorganisms, 9*(2), 251. https://doi.org/10.3390/microorganisms9020251

Zhao, Z., Wang, L., Miao, J., Zhang, Z., Ruan, J., Xu, L., Guo, H., Zhang, M., & Qiao, W. (2021). Regulation of the formation and structure of biofilms by quorum sensing signal molecules packaged in outer membrane vesicles. *The Science of the Total Environment, 806*, 151403. https://doi.org/10.1016/j.scitotenv.2021.151403

Zhao, Z., Gao, B., Yang, C., Wu, Y., Sun, C., Jiménez, N., Zheng, L., Huang, F., Ren, Z., Yu, Z., Yu, C., Zhang, J., & Cai, M. (2024). Stimulating the biofilm formation of Bacillus populations to mitigate soil antibiotic resistome during insect fertilizer application. *Environment International, 190*, 108831. https://doi.org/10.1016/j.envint.2024.108831

Zielińska, A., Savietto, A., De Sousa Borges, A., Martinez, D., Berbon, M., Roelofsen, J. R., Hartman, A. M., De Boer, R., Van Der Klei, I. J., Hirsch, A. K., Habenstein, B., Bramkamp, M., & Scheffers, D. (2020). *Flotillin-mediated membrane fluidity controls peptidoglycan synthesis and MreB movement. eLife, 9.* https://doi.org/10.7554/elife.57179

Culturomics and Microbial Biofilms 13

Zarin Taj, Kamalavannan Raashika, Vinoth Kumarasamy, and Indranil Chattopadhyay

13.1 Introduction

Biofilms consist of groups of microorganisms that attach to surfaces and are surrounded by a self-generated extracellular matrix made up of polysaccharides, proteins, and nucleic acids (Donlan, 2002). These intricate formations can develop on both living and non-living surfaces and are observed in a diverse array of settings, spanning natural ecosystems, such as rivers and oceans, to artificial surface coatings like those seen in medical equipment and industrial machinery. The capacity of microorganisms to develop biofilms serves as an essential survival strategy, offering defense against environmental challenges, antibiotics, and the immune response of the host. The investigation of biofilms holds significant relevance because of their involvement in numerous clinical, environmental, and industrial scenarios. In the field of healthcare, biofilms play a significant role in chronic infections, including those related to cystic fibrosis, urinary tract infections, and catheter-associated infections (Jordan & Nicolle, 2014). Their persistence is closely associated with antibiotic tolerance, which complicates the eradication of these infections. In industrial environments, the development of biofilms can cause equipment corrosion, fouling, and clogging, leading to considerable economic losses (Vishwakarma, 2020). Grasping the molecular processes that drive biofilm formation, maturation, and dispersal may lead to advancements in therapeutic strategies and management practices for addressing biofilm-related challenges (Wang et al., 2023).

Z. Taj · K. Raashika · I. Chattopadhyay (✉)
Department of Biotechnology, School of Integrative Biology, Central University of Tamil Nadu, Neelakudi, Thiruvarur, Tamil Nadu, India
e-mail: indranil@cutn.ac.in

V. Kumarasamy
Department of Parasitology and Medical Entomology, Faculty of Medicine, Universiti Kebangsaan Malaysia, Kuala Lumpur, Malaysia

S. Busi et al. (eds.), *Omics Approaches in Biofilm Research*,
https://doi.org/10.1007/978-3-031-91863-6_13

Culturomics is an innovative methodology in microbiology, emphasizing the cultivation and identification of a diverse range of microorganisms, including many that were once deemed uncultivable through conventional techniques (Lagier et al., 2012). Culturomics represents a significant methodology that integrates high-throughput culturing techniques with sophisticated identification methods to reveal microbial species that were previously unnoticed (Ramasamy et al., 2014). This novel approach utilizes high-throughput culturing and sophisticated identification methods, greatly improving our comprehension of microbial diversity and ecology (Greub, 2012).

The application of diverse culture conditions significantly improves our comprehension of microbial diversity and serves as a valuable complement to metagenomics, which is centered on genetic material. In contrast to metagenomics, culturomics facilitates the functional examination of microbes through cultivation. This method has broadened understanding of microbial ecosystems, carrying important consequences for human health and environmental microbiomes. Regarding the health of humans, culturomics displays the microbiome's influence on health and disease, presenting opportunities for innovative therapeutic approaches. Meanwhile, in environmental studies, it enhances our comprehension of ecosystem dynamics and microbial relationships (Nam et al., 2023). This chapter provides an in-depth analysis of the impacts of biofilms and explores the application of culturomics-based approaches in advancing biofilm research.

13.1.1 Intersections of Biofilms and Culturomics

Biofilms represent a fundamental component of microbial existence both in natural environments and within the human body, significantly influencing microbial interactions. The interactions, encompassing both competitive and cooperative behaviors, play a crucial role in influencing microbial metabolism, diversity, and evolution (Lechón-Alonso et al., 2021). Competition between species has been acknowledged as a key factor in microbial ecology for a considerable time, yet recent studies have brought to light the importance of microbial cooperation. The intricate interactions significantly affect biofilm functionality, biomass, diversity, and pathogenicity (Li & Tian, 2016). In the human body, biofilms constitute an integral component of the personalized microbiome, essential for sustaining host health. The native species within microbial communities generally uphold a consistent, non-threatening interaction with their hosts, even when faced with frequent environmental disturbances and host defense strategies (Luo et al., 2021). This balance, or homeostasis, represents a dynamic equilibrium involving interactions among microbes as well as between microbes and their hosts. Nonetheless, disturbances in this equilibrium may make the host more susceptible to illnesses. Research indicates that biofilm-related diseases may be prevented or treated by focusing on potential pathogens but also on addressing the processes that disturb homeostasis within biofilms (Dhiman et al., 2024).

The hyperlink between biofilms and culturomics is crucial for enhancing our comprehension of biofilm composition, structure, and function (Lagier et al., 2012). Culturomics employs high-throughput culturing techniques to cultivate a wide variety of microbial species under various conditions, significantly contributing to the discovery of microorganisms that constitute biofilms (Ramasamy et al., 2014). In contrast to metagenomics, which reveals the genetic material found in biofilm samples, culturomics facilitates the cultivation and functional examination of these microbes, providing a deeper understanding of their behavior, interactions, and physiological responses within biofilms. Furthermore, culturomics has the potential to uncover species that were previously unnoticed or challenging to culture, which play a role in the complexity and resilience of biofilms. The study of microbial diversity and biofilm dynamics through culturomics plays a crucial role in enhancing our comprehension of biofilm-related pathogenesis and exploring potential treatment strategies (Clagnan et al., 2024).

13.2 Biofilms

Microorganisms develop biofilms by adhering to surfaces and integrating an extracellular polymeric substance (EPS) matrix into their composition. The dual role of this structure as both a physical barrier and a facilitator of resistance to antimicrobial drugs exacerbates persistent and chronic infections. Bacteria can endure antibiotics through various mechanisms, such as entering a state of metabolic dormancy within biofilms, exhibiting poor penetration of the antibiotics, and the presence of persister cells that manage to survive treatments that would typically eradicate them.

13.2.1 Formation and Composition

13.2.1.1 Biofilm Formation

The formation of bacterial biofilms is a sequential and dynamic process. The initial stage of biofilm formation is the adhesion of microorganisms to a surface. This process is mainly driven by bacterial surface structures such as flagella and type IV pili, which play a crucial role in initial attachment and ultimately result in irreversible adhesion (O'Toole & Kolter, 1998). After this attachment, biofilm maturation takes place, leading to the development of a three-dimensional structure. Ultimately, individual bacterial cells can be released from the biofilm, returning to a planktonic state. The durability and harmfulness of microbial attachment to material surfaces, along with biofilm formation, contribute significantly to the emergence of strong resistance to antibiotics thereby promoting the spread of infectious diseases (Flemming & Wingender, 2010; Mah & O'Toole, 2001).

Five major steps may be distinguished in the intricate process of bacterial biofilm formation: (i) Reversible attachment: Bacteria cling loosely to surfaces during this first stage via general interactions. (ii) Irreversible attachment: This phase involves the firm adhesion of bacteria to the surface through adhesins produced by bacteria, including lipopolysaccharide (LPS) and fimbriae (iii) Extracellular polymeric substance (EPS) Production: The bacteria within the biofilm begin synthesizing and secreting EPS, that supports the morphological resilience of biofilm iv) Biofilm maturation: In this phase, bacterial cells generate and secrete signaling molecules, which allow them to sense each other and synchronize their activities, ultimately resulting in the development of microcolonies and the advanced maturation of the biofilm; and (v) Dispersal/Detachment: In the final phase, bacterial cells are released from the biofilm, reverting to an independent, planktonic lifestyle (Sharma et al., 2023) (Fig. 13.1).

13.2.1.2 Biofilm Composition

The EPS matrix is a dynamic and complex structure that functions as both a protective barrier and a facilitator for bacterial communication, motility, and resistance mechanisms. The various elements of the biofilm matrix contribute collaboratively, reinforcing the durability and pathogenicity of bacterial communities, particularly in hostile settings. A comprehensive examination of the key components of EPS and their functions in biofilm dynamics is explained below.

(a) *Polysaccharides*

Types: The polysaccharides found in biofilms are mainly heterogeneous, although there are some homogeneous examples, such as cellulose, fructans, and glucans.

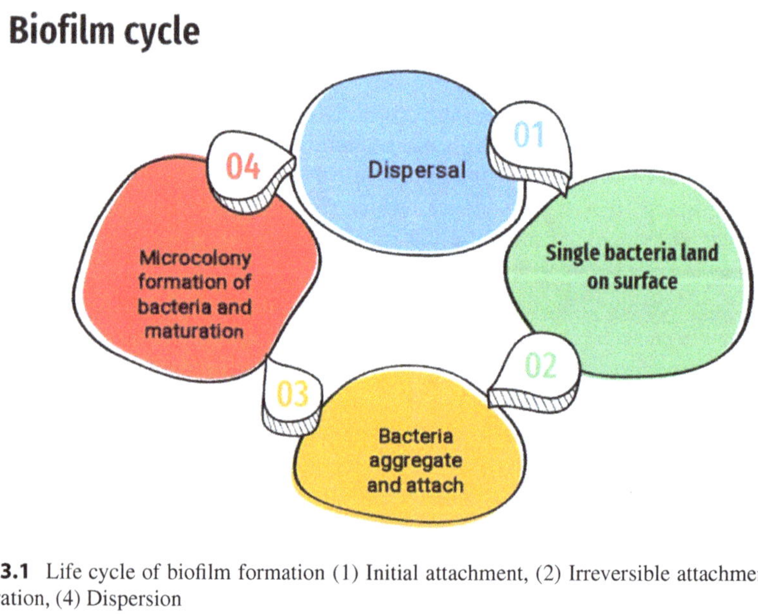

Fig. 13.1 Life cycle of biofilm formation (1) Initial attachment, (2) Irreversible attachment, (3) Maturation, (4) Dispersion

Roles: These polysaccharides serve as molecular adhesive agents, promoting the initial adhesion of bacteria to surfaces that has life and no life (Flemming et al., 2007). Their role is crucial in forming the biofilm's structural framework, which enhances its stability and ability to withstand environmental stresses, such as attacks from the immune system. In *Pseudomonas aeruginosa*, notable polysaccharides that contribute to the role of Biofilm formation include Pel, Psl, and alginate. The presence of these polysaccharides significantly boosts the biofilm's resilience against antimicrobial agents while also playing a crucial role in maintaining its structural integrity (Ghafoor et al., 2011).

(b) *Extracellular Proteins*

Types: This classification encompasses secreted proteins, along with proteins linked to pili, flagella, adhesins, and outer membrane vesicles (OMVs).

Roles: Extracellular proteins are essential for stabilizing the biofilm matrix, aiding in bacterial colonization and the maturation of biofilms. Certain proteins play a role in the degradation and dispersal of the matrix, including proteases, DNases, and dispersin B, which assist in the detachment of cells from the biofilm. Additionally, these proteins encompass enzymes that play a role in metabolic processes, chaperones, and proteins originating from lysed cells. About 30% of the EPS proteins in *Pseudomonas aeruginosa* are associated with membranes, frequently located within outer membrane vesicles (OMVs), which play a vital role in biofilm communication and maintenance (Toyofuku et al., 2012).

(c) *Extracellular DNA (eDNA)*

Origin: eDNA is released through various mechanisms, including secretion systems, cell lysis, quorum sensing, or through outer membrane vesicles (OMVs) that contain DNA.

Roles: The role of eDNA is crucial in maintaining the structural stability of biofilms and facilitating microbial aggregation, both of which are vital for the growth of biofilms. This also enhances the movement of bacteria within the biofilm. Furthermore, eDNA plays a role in enhancing the biofilm's resistance to antibiotics through the chelation of cations, thereby mitigating antimicrobial activity. This process improves the attachment to surfaces and facilitates horizontal gene transfer, thereby promoting genetic diversity and the possible dissemination of virulence factors. Furthermore, eDNA lowers the pH of the biofilm matrix, which additionally enhances antimicrobial resistance in organisms such as *Pseudomonas aeruginosa* (Sharma & Rajpurohit, 2024).

(d) *Surfactants and Lipids*

Roles: Significant surfactants and lipids that play a role in biofilm formation are surfactin, emulsan, viscosin, and rhamnolipids. These compounds enable the adhesion of bacteria to hydrophobic surfaces, including Teflon or waxy materials.

Surfactants and lipids have a significant impact on the structure of the biofilm, enabling dispersion, and binding heavy metals, which enhances the biofilm's adaptability across various environments (Satputea et al., 2016).

(e) *Water*

Significance: Water represents the predominant component of the EPS matrix, essential for providing the hydration required to sustain biofilm integrity.

The water content plays a crucial role in preventing desiccation, thereby ensuring the biofilm's viability over time. Water plays a crucial role in the movement of nutrients and metabolic byproducts, thereby supporting the nutritional requirements of the biofilm and improving its ability to adapt to varying environmental conditions (Flemming et al., 2007).

13.2.2 Microbial Interaction Within Biofilm

Biofilms are structured communities of microorganisms that can involve not only bacteria but also fungi and other eukaryotes. Candida species, for instance, are well-studied biofilm-forming fungi that provide valuable insights into the behavior of polymicrobial communities (Tournu & Van Dijck, 2012). While monomicrobial biofilms (biofilms formed by a single microbial species) are theoretically possible, they are rare. Most biofilms are polymicrobial, comprising multiple species that interact in complex ways. These interactions, which span across species and biological kingdoms, are essential for the stability and homeostasis of the biofilm community.

13.2.2.1 Cooperative and Antagonistic Interactions in Biofilms

Biofilms are dynamic ecosystems where cooperation and competition coexist. In polymicrobial communities, microorganisms often engage in collaborative interactions that enhance their survival, but competitive or antagonistic relationships can also shape biofilm formation and function. Environmental factors and the order of colonization influence whether interactions between species are cooperative or competitive (Flemming et al., 2024). One example of synergistic cooperation in biofilms is the relationship between *Staphylococcus aureus* and *Candida albicans.* While *S. aureus* alone struggles to form stable biofilms, its co-existence with *C. albicans* biofilm development provides vancomycin resistance in a similar vein, *C. albicans*, which shows heightened resistance to the antifungal miconazole in the presence of *S. aureus* (Hu et al., 2021). Despite these cooperative interactions, antagonistic competition between species can still influence biofilm pathogenesis and stability (Luo et al., 2022).

13.2.2.2 Mechanisms of Cooperation and Synergy in Polymicrobial Biofilms

In natural environments, polymicrobial biofilms are the norm, and understanding how cooperation and synergy maintain the coexistence of multiple species is key to understanding biofilm dynamics. Cooperation between species can provide significant advantages by outcompeting antagonistic interactions and promoting community stability. These cooperative behaviors are most likely to evolve when the interests of different species align and benefit mutual survival (Flemming et al., 2023).

13.2.2.3 Cohesion and Coaggregation in Polymicrobial Biofilms

The capacity of various microorganisms to cohabit and physically clump is a crucial requirement for the formation of mixed-species biofilms. A study investigating 30 microbial strains from various environments found that 77% co-aggregated with at least one other strain, with 70% forming co-aggregates with other archetypal strains. This coaggregation is primarily mediated by lectin–sugar interactions on microbial surfaces (Stevens et al., 2015). The oral biofilm is one of the best-characterized models of cohesion and coaggregation. For instance, *Porphyromonas gingivalis* and *Streptococcus gordonii*, the two bacteria that primarily colonize dental plaque, combine to produce a polymicrobial biofilm. Surface polypeptides like SspB on *S. gordonii*, which attaches to fimbriae on *P. gingivalis*, enable this association (Lamont et al., 2002). Similarly, glyceraldehyde-3-phosphate dehydrogenase, a coadhesin that binds to *P. gingivalis* fimbriae, is secreted by *Streptococcus oralis*, another early coloniser (Maeda et al., 2004).

Along with these adhesive interactions, movement of organelles also play a significant role in the cohesion between early colonizers, such as *S. gordonii* and *Veillonella parvula*. Dual RNA sequencing of these species during coaggregation revealed that oxidative stress responses dominate the coaggregation behavior of *V. parvula*, while *S. gordonii* primarily reinforces the metabolism of carbohydrates throughout this period of time (Mutha et al., 2019). These findings highlight the intricate molecular mechanisms behind biofilm formation, where cooperative and competitive dynamics interact to shape the biofilm's structure and function.

13.2.3 Biofilm Pathogenesis

13.2.3.1 Biofilm Dispersal and Infection Propagation

Established biofilms can release single cells or groups, a mechanism that aids in spreading infections to different regions of the body (Hall-Stoodley et al., 2004). Detached clusters from biofilms can result in systemic infections, including endocarditis. The capacity to disseminate within the host underscores the harmful potential of biofilms in leading to extensive tissue damage and complicating the management of infections.

13.2.3.2 Endotoxin Production and Inflammatory Response

Pathogenic bacteria in biofilms, including *Pseudomonas aeruginosa* and *Staphylococcus aureus,* frequently generate endotoxins that elicit significant inflammatory responses in the host (Bjarnsholt, 2013). The presence of these toxins intensifies tissue damage, playing a significant role in the persistent nature of infections associated with biofilms. The release of endotoxins significantly hinders the body's healing process, establishing a detrimental cycle of inflammation and injury.

13.2.3.3 Immune Evasion

The matrix of extracellular polymeric substances (EPS) and the physical architecture of biofilms are essential in protecting microbial communities from the host's immune response. The dense matrix of the biofilm hinders immune cell access, thereby diminishing the effectiveness of phagocytosis and the neutralization of antimicrobial peptides (Sharma et al., 2023). This immune evasion enables biofilm-associated pathogens to endure even in the presence of the host's immune defenses, complicating efforts to eliminate infections.

13.2.3.4 Genetic Diversity and Horizontal Gene Transfer in Biofilms

Biofilms facilitate genetic diversity, especially via horizontal gene transfer (HGT), enabling the swift dissemination of resistance genes among microorganisms (Michaelis & Grohmann, 2023). This process raises considerable concern, as it may result in the swift development of antimicrobial resistance within biofilm communities, thereby complicating infection treatment. The proximity of cells in biofilms and their capacity for genetic material exchange enables the rapid spread of traits that improve survival and virulence.

13.3 Biofilm Applications and Implications

13.3.1 Biofilms in Clinical Settings

Human health is intricately linked to the presence of microorganisms, particularly in the gut microbiota, which plays an important part in maintaining functions of the body. Pathogenic microorganisms, particularly those that can form biofilms, represent considerable risks to human health. Biofilm-associated infections (BAIs) include a diverse array of chronic and acute diseases, such as dental plaques, cystic fibrosis, infective endocarditis, urinary tract infections, ventilator-associated pneumonia, chronic wounds, chronic sinusitis, otitis media, and periodontitis (Gondil & Subhadra, 2023; Sahoo & Meshram, 2024). The capacity of bacteria to create biofilms, whether consisting of a single species or multiple species, significantly complicates the management of these infections.

The formation of biofilms plays a significant role in how microorganisms resist conventional antimicrobial therapies, such as antibiotics, antiseptics, and immune system responses. The extracellular polymeric substances (EPS) generated by biofilm-forming bacteria establish a protective barrier that hinders the effective penetration of therapeutic agents, resulting in biofilms being inherently resistant to conventional treatments (Hall & Mah, 2017). The resistance observed can be attributed primarily to genetic factors, especially the existence of Multidrug Resistance (MDR) genes. These genes empower bacteria to evade the mechanisms of drug action, thereby aiding in the continuation of infections (Cepas et al., 2019; Zafer et al., 2024). Due to the persistence of biofilms, eradicating infections is challenging, as bacteria within biofilms can withstand antimicrobial treatments, including antibiotics, by developing resistance to their inhibitory effects. The mechanisms of resistance found in biofilms are complex and varied. One important factor is the difficulty of antibiotics in reaching the planktonic cells entombed in the biofilm by permeating the extracellular polymeric substances (EPS) (Alav et al., 2018). Furthermore, bacteria in biofilms display modified gene expression when faced with environmental stressors, and the irregular surfaces of biofilms can form microenvironments that hinder drug diffusion (Chen et al., 2013). Moreover, antimicrobial agents frequently become inactivated within the biofilm matrix, and the environmental conditions present in biofilms like low oxygen levels and nutrient gradients—significantly diminish the effectiveness of treatment (Hall & Mah, 2017). The increased activity of efflux pumps, which play a crucial role in expelling antibiotics from bacterial cells, adds to the resistance seen in bacteria associated with biofilms. Numerous families of bacterial efflux pumps have been characterized, such as the major facilitator superfamily (MF), the resistance-nodulation-division family (RND), the small multidrug-resistance family (SMR), the ATP-binding cassette family (ABC), and the multidrug and toxic compound extrusion family (MATE). Each of these families contributes to biofilm-mediated antimicrobial resistance (Steenackers et al., 2016).

The formation of dormant and persister cells within biofilms serves as a crucial strategy for bacterial survival in the face of stress, complementing these mechanisms. The metabolic rate of these cells is diminished, making them less vulnerable to antimicrobial agents and playing a role in the persistent nature of infections linked to biofilms. The presence of these dormant cells significantly contributes to the challenge of eliminating biofilm-related infections, resulting in extended infections and associated complications (Liu et al., 2024). Biofilm-related infections and their resistance to antibiotics present significant worldwide health concerns. Antimicrobial resistance's increasing prevalence renders conditions more challenging for the treatment of these infections. The development of creative approaches is desperately needed, including the agents that can disrupt the biofilm matrix, prevent bacterial adhesion, or target specific microbial processes within biofilms.

13.3.2 Probiotic Biofilms: Beneficial Microbial Communities

While pathogenic biofilms pose significant risks, it is important to recognize that not all microorganisms that form biofilms are harmful. Probiotic biofilms, made up of beneficial bacteria, perform a crucial part in maintaining a balanced gut flora and improving general immune system function. Microorganisms like *Lactobacillus*,

Bacillus laterosporus, and *Pediococcus acidilactici* create biofilms that play a significant role in gut health by facilitating tissue growth and boosting immune responses (Mgomi et al., 2023). These advantageous biofilms are essential for sustaining a balanced microbial ecosystem, enhancing mucosal health, and bolstering immune system function (Pandit et al., 2020).

13.3.3 Biofilm Associated Infections in Healthcare

Biofilm-associated infections present a major challenge in healthcare, with both device-related and tissue-related biofilms contributing to various diseases and complications. Infections associated with medical devices, like urinary tract infections (UTIs) stemming from uropathogenic *Escherichia coli*, frequently occur in hospital environments, especially in patients who are utilizing catheters. Nosocomial infections, commonly attributed to multidrug-resistant ESKAPE pathogens (*Enterococcus, Staphylococcus, Klebsiella, Acinetobacter, Pseudomonas*, and *Enterobacter*), frequently arise due to biofilm formation on medical devices such as catheters, implants, and prosthetic valves (Bouhrour et al., 2024). Complications associated with breast implants and catheter-related bloodstream infections (CRBSIs) are common, involving pathogens such as *Staphylococcus* species and *Candida* species (Gahlot et al., 2014). These infections result in significant complications, including capsular contracture and, in rare cases, forms of lymphoma like BIA-ALCL (Breast Implant-Associated Anaplastic Large Cell Lymphoma). Biofilms on prosthetic joints play a crucial role in the progression of periprosthetic joint infections (PJI), frequently necessitating revision surgeries because of antimicrobial resistance (Staats et al., 2021). Contact lens-related infections can lead to significant ocular complications due to the presence of bacterial and fungal biofilms, one of the most serious being microbial keratitis (Voinescu et al., 2024).

Tissue-related biofilm infections, alongside device-related infections, represent a considerable issue. Dental biofilms significantly contribute to the development of dental caries and periodontitis, primarily influenced by inadequate oral hygiene and microbial imbalance (Kurtzman et al., 2022). Individuals with cystic fibrosis (CF) experience lung infections associated with biofilms, primarily due to pathogens, such as Staphylococcus aureus and Pseudomonas aeruginosa, leading to worsened respiratory complications (Malhotra et al., 2019). Infective endocarditis (IE), frequently associated with Streptococcus, Staphylococcus, and *Enterococcus* species, features biofilms on heart valves or prosthetic devices, presenting intricate treatment challenges stemming from antimicrobial resistance (Kaushik et al., 2024). Chronic wound infections, often attributed to ESKAPE pathogens and fungi, impede the healing process by keeping the wound in a prolonged inflammatory stage. Biofilms, developed in nutrient-dense settings, assimilate with host tissues and resist standard treatment methods (Cavallo et al., 2024). Strategies for managing these infections focus on strict hygiene protocols, the application of antimicrobial coatings, the innovation of technologies that disrupt biofilms, and, in more critical situations, surgical procedures. Addressing biofilm-associated infections through effective prevention and treatment strategies is crucial in tackling the increasing challenge of antimicrobial resistance within healthcare environments.

13.3.4 Biofilms in Food Industries

Biofilms represent intricate microbiological environments created by single or collective species embedded within an extracellular matrix, with their constituents differing according to the food manufacturing environment and the species that colonize them. Biofilms in the food industry present various challenges, such as the contamination of food products and equipment. Pathogenic microorganisms present in biofilms can potentially lead to foodborne illnesses, particularly affecting vulnerable groups, such as the elderly, infants, and those with weakened immune systems (Olanbiwoninu & Popoola, 2023). Food products that are contaminated, including dairy, meat, and vegetables, can contain bacteria that are challenging to eradicate because of the protective characteristics of biofilms. Furthermore, biofilms participate in spoilage by supporting the proliferation of spoilage microorganisms, which in turn diminishes the shelf life and quality of food products (Carrascosa et al., 2021). Biofilms develop in food production settings when microorganisms like *Listeria monocytogenes*, *Salmonella*, *Escherichia coli*, and *Pseudomonas* attach to surfaces that come into contact with food, water, or air. These microorganisms can flourish in damp environments, particularly in difficult-to-clean locations like cracks, crevices, and joints found in machinery, conveyor belts, and storage tanks (Ivers et al., 2024). The development of biofilms allows bacteria to withstand cleaning agents and antimicrobial treatments, resulting in ongoing contamination. A significant issue regarding biofilms in the food industry is their ability to resist antimicrobial agents. The thick extracellular polymeric substances (EPS) surrounding biofilms serve as a protective barrier, hindering the ability of disinfectants, sanitizers, and antibiotics to penetrate and successfully eliminate the microorganisms.

This matrix can adhere to biological structures, such as fruits, vegetables, meat, and bones and hard surfaces like food industry equipment and surfaces used for transportation, dispense, and preservation (Galié et al., 2018). The structural function of this matrix plays a significant role in the enduring presence of biofilms within the food industry. This system generates intricate gradients that enable the diffusion of nutrients and oxygen, contains extracellular enzymes that serve nutritional functions, support the transfer of cell communication molecules, and safeguards the embedded cells from harmful compounds (Liu et al., 2023).

13.3.5 Biofilms in Environment

Biofilms are present in a variety of environments, such as aquatic ecosystems, soils, and both natural and artificial surfaces (Fig. 13.2). Environmental biofilms are crucial in numerous ecological processes, yet their existence can also lead to considerable environmental and health-related consequences. In natural environments,

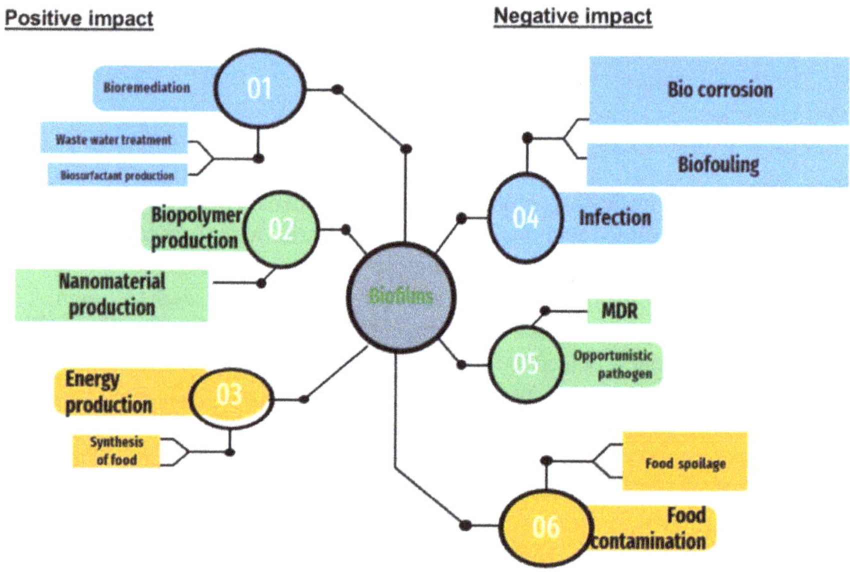

Fig. 13.2 Differential impacts of biofilms: 1–3 describes the positive impact of biofilms on applicative industry. 4–6 describes the negative impact of biofilms on infections

biofilm forming microorganisms including bacteria, algae, fungi, and protozoa that adhere to solid surfaces, such as rocks, sediments, and plant roots in aquatic settings, or to soil particles in terrestrial environments (Yadav, 2017). Biofilms can develop on both submerged and dry surfaces, provided there are suitable moisture levels, nutrients, and temperature conditions (Krishnan et al., 2023). In aquatic environments, biofilms flourish on the surfaces of water bodies, such as rivers, lakes, and oceans, whereas in soils, they form on plant roots, decaying organic matter, and soil aggregates.

13.3.6 Biofilms in Ecological Process

Biofilms play a vital role in numerous biogeochemical cycles, aiding in nutrient cycling and assisting in the decomposition of organic matter. In aquatic ecosystems, biofilms are essential in the nitrogen and carbon cycles by fostering microbial communities that convert organic compounds into more accessible forms for other organisms (Krishnan et al., 2023). Biofilms in aquatic environments also help in the oxidation of nitrogen compounds, playing a key role in nutrient regulation and maintaining ecosystem balance. In soils, biofilms associated with plants are crucial for nutrient uptake and protection, contributing significantly to plant growth and enhancing resistance to pathogens. These biofilms support plant roots by facilitating the uptake of essential nutrients like nitrogen and phosphorus while providing protection from harmful soil-borne pathogens. Biofilms also play a role in detoxification processes, effectively breaking down environmental pollutants, such as heavy metals and organic contaminants, through microbial metabolic activities (Balcázar et al., 2015).

13.3.7 Biofilms in Contaminated Environments

In polluted environments, biofilms can play a dual role in the degradation of contaminants, either facilitating or obstructing the process. Microorganisms within biofilms can decompose hazardous pollutants, including heavy metals, pesticides, and petroleum products, via bioremediation processes (Muhammad et al., 2020). Biofilms hold significant potential for applications in environmental remediation initiatives. Nevertheless, biofilms have the capacity to accumulate toxic substances within their EPS matrix, which may subsequently be released into the surrounding environment, thereby presenting additional risks. The capacity of biofilms to accumulate pollutants, coupled with their heightened resistance to chemical treatments, presents significant challenges in effectively tackling contamination problems.

13.4 Challenges and Current Interventions

The management of biofilm-associated infections presents a significant challenge because of the intrinsic resilience and complexity of biofilms. These microbial communities demonstrate significant resilience against host immune responses and standard antimicrobial therapies. Current therapeutic strategies focus on tackling the different phases of biofilm formation and persistence, utilizing a range of methodologies.

13.4.1 Prevention of Initial Colonization

A key approach in reducing biofilm formation is to inhibit the initial adhesion of microorganisms to surfaces, especially on medical devices. Coatings that prevent adhesion and modifications to surfaces have been created to minimize the attachment of microbes, thereby obstructing the development of biofilms on devices, such as catheters, prostheses, and implants (Yin et al., 2021). The strategies aim to modify the physical or chemical characteristics of the surface to reduce its suitability for microbial attachment, thus hindering the development of biofilm-associated infections.

13.4.2 Investigating Microbial Attachment Mechanisms

A different strategy for preventing biofilms includes employing agents that specifically disrupt the mechanisms by which bacteria adhere. These agents diminish the probability of biofilm formation on host tissues or medical devices by interfering with the molecular interactions that facilitate microbial attachment. These strategies have demonstrated potential in limiting the initial phases of biofilm formation, thereby decreasing the likelihood of later infections (Asma et al., 2022).

13.4.3 Disrupting the Biofilm Structure

Once biofilms are established, they present significant challenges for treatment because of the protective extracellular polymeric substance (EPS) matrix enveloping the microbial cells. To address this challenge, enzymatic treatments, or surfactants have been investigated to break down or interfere with the EPS matrix, thereby improving the infiltration of antibiotics and other therapeutic agents into the biofilm (Jiang et al., 2020). By interfering with the strength of the biofilm structure that can enhance vulnerability to biofilm-associated pathogens to treatment, presenting potential strategies for addressing chronic infections.

13.4.4 Procedure: Extraction of Device

In instances where infections associated with biofilms prove resistant to medical interventions, it may become essential to remove the infected medical devices. Infections associated with implanted devices like catheters, pacemakers, or prostheses present significant challenges for resolution, primarily due to the enduring characteristics of the biofilm. In these cases, the only feasible approach to eliminate the infection may involve surgical procedures to either replace or remove the infected devices (Abdelhamid & Yousef, 2023). This underscores the necessity for efficient preventive strategies to control biofilm formation on medical devices from occurring initially.

13.5 Culturomics

Culturomics is an innovative and efficient culturing technique that has played a crucial role in revitalizing research focused on microbial cultures. Metagenomics has significantly advanced our comprehension of microbiota, while culturomics has greatly enhanced our insights into the bacterial gut microbiome, especially through the identification and characterization of numerous previously unrecognized taxa. The notion of culturomics was initially presented by Lagier et al. (2012) to investigate the human gastrointestinal microbiota. This innovative study utilized 212 unique culturing conditions, leading to the successful isolation of 32,000 bacterial colonies from human fecal samples, which included 340 distinct bacterial species. In the subsequent year, Pfleiderer et al. (2013) enhanced the methodology by evaluating and confirming around seventy optimal culture conditions, thereby broadening the repertoire of the human gut microbiota.

Culturomics has been integrated into clinical and environmental microbiology laboratories for regular application. Over the past 10 years, metagenomics-assembled genomes (MAGs)—which are reconstructed microbial genomes obtained from metagenomic samples—have facilitated the discovery of new, non-cultivable microorganisms in various environments, such as the human microbiota (Sood et al., 2021). While these molecular approaches hold significant potential, they frequently demand considerable resources and time. Culturomics was once regarded as the "gold standard" until the rise of more advanced techniques.

Recent technological advancements, including Sanger sequencing of the 16S rDNA gene and MALDI-TOF mass spectrometry, have greatly minimized the labor and time needed for identifying bacterial isolates, thereby enhancing the exploration of previously challenging-to-culture organisms. The emergence of next-generation sequencing has increasingly supplanted traditional culture-based microbiology in various contexts, greatly improving our comprehension of microbial communities. Nonetheless, the significant importance of microbial culture is being acknowledged once more (Bielen et al., 2024). With the expansion of our understanding of microbial diversity, innovative culture methods that utilize a blend of technologies have been developed. Currently, culturomics utilizes a diverse range of culture conditions, which encompass different media compositions, pretreatment methods, oxygen levels, and temperatures. This is combined with advanced techniques like MALDI-TOF MS and full-length 16S rRNA gene sequencing for the precise identification of microbial isolates (Seng et al., 2010).

13.5.1 Work Flow of Culturomics

MALDI-TOF mass spectrometry serves as an essential technique for identifying bacterial species in culturomics research, owing to its efficiency, specificity, and cost-effectiveness. The process of culturomics for unidentified microbes consists of several essential steps (Fig. 13.3): isolating bacteria, identifying them through MALDI-TOF mass spectrometry, conducting 16S rRNA sequencing, and announcing any new species discovered. The colony derived from bacterial isolates must undergo multiple analyses using MALDI-TOF MS initially. If the isolate's score is less than 1.9, it may be an undetermined bacterial species (Seng et al., 2010). The full 16S sequencing process is then carried out. There is a significant likelihood that the isolate is a novel species if the sequence similarity between it and recognized species is less than 98.65%.

13.5.1.1 Sample Collection and Preparation

The initial phase entails the meticulous gathering of biological samples, frequently sourced from human origins, such as feces, oral cavity, skin, or various body sites, as well as from environmental sources like soil or water.

Sample Processing and Pre-treatment: Samples undergo various pretreatment protocols, such as centrifugation, filtration, or dilution, to effectively isolate

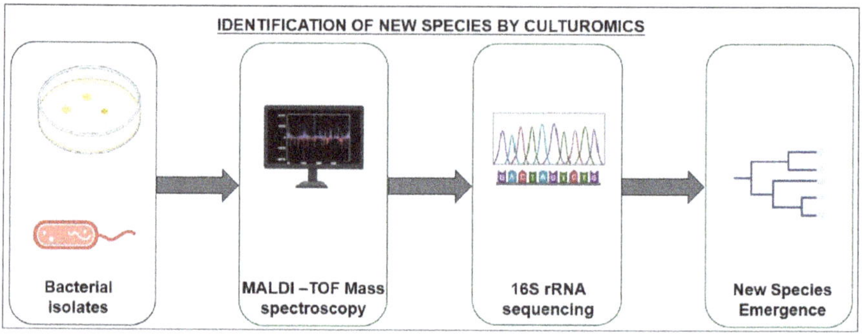

Fig. 13.3 Workflow of Culturomics: Culturomics workflow for identifying novel bacteria. Bacterial isolates are first obtained and cultured on agar plates. These isolates are then analyzed using MALDI-TOF mass spectrometry to determine their protein profiles. Next, 16S rRNA sequencing is performed to analyze genetic similarities and differences among the isolates. Finally, phylogenetic analysis is conducted, leading to the identification and classification of new bacterial species

microbial components and minimize contaminants. In certain instances, enrichment cultures can be employed to facilitate the proliferation of particular microorganisms.

13.5.1.2 High-Throughput Culturing Conditions

With the use of numerous characteristics, such as media composition, oxygen availability, temperature and incubation duration, and sample pretreatment, culturomics focuses on high-throughput screening of a variety of culturing conditions. In order to maintain high-throughput capacities, acquire the entire spectrum of microbial diversity, and promote the growth of a wide variety of microorganisms, these factors are examined. Automated culturing systems are used to handle the complexity and volume of conditions tested, ensuring efficient and effective culturing of diverse taxa.

13.5.1.3 Isolation of Bacterial Colonies

The process of colony isolation and streaking is a fundamental technique in microbiology, employed after the incubation of microbial samples on agar plates. This method allows researchers to isolate individual microbial colonies, which are then meticulously streaked to obtain pure cultures, essential for accurate identification and study. The selection of colonies is guided by their distinct morphological characteristics.

13.5.1.4 Microbial Identification

MALDI-TOF MS: Matrix-assisted laser desorption ionization time-of-flight mass spectrometry serves as an essential instrument for the rapid and accurate identification of microbial species. This method examines the protein profiles of bacterial colonies to create a distinct spectral fingerprint, which is then matched with databases for species identification.

16S rRNA Gene Sequencing: To provide further validation and to detect novel or unculturable microorganisms, the complete 16S rRNA gene is sequenced. This approach offers a detailed phylogenetic examination of bacterial isolates.

Genomic Sequencing: In certain scenarios, whole-genome sequencing may be carried out to improve knowledge of the genetic makeup of the isolated microorganisms, particularly for recently discovered species.

13.5.1.5 Data Integration and Analysis

Bioinformatics Tools: Data produced from MALDI-TOF MS and 16S rRNA sequencing undergo analysis through a range of bioinformatics tools to determine microbial species and evaluate their phylogenetic relationships.

Microbial Diversity Assessment: Sophisticated software is employed to analyze the diversity of microbial communities derived from the cultures collected, charting the diversity present within the particular environment or microbiota under investigation.

Comparative analysis: It involves comparing the obtained data with existing microbial databases and literature to identify novel species or strains.

13.5.1.6 Reporting and Further Characterization

Microbial Cataloguing: The outcomes of culturomic studies are systematically organized, encompassing the identification of novel taxa, uncommon species, or challenging-to-culture microorganisms, thereby enhancing the existing knowledge of microbial diversity.

Functional Characterization: Further analyses could include assessments of particular microbial functions, including metabolic pathways, antibiotic resistance profiles, or virulence factors.

Long-term Cultivation and Repository: Isolates that signify novel or significant species are frequently maintained in culture collections or microbial repositories for prospective studies and applications.

13.6 Applications of Culturomics

13.6.1 Medical Microbiology

Culturomics has emerged as a crucial method in medical microbiology, especially for identifying and characterizing microbial communities that affect human health and wellbeing.

The high-throughput nature facilitates the isolation, identification, and examination of previously unculturable pathogens, providing fresh perspectives on a range of medical conditions.

13.6.1.1 Discovery of Novel Pathogens

Culturomics facilitates the extraction of previously unculturable or rare pathogens from clinical samples, encompassing those linked to chronic infections, biofilm-associated infections, and antibiotic-resistant organisms. This ability improves the precision of diagnostics, especially in the case of intricate or polymicrobial infections (Diakite et al., 2020).

13.6.1.2 Improved Diagnosis of Infectious Diseases

Diverse culturing techniques enable culturomics to identify a wider range of pathogens compared to traditional microbiological methods. This is especially beneficial for identifying infections caused by unique, slow-growing, or difficult-to-culture bacteria that might be overlooked by standard culture methods.

13.6.1.3 Antimicrobial Resistance Surveillance

The application of culturomics is essential for identifying antibiotic-resistant strains in clinical environments. Isolating resistant bacteria from diverse patient samples allows for the tracking of resistance spread, enabling timely interventions and more precise treatment strategies.

13.6.1.4 Investigating the Role of Microbiomes in Disease

Culturomics facilitates a comprehensive examination of the microbiome across different disease conditions, including inflammatory bowel disease (IBD), obesity, diabetes, and cancer. The link between microbial communities and disease pathogenesis can be uncovered by culturomics, which isolates and characterizes gut microbiota or other body-site microorganisms.

13.6.1.5 Personalized Medicine

Culturomics contributes to the development of personalized medicine by identifying microbial communities that may influence disease outcomes or responses to treatment. This includes tailoring antibiotic therapies or probiotic treatments based on the specific microbial profiles of individual patients.

13.6.1.6 Infection Control and Hospital Epidemiology

In hospital environments, culturomics plays a crucial role in tracking healthcare-associated infections, particularly those outbreaks linked to multidrug-resistant organisms (MDROs). Tracking microbial diversity and identifying potential sources of infection, culturomics provides valuable insights for informing infection control strategies (Werno et al., 2012).

13.6.1.7 Identification of Uncultured and Novel Pathogens

Culturomics contributes to the discovery of emerging infectious agents by expanding the range of culturable microorganisms in clinical samples. This allows for the early detection of pathogens that may not be well-characterized or are not routinely tested for in diagnostic laboratories (Verroken et al., 2010).

13.6.2 Environmental Microbiology

Culturomics has significantly advanced the field of environmental microbiology, offering new insights into the diversity, functionality, and dynamics of microbial communities across various ecosystems. Conventional approaches for microbial analysis typically depend on metagenomics or environmental sampling. Although these methods are useful, they may not fully represent the complete diversity of microbes that exist. The isolation of uncultured microorganisms presents a significant challenge in environmental microbiology, as many microbes in natural ecosystems remain difficult to cultivate under standard laboratory conditions. This issue is addressed through the innovative approach of culturomics, which employs a diverse array of culture conditions, such as variations in media composition, temperature, and oxygen levels to successfully isolate a broader spectrum of microorganisms, including those previously deemed non-culturable. This approach improves our knowledge of microbial diversity beyond what sequencing-based approaches can offer, in addition to identifying hitherto undiscovered species with possible ecological and biotechnological value. Although metagenomics provides information on the genetic makeup of microbial communities, it frequently misses the live, metabolically active microorganisms that culturomics can isolate (Kapinusova et al., 2023). This functional approach allows researchers to assess microbial diversity in its active state, facilitating the discovery of rare or fastidious microorganisms that may be overlooked in metagenomic datasets. Furthermore, culturomics improves the representation of microbial functional diversity by enabling the isolation of microbes based on specific metabolic capabilities, which is essential for understanding ecological processes such as nutrient cycling and pollutant degradation (Clagnan et al., 2024). The dynamic nature of microbial communities necessitates long-term monitoring, and culturomics provides the tools to track changes in microbial diversity over time and across different environments, thereby elucidating how these communities adapt to environmental stressors like pollution and climate change (Qian et al., 2021). By integrating culturomics with metagenomics, researchers can achieve a comprehensive understanding of the diversity of microbes, revealing not only the identities of microorganisms present but also their interactions and contributions to ecosystem processes. This integrated approach is particularly valuable in exploring extremophilic microorganisms that thrive in harsh environments, as well as in conserving microbial resources through the establishment of culture collections that preserve unique strains with effective utilization in ecology, agriculture, and industry (Sood et al., 2021). Ultimately, culturomics enhances ecological studies by providing insights into microbial interactions and community dynamics, which are crucial for effective biodiversity conservation and ecosystem management.

13.6.3 Biotechnology and Industrial Applications

Culturomics is becoming increasingly important in biotechnology and industrial applications, facilitating the isolation and characterization of diverse microorganisms that possess valuable metabolic and enzymatic properties. Through the application of high-throughput culturing techniques, culturomics can identify microorganisms with distinctive biochemical capabilities, including those that contribute to the production of biofuels, bioplastics, pharmaceuticals, enzymes, and various other bioproducts.

13.6.3.1 Discovery of Novel Enzymes for Industrial Applications

A variety of industrial processes depend on enzymes to facilitate specific reactions, including those in food production, detergents, biofuel synthesis, and paper manufacturing. Conventional approaches to enzyme discovery typically emphasize established or well-studied enzymes; however, culturomics facilitates the detection of new enzymes from microorganisms that have not been cultured or are less explored. The isolation and characterization of these microbes can facilitate the discovery of enzymes with unique properties, including thermostability, pH tolerance, or specificity, positioning them as ideal candidates for industrial applications. Recent studies highlight the successful application of culturomics in isolating enzymes from previously unculturable microorganisms. For instance, culturomic techniques were employed to discover novel thermostable proteases, which showed high stability and activity at elevated temperatures, making them valuable for biofuel production processes (Zhang et al., 2022a, b).

13.6.3.2 Production of Biofuels and Bio-based Chemicals

Culturomics participate in the discovery of microorganisms capable of producing biofuels, including ethanol, biodiesel, and methane, more efficiently. By isolating and studying microorganisms from diverse environments, including extreme environments like hot springs, deep-sea habitats, and other harsh conditions, culturomics identifies novel strains of bacteria, algae, and fungi that have optimized metabolic pathways for biofuel production. For instance, recent culturomic studies have isolated new strains of *Clostridium* species from hot springs that demonstrate enhanced ethanol production under extreme conditions (Liu et al., 2020). Culturomic techniques have been instrumental in discovering microorganisms capable of producing valuable bio-based chemicals, such as lactic acid, succinic acid, and bioethanol, offering sustainable alternatives to petroleum-based chemicals.

13.6.3.3 Microbial Production of Bioplastics

The increasing concern regarding plastic pollution has led to a heightened demand for bioplastics, which are derived from renewable biomass. Microorganisms that synthesize bioplastics, such as polyhydroxyalkanoates (PHA), offer an environmentally sustainable alternative to conventional plastics. Culturomics plays a crucial role in identifying new bacterial or algal strains that can produce these biopolymers more efficiently and under diverse environmental conditions. For example, recent studies have utilized culturomics to isolate *Cupriavidus necator* strains with improved PHA production capabilities (Zhang et al., 2022a, b). These microorganisms have been optimized for better carbon source utilization and higher polymer yields, demonstrating their industrial potential. By isolating and optimizing these microbial strains, culturomics enhances the production processes for bioplastics, thereby improving their cost-effectiveness and scalability for industrial use (Dementyeva et al., 2021).

13.6.3.4 Pharmaceutical Production

The pharmaceutical industry heavily relies on microorganisms for the production of antibiotics, vaccines, and other therapeutic compounds. Culturomics plays an essential role in the discovery of novel microorganisms that produce bioactive compounds with potential pharmaceutical applications. By isolating microorganisms from diverse ecosystems, including extreme environments where novel bioactive molecules may be present, culturomics helps to expand the pool of microorganisms that can be studied for drug discovery. For example, culturomics has led to the identification of novel strains of fungi and bacteria that produce antibiotics, antivirals, or anticancer agents, offering new avenues for pharmaceutical development (Chunduri et al., 2022).

13.6.3.5 Synthetic Biology and Metabolic Engineering

Culturomics can enhance fermentation processes in various industries by identifying new strains of microorganisms with enhanced capabilities. This can optimize production in food, beverage, and other fermentation-based industries. Culturomics can also help develop sustainable energy production methods in microbial fuel cells (MFCs), which convert chemical energy into electrical energy. It can also aid in synthetic biology and metabolic engineering by identifying microorganisms that can be engineered to produce specific products, such as drugs, chemicals, or biofuels. This promotes sustainable production practices and reduces reliance on traditional chemical synthesis methods (Walsh et al., 2023).

13.6.4 Probiotic Research and Gut Microbiota Exploration

Culturomics has revolutionized the study of the human microbiota and microbial diversity, making it an invaluable tool in probiotic research and development. Probiotics, defined as live microorganisms that confer health benefits to the host when administered in adequate amounts, are of significant interest for their potential therapeutic applications in gut health, immunity, and even metabolic disorders. Lagier et al. (2012) isolated novel probiotic bacteria from the gut using culturomics. Over 200 growth conditions were tested to find new *Lactobacillus* and *Bifidobacterium* strains, two probiotic genera. This high-throughput culture method helped uncover probiotic strains that tolerate bile salts and acidic conditions, which are essential for successful probiotics. A study by Huang et al. (2023) employed culturomics to isolate *Bifidobacterium* strains from human feces and investigate their interactions with gut epithelial cells. They identified specific strains that not only survived under low pH but also promoted the growth of beneficial gut microbiota.

13.6.5 Bioremediation and Wastewater Treatment

Bioremediation and wastewater treatment are two crucial environmental processes that leverage microorganisms to degrade pollutants, treat contaminants, and restore ecological balance. While bioremediation refers to the use of living organisms, predominantly microbes, to clean up polluted environments, wastewater treatment involves the removal of harmful substances from water to make it safe for human consumption or to return it to natural water bodies. Culturomics, a high-throughput culturing approach, has significantly advanced these fields by improving the identification, isolation, and understanding of microorganisms with specific capabilities to degrade pollutants, tolerate extreme conditions, and facilitate cleaner environments. Culturomics has been pivotal in discovering extremophilic microbes that can thrive in harsh environmental conditions, such as high salinity, high temperatures, or acidic environments. These extremophiles are invaluable in bioremediation efforts, as they can be utilized in environments where conventional microorganisms would fail. The *Halomonas* species, isolated from saline wastewater, are capable of degrading pollutants like phenol and heavy metals, demonstrating their potential in bioremediation applications (Mainka et al., 2021). By expanding the range of known microbial species capable of degrading pollutants or facilitating wastewater treatment, culturomics plays a pivotal role in enhancing the efficiency and sustainability of environmental remediation processes.

13.6.6 Veterinary and Agricultural Microbiology

Animal Health: The application of microbial culturomics in veterinary medicine allows for the exploration of the microbiomes present in various animals, including livestock, poultry, and pets. Exploring the influence of microbial communities on animal health can enhance disease management approaches and facilitate the creation of probiotics for animals (Howe et al., 2024).

In the realm of agriculture, the study of microbial culturomics is employed to investigate the microbial communities present in soils and plants. Through the examination of microbes associated with plants, it is possible to support crop resilience, enhance soil health, and create sustainable agricultural methods (Jagadesh et al., 2024). Microbial culturomics offers robust methodologies for exploring the intricate nature of microbial existence in diverse settings, ranging from the human body to extreme natural ecosystems. The range of applications includes fields such as medicine, agriculture, environmental science, and biotechnology, facilitating notable progress in health diagnostics, disease prevention, ecosystem monitoring, and industrial processes. Utilizing extensive culturing and sequencing methods, microbial culturomics paves the way for deeper insights and the exploration of the vast potential within the microbial realm (Seng et al., 2010).

13.7 Interaction Between the Biofilms and Culturomics

Culturomics provides a powerful and comprehensive approach to biofilm research, allowing scientists to study microbial diversity, biofilm formation mechanisms, antibiotic resistance, and microbial interactions in greater depth. The interplay between biofilms and culturomics offers a range of challenges alongside promising opportunities for uncovering new microbial insights.

13.7.1 Biofilm Formation Mechanisms and Diversity

The study of culturomics facilitates an exploration into the diversity and mechanisms underlying biofilm formation through the cultivation of various microorganisms, allowing for an analysis of their growth patterns, morphology, and matrix production. The culturomics approach facilitated the identification of previously overlooked bacterial species within biofilm communities. A research study was carried out by Mangiaterra et al. (2018) employed culturomics to examine Pseudomonas aeruginosa strains obtained from patients with cystic fibrosis. The study focused on the genetic and phenotypic variations in biofilm formation among different strains, pinpointing essential genes linked to biofilm development. This work contributes to a deeper understanding of how specific strains behave in biofilm formation, particularly about chronic infections.

13.7.2 Identification of Biofilm-Forming Pathogens

Caputo et al. (2018) conducted a study utilizing culturomics to investigate the biofilm-forming capabilities of microorganisms present in medical equipment such as catheters. They employed a sophisticated culturing approach that enabled the isolation of a wider range of biofilm-forming bacteria, *Staphylococcus epidermidis*, from non-biofilm-producer strains. Their research illustrated how culturomics can assist in the identification of pathogenic bacteria. Boyapati et al. (2024) utilized culturomics to investigate biofilm communities in dental plaque from patients with periodontal disease.

13.7.3 Antibiotic Resistance in Biofilms

Culturomics facilitates extensive evaluation of antibiotics against diverse biofilm-forming microorganisms, aiding in the identification of resistance mechanisms at both species and strain levels. Kabir et al. (2024) employed culturomics to evaluate the antibiotic resistance profiles of *Pseudomonas aeruginosa* strains that form biofilms in patients with cystic fibrosis. Through the cultivation of various strains and their exposure to different antibiotics, it was observed that biofilm-associated *P. aeruginosa* strains demonstrated markedly greater resistance to antibiotics in comparison to planktonic cells. The culturomic approach facilitated the identification of genetic and phenotypic heterogeneity within these biofilms, potentially aiding in the development of more effective treatment strategies. In a study conducted by Sharan et al. (2024), a culturomic approach was employed to investigate biofilm resistance in *Staphylococcus aureus*. Through the culture of clinical isolates and subsequent antibiotic exposure, it was determined that specific biofilm-forming strains of *S. aureus* demonstrated multidrug resistance. This resistance was associated with the production of extracellular polymeric substances (EPS) and modifications in the expression of efflux pumps. This finding demonstrates the application of culturomics in investigating antibiotic resistance within biofilms and examining the genetic determinants of resistance.

13.7.4 Microbial Interactions in Biofilms

Biofilms are often polymicrobial, consisting of multiple species that interact either synergistically or antagonistically. Culturomics aids in identifying and quantifying interactions that are essential for comprehending biofilm behavior and pathogenesis. Mashima and Nakazawa (2014) investigated microbial interactions within oral biofilms through the cultivation of bacteria derived from plaque samples. Several synergistic interactions were identified between oral pathogens, including *Streptococcus* sp. and *Veillonella* sp., both of which are implicated in the progression of periodontal disease. The research demonstrated the application of culturomics in elucidating complex microbial dynamics within biofilms, highlighting its potential for informing the design of therapies aimed at microbial interactions.

13.7.5 Biofilm Disruption and Novel Therapeutic Strategies

Culturomics is essential for the development and evaluation of novel therapeutic strategies aimed at disrupting biofilms. Researchers can identify potential therapies to disrupt biofilms and restore antimicrobial efficacy by culturing biofilm forming pathogens and testing various biofilm-targeting agents, such as enzymes, antimicrobial peptides, or nanoparticles. Zhao et al. (2024) utilized culturomics to examine the effects of antimicrobial peptides (AMPs) on biofilms produced by *Escherichia coli* and *Klebsiella pneumoniae*. The study demonstrated that culturing biofilm forming strains and exposing them to antimicrobial peptides (AMPs) allowed certain peptides to penetrate biofilms and disrupt their structure, thereby enhancing the effectiveness of antibiotics.

13.7.6 Biofilm Detection and Imaging

Culturomics can enhance biofilm detection methods by integrating culturing with advanced imaging techniques, such as confocal laser scanning microscopy and atomic force microscopy, to visualize and quantify biofilm formation in real time.

Gulati et al. (2018) combined culturomics with confocal microscopy to monitor biofilm formation in *Candida albicans*. The researchers utilized culturing techniques to cultivate biofilms and applied advanced imaging methods to examine the three-dimensional structure of the biofilm. This enabled the evaluation of the correlation between biofilm thickness and density with antifungal resistance, illustrating the effectiveness of culturomics in enhancing biofilm detection methods. Navarro et al. (2014) employed a culturomic approach alongside high-throughput biofilm assays to identify biofilm inhibitors. By cultivating extensive microbial strain libraries and applying automated imaging techniques, researchers identified novel compounds effective in inhibiting biofilm formation. This work illustrates the potential of integrating culturomics with high-throughput screening for drug discovery aimed at biofilm targets. The studies mentioned above demonstrate how culturomics is advancing biofilm research and enabling the development of more effective treatments and interventions.

13.8 Future Challenges and Prospectives

Culturomics integrates high-throughput culturing, genomic sequencing, and omics technologies, offering significant potential for elucidating microbial diversity and functional roles. Nonetheless, various challenges need to be resolved to realize its complete potential. A significant challenge in culturomics is the identification and cultivation of the majority of microorganisms that are unculturable in laboratory settings. Approximately 50% of the genes in metagenome-assembled genomes (MAGs) are unannotated, largely because their functions are not yet understood. This limitation impedes the prediction of metabolic pathways and functional roles, thereby obstructing the thorough investigation of microbial communities. Incorporating meta-transcriptomics, meta-proteomics, and metabolomics into culturomics workflows offers insights into active microbial populations, their metabolic activities, and nutrient requirements, facilitating the cultivation of previously unculturable microbes. Integrating these technologies with culturomic methods enables researchers to pinpoint essential factors influencing microbial activity, thereby aiding in the formulation of culture media that replicate natural environments and promote the growth of hard-to-culture microbes. A significant challenge in culturomics is the annotation of novel genomes, frequently classified as "hypothetical" due to their ambiguous functional roles. Enhancing bioinformatics and computational biology is crucial for bridging these gaps. Establishing robust taxonomic frameworks ensures consistent classification of newly discovered microbes, facilitating collaborative research and enhancing reproducibility. Advancements in high-throughput culturing systems, automation, and microfluidics are expected to enhance microbial isolation and increase the reproducibility of experiments. The incorporation of artificial intelligence and machine learning has the potential to transform culturomics through the automation of growth conditions, optimization of culture media formulations, and prediction of metabolic pathways in uncultured microbes.

13.9 Conclusion

The integration of culturomics into biofilm research signifies a notable advancement in comprehending the complexity and diversity of microbial communities within biofilms. The integration of large-scale culturing techniques with genomic, transcriptomic, proteomic, and metabolomic analyses facilitates a thorough investigation of biofilm formation, microbial interactions, and resistance mechanisms. This method improves the identification of new biofilm-forming microorganisms and offers insights into the metabolic pathways and ecological functions that support biofilm behavior across different environments. Despite existing challenges, including the enhancement of culturing techniques for unculturable microbes and addressing deficiencies in functional annotation, the potential of culturomics in biofilm research is substantial. This has the potential to transform our comprehension of microbial ecosystems, provide novel therapeutic approaches for biofilm-associated infections, and facilitate biotechnological advancements, leading to improved treatments and innovations in environmental and industrial microbiology.

References

Abdelhamid, A. G., & Yousef, A. E. (2023). Combating bacterial biofilms: Current and emerging antibiofilm strategies for treating persistent infections. *Antibiotics, 12*(6), 1005. https://doi.org/10.3390/antibiotics12061005

Alav, I., Sutton, J. M., & Rahman, K. M. (2018). Role of bacterial efflux pumps in biofilm formation. *The Journal of Antimicrobial Chemotherapy, 73*(8), 2003–2020. https://doi.org/10.1093/jac/dky042

Asma, S. T., Imre, K., Morar, A., Herman, V., Acaroz, U., Mukhtar, H., Arslan-Acaroz, D., Shah, S. R. A., & Gerlach, R. (2022). An overview of biofilm formation-combating strategies and mechanisms of action of antibiofilm agents. *Life (Basel, Switzerland), 12*(8), 1110. https://doi.org/10.3390/life12081110

Balcázar, J. L., Subirats, J., & Borrego, C. M. (2015). The role of biofilms as environmental reservoirs of antibiotic resistance. *Frontiers in Microbiology, 6*, 1216. https://doi.org/10.3389/fmicb.2015.01216

Bielen, A., Babić, I., VukSurjan, M., Kazazić, S., Šimatović, A., Lajtner, J., Udiković-Kolić, N., Mesić, Z., & Hudina, S. (2024). Comparison of MALDI-TOF mass spectrometry and 16S rDNA sequencing for identification of environmental bacteria: A case study of cave mussel-associated culturable microorganisms. *Environmental Science and Pollution Research International, 31*(14), 21752–21764. https://doi.org/10.1007/s11356-024-32537-1. Epub 2024 Feb 23. PMID: 38393570.

Bjarnsholt, T. (2013). The role of bacterial biofilms in chronic infections. *APMIS. Supplementum, 136*, 1–51. https://doi.org/10.1111/apm.12099

Bouhrour, N., Nibbering, P. H., & Bendali, F. (2024). Medical device-associated biofilm infections and multidrug-resistant pathogens. *Pathogens, 13*(5), 393. https://doi.org/10.3390/pathogens13050393

Boyapati, R., Lanke, R. B., Mudaliyar, M. C., Gaddam, B., Babu Dasari, A., & Dhulipalla, R. (2024). Exploring the microbiome landscape of dental plaque: A cross-sectional analysis in periodontal health and disease. *Cureus, 16*(3), e57334. https://doi.org/10.7759/cureus.57334

Caputo, P., Di Martino, M. C., Perfetto, B., Iovino, F., & Donnarumma, G. (2018). Use of MALDI-TOF MS to discriminate between biofilm-producer and non-producer strains of Staphylococcus epidermidis. *International Journal of Environmental Research and Public Health, 15*, 1695. https://doi.org/10.3390/ijerph15081695

Carrascosa, C., Raheem, D., Ramos, F., Saraiva, A., & Raposo, A. (2021). Microbial biofilms in the food industry-A comprehensive review. *International Journal of Environmental Research and Public Health, 18*(4), 2014. https://doi.org/10.3390/ijerph18042014

Cavallo, I., Sivori, F., Mastrofrancesco, A., Abril, E., Pontone, M., Di Domenico, E. G., & Pimpinelli, F. (2024). Bacterial biofilm in chronic wounds and possible therapeutic approaches. *Biology, 13*(2), 109. https://doi.org/10.3390/biology13020109

Cepas, V., López, Y., Muñoz, E., Rolo, D., Ardanuy, C., Martí, S., Xercavins, M., Horcajada, J. P., Bosch, J., & Soto, S. M. (2019). Relationship between biofilm formation and antimicrobial resistance in gram-negative bacteria. *Microbial Drug Resistance (Larchmont, N.Y.), 25*(1), 72–79. https://doi.org/10.1089/mdr.2018.0027

Chen, M., Yu, Q., & Sun, H. (2013). Novel strategies for the prevention and treatment of biofilm related infections. *International Journal of Molecular Sciences, 14*(9), 18488–18501. https://doi.org/10.3390/ijms140918488. PMID: 24018891; PMCID: PMC3794791.

Chunduri, A., Donthula, N., Jahanavi, M., Golla, S., Aich, P., Sahithya, K., Mishra, B., Mandal, S. K., & Anuranjeeta. (2022). Role of microbes in the pharmaceutical industry. In S. Kumar, N. Kumar, & Shahid-ul-Islam (Eds.), *Role of microbes in industrial products and processes*. https://doi.org/10.1002/9781119901198.ch1

Clagnan, E., Costanzo, M., Visca, A., Di Gregorio, L., Tabacchioni, S., Colantoni, E., Sevi, F., Sbarra, F., Nolfi, L., Magarelli, R. A., Trupo, M., & Bevivino, A. (2024). Culturomics- and metagenomics-based insights into the soil microbiome preservation and application for sustainable agriculture. *Frontiers in Microbiology, 15*, 1473666. https://doi.org/10.3389/fmicb.2024.1473666

Dementyeva, P., Freudenberg, R. A., Baier, T., Rojek, K., Wobbe, L., Weisshaar, B., & Kruse, O. (2021). A novel, robust and mating-competent Chlamydomonas reinhardtii strain with an enhanced transgene expression capacity for algal biotechnology. *Biotechnology Reports, 31*, e00644. https://doi.org/10.1016/j.btre.2021.e00644

Dhiman, S., Kumar, A., Kaur, G., Mukherjee, G., Rustagi, S., Shreaz, S., Negi, R., & Yadav, A. N. (2024). Bacterial biofilms: Pathogenesis, monitoring, treatment approaches and associated challenges. *Biologia, 79*, 3161–3181. https://doi.org/10.1007/s11756-024-01767-6

Diakite, A., Dubourg, G., Dione, N., et al. (2020). Optimization and standardization of the culturomics technique for human microbiome exploration. *Scientific Reports, 10*, 9674. https://doi.org/10.1038/s41598-020-66738-8

Donlan, R. M. (2002). Biofilms: Microbial life on surfaces. *Emerging Infectious Diseases, 8*(9), 881–890. https://doi.org/10.3201/eid0809.020063

Flemming, H. C., & Wingender, J. (2010). The biofilm matrix. *Nature Reviews Microbiology, 8*(9), 623–633. https://doi.org/10.1038/nrmicro2415

Flemming, H. C., Neu, T. R., & Wozniak, D. J. (2007). The EPS matrix: The "house of biofilm cells". *Journal of Bacteriology, 189*(22), 7945–7947. https://doi.org/10.1128/JB.00858-07

Flemming, H.-C., Wuertz, S., & Wingender, J. (2023). Polymicrobial biofilms: Cooperation and competition in natural and engineered systems. *Nature Reviews Microbiology, 21*(10), 670–685.

Flemming, H.-C., Wuertz, S., & Wingender, J. (2024). Microbial interactions in biofilms: Balancing cooperation and competition. *Trends in Microbiology, 32*(3), 210–221. https://doi.org/10.1016/j.tim.2024.01.005

Gahlot, R., Nigam, C., Kumar, V., Yadav, G., & Anupurba, S. (2014). Catheter-related bloodstream infections. *International Journal of Critical Illness and Injury Science, 4*(2), 162–167. https://doi.org/10.4103/2229-5151.134184

Galié, S., García-Gutiérrez, C., Miguélez, E. M., Villar, C. J., & Lombó, F. (2018). Biofilms in the food industry: Health aspects and control methods. *Frontiers in Microbiology, 9*, 898. https://doi.org/10.3389/fmicb.2018.00898

Ghafoor, A., Hay, I. D., & Rehm, B. H. A. (2011). Role of exopolysaccharides in Pseudomonas aeruginosa biofilm formation and architecture. *Applied and Environmental Microbiology, 77*(15), 1–11. https://doi.org/10.1128/AEM.00637-11

Gondil, V. S., & Subhadra, B. (2023). Biofilms and their role on diseases. *BMC Microbiology, 23*(1), 203. https://doi.org/10.1186/s12866-023-02954-2

Greub, G. (2012). Culturomics: A new approach to study the human microbiome. *Clinical Microbiology and Infection, 18*, 1157–1159. https://doi.org/10.1111/1469-0691.12032

Gulati, M., Lohse, M. B., Ennis, C. L., Gonzalez, R. E., Perry, A. M., Bapat, P., Arevalo, A. V., Rodriguez, D. L., & Nobile, C. J. (2018). In vitro culturing and screening of candida albicans biofilms. *Current Protocols in Microbiology, 50*(1), e60. https://doi.org/10.1002/cpmc.60

Hall, C. W., & Mah, T.-F. (2017). Molecular mechanisms of biofilm-based antibiotic resistance and tolerance in pathogenic bacteria. *FEMS Microbiology Reviews, 41*(3), 276–301.

Hall-Stoodley, L., Costerton, J. W., & Stoodley, P. (2004). Bacterial biofilms: From the natural environment to infectious diseases. *Nature Reviews Microbiology, 2*(2), 95–108. https://doi.org/10.1038/nrmicro821

Howe, S., Liu, Z., Zuo, B., & Zhao, J. (2024). Culturomics: A critical approach in studying the roles of human and animal microbiota. *Animal Nutriomics, 1*(e6). https://doi.org/10.1017/anr.2024.6

Hu, Y., Niu, Y., Ye, X., Zhu, C., Tong, T., Zhou, Y., Zhou, X., Cheng, L., & Ren, B. (2021). Staphylococcus aureus synergized with Candida albicans to increase the pathogenesis and drug resistance in cutaneous abscess and peritonitis murine models. *Pathogens, 10*(8), 1036. https://doi.org/10.3390/pathogens10081036

Huang, Y., Sheth, R. U., Zhao, S., et al. (2023). High-throughput microbial culturomics using automation and machine learning. *Nature Biotechnology, 41*, 1424–1433. https://doi.org/10.1038/s41587-023-01674-2

Ivers, C., Kaya, E. C., Yucel, U., et al. (2024). Evaluation of Salmonella biofilm attachment and hydrophobicity characteristics on food contact surfaces. *BMC Microbiology, 24*, 387. https://doi.org/10.1186/s12866-024-03556-2

Jagadesh, M., Dash, M., Kumari, A., Singh, S. K., Verma, K. K., Kumar, P., et al. (2024). Revealing the hidden world of soil microbes: Metagenomic insights into plant, bacteria, and fungi interactions for sustainable agriculture and ecosystem restoration. *Microbiological Research, 285*, 127764. https://doi.org/10.1016/j.micres.2024.127764

Jiang, Y., Geng, M., & Bai, L. (2020). Targeting biofilms therapy: Current research strategies and development hurdles. *Microorganisms, 8*(8), 1222. https://doi.org/10.3390/microorganisms8081222

Jordan, R. P., & Nicolle, L. E. (2014). Preventing infection associated with urethral catheter biofilms. In *Biofilms in infection prevention and control: A healthcare handbook* (pp. 287–309). Elsevier. https://doi.org/10.1016/b978-0-12-397043-5.00016-5

Kabir, R. B., Ahsan, T., Rahman, M. F., Jobayer, M., & Shamsuzzaman, S. M. (2024). Biofilm-producing and specific antibiotic resistance genes in Pseudomonas aeruginosa isolated from patients admitted to a tertiary care hospital, Bangladesh. *IJID Regions, 11*, 100369. https://doi.org/10.1016/j.ijregi.2024.100369

Kapinusova, G., Lopez Marin, M. A., & Uhlik, O. (2023). Reaching unreachables: Obstacles and successes of microbial cultivation and their reasons. *Frontiers in Microbiology, 14*, 1089630. https://doi.org/10.3389/fmicb.2023.1089630

Kaushik, A., Kest, H., Sood, M., Thieman, C., Steussy, B. W., Padomek, M., & Gupta, S. (2024). Infective endocarditis by biofilm-producing methicillin-resistant staphylococcus aureus—Pathogenesis, diagnosis, and management. *Antibiotics, 13*(12), 1132. https://doi.org/10.3390/antibiotics13121132

Krishnan, S., Patil, S. A., & Nancharaiah, Y. V. (2023). Environmental microbial biofilms. In *Material-microbes interactions* (pp. 3–45). Elsevier. https://doi.org/10.1016/B978-0-323-95124-1.00013-9

Kurtzman, G. M., Horowitz, R. A., Johnson, R., Prestiano, R. A., & Klein, B. I. (2022). The systemic oral health connection: Biofilms. *Medicine, 101*(46), e30517. https://doi.org/10.1097/MD.0000000000030517

Lagier, J. C., Armougom, F., Million, M., Hugon, P., Pagnier, I., Robert, C., Bittar, F., Fournous, G., Gimenez, G., Maraninchi, M., Trape, J. F., Koonin, E. V., La Scola, B., & Raoult,

D. (2012). Microbial culturomics: Paradigm shift in the human gut microbiome study. *Clinical Microbiology and Infection, 18*(12), 1185–1193. https://doi.org/10.1111/1469-0691.12023

Lamont, R. J., El-Sabaeny, A., Park, Y., Cook, G. S., Costerton, J. W., & Demuth, D. R. (2002). Role of the Streptococcus gordonii SspB protein in the development of Porphyromonas gingivalis biofilms on streptococcal substrates. *Microbiology, 148*(6), 1627–1636. https://doi.org/10.1099/00221287-148-6-1627

Lechón-Alonso, P., Clegg, T., Cook, J., Smith, T. P., & Pawar, S. (2021). The role of competition versus cooperation in microbial community coalescence. *PLoS Computational Biology, 17*(11), e1009584. https://doi.org/10.1371/journal.pcbi.1009584

Li, Y. H., & Tian, X. L. (2016). *Microbial interactions in biofilms: Impacts on homeostasis and pathogenesis*. InTech. https://doi.org/10.5772/62942

Liu, L., Jiao, J. Y., Fang, B. Z., Lv, A. P., Ming, Y. Z., Li, M. M., Salam, N., & Li, W. J. (2020). Isolation of Clostridium from Yunnan-Tibet hot springs and description of Clostridium thermarum sp. nov. with lignocellulosic ethanol production. *Systematic and Applied Microbiology, 43*(5), 126104. https://doi.org/10.1016/j.syapm.2020.126104

Liu, X., Yao, H., Zhao, X., & Ge, C. (2023). Biofilm formation and control of foodborne pathogenic bacteria. *Molecules (Basel, Switzerland), 28*(6), 2432. https://doi.org/10.3390/molecules28062432

Liu, H. Y., Prentice, E. L., & Webber, M. A. (2024). Mechanisms of antimicrobial resistance in biofilms. *npj Antimicrobials and Resistance, 2*, 27. https://doi.org/10.1038/s44259-024-00046-3

Luo, A., Wang, F., Sun, D., Liu, X., & Xin, B. (2021). Formation, development, and cross-species interactions in biofilms. *Frontiers in Microbiology, 12*, 757327. https://doi.org/10.3389/fmicb.2021.757327

Luo, A., Wang, F., Sun, D., Liu, X., & Xin, B. (2022). Formation, development, and cross-species interactions in biofilms. *Frontiers in Microbiology, 12*, 757327. https://doi.org/10.3389/fmicb.2021.757327

Maeda, K., Nagata, H., Yamamoto, Y., Tanaka, M., Tanaka, J., Minamino, N., & Shizukuishi, S. (2004). Glyceraldehyde-3-phosphate dehydrogenase of Streptococcus oralis functions as a coadhesin for Porphyromonas gingivalis major fimbriae. *Infection and Immunity, 72*(3), 1341–1348. https://doi.org/10.1128/IAI.72.3.1341-1348.2004

Mah, T. F., & O'Toole, G. A. (2001). Mechanisms of biofilm resistance to antimicrobial agents. *Trends in Microbiology, 9*(1), 34–39. https://doi.org/10.1016/S0966-842X(00)01913-2

Mainka, T., Weirathmüller, D., Herwig, C., & Pflügl, S. (2021). Potential applications of halophilic microorganisms for biological treatment of industrial process brines contaminated with aromatics. *Journal of Industrial Microbiology & Biotechnology, 48*(1–2), kuab015. https://doi.org/10.1093/jimb/kuab015

Malhotra, S., Hayes, D., Jr., & Wozniak, D. J. (2019). Cystic fibrosis and Pseudomonas aeruginosa: The host-microbe interface. *Clinical Microbiology Reviews, 32*(3), e00138–e00118. https://doi.org/10.1128/CMR.00138-18

Mangiaterra, G., Amiri, M., Di Cesare, A., Pasquaroli, S., Manso, E., Cirilli, N., Citterio, B., Vignaroli, C., & Biavasco, F. (2018). Detection of viable but non-culturable Pseudomonas aeruginosa in cystic fibrosis by qPCR: A validation study. *BMC Infectious Diseases, 18*(1), 701. https://doi.org/10.1186/s12879-018-3612-9

Mashima, I., & Nakazawa, F. (2014). The influence of oral Veillonella species on biofilms formed by Streptococcus species. *Anaerobe, 28*, 54–61. https://doi.org/10.1016/j.anaerobe.2014.05.003

Mgomi, F. C., Yang, Y. R., Cheng, G., & Yang, Z. Q. (2023). Lactic acid bacteria biofilms and their antimicrobial potential against pathogenic microorganisms. *Biofilms, 5*, 100118. https://doi.org/10.1016/j.bioflm.2023.100118

Michaelis, C., & Grohmann, E. (2023). Horizontal gene transfer of antibiotic resistance genes in biofilms. *Antibiotics (Basel, Switzerland), 12*(2), 328. https://doi.org/10.3390/antibiotics12020328

Muhammad, M. H., Idris, A. L., Fan, X., Guo, Y., Yu, Y., Jin, X., Qiu, J., Guan, X., & Huang, T. (2020). Beyond risk: Bacterial biofilms and their regulating approaches. *Frontiers in Microbiology, 11*, 928. https://doi.org/10.3389/fmicb.2020.00928

Mutha, N. V. R., Mohammed, W. K., Krasnogor, N., Tan, G. Y. A., Wee, W. Y., Li, Y., Choo, S. W., & Jakubovics, N. S. (2019). Transcriptional profiling of coaggregation interactions between Streptococcus gordonii and Veillonellaparvula by Dual RNA-Seq. *Scientific Reports, 9*(1), 7664. https://doi.org/10.1038/s41598-019-43979-w

Nam, N. N., Do, H. D. K., Loan Trinh, K. T., & Lee, N. Y. (2023). Metagenomics: An effective approach for exploring microbial diversity and functions. *Food, 12*(11), 2140. https://doi.org/10.3390/foods12112140

Navarro, G., Cheng, A. T., Peach, K. C., Bray, W. M., Bernan, V. S., Yildiz, F. H., & Linington, R. G. (2014). Image-based 384-well high-throughput screening method for the discovery of skyllamycins A to C as biofilm inhibitors and inducers of biofilm detachment in Pseudomonas aeruginosa. *Antimicrobial Agents and Chemotherapy, 58*(2), 1092–1099. https://doi.org/10.1128/AAC.01781-13

O'Toole, G., & Kolter, R. (1998). Initiation of biofilm formation in Pseudomonas fluorescens WCS365 proceeds via multiple, convergent signaling pathways: A genetic analysis. *Molecular Microbiology, 28*(3), 449–461. https://doi.org/10.1046/j.1365-2958.1998.00787.x

Olanbiwoninu, A. A., & Popoola, B. M. (2023). Biofilms and their impact on the food industry. *Saudi Journal of Biological Sciences, 30*(2), 103523. https://doi.org/10.1016/j.sjbs.2022.103523

Pandit, A., Adholeya, A., Cahill, D., Brau, L., & Kochar, M. (2020). Microbial biofilms in nature: Unlocking their potential for agricultural applications. *Journal of Applied Microbiology, 129*(2), 199–211. https://doi.org/10.1111/jam.14609

Pfleiderer, A., Lagier, J. C., Armougom, F., Robert, C., Vialettes, B., & Raoult, D. (2013). Culturomics identified 11 new bacterial species from a single anorexia nervosa stool sample. *European Journal of Clinical Microbiology and Infectious Diseases, 32*, 1471–1481. https://doi.org/10.1007/s10096-013-1900-2

Qian, Y., Lan, F., & Venturelli, O. S. (2021). Towards a deeper understanding of microbial communities: Integrating experimental data with dynamic models. *Current Opinion in Microbiology, 62*, 84–92. https://doi.org/10.1016/j.mib.2021.05.003

Ramasamy, D., Mishra, A., Lagier, J. C., Padhmanabhan, R., Rossi-Tamisier, M., Sentausa, E., et al. (2014). A polyphasic strategy incorporatinggenomic data for the taxonomic description of newbacterial species. *International Journal of Systematicand Evolutionary Microbiology, 64*, 384–391.

Sahoo, K., & Meshram, S. (2024). Biofilm formation in chronic infections: A comprehensive review of pathogenesis, clinical implications, and novel therapeutic approaches. *Cureus, 16*(10), e70629. https://doi.org/10.7759/cureus.70629

Satputea, S. K., Banpurkar, A. G., Banat, I. M., Sangshetti, J. N., Patil, R. H., & Gade, W. N. (2016). Multiple roles of biosurfactants in biofilms. *Current Pharmaceutical Design, 22*(11), 1429–1448. https://doi.org/10.2174/1381612822666160120152704

Seng, P., Rolain, J. M., Fournier, P. E., La Scola, B., Drancourt, M., & Raoult, D. (2010). MALDI-TOF-mass spectrometry applications in clinical microbiology. *Future Microbiology, 5*(11), 1733–1754. https://doi.org/10.2217/fmb.10.127

Sharan, M., Dhaka, P., Bedi, J. S., et al. (2024). Assessment of biofilm-forming capacity and multidrug resistance in Staphylococcus aureus isolates from animal-source foods: Implications for lactic acid bacteria intervention. *Annales de Microbiologie, 74*, 22. https://doi.org/10.1186/s13213-024-01768-5

Sharma, D. K., & Rajpurohit, Y. S. (2024). Multitasking functions of bacterial extracellular DNA in biofilms. *Journal of Bacteriology, 206*(4), e00006-24. https://doi.org/10.1128/jb.00006-24

Sharma, S., Mohler, J., Mahajan, S. D., Schwartz, S. A., Bruggemann, L., & Aalinkeel, R. (2023). Microbial biofilm: A review on formation, infection, antibiotic resistance, control measures, and innovative treatment. *Microorganisms, 11*(6), 1614. https://doi.org/10.3390/microorganisms11061614

Sood, U., Kumar, R., & Hira, P. (2021). Expanding culturomics from gut to extreme environmental settings. *mSystems, 6*(4), e00848–e00821. https://doi.org/10.1128/msystems.00848-21

Staats, A., Li, D., Sullivan, A. C., & Stoodley, P. (2021). Biofilm formation in periprosthetic joint infections. *Annals of Joint, 6*, 43. https://doi.org/10.21037/aoj-20-85

Steenackers, H. P., Parijs, I., Foster, K. R., & Vanderleyden, J. (2016). Experimental evolution in biofilm populations. *FEMS Microbiology Reviews, 40*(3), 373–397. https://doi.org/10.1093/femsre/fuw002

Stevens, M. R. E., Luo, T. L., Vornhagen, J., Jakubovics, N. S., Gilsdorf, J. R., Marrs, C. F., Møretrø, T., & Rickard, A. H. (2015). Coaggregation occurs between microorganisms isolated from different environments. *FEMS Microbiology Ecology, 91*(11), fiv123. https://doi.org/10.1093/femsec/fiv123

Tournu, H., & Van Dijck, P. (2012). Candida biofilms and the host: Models and new concepts for eradication. *International Journal of Microbiology, 2012*, 845352. https://doi.org/10.1155/2012/845352

Toyofuku, M., Roschitzki, B., Riedel, K., & Eberl, L. (2012). Identification of proteins associated with the Pseudomonas aeruginosa biofilm extracellular matrix. *Journal of Proteome Research, 11*(10), 4906–4915. https://doi.org/10.1021/pr300395j

Verroken, A., Janssens, M., Berhin, C., Bogaerts, P., Huang, T. D., Wauters, G., et al. (2010). Evaluation of matrix-assisted laser desorption ionization–time of flight mass spectrometry for identification of Nocardia species. *Journal of Clinical Microbiology, 48*, 4015–4021. https://doi.org/10.1128/jcm.01234-10

Vishwakarma, V. (2020). Impact of environmental biofilms: Industrial components and its remediation. *Journal of Basic Microbiology, 60*, 198–206. https://doi.org/10.1002/jobm.201900569

Voinescu, A., Licker, M., Muntean, D., Musuroi, C., Musuroi, S. I., Izmendi, O., Vulpie, S., Jumanca, R., Munteanu, M., & Cosnita, A. (2024). A comprehensive review of microbial biofilms on contact lenses: Challenges and solutions. *Infection and Drug Resistance, 17*, 2659–2671. https://doi.org/10.2147/IDR.S463779

Walsh, A. M., Leech, J., Huttenhower, C., Delhomme-Nguyen, H., Crispie, F., Chervaux, C., & Cotter, P. D. (2023). Integrated molecular approaches for fermented food microbiome research. *FEMS Microbiology Reviews, 47*(2), fuad001. https://doi.org/10.1093/femsre/fuad001

Wang, X., Liu, M., Yu, C., et al. (2023). Biofilm formation: Mechanistic insights and therapeutic targets. *Molecular Biomedicine, 4*, 49. https://doi.org/10.1186/s43556-023-00164-w

Werno, A. M., Christner, M., Anderson, T. P., & Murdoch, D. R. (2012). Differentiation of Streptococcus pneumoniae from nonpneumococcal streptococci of the Streptococcus mitis group by matrix-assisted laser desorption ionization-time of flight mass spectrometry. *Journal of Clinical Microbiology, 50*, 2863–2867. https://doi.org/10.1128/jcm.00508-12

Yadav, M. K. (2017). Role of biofilms in environment pollution and control. In J. Patra, C. Vishnuprasad, & G. Das (Eds.), *Microbial biotechnology*. Springer. https://doi.org/10.1007/978-981-10-6847-8_16

Yin, W., Xu, S., Wang, Y., Zhang, Y., Chou, S. H., Galperin, M. Y., & He, J. (2021). Ways to control harmful biofilms: Prevention, inhibition, and eradication. *Critical Reviews in Microbiology, 47*(1), 57–78. https://doi.org/10.1080/1040841X.2020.1842325

Zafer, M. M., Mohamed, G. A., Ibrahim, S. R. M., et al. (2024). Biofilm-mediated infections by multidrug-resistant microbes: A comprehensive exploration and forward perspectives. *Archives of Microbiology, 206*, 101. https://doi.org/10.1007/s00203-023-03826-z

Zhang, K., Huang, Q., Li, Y., Liu, L., Tang, X. F., & Tang, B. (2022a). Maturation process and characterization of a novel thermostable and halotolerant subtilisin-like protease with high collagenolytic activity but low gelatinolytic activity. *Applied and Environmental Microbiology, 88*(3), e0218421. https://doi.org/10.1128/aem.02184-21

Zhang, L., Jiang, Z., Tsui, T. H., Loh, K. C., Dai, Y., & Tong, Y. W. (2022b). A review on enhancing cupriavidusnecator fermentation for Poly(3-hydroxybutyrate) (PHB) production from low-cost carbon sources. *Frontiers in Bioengineering and Biotechnology, 10*, 946085. https://doi.org/10.3389/fbioe.2022.946085

Zhao, D., Tang, M., Hu, P., Hu, X., Chen, W., Ma, Z., Chen, H., Liu, H., Cao, J., & Zhou, T. (2024). Antimicrobial peptide Hs02 with rapid bactericidal, anti-biofilm, and anti-inflammatory activity against carbapenem-resistant Klebsiella pneumoniae and Escherichia coli. *Microbiology Spectrum, 13*, e01050-24.

Part III

Omics as Diagnostic and Therapeutic Tools

Multi-omics Approaches in the Detection of Biofilms and Biofilm-Associated Infections

14

Mahima S. Mohan, Siddhardha Busi, Asad Syed, and Rajaneesh Anupam

14.1 Introduction

Generally, microorganisms exist in two forms in the environment, that is, planktonic, which is independently suspended and biofilm, attached to a substrate in an extra polymeric substances (Dutta et al., 2024). Biofilm is a community of microorganisms attached in a matrix, exist as single species or as multi-species. Extracellular polymeric matrix acts as a shield to protect microbes from environmental stress, especially antibiotics (Sharma et al., 2023). In the environment, biofilm exists as multispecies biofilm where they cooperate through inter and intra-species interactions (Azevedo et al., 2009). Most of the microorganisms produces biofilm and are often found in hospital acquired and chronic infections. Due to the occurrence of antibiotic resistance, biofilm-associated infections are challenging to treat (Sharma et al., 2023).

Microbial biofilm formation occurs in several stages, such as attachment, maturation, and dispersion. In the attachment phase, bacteria produce different adhesion proteins for the surface attachment, it may be reversible or irreversible adhesion. EPS matrix was secreted based on the nutrient availability and environmental

M. S. Mohan · S. Busi (✉)
Department of Microbiology, School of Life Sciences, Pondicherry University, Kalapet, Puducherry, India

A. Syed
Department of Botany and Microbiology, College of Science, King Saud University, Riyadh, Saudi Arabia

R. Anupam
Department of Biochemistry and Molecular Biology, School of Life Sciences, Pondicherry University, Puducherry, India
e-mail: ranupam@pondiuni.ac.in

S. Busi et al. (eds.), *Omics Approaches in Biofilm Research*,
https://doi.org/10.1007/978-3-031-91863-6_14

conditions. Once the adhesion becomes stable, next phase of the biofilm cycle initiates, that is, the maturation phase. Here, bacteria communicate through quorum sensing, where autoinducers acts as signaling molecules. Based on the cell density, bacteria switches to initiates biofilm formation through cellular communication in a synchronized mode. Finally, bacteria changes into a planktonic state from the sessile form and dispersed into its surrounding environment; it depends upon the degradation of biofilm matrix constituents. However, still the biofilm development mechanism, genes and metabolites are not completely known, and it needs to be considered for the eradication of biofilm-related infections (Zhang et al., 2023).

Biofilm-producing microorganisms are a public concern not only in the case of the medical field but also in the environment, mainly leads to biofouling. The emergence of antimicrobial resistance in biofilm worsens the patient's condition and the treatment, leading to escalation of morbidity and mortality rates. Since the clinical diagnoses of any microbial diseases are mainly targeting planktonic forms of pathogens, such as susceptibility test, the detection of biofilm is still challenging. The conventional ways of diagnosis and detection of microorganisms are growing them in a laboratory, serological tests, and DNA/RNA detection using PCR techniques. Molecular detection of microorganisms is limited to a few groups of pathogens. Omics approaches permit the analysis of the microbial community in the level of transcriptome, proteome, and metabolites. Moreover, using omics methods, the differences in the microbial state or the stress due to environmental factors can be identified (Silva et al., 2021). This chapter mainly focuses on the different ways of detection and diagnosis of biofilm through omics approaches. Moreover, several techniques involved in biofilm diagnosis at the transcriptomic, proteomic, and metabolomic levels were also described in this section.

14.2 Multi-omics Technology in Biofilm Studies

The development of omics approaches in microbiology has deepened our knowledge into the complex physiological mechanisms of microbial life. Through omics technology, the differential expression of genes and biological molecules, such as DNA, RNA, or proteins present in the biofilm can be studied (Vohra et al., 2024). Integration of different omics approaches such as genomics, transcriptomics, proteomics, and metabolomics improves the basic idea about the microbial community structure, functions, and environmental responses (Tian & Wang, 2021). Metagenomics focuses on the analyses of genomes present in the microbial communities. Meta-transcriptomics captures the mRNA content of the sample to analyze the functional properties. It can provide information about the taxonomy, functional characteristics, and genomic expression profile in an organism. On the otherhand, meta proteomics gives insights into the protein present in the sample through which its function in the microbial cells can be determined. Usually, it is performed using protein extraction followed by mass-spectrometric analysis. Whereas, metabolomics emphasizes the detection of molecules or metabolites produced by microorganisms, which determines the health or dysbiosis in the host. It is

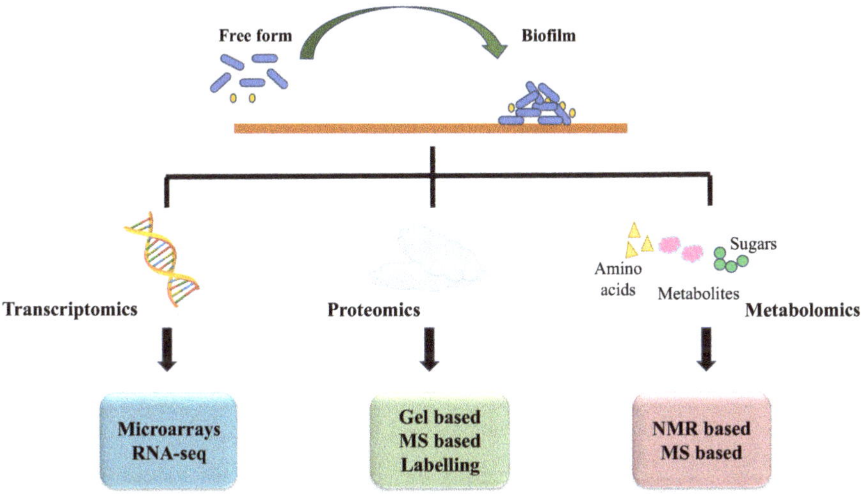

Fig. 14.1 Omics methods, such as transcriptomics, proteomics, and metabolomics employed for studying the biofilm and planktonic phenotypes of microorganism

quantified using chromatographic techniques, mass spectrometry, and nuclear magnetic resonance methods. Therefore, using these techniques, the mechanism and markers associated with diseases can be identified (Lin et al., 2024).

Omics techniques are commonly used for studying the mechanism behind biofilm formation. Integration of different omics approaches can provide information at genetic, transcriptional, and translational level, and also about cellular metabolism, and its regulation compared to the single omics approaches (Fig. 14.1). Zhang et al. used multi-omics methods for the identification of genes and metabolites involved in the biofilm formation of *Bifidobacterium pseudocatenulatum.* They cultured the bacterium in wheat fibers and using multi-omics tools, the genes and metabolites that take part in biofilm formation were identified. Transcriptomics results showed the expression of genes contained in stress response, quorum sensing, and polysaccharide metabolism, which may induce biofilm formation. Metabolomic data showed that metabolites, for instance amino acids, amides, organic acids, sugars, and alcohol are synthesized in the biofilm. Hence, they concluded that stress response, QS, and EPS secretion are the important factors in biofilm formation of *B. pseudocatenulatum* (Zhang et al., 2023). Table 14.1 enlisted some of the omics and multi-omics approaches involved in the biofilm studies.

14.3 Transcriptomics in Biofilm Detection/Diagnosis

Transcriptomics, also known as meta-transcriptomics, is used to detect the genes that are differently expressed at a given time. The two major tools used in transcriptomics to characterize various roles of RNA are microarray and RNA-sequencing (RNA-seq) techniques. Through these methods, the mechanism in biofilm

Table 14.1 Different omics, their integration, and techniques used for biofilm studies

Omics	Study	Technique used	References
Transcriptomics	Analysis of *S. epidermidis* biofilm interface with human immune cells and soluble factors	RNA-sequencing	França et al. (2016)
Transcriptomics	Identification of markers associated with *Clostridium difficile* biofilm linked recurrent and non-recurrent infections	Microarrays	Rubio-Mendoza et al. (2023)
Transcriptomics	MRSA biofilm inhibition via ursolic acid and resveratrol	RNA-seq	Qin et al. (2014)
Transcriptomics	Analysis of *Haemophilus parasuis* biofilm	RNA-seq	Jiang et al. (2021)
Transcriptomics	*P. aeruginosa* biofilm lung infection in ex vivo pig model	RNA-seq	Harrington et al. (2022)
Transcriptomics	Meta transcriptomic analysis on river biofilm exposed to pharmaceutical products	DNA microarrays	Yergeau et al. (2010)
Proteomics	*Corynebacterium pseudotuberculosis* Biofilm forming and non-biofilm forming strains from goat were analyzed	LC-MS/MS	de Sá et al. (2021)
Proteomics	Protein expression levels of *S. aureus* biofilm subjected to shear stress conditions	2DE and nano-LC-ESI-MS/MS	Islam et al. (2014)
Proteomics	Proteome analysis of *S. epidermidis* biofilm like behavior of planktonic cells under stress conditions	MALDI-TOF/TOF LC-high-definition MS	Bottagisio et al. (2019)
Proteomics	Identification of markers for invasive candidiasis by *C. albicans* through proteomic and immunoproteomics	LC-MS/MS	Vaz et al. (2021)
Proteomics	Proteomic variation in polymicrobial biofilm formed by *Anaeroglobus geminatus*	LC-MS/MS	Bao et al. (2017)
Metabolomics	Metabolite quantification in *Pseudomonas aeruginosa* biofilm	GC-MS	Yang et al. (2021)
Metabolomics	Comparison of planktonic and biofilm forms of *S. aureus*	HILIC-MS	Stipetic et al. (2016)
Metabolomics	Metabolic variation in *P. aeruginosa* planktonic and biofilm in bovine synovial fluid	2D NMR	Leggett et al. (2022)
Metabolomics	Metabolomic analyses of single and co-cultured *C. albicans* and *S. aureus* biofilms	GC-Orbitrap MS	Weidt et al. (2016)
Metabolomics	Metabolic differences between virulent and avirulent *P. aeruginosa* clinical strains	LC-MS	Depke et al. (2020)
Metabolomics	Metabolomics in planktonic and biofilms of *Mycoplasma fermentans* and *Mycoplasma pneumoniae*	NMR	Awadh et al. (2021)

(continued)

Table 14.1 (continued)

Omics	Study	Technique used	References
Metabolomics	Metabolomic analyses of *Pseudomonas putida* in drinking water	LC-TOF-MS	Kouremenos et al. (2014)
Multi-omics	Carbohydrate metabolism in mixed species biofilm of *Streptococcus mutans* and *C. albicans*	RNA-seq and iTRAQ	Ellepola et al. (2019)
Multi-omics	Identification of pathways monitoring *Aspergillus fumigatus* and *Klebsiella pneumoniae* interspecies interactions.	RNA nanochips Orbitrap MS HILIC-MS	Bitencourt et al. (2024)
Multi-omics	Identification of molecular mechanisms and factors contained in *Leptospira* pathogenesis	RNA-seq LC-MS/MS	Kavela et al. (2023)

formation, biofilm stress response, and effectiveness of control procedures can be studied (Yuan et al., 2023; Nguyen et al., 2020). This method is an indispensable approach for exploring the metabolism and genes associated with microorganisms and also offers insights into the functional aspects of microbial communities. It provides numerous benefits compared to DNA amplicon sequencing. Metatranscriptomic approaches capture the live microbes and identify microbial species and subspecies in the mixed biofilm without any amplification biases. Moreover, this method gives an idea about the species involved in the metabolic activity within a microbial population, whether in a healthy or dysbiosis condition. Additionally, this approach can be used to assess the biofilm effectors associated with dairy products, thereby improving food safety concerns (Almatroudi, 2024).

DNA microarrays consist of numerous short oligonucleotide probes, unique to thousands of known transcripts known as probes that are anchored to a surface. This method is extensively employed for the analysis of biofilm-associated gene expression (Yuan et al., 2023). A complementary DNA chip can be used to detect specific genes from the gene pool. The crude samples of RNA from cell culture can be directly combined with these chips, and the expression profiles of complementary binding with fluorescent probes can be detected. Moreover, with the help of microarrays, biofilm regulators can be identified (Kumar & Kumar, 2023). Microarrays are prepared on a glass slides or chips coated with fragments of DNA or nucleotides in a precise form. A typical diagnostic microarray composed of three probes includes short and long oligo (dT) and PCR amplicons (Panwar et al., 2023). Designing the probes is an important part of this method as it is necessary for the proper detection of pathogens. Over the years, this method was modified, where a planar glass slide with different surface alterations was made, which was commonly consented by the researchers. The probes are designed to have a distinct region of the gene. The detection process consists of two steps. Initially, the fragments were tagged with a fluorescent dye and produces single-stranded fragments by denaturation. Further, these denatured filaments undergo hybridization with their corresponding probes in the array. After binding, the probe-ssDNA produces fluorescence, which was detected to know the concentration of DNA fragments. First commercial microarrays for the identification of methicillin resistant *S. aureus* (MRSA) were established by Affymetrix and collaborators known as Staphy Chips; which can identify upto 15 species of *Staphylococcus* (Panwar et al., 2023). Moreover, microarrays are comparatively lower cost however, the limitation of this method is which cannot be used for unknown transcripts due to its unavailability, creates background noises during hybridization, and low accuracy (Yuan et al., 2023).

In a study, early (3 days) and late (31 days) oral biofilm formation on different dental implants were studied using microarray technique. After 3 days, 16 species of bacteria capable of forming biofilm were identified, whereas 17 species were detected after 31 days of biofilm formation. The topmost periodontal pathogens, such as *Porphyromonas gingivalis, Prevotella intermedia*, and *Aggregatibacter actinomycetemcomitans*, are absent in the samples. However, other pathogens, such as *Streptococcus mitis* and *Parvimonas micra* were detected. Moreover, healthly microorganisms associated, includes *Actinomyces odontolyticus, Veillonella parvula*, and *Capnocytophaga* species were also detected in all samples (Herrmann et al., 2020). Using microarray analysis, *Staphylococcus aureus* isolates from humans, animals, and food were assessed for biofilm-forming ability and associated genes. Around 41.8% of isolates are biofilm producers, and 71% of isolates are slime producers. All the isolates identified from food samples are good biofilm producers. Around 17.9% of human isolates and 52.6% of isolates obtained from sheep diseased with mastitis are biofilm formers. Human *S. aureus* isolates (4) showed the *mecA* gene encodes for methicillin resistance (Achek et al., 2020). In another study, the genes connected with the *Streptococcus mutans* biofilm thickness at different depths were analyzed using the DNA-microarray technique. They found out 29 genes were differentially expressed at 400 microns and 100 microns depth and 39 genes at 200- and 100-microns depth (Shemesh et al., 2008).

The RNA-seq technique sequences the RNA indirectly after its conversion into cDNA. Through this sequencing, the real-time expression of host genes as well as the pathogen can be determined. The presence of RNA in the system represents the transcription process occurring in the system or the incidence of RNA genome. RNA molecules after transcription are short lived than DNA and serves as an ideal marker to understand the current biological activity happening inside the host system. The high number of RNA copies during transcription can be detected easily. Moreover, RNA has less repetitive and exons than DNA. RNA-seq directly from the RNA called direct RNA sequencing (DRS) can avoid the inaccuracies throughout cDNA amplification and RNA-RNA chimera production in the cDNA construction process. DRS is in its early stages, which is a quick method, and also it can bypass the amplification steps (Spatz & Afonso, 2024). RNA-seq is a potential method used for the identification of known and unknown transcripts. It offers several merits, such as sensitivity, reproducibility, and good capacity. The cost and labor for performing RNA-seq is relatively expensive; however, it can be considered as an ideal tool for biofilm studies (Yuan et al., 2023). It has become a preferable method for the identification of genes related to biofilms. RNA-seq can analyses multiple number of samples simultaneously makes it a cost-effective method (Almatroudi, 2024). Compared to microarrays, RNA-seq has more sensitivity to detect low RNA levels, which helps in studying about persisters and biofilms (Almatroudi, 2024).

Transcriptome studies were used to identify the transcriptome of *Gardenella vaginalis, Fannyhessea vaginae*, and *Prevotella bivia* on polymicrobial biofilm associated with bacterial vaginosis (BV). RNA-seq exposed several differentially expressed genes of 432, 126, and 39 for *G. vaginalis, F. vaginae*, and *P. bivia*, respectively, on polymicrobial biofilm forms. Study revealed that the three BV organisms are linked in their transcriptome and the molecular mechanisms behind its interaction in the triple species biofilm leading to BV (Sousa et al., 2024).

Furthermore, RNA-seq can be used to detect the molecular mechanism behind the action of antibiofilm agents. In a study, the mechanism of DNase action on biofilm formation of *Pseudomonas aeruginosa* and *S. aureus* was studied using RNA-seq transcriptome analysis. The study proposed that DNase action may suppress biofilm development by reducing the gene expression of T4SS, capsular polysaccharides, acid metabolism, and flagellar arrangement (Deng et al., 2024). Single-cell RNA-sequencing (scRNA-seq) is a broadly accepted tool to understand eukaryotic transcriptome profiles from crude samples. This method can be used to study immune response during microbial infections such as biofilm, reactive oxygen production, and inflammatory signaling. Nevertheless, this method was limited to prokaryotic cells due to low amount of mRNA and complexity in cell wall. Bulk RNA-seq methods are mainly focused to sense about biofilm and pathogens; however, it fails to detect the heterogeneity (Korshoj & Kielian, 2024). scRNA-seq helps in identifying the individual cells and heterogeneity in the biofilm, especially in the mixed species community. However, the development of scRNA-seq sheds the light on studying the host-bacterial cell interaction and cell diversity at the single cell level. Currently, scRNA-seq studies in prokaryotic organisms are in its early stages. Bacterial single-cell sequencing offers several advantages such as identification of phenotypic heterogeneity in the same group, rare cell type identification, individual gene expression pattern detection, metabolic activity, and host–bacterial interactions. Moreover, it can also simplify the recognition of antibiotic resistant and also persister cells. Identification of such cells provides better knowledge about the underlying diseases and personalized therapeutic methods for biofilm-associated infections (Moon et al., 2023). The initial phase in the single-cell transcriptomics is the separation and lysis of the individual microbial cells. Fluorescent-activated cell sorting (FACS) is a rapid and high throughput technique for the separation of individual microbial cells from the bulk samples. The cell subsets can be selected based on the fluorescent signals emitted from the cells. Additionally, Raman micro spectroscopy also called as Raman-activated cell sorting, RACS in combination with microfluidics and optical tweezers are used to discrete bacteria from the diverse environments on the basis of the deuterium metabolizing ability. Both techniques lack the microbial phenotypic imaging to categorize based on size and shape. Multiple annealing and tailing based quantitative single-cell RNA sequencing (MATQ-seq) is a type of scRNA sequencing method, that can captures the rare transcripts by commencing reverse transcription using random primers followed by circumventing the melting stages between each cycle with several annealing stages (Imdahl & Saliba, 2020). Transcriptomic analysis along with computational tools helps in the identification of biomarkers for the detection of underlying pathologies.

The transcriptomic data availability in the NCBI Gene Expression Omnibus (GEO) database allows to analyze multiple datasets at a time (Moreno et al., 2022).

14.4 Proteomics in Identification/Diagnosis of Biofilm

Unlike transcriptomics, proteomics deals with scrutinizing the proteins secreted by an organism or a community, which can offer a clearer picture about the biological modifications happening in an organism. Microorganisms are able to sense and tolerate severe environmental conditions by altering their protein expressions (Seneviratne et al., 2020). Meta-proteomics is useful for revealing the microbial interactions and functions in a community. By tracing the proteomic expression, the response of the microbe under physiological condition, and biofilm phenotypic features can be uncovered (Zhao et al., 2023).

Mass spectrometry (MS) has been widely used for investigating protein in the crude samples. The proteome present in the living organism is dynamic due to the activity, conformation, expression levels, location, etc. which makes differences in the organism's response to stimuli. Proteomics, in combination with mass spectrometry, upgrades the quantification of genes and cell functions at the proteome level. Since the integration of MS-based approaches (mainly tandem MS/MS) in proteomics, the accuracy and resolution for proteome quantification have increased on a large scale. MS-proteomics methods usually follow bottom-up or top-down approaches. Top-down deals with the analysis of complete protein, whereas bottom-up methods analyze proteins in the form of peptides after enzymatic digestion process (Torres-Sangiao et al., 2022). The top-down approach analyses the proteins using high-resolution MS and allows the protein characterization at the proteoform level with good precision. The proteins can be intactly subjected to LC-MS/MS without enzymatic digestion. In the LC step, the proteoforms were separated before its introduction into the mass spectrometer, followed by fragmentation of the protein, where the proteoform can be identified through databases. However, identification of proteins with single amino acid change is quite challenging in top-down approach than the bottom up. Moreover, the same complexity of the protein cannot be maintained in top-down method. Despite its limitations, bacterial genomes are suitable for top-down method because of its less complexity and the proteins are relatively small (less than 30 kDa) (Gant & Chamot-Rooke, 2024). The bottom-up method characterizes the protein based on peptide fragments. The extracted protein was digested using trypsin, and the obtained peptides were subjected to LC, separated, and detected by MS. Identification of peptide fragments was done by matching the mass spectra attained from the fragmentation process with theoretical spectra from the protein database (Gant & Chamot-Rooke, 2024).

Proteome analyses can be gel-free or gel-based methods. The former method primarily involves the two-dimensional gel electrophoresis of protein samples and its analysis (Kalpana et al., 2023). In gel-dependent techniques, the proteins were isolated by 2D-gel electrophoresis and are identified using MALDI-TOF MS/MS, whereas in gel-independent methods, the protein was directly identified by MS

(Katsarou et al., 2021). Sodium dodecyl sulfate polyacrylamide gel electrophoresis (SDS-PAGE) is used for the electrophoretic separation of proteins based on the molecular weight in an electric field. Sodium dodecyl sulfate is an anionic detergent used for the denaturation of protein; however, the functions of the protein cannot be studied using this method. To overcome this problem, native PAGE gels can be employed without any denaturation process. In order to maintain the functionality of protein and better separation native, SDS-PAGE are widely used. Two-dimensional gel electrophoresis is able to separate proteins from a mixture and can explore the variations of proteome in the samples. Here, the proteins are initially parted based on the isoelectric point followed by mass separation using SDS-PAGE. Proteins can be detected from the gel using dyes, such as coomassie blue, silver nitrate, and deep purple dye; gel images can be analyzed using different software. This method enables the separation of thousands of samples to study the differential expression of proteins; which can be isolated and subjected for protein sequencing. Differential-in-gel electrophoresis (DIGE) used for the detection of two different protein samples labeled with fluorophores, such as Cy2, Cy3, and Cy5 and subjected to isoelectric separation. These dyes bind to the lysine amino acids in the amino group and the positive charge of the lysine was replaced by the single cationic charge of the dye at neutral and acidic pH and maintains equal charge. Further, the samples were separated by 2D electrophoresis and the proteins were detected upon fluorophore excitation. This method can be used for the simultaneous detection of protein samples, and thereby minimizes the usage of gels (Shiny Matilda et al., 2020).

Shotgun proteomics includes the enzymatic cleavage of protein into fragments with equal characteristics, which are subjected to LC-MS/MS. Techniques such as MALDI-TOF/MS, gas chromatography MS (GC-MS), and liquid chromatography MS (LC-MS) are widely used for the detection and diagnosis of diseases (Tsuchida et al., 2018). MS is an analytical technique that discovers the mass-to-charge ratio (*m/z*) and molecular weight of unknown molecules. The most frequent ionization methods used in MS are laser desorption/ionization (LDI), matrix-assisted laser desorption ionization (MALDI) or surface enhanced laser desorption/ionization (SELDI) and electrospray ionization (ESI). In the late 1990s, MALDI supported with time of flight (TOF) analyzer was developed to distinguish bacterial cells. MALDI-TOF mass spectrometer can be used to analyze both digested as well as complete proteins. It consists of three parts, namely an ionization source, commonly laser, mass analyzer (TOF), and an ion detector. Initially, the samples were combined with a matrix, which is an energy-absorbent and organic solution where the combined sample was subjected to drying to form a crystalline sample, followed by ionization using laser. During ionization and desorption, the samples switch into gaseous state charged ions that are affected by the electric and the magnetic fields. Here, MALDI is working on the basis of a soft ionization process, conserves the stability of the samples, such as proteins, peptides, and other biomolecules, that are delicate and tend to fragment easily on other ionization techniques. After ionization, the mass analyzer differentiates the samples based on their *m/z* ratios. The detector analyses the samples and generates a mass spectrum with *m/z* ratios and relative abundance. This method sorts the microorganisms based on the mass spectra of protein using MS methods. It is widely used for disease diagnosis in humans or animals by analyzing the infected tissue samples as well as bacterial biofilm (Torres-Sangiao et al., 2021). MALDI-TOF MS can also be utilized for the structural analysis and chemical composition, such as proteins, nucleic acids, organic compounds in the biofilm. It ensures rapid analysis of samples, minimal samples are required, and shows high sensitivity; however, it requires expertise and specialized equipment, ion suppression effects, expensive consumables, and instruments which are considered as limitations to this technique (Mir & Rather, 2024).

In a study, the differential expression of protein in both biofilm and planktonic forms of *Mycobacterium tuberculosis* was analyzed using proteomics approaches, such as 2D gel electrophoresis succeeded by LC-MS/MS. In biofilm forms, around 47 different proteins were expressed compared to planktonic counterparts, where 37 were nonredundant (Wang et al., 2019). Researchers carried out a comparative proteomic experiment to identify the biofilm infection-associated immunogenic proteins in *Streptococcus suis* biofilm and planktonic cells. Almost 13 proteins were identified, where six and seven proteins were upregulated and downregulated, respectively, in biofilms than the planktonic forms. Followed by challenging of *S. suis* with serum, proteins were visualized using Western blot analysis, which was previously separated by 2DE. Around 10 immunogenic protein spots were matching with the nine distinct proteins analyzed by MALDI-TOF/TOF-MS. Out of nine, five proteins are common to both forms, which can be considered as a target for the development of vaccine against acute and biofilm infections of *S. suis* (Wang et al., 2012). Similarly, the protein expression associated with antimicrobial resistance in *S. aureus* biofilm was studied using proteomics. Protein extraction followed by MALDI-TOF/TOF MS showed 14 differentially expressed proteins in both biofilm and free forms (Piras et al., 2021).

Protein can be quantified using mass spectrometer with label or label-free methods. In stable isotope labeling, the proteins or peptides from different samples can be labeled to create mass difference. Such kind of stable isotopic labels includes isotope-coded affinity tag (ICAT), tandem mass tag (TMT), stable isotopic labeling with amino acids (SILAC), mass differential tags for relative and absolute quantifications (mTRAQ), and isobaric tag for relative and absolute quantification (iTRAQ) (Rizal et al., 2020). Label-free methods (shotgun proteomics) in MS quantifies the proteins and peptides without labeling with isotopes. LC-MS is a label free quantification method (Rizal et al., 2020). Label-free quantitative proteomics, such as LC-MS/MS is exploited to analyze the interactions of *Fusobacterium nucleatum* and *Porphyromonas gingivalis* in oral biofilm formation and periodontitis. The study implies that *P. gingivalis* biofilm and secretion of protein were reduced in the dual culture with *F. nucleatum* (Ali Mohammed et al., 2021).

Labeling techniques were used to derivatize the protein or peptides in the samples for analysis. Initially, isotope-coded affinity tagging (ICAT) was used for the differential expression of proteins in *Saccharomyces cerevisiae* in presence of galactose/ethanol as a carbon source. The ICAT setup comprised of an avidin column, and the peptides tagged with biotin (affinity tag) present in the mixture bind with the avidin column. Here, an isotopically labeled linker molecule was used, which is either light with eight hydrogen atom or heavy with eight deuterium atom and a thiolate group that labels the cysteine amino acids present in the peptide. The protein mixtures were labeled separately with both heavy and light reagents; the two samples were mixed, digested with trypsin, and the mixture was allowed to run through the column for affinity binding. After running the column, biotin was removed and the separated labeled protein was removed from the column and was analyzed using MS. The relative abundance of the labeled (light and heavy) peptides is equated and the differential expression of proteins can be detected. The currently used form of reagent was modified from the original, where the biotin was attached to an acid-cleavable linker for easier removal; the light and heavy regions consist of either nine C_{12} or C_{13} atoms, in order to overcome the elution issues observed with hydrogen and deuterium atoms (Graham et al., 2007).

Proteomics technology based on iTRAQ (Isobaric Tags for Relative and Absolute Quantification) can be used for the detection of biofilm and also for searching potential targets for anti-biofilm activity. It is a mass-spectrometric technique useful for the quantification of protein present in the samples. The peptides from different sources were labeled with isobaric tags (reagents primarily reacting with the amine group) permits the simultaneous detection. The typical model of iTRAQ-based biofilm proteome analyses includes sample collection (biofilm), extraction of protein, digestion, labeling of peptides, peptide fractionation, mass spectrometry, and data interpretation. For effective protein extraction techniques such as sonication and enzymatic digestion of biofilm samples can be exercised. This method able to identify the proteins expressed at different stages of biofilm includes formation, maturation, and dispersion. Moreover, with the expressed protein levels, the mechanism of biofilm formation, host–cell interaction, and antibiotic resistance are explored. Finally, it helps to compare the proteomes of planktonic forms with their biofilm counterparts, in this way the development of potential therapeutics is possible. Additionally, this method can be used for the detection of biofilm proteins associated with caries and other problems. It can be considered as an ideal tool for the detection of protein and metabolic pathway targets for the eradication of biofilm (Zafer et al., 2024; Cruz et al., 2021).

Tandem mass tags (TMT) are an MS/MS-based approach for the quantification of proteins and peptides (Tsuchida et al., 2018), which is an alternative to iTRAQ method. The reactive groups in TMT is an amino group, which modifies the primary amine (Grocholska et al., 2023). This method can be used to measure the protein in tissue, plasma, or other fluids from the diseased part. It is a reliable method for the identification of proteins expressed at various times of disease, and helps in tracking pathophysiology (Zubair et al., 2022). Isotopic variants vary in reporter mass but possess equal overall mass, which aids in quantification with great accuracy, which can be applied for low-mass proteins as well. TMT allows simultaneous detection of ten samples with a simplified experimental setup and time-saving method. A TMT-based proteomic approach was used to detect the proteome variations in the dry and wet biofilm of *S. aureus*. Isobaric tags are ideal tools for the detection and quantification of protein in samples by MS (Rahman et al., 2022; Tsuchida et al., 2018).

SILAC is a labeling procedure commonly used in the quantification of proteins. In this technique, amino acids are labeled with isotopes supplemented into the growth medium with the living cells. The specific residues are labeled and are used for quantification, showed potential advantages over other metabolic labeling techniques, such as N^{15} labeling, where quantification is depends on the nitrogen numbers in the protein samples. The isotopes used in amino acid labeling are mainly C^{13} or N^{15}-labeled arginine or lysine, which was incorporated into the proteome during cellular metabolism, especially in cell culturing. After labeling the amino acids in light and heavy medium, the samples undergo digestion, and the peptides are analyzed using LC-MS/MS. Quantification is on the basis of labeled peptides and unlabeled peptides ratio in the given sample. It is used for the analyses of proteomic expression in two different biological samples. Moreover, SILAC can be applied to identify the membrane protein biomarkers for the diagnosis of cancer and also as drug targets (Chen et al., 2015). Moreover, SILAC can be used to study about biofilm and planktonic proteome. In a study, the proteomic expression of both planktonic and biofilm forms of *Neisseria gonorrhoeae* was studied using SILAC. Two sets of arginine auxotroph of the strain were grown, the planktonic population was added with C^{13}-Arg, and the biofilm form was kept unlabeled. After incubation, both cells were harvested, digested, separated by SDS-PAGE gels and further analyzed using LC-MS (Phillips et al., 2012). Labeling methods have a good accuracy rate; however, it cannot be subjected to large-scale applications. Small-scale use of labeling methods consumes time, labor, and expensive (Rizal et al., 2020).

14.5 Metabolomics in Biofilm Studies

Bacterial biofilms are well known for their minimal metabolism and lack of cell division, where metabolic activities are almost equivalent to that of dormant bacteria. However, recently one study reported that biofilms under nutrient limitation showed enhanced metabolic activities. Metabolomic profiling has become an influential tool for investigating biofilm metabolism and molecular composition. Metabolites are the small molecules secreted by a cell, categorized into two types, such as primary and secondary metabolites. The former are necessary for normal cellular development and are secreted during active cell growth, whereas secondary metabolites are produced for the survival and adaptability of the cell, for instance, antibiotics, but not for growth (Malviya et al., 2023; Seneviratne et al., 2020). The metabolites are the end products of transcription, translation, protein synthesis, and the enzymatic processes (Washio & Takahashi, 2016). Microorganisms show heterogeneity in metabolism and composition with the same biofilm leads to differences in metabolic spectrum. Both biofilm and planktonic forms differ in their metabolic activities. For instance, the glycolytic enzyme, glyceraldehyge-3-phosphate dehydrogenase was highly active in biofilm than in the suspended forms. The physical interaction of bacteria with the environment establishes matrix production, biofilm structure, and also diverseness in metabolism. It induces metabolic cross-feeding and promotes the emergence of metabolically diverse microorganisms in the same biofilm (Malviya et al., 2023; Mirzaei et al., 2024; Joshi et al., 2021). Metabolomics, are also known as metabolic fingerprinting, can be used to categorize biological samples. The combination of multiple approaches makes metabolic studies versatile for the analysis of biofilms (Zhang & Powers, 2012). The metabolomics technique allows the detection of these small molecules and its interaction in the biofilm (Malviya et al., 2023).

Analytical methods, including nuclear magnetic resonance (NMR) spectroscopy and mass spectrometry (MS), provide information about the biofilm metabolism (Malviya et al., 2023). NMR-based approaches can be employed for the identification of novel drugs against bacterial biofilms. Characterization of biofilms and planktonic forms provides a vast idea about the biological processes linked with biofilm development. This spectroscopic method can analyze the samples both quantitatively and qualitatively. Metabolic studies using NMR in bacterial biofilms comprised of several steps, which includes culturing of bacteria, sampling and extraction of metabolites, NMR analysis, statistical analysis, metabolic identification using different databases, metabolic pathway construction, and integration with other omics approaches (Zhang & Powers, 2012). NMR spectroscopy can provide overall metabolic profiles of the crude sample. Principal component analysis was often used to analyze the NMR spectra due to distinct peaks of chemical molecules. High-resolution magic angle spinning (HRMAS) NMR spectra can be used for the metabolomic studies in semisolid samples like cells. H^1 HRMAS NMR approach was studied to compare the metabolites in the actively grown culture and supernatants from batch and chemostat planktonic as well as biofilm cultures of *P. aeruginosa* (Gjersing et al., 2007). Moreover, NMR metabolomics uses unsupervised statistical analysis called as principal component analysis (PCA), which provides understanding about the metabolic relation between bacterial species without predetermining the statistical clustering results of those phenotypes. In a study, NMR was used to distinguish the metabolic profiles of biofilm and suspended forms of methicillin-resistant and methicillin-susceptible *S. aureus*. PCA differentiates biofilm from free-form phenotypes and proposes the main characteristics that distinguish biofilm from free forms includes selective amino acid uptake, lipid catabolism, butanediol fermentation, and metabolic shift in energy for matrix formation and cell wall assemblage. These metabolites or the pathways can be targeted against *S. aureus* wound biofilm infections (Ammons et al., 2014). The benefits of NMR include non-destructive, non-invasive, minimal sample required, and is applicable to both static as well as free forms; behavioral or physical characteristics can be measured. Limitations of NMR include low sensitivity, and careful optimization is required to lower the number of measurements and lower concentration detection thresholds (Abreu & Fernández, 2020).

Usually, the metabolomics studies were done using NMR and GC-MS techniques. Recently, LC-MS has become a robust platform for the non-targeted metabolomics. Owing to the use of a stationary phase with a particle size of 2 μm (ultra-high performance liquid chromatography, UPLC) and also the separation of a broader array of molecules in a shorter time. The combination of quadrupole time-of-flight mass spectrometry (QToF-MS) with UPLC is used for targeted and non-targeted studies. Moreover, the addition of MS/MS experiments onto this MS yields substantial information about metabolites (Favre et al., 2018). An in vitro biofilm model was prepared from the saliva of normal individuals and investigates the commensal as well as cariogenic organisms by supplementing sucrose. Metabolomic approaches using UHPLC/QToF-MS analyses were conducted, where 42 metabolites were abundant in cariogenic biofilms, 17 metabolites were connected with commensal biofilms, specifically linked with protein metabolism (Sánchez et al., 2022).

Metabolomics using the GC-MS technique was employed to investigate biofilm in the water distribution system. As biofilms form on the water pipeline can cause serious waterborne illness, corrosion of the pipelines, and changes in the odor and taste of drinking water. The metabolites produced by the biofilm-forming organism can be identified using GC-MS, and thereby targeting such pathways can manage biofouling in the water system (Beale et al., 2013).

Gas chromatography-mass spectrometry (GC-MS) is a diagnostic procedure because of its good reproducibility, separation capacity, selectivity, and high sensitivity. In GC-MS there are two forms of ionization, includes electron ionization (EI) and chemical ionization (CI), where EI is generally used in GC-MS metabolomic studies. Spectral databases/libraries are available for the detection of metabolites corresponding to the GC-MS spectra. Metabolite identification is through GC system combined with single quadrupole LC-MS system. Besides, GC-MS can overcome the issues compared to LC-MS, such as matrix effects and ion suppression by co-eluting compounds and thereby attain greater resolution in the separation of compounds. Detection of low-mass (50–600 Da) metabolites and volatile compounds is only possible with GC-MS, which remains as a limitation to this technique. Before the analysis of polar, thermolabile, and non-volatile compounds, chemical derivatization should be performed; with this sensitivity, and volatility of the compounds can be enhanced for the detection of signals, and thereby the problems that arise during derivative measurement for target metabolite identification can be avoided. The general metabolites that can be analyzed using GC-MS includes amino acids, fatty acids, and mono-, di-, and tri-saccharides (Beale et al., 2018). The workflow of GC-MS method starts with sample preparation/extraction, sample preparation based on volatile and non-volatile nature of the metabolites, separation using chromatographic techniques followed by data analysis through bioinformatics platforms and finally biological assessment of the interpreted data.

With metabolomic techniques, small metabolites like amino acids, carbonic acids, and sugars in the sample are difficult to be analyzed simultaneously. In the last decade, introduction of capillary electrophoresis (CE) and time-of-flight mass spectrometer (TOFMS) development can overcome this hurdle in analysis. Capillary electrophoresis comprised of capillary tube with a diameter less than or equal to 100 μm connected to an electrode. High voltage of current was passed at the ends of the capillary tube after loading the samples results in the movement of ionic metabolites toward anode or cathode. Each metabolite in the samples have different electric charge, ionic radius and different velocities, hence the molecules can be separated. Compared to the liquid chromatography, CE have good separation capacity and there is no interference of molecular mass of the metabolites. Using TOFMS, the ionic metabolite's mass can be determined and separated based on the flight time. By combining CE and TOFMS, the ionic molecules can be parted on the basis of their electric charge and the mass can be determined. Analysing *m/z* ratio and the molecular weight the metabolite can be identified precisely with small quantity of samples. CE-TOFMS was widely employed in metabolomic research for studying metabolic pathways, such as glycolysis, pentose phosphate pathway, TCA, and amino acid metabolic pathways. CE-TOFMS was widely used to study metabolomics in oral biofilms. Caries preventive agents, such as fluoride and xylitol effects on oral biofilm metabolites were studied using metabolomics (Washio & Takahashi, 2016). Similarly, the effects of erythritol on periodontopathic biofilm of *Streptococcus gordonii* and *Porphyromonas gingivalis* was studied using CE-TOFMS metabolomic approaches. Study showed that the production of extracellular matrix and nucleic acid intermediates were decreased after exposure with erythritol in a dose dependent manner. Moreover, erythritol negatively affects the polymicrobial biofilm through several signaling pathways such as DNA and RNA depletion, matrix production, and amino acid metabolism (Hashino et al., 2013).

Untargeted metabolomics needs reference compounds for the identification of metabolites and provides information about metabolic phenotypes only, not about the functional metabolism without the purified compounds (Guo et al., 2021). Untargeted metabolomics are non-selective screening of metabolite characterization. Analysis of wider metabolites needs longer separation methods for its identification. High-resolution mass spectrometers (HRMS) are able to perform high-throughput profiling of metabolites without any additional equipment and also with shorter analysis time. HRMS are mass analyzers that includes time-of-flight, Orbitrap, and Fourier-transform ion cyclotron resonance system that possess high-mass accuracy and separates metabolites with a resolution of 0.001 amu. This enables metabolite identification using HRMS and permits the investigation of metabolic changes within less time and also compares the results with mass spectral libraries. It can also detect the metabolic alteration in antimicrobial resistance and biofilm development (Kok et al., 2022). Some researchers used untargeted metabolomics to study the metabolism in biofilm and planktonic states of *Bifidobacterium bifidum*. Both states were analyzed using GC-MS to reveal the differences in carbohydrates, lipids, and amino acids, essential for the cell regulatory processes, and also functions as intermediates in various metabolic processes. A total of 173 metabolites were identified in both states, where 64 showed substantial difference in both states; 16 metabolites were comparatively expressed in biofilm states. In planktonic states, 48 metabolites were significantly upregulated when compared to biofilm. The five metabolic pathways that are downregulated in biofilm are aminoacyl-tRNA biosynthesis, alanine, aspartate and glutamate metabolism, TCA cycle, nitrogen metabolism, and arginine and proline metabolic pathways (Sadiq et al., 2020). Researchers used untargeted LC-MS metabolomic techniques to discriminate four marine biofilm formers and its culture parameters.

They have used marine bacteria, such as *Persicivirgia mediterranea* TC4 and TC7, *Pseudoalteromonas lipolytica* TC8 and *Shewanella* sp. TC11 as model organisms. From the MS/MS analysis, novel markers for the differentiation of these bacteria were identified, where TC8 strain was distinguished by ornithine lipids, and TC4 and TC7 produced hydroxylated ornithine and glycine lipids. Moreover, TC7 strain produces diamine derivatives like putrescine amides (Favre et al., 2017). In another study, a GC-MS based untargeted metabolomic method was used to study about the mono- and dual-species biofilm of *Candida albicans* and *Klebsiella pneumoniae*. Both bacteria often form dual-species biofilms in humans and cause high morbidity and mortality rates in patients. The study demonstrated that strict association between two organisms is mainly due to the stress-related metabolites, such as vitamin B6 and trehalose, secreted by *C. albicans* in the mature biofilm. The metabolites produced in the single biofilm of both bacteria and fungus is different from the dual species biofilm. Studies related to the exploration of metabolites in the biofilm can figure out the strategies to combat such dual-species biofilm infections (Galdiero et al., 2021).

Targeted metabolomics are widely used to investigate about functional metabolism (Guo et al., 2021). Targeted metabolomics needs previous knowledge of the metabolites to be extracted from the cells. The high sensitivity and selective quantification of the metabolites allow us to characterize the changes in metabolites. This needs a more comprehensive and expensive platform, and also the validation process is more time-consuming (Kok et al., 2022). Due to the commercial accessibility of compounds, understanding the functional regulatory genes and the catalyzing enzymes involved in the synthesis can accurately ascertain the mechanism behind the altered metabolism in a regulatory pathway. High-throughput LC-MS-based tools can be used for the large-scale analysis of metabolomes, which eases the metabolomic investigation in diseases and physiological studies (Guo et al., 2021). Recently, LC/TQ-based targeted metabolomics approaches using reverse-phase liquid chromatography (RPLC) and hydrophilic interaction liquid chromatography (HILIC) were used for the detection of non-polar and polar metabolomes. From this investigation, more than 100 important microbial metabolites were determined. Moreover, five functional metabolites were detected in the regulation of *E. coli* biofilm formation (Guo et al., 2021). In a study, targeted metabolomics using LC-MS/MS Orbitrap was applied to reveal the pathways allied with biofilm establishment in clinical isolates of *Enterococcus faecalis*. Researchers have identified that hypoxanthine and aspartate metabolic pathways are the key markers to target *E. faecalis* infections through semi-targeted and targeted metabolomics (Suriyanarayanan et al., 2024).

Selected ion flow tube-mass spectroscopy (SIFT-MS), a technique recognized for the identification and quantification of trace gases as low as in the range of parts-per-trillion by volume in real time. This method offers several advantages such as minimal samples, especially volatile samples, high sensitivity, specificity, high repeatability, and simultaneous detection of wide range of samples. In a study, in vitro collagen wound biofilm model of *Pseudomonas aeruginosa* was investigated using SIFT-MS to detect ammonia and hydrogen cyanide production in real time. Around 12 h of incubation, the level of ammonia and HCN was at its extreme level of 2273 ppb and 138 ppb, respectively, and are noticeable throughout the 72 h of sampling. Hence, detection of these compounds can be considered as a marker for the rapid diagnosis of *P. aeruginosa* biofilm (Slade et al., 2019).

14.6 Limitations of Omics Approaches

With the help of omics, differentiation of two forms of microbial form, that is, planktonic and biofilm, specifically in terms of matrix components, structure and physiological properties can be studied. Omics such as transcriptomics, proteomics, and metabolomics can deliver idea about the gene expression, proteins, and metabolites produced in a living system. However, these approaches are not connected for the evaluation of data. Coupling of omics approaches provides a clear picture behind the mechanism of biofilm development, and the biofilm infection (Azevedo et al., 2009). Additionally, the instruments commonly used in omics are quite expensive, considered as a great challenge for the study. In transcriptome analyses, RNA-sequencing explores the differential expression of genes in two biological conditions. Microarrays are developed for the expression levels of large numbers of transcriptomes, possibly thousands of samples simultaneously. This method provides expression levels on the basis of high-throughput and semi quantitative analysis of fluorescent signals. However, microarray technique has limitations in the hybridization process that includes cross-hybridization, dye binding variations, and background signal interference. Due to this issue, the quantification of mRNA may not be accurate and difficult in quantifying least available samples as the fluorescent signals are not enough to be detected from the background noise (Muszkieta, 2013). Hence, Microarrays are not widely used in microbiology because of complexity of the procedure and high cost for an experiment (Gerace et al., 2022). Proteomics delivers insights into the proteomes associated with host–pathogen interactions, and pathogenic profiles of the microorganisms; moreover, it explains about the antimicrobial resistance associated with biofilm, helps in designing of novel drugs; distinguishing biofilm forms from the planktonic forms in terms of physiology. The coupling of proteomics with MS extends its application in the diagnosis of diseases linked with microbial biofilms and its interaction with the host (Magana et al., 2018). Although, proteomics technology offers numerous advantages in the biological studies, it has some limitations. First one is the low abundance of protein makes the detection difficult, especially in the dental caries. High-protein content in the saliva and plaques interferes with the recognition of low abundance proteins, which affects the results. Second one is the post-translational modifications alters protein function. Identification and characterization of these protein modifications are still challenging for proteomic technology. Third one is data analysis and interpretation, requires specialized and intensive tools, and also proficiency in data analysis. The dental biofilm sample collection and processing without any contamination or metabolic loss is still a challenging process in metabolomics. Moreover, diverseness, rapid metabolic turnover, and chemical properties of the metabolites are challenging in metabolomic studies. Data interpretation and understanding may be complicated due to large datasets (Zhang et al., 2024).

14.7 Conclusion

Several high-throughput technologies were developed recently for the analysis of varying aspects of cellular functions, includes transcriptome, proteome, and metabolomics. Integrative analyses of different stages of cellular processes using omics approaches provide a better understanding of the functioning and mechanism of the biological process. However, the comparison of data from the combination studies is still challenging to understand. Transcriptomics provides information about the gene expression of the target. Proteomic analysis offers the identification of proteins that are the end product of gene expression. Nevertheless, the protein abundance may be limited due to its high variability and complications in detecting membrane-associated proteins. Metabolomics approaches provide information about the metabolites produced in biofilm. However, our current knowledge about biofilm, developmental processes, metabolic modifications, and variations in expression levels during diseased or normal states is not adequate to tackle biofilm and associated infections. The area of biofilm research should be explored much more for the detection and diagnosis of biofilm and also in the development of novel therapeutic agents.

References

Abreu, A. C., & Fernández, I. (2020). Nuclear magnetic resonance to study bacterial biofilm's structure, formation, and resilience. In *Recent trends in biofilm science and technology* (pp. 23–70). Elsevier.

Achek, R., Hotzel, H., Nabi, I., Kechida, S., Mami, D., Didouh, N., Tomaso, H., Neubauer, H., Ehricht, R., Monecke, S., & El-Adawy, H. (2020). Phenotypic and molecular detection of biofilm formation in Staphylococcus aureus isolated from different sources in Algeria. *Pathogens, 9*(2), 153. https://doi.org/10.3390/pathogens9020153

Ali Mohammed, M. M., Pettersen, V. K., Nerland, A. H., Wiker, H. G., & Bakken, V. (2021). Label-free quantitative proteomic analysis of the oral bacteria Fusobacterium nucleatum and Porphyromonas gingivalis to identify protein features relevant in biofilm formation. *Anaerobe, 72*, 102449. https://doi.org/10.1016/j.anaerobe.2021.102449

Almatroudi, A. (2024). Investigating biofilms: Advanced methods for comprehending microbial behavior and antibiotic resistance. *Frontiers in Bioscience-Landmark, 29*(4). https://doi.org/10.31083/j.fbl2904133

Ammons, M. C. B., Tripet, B. P., Carlson, R. P., Kirker, K. R., Gross, M. A., Stanisich, J. J., & Copié, V. (2014). Quantitative NMR metabolite profiling of methicillin-resistant and methicillin-susceptible Staphylococcus aureus discriminates between biofilm and planktonic phenotypes. *Journal of Proteome Research, 13*(6), 2973–2985. https://doi.org/10.1021/pr500120c

Awadh, A. A., Le Gresley, A., Forster-Wilkins, G., Kelly, A. F., & Fielder, M. D. (2021). Determination of metabolic activity in planktonic and biofilm cells of Mycoplasma fermentans and Mycoplasma pneumoniae by nuclear magnetic resonance. *Scientific Reports, 11*(1), 5650. https://doi.org/10.1038/s41598-021-84326-2

Azevedo, N. F., Lopes, S. P., Keevil, C. W., Pereira, M. O., & Vieira, M. J. (2009). Time to "go large" on biofilm research: Advantages of an omics approach. *Biotechnology Letters, 31*(4), 477–485. https://doi.org/10.1007/s10529-008-9901-4

Bao, K., Bostanci, N., Thurnheer, T., & Belibasakis, G. N. (2017). Proteomic shifts in multi-species oral biofilms caused by Anaeroglobus geminatus. *Scientific Reports, 7*(1), 4409. https://doi.org/10.1038/s41598-017-04594-9

Beale, D. J., Barratt, R., Marlow, D. R., Dunn, M. S., Palombo, E. A., Morrison, P. D., & Key, C. (2013). Application of metabolomics to understanding biofilms in water distribution systems: A pilot study. *Biofouling, 29*(3), 283–294. https://doi.org/10.1080/08927014.2013.772140

Beale, D. J., Pinu, F. R., Kouremenos, K. A., Poojary, M. M., Narayana, V. K., Boughton, B. A., Kanojia, K., Dayalan, S., Jones, O. A. H., & Dias, D. A. (2018). Review of recent developments in GC–MS approaches to metabolomics-based research. *Metabolomics, 14*(11), 152. https://doi.org/10.1007/s11306-018-1449-2

Bitencourt, T., Nogueira, F., Jenull, S., Phan-Canh, T., Tscherner, M., Kuchler, K., & Lion, T. (2024). Integrated multi-omics identifies pathways governing interspecies interaction between A. fumigatus and K. pneumoniae. *Communications Biology, 7*(1), 1496. https://doi.org/10.1038/s42003-024-07145-x

Bottagisio, M., Soggiu, A., Piras, C., Bidossi, A., Greco, V., Pieroni, L., Bonizzi, L., Roncada, P., & Lovati, A. B. (2019). Proteomic analysis reveals a biofilm-like behavior of planktonic aggregates of Staphylococcus epidermidis grown under environmental pressure/stress. *Frontiers in Microbiology, 10*. https://doi.org/10.3389/fmicb.2019.01909

Chen, X., Wei, S., Ji, Y., Guo, X., & Yang, F. (2015). Quantitative proteomics using SILAC: Principles, applications, and developments. *Proteomics, 15*(18), 3175–3192. https://doi.org/10.1002/pmic.201500108

Cruz, A., Condinho, M., Carvalho, B., Arraiano, C. M., Pobre, V., & Pinto, S. N. (2021). The two weapons against bacterial biofilms: Detection and treatment. *Antibiotics, 10*(12), 1482. https://doi.org/10.3390/antibiotics10121482

de Sá, M. C. A., da Silva, W. M., Rodrigues, C. C. S., Rezende, C. P., Marchioro, S. B., Rocha Filho, J. T. R., Sousa, T. D. J., de Oliveira, H. P., da Costa, M. M., Figueiredo, H. C. P., Portela, R. D., Castro, T. L. D. P., Azevedo, V., Seyffert, N., & Meyer, R. (2021). Comparative proteomic analyses between biofilm-forming and non-biofilm-forming strains of Corynebacterium pseudotuberculosis isolated from goats. *Frontiers in Veterinary Science, 8*. https://doi.org/10.3389/fvets.2021.614011

Deng, W., Zhou, C., Qin, J., Jiang, Y., Li, D., Tang, X., Luo, J., Kong, J., & Wang, K. (2024). Molecular mechanisms of DNase inhibition of early biofilm formation Pseudomonas aeruginosa or Staphylococcus aureus: A transcriptome analysis. *Biofilms, 7*, 100174. https://doi.org/10.1016/j.bioflm.2023.100174

Depke, T., Thöming, J. G., Kordes, A., Häussler, S., & Brönstrup, M. (2020). Untargeted LC-MS metabolomics differentiates between virulent and avirulent clinical strains of Pseudomonas aeruginosa. *Biomolecules, 10*(7), 1041. https://doi.org/10.3390/biom10071041

Dutta, B., Chatterjee, D., Sarkar, N., Lahiri, D., Nag, M., & Ray, R. R. (2024). Multi-omics technology in detection of multispecies biofilm. *The Microbe, 4*, 100128. https://doi.org/10.1016/j.microb.2024.100128

Ellepola, K., Truong, T., Liu, Y., Lin, Q., Lim, T. K., Lee, Y. M., Cao, T., Koo, H., & Seneviratne, C. J. (2019). Multi-omics analyses reveal synergistic carbohydrate metabolism in Streptococcus mutans-Candida albicans mixed-species biofilms. *Infection and Immunity, 87*(10). https://doi.org/10.1128/IAI.00339-19

Favre, L., Ortalo-Magné, A., Greff, S., Pérez, T., Thomas, O. P., Martin, J.-C., & Culioli, G. (2017). Discrimination of four marine biofilm-forming bacteria by LC–MS metabolomics and influence of culture parameters. *Journal of Proteome Research, 16*(5), 1962–1975. https://doi.org/10.1021/acs.jproteome.6b01027

Favre, L., Ortalo-Magné, A., Pichereaux, C., Gargaros, A., Burlet-Schiltz, O., Cotelle, V., & Culioli, G. (2018). Metabolome and proteome changes between biofilm and planktonic phenotypes of the marine bacterium Pseudoalteromonas lipolytica TC8. *Biofouling, 34*(2), 132–148. https://doi.org/10.1080/08927014.2017.1413551

França, A., Pier, G. B., Vilanova, M., & Cerca, N. (2016). Transcriptomic analysis of Staphylococcus epidermidis biofilm-released cells upon interaction with human blood circulating immune cells and soluble factors. *Frontiers in Microbiology, 7*. https://doi.org/10.3389/fmicb.2016.01143

Galdiero, E., Salvatore, M. M., Maione, A., de Alteriis, E., Andolfi, A., Salvatore, F., & Guida, M. (2021). GC-MS-based metabolomics study of single- and dual-species biofilms of Candida albicans and Klebsiella pneumoniae. *International Journal of Molecular Sciences, 22*(7), 3496. https://doi.org/10.3390/ijms22073496

Gant, M. S., & Chamot-Rooke, J. (2024). Present and future perspectives on mass spectrometry for clinical microbiology. *Microbes and Infection, 26*(7), 105296. https://doi.org/10.1016/j.micinf.2024.105296

Gerace, E., Mancuso, G., Midiri, A., Poidomani, S., Zummo, S., & Biondo, C. (2022). Recent advances in the use of molecular methods for the diagnosis of bacterial infections. *Pathogens, 11*(6), 663. https://doi.org/10.3390/pathogens11060663

Gjersing, E. L., Herberg, J. L., Horn, J., Schaldach, C. M., & Maxwell, R. S. (2007). NMR metabolomics of planktonic and biofilm modes of growth in Pseudomonas aeruginosa. *Analytical Chemistry, 79*(21), 8037–8045. https://doi.org/10.1021/ac070800t

Graham, R. L., Graham, C., & McMullan, G. (2007). Microbial proteomics: A mass spectrometry primer for biologists. *Microbial Cell Factories, 6*(1), 26. https://doi.org/10.1186/1475-2859-6-26

Grocholska, P., Kowalska, M., & Bąchor, R. (2023). Qualitative and quantitative mass spectrometry in salivary metabolomics and proteomics. *Metabolites, 13*(2), 155. https://doi.org/10.3390/metabo13020155

Guo, R., Luo, X., Liu, J., & Lu, H. (2021). Mass spectrometry based targeted metabolomics precisely characterized new functional metabolites that regulate biofilm formation in Escherichia coli. *Analytica Chimica Acta, 1145*, 26–36. https://doi.org/10.1016/j.aca.2020.12.021

Harrington, N. E., Littler, J. L., & Harrison, F. (2022). Transcriptome analysis of Pseudomonas aeruginosa biofilm infection in an ex vivo pig model of the cystic fibrosis lung. *Applied and Environmental Microbiology, 88*(3). https://doi.org/10.1128/aem.01789-21

Hashino, E., Kuboniwa, M., Alghamdi, S. A., Yamaguchi, M., Yamamoto, R., Cho, H., & Amano, A. (2013). Erythritol alters microstructure and metabolomic profiles of biofilm composed of S treptococcus gordonii and P orphyromonas gingivalis. *Molecular Oral Microbiology, 28*(6), 435–451. https://doi.org/10.1111/omi.12037

Herrmann, H., Kern, J., Kern, T., Lautensack, J., Conrads, G., & Wolfart, S. (2020). Early and mature biofilm on four different dental implant materials: An in vivo human study. *Clinical Oral Implants Research, 31*(11), 1094–1104. https://doi.org/10.1111/clr.13656

Imdahl, F., & Saliba, A.-E. (2020). Advances and challenges in single-cell RNA-seq of microbial communities. *Current Opinion in Microbiology, 57*, 102–110. https://doi.org/10.1016/j.mib.2020.10.001

Islam, N., Kim, Y., Ross, J. M., & Marten, M. R. (2014). Proteomic analysis of Staphylococcus aureus biofilm cells grown under physiologically relevant fluid shear stress conditions. *Proteome Science, 12*(1), 21. https://doi.org/10.1186/1477-5956-12-21

Jiang, R., Xiang, M., Chen, W., Zhang, P., Wu, X., Zhu, G., Tu, T., Jiang, D., Yao, X., Luo, Y., Yang, Z., Chen, D., & Wang, Y. (2021). Biofilm characteristics and transcriptomic analysis of Haemophilus parasuis. *Veterinary Microbiology, 258*, 109073. https://doi.org/10.1016/j.vetmic.2021.109073

Joshi, R. V., Gunawan, C., & Mann, R. (2021). We are one: Multispecies metabolism of a biofilm consortium and their treatment strategies. *Frontiers in Microbiology, 12*. https://doi.org/10.3389/fmicb.2021.635432

Kalpana, S., Lin, W.-Y., Wang, Y.-C., Fu, Y., Lakshmi, A., & Wang, H.-Y. (2023). Antibiotic resistance diagnosis in ESKAPE pathogens—A review on proteomic perspective. *Diagnostics, 13*(6), 1014. https://doi.org/10.3390/diagnostics13061014

Katsarou, E. I., Billinis, C., Galamatis, D., Fthenakis, G. C., Tsangaris, G. T., & Katsafadou, A. I. (2021). Applied Proteomics in 'One Health.'. *Proteomes, 9*(3), 31. https://doi.org/10.3390/proteomes9030031

Kavela, S., Vyas, P., Cp, J., Kushwaha, S. K., Majumdar, S. S., & Faisal, S. M. (2023). Use of an integrated multi-omics approach to identify molecular mechanisms and critical factors involved in the pathogenesis of Leptospira. *Microbiology Spectrum, 11*(2). https://doi.org/10.1128/spectrum.03135-22

Kok, M., Maton, L., van der Peet, M., Hankemeier, T., & van Hasselt, J. G. C. (2022). Unraveling antimicrobial resistance using metabolomics. *Drug Discovery Today, 27*(6), 1774–1783. https://doi.org/10.1016/j.drudis.2022.03.015

Korshoj, L. E., & Kielian, T. (2024). Bacterial single-cell RNA sequencing captures biofilm transcriptional heterogeneity and differential responses to immune pressure. *Nature Communications, 15*(1), 10184. https://doi.org/10.1038/s41467-024-54581-8

Kouremenos, K. A., Beale, D. J., Antti, H., & Palombo, E. A. (2014). Liquid chromatography time of flight mass spectrometry based environmental metabolomics for the analysis of Pseudomonas putida Bacteria in potable water. *Journal of Chromatography B, 966*, 179–186. https://doi.org/10.1016/j.jchromb.2014.02.058

Kumar, D., & Kumar, A. (2023). High throughput bioanalytical techniques for elucidation of Candida albicans biofilm architecture and metabolome. *Rendiconti Lincei. Scienze Fisiche e Naturali, 34*(1), 117–129. https://doi.org/10.1007/s12210-022-01115-3

Leggett, A., Li, D.-W., Bruschweiler-Li, L., Sullivan, A., Stoodley, P., & Brüschweiler, R. (2022). Differential metabolism between biofilm and suspended Pseudomonas aeruginosa cultures in bovine synovial fluid by 2D NMR-based metabolomics. *Scientific Reports, 12*(1), 17317. https://doi.org/10.1038/s41598-022-22127-x

Lin, Y., Liang, X., Li, Z., Gong, T., Ren, B., Li, Y., & Peng, X. (2024). Omics for deciphering oral microecology. *International Journal of Oral Science, 16*(1), 2. https://doi.org/10.1038/s41368-023-00264-x

Magana, M., Sereti, C., Ioannidis, A., Mitchell, C. A., Ball, A. R., Magiorkinis, E., Chatzipanagiotou, S., Hamblin, M. R., Hadjifrangiskou, M., & Tegos, G. P. (2018). Options and limitations in clinical investigation of bacterial biofilms. *Clinical Microbiology Reviews, 31*(3). https://doi.org/10.1128/CMR.00084-16

Malviya, J., Alameri, A. A., Al-Janabi, S. S., Fawzi OF, Azzawi, A. L., Obaid, R. F., Alsudani, A. A., Alkhayyat, A. S., Gupta, J., Mustafa, Y. F., Karampoor, S., & Mirzaei, R. (2023). Metabolomic profiling of bacterial biofilm: Trends, challenges, and an emerging antibiofilm target. *World Journal of Microbiology and Biotechnology, 39*(8), 212. https://doi.org/10.1007/s11274-023-03651-y

Mir, D. H., & Rather, M. A. (2024). Advantages and limitations of the biofilm study methods. *Applied Biochemistry and Microbiology, 60*(2), 264–279. https://doi.org/10.1134/S000368382402011X

Mirzaei, R., Campoccia, D., Ravaioli, S., & Arciola, C. R. (2024). Emerging issues and initial insights into bacterial biofilms: From orthopedic infection to metabolomics. *Antibiotics, 13*(2), 184. https://doi.org/10.3390/antibiotics13020184

Moon, J.-H., Roh, D.-H., Kwack, K. H., & Lee, J.-H. (2023). Bacterial single-cell transcriptomics: Recent technical advances and future applications in dentistry. *Japanese Dental Science Review, 59*, 253–262. https://doi.org/10.1016/j.jdsr.2023.08.001

Moreno, C., Bybee, E., Tellez Freitas, C. M., Pickett, B. E., & Weber, K. S. (2022). Meta-analysis of two human RNA-seq datasets to determine periodontitis diagnostic biomarkers and drug target candidates. *International Journal of Molecular Sciences, 23*(10), 5580. https://doi.org/10.3390/ijms23105580

Muszkieta, L. (2013). Investigation of Aspergillus fumigatus biofilm formation by various "omics" approaches. *Frontiers in Microbiology, 4*. https://doi.org/10.3389/fmicb.2013.00013

Nguyen, T., Sedghi, L., Ganther, S., Malone, E., Kamarajan, P., & Kapila, Y. L. (2020). Host-microbe interactions: Profiles in the transcriptome, the proteome, and the metabolome. *Periodontology 2000, 82*(1), 115–128. https://doi.org/10.1111/prd.12316

Panwar, S., Duggirala, K. S., Yadav, P., Debnath, N., Yadav, A. K., & Kumar, A. (2023). Advanced diagnostic methods for identification of bacterial foodborne pathogens: Contemporary and upcoming challenges. *Critical Reviews in Biotechnology, 43*(7), 982–1000. https://doi.org/10.1080/07388551.2022.2095253

Phillips, N. J., Steichen, C. T., Schilling, B., Post, D. M. B., Niles, R. K., Bair, T. B., Falsetta, M. L., Apicella, M. A., & Gibson, B. W. (2012). Proteomic analysis of Neisseria gonorrhoeae biofilms shows shift to anaerobic respiration and changes in nutrient transport and outermembrane proteins. *PLoS One, 7*(6), e38303. https://doi.org/10.1371/journal.pone.0038303

Piras, C., Di Ciccio, P. A., Soggiu, A., Greco, V., Tilocca, B., Costanzo, N., Ceniti, C., Urbani, A., Bonizzi, L., Ianieri, A., & Roncada, P. (2021). S. aureus biofilm protein expression linked to antimicrobial resistance: A proteomic study. *Animals, 11*(4), 966. https://doi.org/10.3390/ani11040966

Qin, N., Tan, X., Jiao, Y., Liu, L., Zhao, W., Yang, S., & Jia, A. (2014). RNA-Seq-based transcriptome analysis of methicillin-resistant Staphylococcus aureus biofilm inhibition by ursolic acid and resveratrol. *Scientific Reports, 4*(1), 5467. https://doi.org/10.1038/srep05467

Rahman, M. A., Amirkhani, A., Parvin, F., Chowdhury, D., Molloy, M. P., Deva, A. K., Vickery, K., & Hu, H. (2022). One step forward with dry surface biofilm (DSB) of Staphylococcus aureus: TMT-based quantitative proteomic analysis reveals proteomic shifts between DSB and hydrated biofilm. *International Journal of Molecular Sciences, 23*(20), 12238. https://doi.org/10.3390/ijms232012238

Rizal, M. I., Soeroso, Y., Sulijaya, B., Assiddiq, B. F., Bachtiar, E. W., & Bachtiar, B. M. (2020). Proteomics approach for biomarkers and diagnosis of periodontitis: Systematic review. *Heliyon, 6*(6), e04022. https://doi.org/10.1016/j.heliyon.2020.e04022

Rubio-Mendoza, D., Córdova-Fletes, C., Martínez-Meléndez, A., Morfín-Otero, R., Maldonado-Garza, H. J., & Garza-González, E. (2023). Transcriptomic analysis of biofilm formation in strains of Clostridioides difficile associated with recurrent and non-recurrent infection reveals potential candidate markers for recurrence. *PLoS One, 18*(8), e0289593. https://doi.org/10.1371/journal.pone.0289593

Sadiq, F. A., Yan, B., Zhao, J., Zhang, H., & Chen, W. (2020). Untargeted metabolomics reveals metabolic state of Bifidobacterium bifidum in the biofilm and planktonic states. *LWT, 118*, 108772. https://doi.org/10.1016/j.lwt.2019.108772

Sánchez, M. C., Velapatiño, A., Llama-Palacios, A., Valdés, A., Cifuentes, A., Ciudad, M. J., & Collado, L. (2022). Metataxonomic and metabolomic evidence of biofilm homeostasis disruption related to caries: An in vitro study. *Molecular Oral Microbiology, 37*(2), 81–96. https://doi.org/10.1111/omi.12363

Seneviratne, C. J., Suriyanarayanan, T., Widyarman, A. S., Lee, L. S., Lau, M., Ching, J., Delaney, C., & Ramage, G. (2020). Multi-omics tools for studying microbial biofilms: Current perspectives and future directions. *Critical Reviews in Microbiology, 46*(6), 759–778. https://doi.org/10.1080/1040841X.2020.1828817

Sharma, S., Mohler, J., Mahajan, S. D., Schwartz, S. A., Bruggemann, L., & Aalinkeel, R. (2023). Microbial biofilm: A review on formation, infection, antibiotic resistance, control measures, and innovative treatment. *Microorganisms, 11*(6), 1614. https://doi.org/10.3390/microorganisms11061614

Shemesh, M., Tam, A., Kott-Gutkowski, M., Feldman, M., & Steinberg, D. (2008). DNA-microarrays identification of Streptococcus mutans genes associated with biofilm thickness. *BMC Microbiology, 8*(1), 236. https://doi.org/10.1186/1471-2180-8-236

Shiny Matilda, C., Madhusudan, I., Gaurav Isola, R., & Shanthi, C. (2020). Potential of proteomics to probe microbes. *Journal of Basic Microbiology, 60*(6), 471–483. https://doi.org/10.1002/jobm.201900628

Silva, N. B. S., Marques, L. A., & Röder, D. D. B. (2021). Diagnosis of biofilm infections: Current methods used, challenges and perspectives for the future. *Journal of Applied Microbiology, 131*(5), 2148–2160. https://doi.org/10.1111/jam.15049

Slade, E. A., Thorn, R. M. S., Young, A., & Reynolds, D. M. (2019). An in vitro collagen perfusion wound biofilm model; with applications for antimicrobial studies and microbial metabolomics. *BMC Microbiology, 19*(1), 310. https://doi.org/10.1186/s12866-019-1682-5

Sousa, L. G. V., Novak, J., França, A., Muzny, C. A., & Cerca, N. (2024). Gardnerella vaginalis, Fannyhessea vaginae, and Prevotella bivia strongly influence each other's transcriptome in triple-species biofilms. *Microbial Ecology, 87*(1), 117. https://doi.org/10.1007/s00248-024-02433-9

Spatz, S., & Afonso, C. L. (2024). Non-targeted RNA sequencing: Towards the development of universal clinical diagnosis methods for human and veterinary infectious diseases. *Veterinary Sciences, 11*(6), 239. https://doi.org/10.3390/vetsci11060239

Stipetic, L. H., Dalby, M. J., Davies, R. L., Morton, F. R., Ramage, G., & Burgess, K. E. V. (2016). A novel metabolomic approach used for the comparison of Staphylococcus aureus planktonic cells and biofilm samples. *Metabolomics, 12*(4), 75. https://doi.org/10.1007/s11306-016-1002-0

Suriyanarayanan, T., Lee, L. S., Han, S. H. Y., Ching, J., & Seneviratne, C. J. (2024). Targeted metabolomics analysis approach to unravel the biofilm formation pathways of Enterococcus faecalis clinical isolates. *International Endodontic Journal, 57*(10), 1505–1520. https://doi.org/10.1111/iej.14110

Tian, L., & Wang, L. (2021). Multi-omics analysis reveals structure and function of biofilm microbial communities in a pre-denitrification biofilter. *Science of the Total Environment, 757*, 143908. https://doi.org/10.1016/j.scitotenv.2020.143908

Torres-Sangiao, E., Leal Rodriguez, C., & García-Riestra, C. (2021). Application and perspectives of MALDI–TOF mass spectrometry in clinical microbiology laboratories. *Microorganisms, 9*(7), 1539. https://doi.org/10.3390/microorganisms9071539

Torres-Sangiao, E., Giddey, A. D., Leal Rodriguez, C., Tang, Z., Liu, X., & Soares, N. C. (2022). Proteomic approaches to unravel mechanisms of antibiotic resistance and immune evasion of bacterial pathogens. *Frontiers in Medicine (Lausanne), 9.* https://doi.org/10.3389/fmed.2022.850374

Tsuchida, S., Satoh, M., Takiwaki, M., & Nomura, F. (2018). Current status of proteomic technologies for discovering and identifying gingival crevicular fluid biomarkers for periodontal disease. *International Journal of Molecular Sciences, 20*(1), 86. https://doi.org/10.3390/ijms20010086

Vaz, C., Pitarch, A., Gómez-Molero, E., Amador-García, A., Weig, M., Bader, O., Monteoliva, L., & Gil, C. (2021). Mass spectrometry-based proteomic and immunoproteomic analyses of the Candida albicans hyphal secretome reveal diagnostic biomarker candidates for invasive candidiasis. *Journal of Fungi, 7*(7), 501. https://doi.org/10.3390/jof7070501

Vohra, M., Kour, A., Kalia, N. P., Kumar, M., Sharma, S., Jaglan, S., Kamath, N., & Sharma, S. (2024). A comprehensive review of genomics, transcriptomics, proteomics, and metabolomic insights into the differentiation of Pseudomonas aeruginosa from the planktonic to biofilm state: A multi-omics approach. *International Journal of Biological Macromolecules, 257*, 128563. https://doi.org/10.1016/j.ijbiomac.2023.128563

Wang, Y., Yi, L., Wu, Z., Shao, J., Liu, G., Fan, H., Zhang, W., & Lu, C. (2012). Comparative proteomic analysis of Streptococcus suis biofilms and planktonic cells that identified biofilm infection-related immunogenic proteins. *PLoS One, 7*(4), e33371. https://doi.org/10.1371/journal.pone.0033371

Wang, C., Zhang, Q., Wang, Y., Tang, X., An, Y., Li, S., Xu, H., Li, Y., Luan, W., Wang, X., Liu, M., & Yu, L. (2019). Comparative proteomics analysis between biofilm and planktonic cells of Mycobacterium tuberculosis. *Electrophoresis, 40*(20), 2736–2746. https://doi.org/10.1002/elps.201900030

Washio, J., & Takahashi, N. (2016). Metabolomic studies of oral biofilm, oral cancer, and beyond. *International Journal of Molecular Sciences, 17*(6), 870. https://doi.org/10.3390/ijms17060870

Weidt, S., Haggarty, J., Kean, R., Cojocariu, C. I., Silcock, P. J., Rajendran, R., Ramage, G., & Burgess, K. E. V. (2016). A novel targeted/untargeted GC-Orbitrap metabolomics methodology applied to Candida albicans and Staphylococcus aureus biofilms. *Metabolomics, 12*(12), 189. https://doi.org/10.1007/s11306-016-1134-2

Yang, C., Lu, L., Liao, L., Zhang, B., Zeng, M., Zou, K., Liu, X., & Zhang, M. (2021). Establishment of GC–MS method for the determination of Pseudomonas aeruginosa biofilm and its application in metabolite enrichment analysis. *Journal of Chromatography B, 1179*, 122839. https://doi.org/10.1016/j.jchromb.2021.122839

Yergeau, E., Lawrence, J. R., Waiser, M. J., Korber, D. R., & Greer, C. W. (2010). Metatranscriptomic analysis of the response of river biofilms to pharmaceutical products, using anonymous DNA microarrays. *Applied and Environmental Microbiology, 76*(16), 5432–5439. https://doi.org/10.1128/AEM.00873-10

Yuan, L., Dai, H., He, G., Yang, Z., & Jiao, X. (2023). Invited review: Current perspectives for analyzing the dairy biofilms by integrated multiomics. *Journal of Dairy Science, 106*(12), 8181–8192. https://doi.org/10.3168/jds.2023-23306

Zafer, M. M., Mohamed, G. A., Ibrahim, S. R. M., Ghosh, S., Bornman, C., & Elfaky, M. A. (2024). Biofilm-mediated infections by multidrug-resistant microbes: A comprehensive exploration and forward perspectives. *Archives of Microbiology, 206*(3), 101. https://doi.org/10.1007/s00203-023-03826-z

Zhang, B., & Powers, R. (2012). Analysis of bacterial biofilms using NMR-based metabolomics. *Future Medicinal Chemistry, 4*(10), 1273–1306. https://doi.org/10.4155/fmc.12.59

Zhang, T., Liu, Z., Wang, H., Zhang, H., Li, H., Lu, W., & Zhu, J. (2023). Multi-omics analysis reveals genes and metabolites involved in Bifidobacterium pseudocatenulatum biofilm formation. *Frontiers in Microbiology, 14*. https://doi.org/10.3389/fmicb.2023.1287680

Zhang, J. S., Huang, S., Chen, Z., Chu, C.-H., Takahashi, N., & Yu, O. Y. (2024). Application of omics technologies in cariology research: A critical review with bibliometric analysis. *Journal of Dentistry, 141*, 104801. https://doi.org/10.1016/j.jdent.2023.104801

Zhao, A., Sun, J., & Liu, Y. (2023). Understanding bacterial biofilms: From definition to treatment strategies. *Frontiers in Cellular and Infection Microbiology, 13*. https://doi.org/10.3389/fcimb.2023.1137947

Zubair, M., Wang, J., Yu, Y., Faisal, M., Qi, M., Shah, A. U., Feng, Z., Shao, G., Wang, Y., & Xiong, Q. (2022). Proteomics approaches: A review regarding an importance of proteome analyses in understanding the pathogens and diseases. *Frontiers in Veterinary Science, 9*. https://doi.org/10.3389/fvets.2022.1079359

15 Omics Approaches in Diagnosis of Polymicrobial Biofilms and Identification of Novel Therapeutic Drugs

Santhi Latha Pandrangi, Prasanthi Chittineedi, and Suseela Lanka

15.1 Introduction

Biofilms belong to the syntropic community of microorganisms that have tendency to adhere on one another and even on the surface to form three-dimensional structures. Once adherent, these biofilms embed into the slimy extracellular matrix composing extracellular polymeric substances (EPS) which contain extracellular DNA, polysaccharides, lipids, and proteins. Due to the presence of 3D structure and community lifestyle, these biofilms are described as "cities for microbes" (Zhao et al., 2023).

Polymicrobial biofilm (PMBFs) is a very interesting and complex dynamic community of microorganisms that includes bacteria, fungi, archaea, protozoa, and yeast. These species are involved in cell–cell adhesion, quorum sensing, and metabolite exchange (Gabrilska & Rumbaugh, 2015), leading to the development of resistance to antimicrobial agents making them more virulent and leading to sustain even in high environmental stress conditions. These communities further exhibit coaggregation, wherein they stick to each other and express multiple genes dynamically, making them responsive to change (Peters et al., 2012).

Although PMBFs have numerous applications among various fields, their intricate nature in the formation of biofilms, their interaction with other microbial communities, long-term persistence, identification of new therapeutic targets from PMBFs, etc., has not been fully understood. However, these concepts could be deciphered through OMICS technologies. The current chapter is focused on elucidating the current state of omics research and the latest advancements in genomics,

S. L. Pandrangi · P. Chittineedi
Onco-Stem Cell Research Laboratory, Department of Life Sciences, GITAM School of Science, GITAM Deemed to be University, Visakhapatnam, Andhra Pradesh, India

S. Lanka (✉)
Department of Biosciences and Biotechnology, Krishna University, Machilipatnam, Andhra Pradesh, India

S. Busi et al. (eds.), *Omics Approaches in Biofilm Research*,
https://doi.org/10.1007/978-3-031-91863-6_15

transcriptomics, proteomics, and metabolic and their applications in understanding PMBFs formation, highlighting the potential aspects of these technologies to combat recalcitrant infections, thereby leading to better patient survival outcomes.

15.2 Driving Forces Behind PMBF Formation

The infectious microorganism entered into a host organisms face critical stress due to various factors, such as activate immune system, antimicrobial activity of the antibiotics, and limited availability of required nutrients which effect microbial growth triggered antioxidant mechanisms. To overcome this and to adapt to the environmental stress the microbes transform themselves into PMBFs (Zachary, 2017). The formed PMBFs protect the microbes from host immune system by evading the immune molecules through interspecies communication leading to the development of antibiotic resistance.

When compared with mono microbial biofilms (MMBFs), the rate of formation of PMBFs is quiet high and is associated with increased biofilm thickness, biomass, and higher exopolysaccharide content. Additionally, PMBFs exhibit greater elastic modulus, compressive strength, and rigidity, making them more resistant to shear stresses (Rather et al., 2021). For instance, *Candida albicans* and *Staphylococcus aureus* co-infection enhance antibiotic resistance when compared with infection caused by a single organism. Similarly, *Streptococcus mutans* and *Candida albicans*; *Haemophilus somoni* and *Pastourelle multocida* PMBFs increase biomass and exopolysaccharide content and exhibit increased biofilm thickness and biomass, respectively (Lohse et al., 2018). Hence, understanding the differences between mono- and polymicrobial biofilms is crucial for developing effective therapeutic approaches. Elucidating the complex interactions within PMBFs will inform the design of novel antimicrobial strategies, improving treatment outcomes for PMBF-related infections.

15.2.1 Underlying Mechanism of Developing Polymicrobial Infections

The main source of acquiring polymicrobial infections is due to unhygienic practices. Most of which are due to unsterilized environments during surgeries. Matthew P. Cheng et al. (2016) has reported a case presentation of a 66-year-old male who have diagnosed with polymicrobial infection after 10 days of surgery for olecranon fracture. The patient developed green discharge from the site of surgical stitches. The green discharge was found to be *Terrisporobacter glycolicus/mayombei*, a polymicrobial infection (Cheng et al., 2016). Similar kinds of case presentations were observed in 2020 by Markus Rupp et al. They performed a descriptive study in patients who underwent surgery for long bone fractures and non-union from 2010 to 2018. PCR and microbial cultures were performed on 42 patients who were recruited in the study before the surgery and during the follow-up. They observed that 14.3% of the study groups developed polymicrobial infections during their recovery time among which *Staphylococcus aureus* (86.4% of patients) and *Enterococcus* spp. were most commonly detected (Rupp et al., 2020).

15.3 Detection of Multi-Species Biofilms

Due to their remarkable resistance to antibiotic treatments, complex microbial communities, identification of mutli-species biofilms is essential for identifying novel strategies for early detection of resilient biofilm formations. PMBF-associated infections differ from normal microbial infections through the presence of distressing symptoms, such as persistent fever, compromised immune system-associated inflammatory reactions with severe pain. However, one cannot rely on symptom-based diagnosis as it may lead to misdiagnosis or delayed treatment. However, this could be controlled by developing more accurate diagnostic procedures (Zachary, 2017).

The European Society for Clinical Microbiology and Infectious Disease Journal highlights several advanced diagnostic approaches for detecting microbial colonies and inflammatory cells (Martin-Loeches et al., 2023). One notable technique is polymerase chain reaction (PCR), which stands out for its rapid and reliable results, ensuring the detection of ultralow concentrations. Electron microscopy, on the other hand, delivers comprehensive visual insights, enabling researchers to closely examine the ultrafine structures of microorganisms in detail. Conventional techniques, such as agar planting and incubation, face major challenges in accurately detecting specific organisms that form biofilms—communities of microorganisms that adhere to surfaces and are encased in a protective matrix. It offers the detection of biofilm-related genes through specific primer sequences. When the DNA isolated from a patient is subjected to a PCR reaction mixture and biofilm-specific gene primers, the DNA would be amplified with the help of primers leading to the detection of a specific disease along with fold change. A fold change represents how many times the gene expression from the patient DNA is high when compared to healthy persons.

Microscopic techniques not only provide the information related to morphology; however, sophisticated microscopic techniques, such as fluorescent microscopy or fluorescent in-situ hybridization technique (FISH), also enable the scientists/clinicians to understand the spatial information about the distribution of microbial communities. Advances in microscopic techniques enabled researchers to distinguish live and dead cells through fluorescent probe where green fluorescence represents live cells and red fluorescence represents dead cells. This dual-color approach provides a clear visual distinction, assisting scientists to evaluate the overall condition of microbial populations. FISH uses fluorescent-tagged oligonucleotides which enables targeting and binding of these probes to the specific nucleotide sequences that once bound emit fluorescence leading to the morphological identification of various pathogens in a biological sample. Electron microscopy techniques, such as scanning electron microscopy (SEM), confocal laser scanning electron microscopy (CLSM), and transmission electron microscopy (TEM) are more advanced than fluorescent microscopic techniques. These techniques are instrumental in examining the intracellular compartments of diverse cell types, including biofilms. SEM, for instance, yields intricate surface images of biofilm structures, illustrating the microbial interactions with surfaces that lead to cluster formation. However, TEM (transmission electron microscopy), however, offers comprehensive insights into internal structures of biofilms at the nanometer scale, emphasizing the relationship between cells and their extracellular matrix. Although CLSM merges laser scanning techniques with fluorescently labeled probes, it creates 3D reconstructions of biofilms, thus, facilitating an overview of microbial communities in their natural habitats. This convergence of methods enhances our understanding of these complex systems.

Biopsy samples are essential in microbiological research and clinical diagnostics (because they offer crucial insights into microbial ecosystem within tissues) (Mahnic et al., 2021). These samples can reveal presence of proteins associated with formation of extracellular matrix and carbohydrates that serve as cytoskeleton for cells. However, they also provide information about immune cells, which serve as defense mechanism.

Furthermore, these samples aid in identifying clusters of microorganisms (termed microbial aggregates) that can demonstrate cell-cell interactions with one another and their environment. Although this process is intricate, it is vital for understanding microbial dynamics (Selway et al., 2020). Matrix-assisted laser desorption ionization time-of-flight mass spectrometry MALDI-TOF is yet another revolutionary analytical technique that aids in the identification of microbes at a rapid pace by generating intricate protein profiles of the microbial sample to evaluate mass-to-charge ratios by comparing them with the reference strains available in the database. MALDI-TOF MS accurately identifies and differentiates a wide range of microbes, such as biofilms; this advanced technique streamlines the whole diagnostic process and plays an important role in developing targeted strategies toward treatment leading to enhanced understanding of microbial behavior resulting in better patient survival outcomes (Calderaro & Chezzi, 2024). A wide range of in vitro and in vivo biofilm detection techniques have been meticulously developed for biofilm identification. One notable method is the Congo Red agar test, which accurately identifies biofilm-producing Staphylococcus species with impressive precision (Guzmán-Soto et al., 2021). One gold-standard method, the tube biofilm formation test, uses crystal violet for qualitative detection of microbes involved in biofilm formation. While crystal violet is effective, it has some limitations in terms of specificity and quantitative measurement. In contrast, microplate tests adopt a more quantitative approach by using agar-filled wells along with dye staining to provide a more detailed analysis of biofilm development (Kamimura et al., 2022). Each isolated biofilm is categorized as a non-producer, weak producer, moderate producer, or strong producer, depending on absorbance results from the tests. However, this categorization is critical because it informs subsequent research and potential clinical applications. Although the tests vary in methodology, their ultimate goal remains the same: to enhance our comprehension of microbial interactions.

These detection methods are crucial for understanding complexities of biofilm formation and resistance mechanisms. Continuous standardization and development of new techniques are vital for effective diagnosis and treatment of infections (Maeda et al., 2021). Moreover, advanced techniques, such as polymerase chain reaction (PCR), fluorescence in situ hybridization (FISH), electron microscopy, and MALDI-TOF MS significantly enhance detection capabilities; ensuring we remain at cutting edge of microbial research and patient care. However, this advancement is not without challenges, because it requires ongoing commitment and resources. Although progress is evident, the dynamic nature of microorganisms complicates the process.

OMICS represents a scientific discipline that deciphers the roles of biomolecules and their associated molecular pathways. Table 15.1 represents various branches and their applications in medical field (Table 15.1).

Table 15.1 Various branches of OMICS technologies and their applications in healthcare sector

Omics Technology	Instrument	Applications
Genomics	NGS	Biofilm genome analysis, identification of biofilm-forming genes, genetic mutation studies
	Whole-genome sequencing	Comprehensive genome analysis
	Short gun metagenomics	Community-wide genome analysis
Transcriptomics	RNA sequencing	Biofilm gene expression profiling, identification of biofilm-specific genes, gene regulation studies
	Microarray analysis	Gene expression analysis
	qRT –PCR	Targeted gene expression analysis
Proteomics	MS	Biofilm protein profiling identification of biofilm-specific proteins
	LC-MS/MS	Protein separation, identification, and quantification
	MALDI-TOF	Rapid protein identification
Metabolomics	NMR	Biofilm metabolite profiling, identification of biofilm-specific metabolites
	GC-MS	Volatile metabolite analysis, metabolic pathway studies
	LC-MS	Comprehensive metabolite profiling. Biofilm biology understanding, novel target identification
Integrated Omics	Genomics, transcriptomics, proteomics (GTP) analysis	Comprehensive biofilm biology understanding, novel target identification
	Metabolomics-Genomics (MG) analysis	Integration of metabolite and genome data for functional insights.
Bioinformatics tools	Biofilm-specific databases	Data integration and prediction of biofilm formation
	Genome-scale metabolite modeling	Pathway and behavior prediction for biofilms
	Protein–Protein interaction networks	Interaction studies and novel target discovery

Our understanding of PMBF formation and its applications is accelerated due to the current advancements in omics technologies, which include genome profiling, protein profiling, and metabolite profiling. The current advancements in omics technology includes genome profiling, protein profiling, and metabolite profiling, which have fostered our understanding of PMBF formation and its application. For instance, transcriptomic profiling has provided insights into dynamic patterns of gene expression that occur during various stages of biofilm development and persistence under harsh environmental conditions. However, proteomic analysis has revealed complex protein-protein interactions and structural components of biofilms; it helps to understand how these communities maintain their integrity. However, metabolomics focuses on nutritional strategies that allow them to survive and thrive across diverse ecological niches (Micheel et al., 2012).

15.4 Omics and Its Importance in PMBFs

Using omics technologies to study PMBF formation opens up a new door for developing highly effective diagnostic tools and novel therapeutic approaches; however, it is a complex process. Although the potential is great, it requires careful consideration (and adaptation) to be truly effective. This optimization of current treatment regimens leads to improved survival outcomes and reduced mortality rates (Tebani et al., 2016). However, omics research holds significant promise in several key areas. For instance, it aids in discovering novel biomarkers for the early detection of various infectious diseases caused by PMBFs, which leads to timely treatment. Furthermore, it focuses on identifying therapeutics that specifically target PMBFs to address the challenges posed by multidimensional microbial communities. Although understanding the diverse mechanisms of microbial interaction contributes to the formation and persistence of PMBFs, it also provides a clearer understanding of disease progression and resilience. Because of this, there is a substantial improvement in the strategies to develop personalized treatment approaches (Micheel et al., 2012; Tebani et al., 2016).

Although OMICS technology plays a crucial role in understanding the formation of PMBFs and their potential implications in disease diagnosis, progression and resilience, the integration of multi-omics data presents a significant challenge. The underlying reason for this is that multi-omics data analyzes complex information derived from vast datasets; however, acquiring these extensive datasets necessitates advanced bioinformatics tools and computational models (Subramanian et al., 2020). A strong collaboration between clinicians, scientists, and industry experts is necessary to effectively translate the findings obtained from OMICS technology into practical to tackle PMBF-related infections. This collaboration may enhance our understanding to address PMBF-related infections, but it requires robust coordination and effort.

15.4.1 The Role of Multi-omics Technology in Infection Detection

Multi-omics technology has greatly changed the way we study molecular biology and microbial physiology. Over the past 15 years, there has been a rapid increase in the omics data, it enhances our understanding of the complex human body. While the advances are remarkable, the challenges still exist because combining multiple data is complicated. But this growing field continuously provides insights that we couldn't get before, changing our understanding of biological systems. Furthermore, omics science allows us to do in-depth analysis of host–pathogen interactions. Additionally, the identification of specific proteins associated with pathogens. The omics sciences, which include genomics, proteomics, metabolomics, and immunomics, are employed to investigate complex biological systems (Micheel et al., 2012).

Omics-based approaches facilitate system-based evaluations of microbes' physicochemical problems and characterize protein production, carbon, and carbohydrate metabolism. Researchers are using multi-omics information to upgrade microbial risk assessment models. Genomics assists in discovering druggable genomes, while transcriptomics detects specific genes in biofilm formation and antimicrobial resistance (Tebani et al., 2016). Various omics-related techniques are employed in infection detection, including transcriptomics, proteomics, and metabolomics. These techniques utilize express sequence tags, serial gene expression analysis, DNA microarrays, RNA-seq, 2D gels, gel-free quantification, label-free quantification, capillary electrophoresis, and time-of-flight mass spectrometry. The integration of omics sciences with computational methods and post-data analysis has expanded biochemistry, molecular biology, and microbiology to the whole-systems level (Subramanian et al., 2020).

The continued advancements of multi-omics technology hold promise for breakthroughs in infection detection, treatment, and prevention. Its applications in clinical laboratories, combined with traditional methods, will enhance diagnostics accuracy and facilitate the development of targeted therapies (Micheel et al., 2012).

15.4.1.1 Detection of Multispecies Biofilm by Multi-omics Technology

Microorganisms develop biofilms through a multifaceted process that facilitates their attachment to surfaces and enhances their resistance against host defenses and antibiotics. The extracellular matrix serves to provide stability and shields cells from antibiotics, UV radiation, and immune responses; thus, it is a crucial element in the persistence of infections associated with biofilms (Tebani et al., 2016). Biofilms exhibit resistance to antibiotics through various mechanisms. One significant mechanism is the presence of dormant cell subpopulations that persist and contribute to drug resistance, particularly in species such as Staphylococcus and Pseudomonas (Jolivet-Gougeon & Bonnaure-Mallet, 2014). To tackle these challenges, researchers are utilizing omics technologies, including metabolomics, proteomics, and transcriptomics. These approaches provide valuable insights into the regulation of biofilms and their cellular functions. However, because they identify the factors that encourage biofilm formation and drug tolerance, these methods enable the potential development of targeted therapies for multidrug-resistant biofilms. The integration of multi-omics technology not only enhances our understanding of the mechanisms underlying biofilm resistance, but it also aids in identifying essential regulatory factors. This is crucial for fostering the creation of innovative treatments for biofilm-related infections. Although these techniques are promising, challenges remain (Zachary, 2017). Figure 15.1 represents various branches of OMICS that are being studied for identification of PMBFs (Fig. 15.1).

15.4.2 Transcriptomics in Biofilm Research

Transcriptomics, which focuses on the comprehensive analysis of RNA transcripts derived from microbes, offers invaluable insights into the complex processes through which mRNA is synthesized from DNA. This intricate methodology allows researchers to explore the various stages and regulatory mechanisms involved in transcript production, thereby yielding a deeper understanding of gene expression (Dong & Chen, 2013). A prominent application of transcriptomics is it helps us to understand how biofilms form and function. These biofilms are complex structures and play a crucial role in antimicrobial drug resistance. By studying transcriptomic data, researchers can identify the specific genes and regulatory pathways that are activated during biofilm formation. This provides novel insights into the mechanisms that enable host–microbial interactions to survive challenging environments

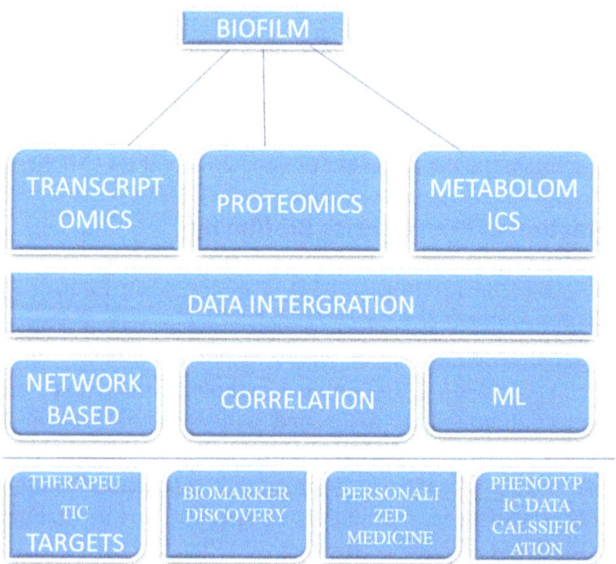

Fig. 15.1 Flow chart representing various branches of OMICS technology that are involved in identification of PMBFs

(Ali et al., 2023). Additionally, transcriptomics is essential for exploring how biofilms and their hosts influence each other.

Moreover, transcriptomics plays a crucial role in examining host–biofilm interactions. Understanding how biofilms engage with host organisms, such as plants, animals, or humans, is essential for developing effective strategies to tackle infections or promote beneficial relationships. Through transcriptomics analyses, researchers can reveal the molecular exchanges between hosts and biofilms, uncovering how biofilms evade immune responses or modify host physiology.

Additionally, transcriptomics sheds light on the mechanisms of drug resistance often seen in biofilm-related infections. Many bacteria within biofilms show increased resistance to antibiotics and other treatments, creating significant challenges in clinical environments. By studying the transcriptomics profiles of biofilm cells, scientists can identify critical genes and pathways linked to resistance, ultimately aiding in the development of more effective treatment strategies (Guo et al., 2023).

Transcriptomics has emerged as a vital and effective tool in microbiology, allowing for a deeper understanding of the intricate nature of biofilm biology and its relevance to health and disease (Jolivet-Gougeon & Bonnaure-Mallet, 2014).

15.4.2.1 Transcriptomics in Detecting Microbial Biofilm-Mediated Drug Resistance

High-throughput transcriptomics techniques are crucial for unraveling the complex mechanisms that contribute to drug resistance in microbial biofilms, which are densely packed aggregates of microorganisms adhering to various surfaces (Subramanian et al., 2020). Importantly, RNA sequencing analyses have shed light on genetic factors associated with biofilm formation and the ability of Enterococcus faecalis and Staphylococcus aureus to endure a range of treatments. These extensive studies have identified numerous genes vital for processes such as sporulation regulation—essential for bacterial survival in extreme environments—arsenic resistance, which enables microbes to thrive in toxic conditions, and multidrug efflux pumps that facilitate the removal of multiple antibiotic agents from bacterial cells (Ali et al., 2023; Marco-Puche et al., 2019). For example, the RNA-seq analysis performed by Seneviratne et al. significantly enhanced our understanding of the relationship between specific genes and biofilm development along with drug resistance in E. faecalis. While these findings deepen our insight into the survival strategies of these bacteria, they also introduce challenges in clinical applications (Guo et al., 2023).

15.4.2.2 Transcriptomics in Biofilm Development and Maturation

Transcriptomic analysis has significantly advanced our comprehension of the mechanisms driving biofilm formation in dental plaque, which is essential for maintaining oral health. Through RNA sequencing, researchers have successfully identified and defined 27 distinct genera of metabolically active microorganisms that collectively constitute the complex and diverse community present in dental plaque. This analysis offers valuable insights into the functional roles these microorganisms fulfill (Marco-Puche et al., 2019). The functional mapping of these genera has indicated that two predominant transcriptional categories—translation and carbohydrate utilization—are particularly significant. Protein translation is vital for the growth and sustenance of microbial cells, while carbohydrate utilization demonstrates how these organisms leverage the carbohydrates available in their oral environment, thereby promoting the stabilization and proliferation of the biofilm. Furthermore, a comprehensive study conducted by Peterson et al. employed transcriptomics to investigate the intricate dynamics of biofilm formation in dental plaque. Utilizing RNA sequencing, the research unveiled gene expression profiles associated with biofilm development, thus illuminating the underlying processes (Sedghi et al., 2021).

15.4.2.3 Single-Cell Transcriptomics

Single-cell transcriptomics is a branch of omics technology that studies the whole RNA sequence of a single cell in a group. This helps to identify heterogeneity and functional diversity among populations. Although single-cell transcriptomics is well employed in eukaryotic systems, due to their resistance toward nucleic acid lysis and low RNA yield, prokaryotic systems pose major challenges. Other challenges faced by the prokaryotic system are reduced mRNA stability due to lack of poly(A) tail transcripts that facilitate rapid sequencing (Duhan et al., 2024). However, advanced technologies such as microSPLiT, par-seqFISH, and PETRI-seq showed more promising results for single-cell transcriptomics in prokaryotic systems (Adil et al., 2021).

A typical single-cell transcriptomics involves isolation of a single cell from a subpopulation followed by RNA extraction, cDNA conversion, and preparation of cDNA libraries. Once the libraries are produced these would be subjected to sequencing. The sequenced data would be further analyzed by aligning, quantifying, normalizing, and clustering the obtained data and comparing the results with existing reference genes that are deposited in the repository (Jovic et al., 2022). Existing literature suggests the role of single-cell transcriptomics in various medicinal fields such as cancer biology, neuroscience, stem cell, and regenerative medicine, etc. (Adil et al., 2021). For instance, mounting evidence suggests that single-cell transcriptomics helps in identifying the tumor stage and grade thereby identifying tumor heterogeneity. It also helps in differentiating a cancer cell from cancer stem cells helping the clinicians to predict tumor relapse/recurrence. In neuroscience, it investigates the diversity of neuronal cell types (Sedghi et al., 2021). Nevertheless, single-cell transcriptomics faces obstacles such as technical variability, low RNA yield, and data complexity. Future developments may focus on establishing microfluidic techniques for cell isolation and library preparation; capturing temporal dynamics and improving the resolution of data analysis are also vital objectives.

The integration of single-cell transcriptomics (in conjunction with other omics technologies), such as proteomics and metabolomics, provides a more holistic understanding of cellular biology. However, this approach can be complex; it requires careful consideration of the interactions between various biological systems. Although the data obtained can be extensive, it is essential to analyze it critically. Because of the richness of information, researchers must navigate the challenges presented by the sheer volume of data, but the insights gained are often invaluable for advancing our knowledge in the field. Additionally, the creation of computational tools (and machine learning algorithms) will enhance data analysis and interpretation (Guo et al., 2023). Thus, single-cell transcriptomics stands out as a powerful approach to exploring cellular heterogeneity and functional diversity. Its applications across various fields—such as microbiology, cancer research, and neuroscience—will further deepen our understanding of biological systems. As research progresses, single-cell transcriptomics will be essential in unraveling the complexities of life at the cellular level; this paves the way for innovative therapeutic strategies and treatments for a range of diseases. However, challenges remain because the integration of these technologies is complex, but the potential benefits are significant (Duhan et al., 2024; Jovic et al., 2022).

15.4.2.4 Applications and Limitations of Transcriptomics

Transcriptomic methods provide significant insights into complex biological processes—especially regarding biofilm formation and antimicrobial resistance. Biofilms (which are organized communities of microorganisms) play a crucial role in the persistence of infections; thus, understanding the genetic mechanisms involved is vital (Dong & Chen, 2013). However, the choice of experimental model—whether in vitro (lab-based) or in vivo (within living organisms)—can greatly influence the gene expression patterns observed. This variability poses challenges for interpreting results, as it may limit the applicability of findings to real-world contexts. To address these issues, researchers are employing advanced machine learning techniques to analyze transcriptomic data. By identifying unique transcriptome signatures, they can distinguish between gene expression profiles associated with human infections and those originating from in vitro cultures. This distinction is essential for developing targeted therapies and comprehending the biological intricacies of infections (Subramanian et al., 2020).

15.4.3 Second Stage of Biofilm Monitoring: Proteomics

Proteomics, the quantitative study of the expression of various proteins produced by an organism under specific conditions, is essential. Unlike transcriptomics, which focuses on gene expression, proteomics offers a more stable and comprehensive perspective on natural processes being within an organism. Microorganisms, including those set up in biofilms, acclimate their protein expression in response to environmental stressors to survive; still, this rigidity is pivotal. Due to advancements in mass spectrometry, proteomics has become an essential tool in biofilm exploration, enabling identification and quantification of a large number of proteins, including complex fragments like membrane proteins. Gel-free mass spectrometry-grounded styles are now preferred because of their effectiveness and perfection, although some traditional ways still have applicability.

Proteomics has revolutionized biofilm research by facilitating high-throughput analyses of microbial communities. Techniques, such as differential isotopic labeling, including tandem mass tag (TMT), isobaric tags for relative and absolute quantitation, as well as stable isotope labeling by amino acids in cell culture, are widely utilized. For instance, the SILAC method was employed to analyze the proteomes of both planktonic and biofilm forms of Neisseria gonorrhoeae; however, TMT labeling combined with iTRAQ was applied to investigate antibiotic-sensitive and resistant biofilms of Aeromonas hydrophilia and Candida. These studies have underscored the significance of proteomics in elucidating differences in protein expression, particularly in antibiotic-resistant strains, because this allows for a deeper understanding of microbial behavior. Although the techniques have evolved, the core objective remains the same: to explore the complexities of microbial life.

15.4.3.1 Historical Context and Methodological Advances

Significant advancements in proteomics have occurred (Zhao et al., 2023) over the past two decades, particularly concerning sessile bacteria like Bacillus cereus and Pseudomonas aeruginosa. These bacteria are known for their ability to form biofilms—intricate communities that attach to surfaces and pose unique challenges in various settings, including clinical environments. The emergence of advanced mass spectrometry techniques has transformed (Gabrilska & Rumbaugh, 2015) how researchers analyze protein compositions found within these biofilms; however, progress in bioinformatics is equally vital. Two prominent methods (SILAC, which stands for stable isotope labeling by amino acids in cell culture, and iTRAQ, or isobaric tags for relative and absolute quantitation) enable scientists to effectively compare protein profiles of bacteria in their biofilm state (where they are fixed to surfaces) against their planktonic state (where they exist as free-floating cells). This comparative analysis is essential for understanding distinct physiological and metabolic differences between these two states and for uncovering mechanisms that enable bacteria to thrive in biofilm-forming environments. Although researchers have made great strides, many challenges remain because the complexity of these systems requires further investigation (Cho, 2007).

15.4.4 Biofilm–Host Interactions

Few studies have investigated biofilm–host interactions through proteomics; however, they have provided significant insights. For example, in a label-free proteomics analysis of a co-culture model involving multi-layered gingival epithelium and a 10-species subgingival biofilm, inflammation and apoptosis pathways were found to be upregulated. Conversely, pathways related to epithelial tissue integrity were downregulated. Over the span of 48 hours, metabolic processes and molecular complexes deteriorated, which suggests that "red-complex" bacteria play a pivotal role in mediating the host immune response. Although the findings are compelling, further research is necessary to fully understand these complex interactions (Chen & Kolodkin-Gal, 2022).

15.4.5 Drug Resistance in Biofilms

Boxed in a defensive extracellular matrix, biofilms are intricate populations of bacteria that stick to shells. They contribute to patient infections and are common in both clinical and natural settings. Because of the special parcels of biofilms, ails become more delicate to cure because of increased medicine forbearance and resistance. Antibiotic resistance in biofilms is caused by a variety of mechanisms, including inheritable, physiological, and structural rudiments. One important element is the EPS matrix, which forms a medium that protects bacteria from environmental stressors and acts as a physical hedge precluding antibiotic penetration. Also, because idle and slow-growing cells are less vulnerable to antibiotics, biofilms' changed metabolic countries and growth rates also contribute to resistance. Antibiotic resistance in biofilms is also significantly told by vertical gene transfer. The transmission of inheritable material, particularly genes that confer resistance to antibiotics, iseased by the close propinquity of cells. Antibiotic resistance is further told by relations between species inside biofilms, with treatment issues being impacted by both collaborative and competitive relations between species. Low temperatures and high swab attention are exemplifications of environmental stresses that might increase HGT rates and encourage resistant strains. Original attachment is followed by unrecoverable attachment, development, and dissipation as the phases of biofilm development progress. It's essential to comprehend these phases in order to produce styles that effectively combat ails linked to biofilms. Reduced penetration, changed microbial metabolism, phenotypic forbearance, inheritable resistance, and efflux pumps are some of the mechanisms of antibiotic resistance in biofilms. The high attention of antibiotics demanded to break through biofilms—which can be 10,000 times advanced than those demanded for planktonic bacteria—make treating infections linked to biofilms delicate. Also, biofilms can affect in defiance, a condition in which bacteria develop more resistant biofilms after surviving antibiotic treatment. These issues are the focus of new remedial approaches. While bacteriophage remedy offers a focused treatment system using bacteriophages, antimicrobial peptides break down bacterial membranes and help the product of biofilms. To enhance treatment issues for cases with patient infections, further biofilm biology exploration and the creation of new remedial approaches are necessary. To effectively combat ails linked to biofilms, it's essential to comprehend the mechanisms underpinning antibiotic resistance in biofilms. To attack the adding problem of antimicrobial resistance in clinical settings, new strategies fastening on biofilm development, conservation, and inheritable underpinnings of resistance are needed. One of the main themes in proteomic exploration has been the significance of metabolic exertion in biofilm development. Interesting trends have been set up in thorough comparisons of Enterococcus faecalis strains with strong and poor biofilm-forming capacities. Oxidative stress proteins, especially OsmC, which is essential for the bacteria's response to external stressors, are significantly downregulated in weaker biofilm formers. On the other hand, compared to their weaker counterparts, robust biofilm formers generally show reduced overall metabolic exertion. This connection implies that reduced metabolic exertion could be a unique index of biofilm adaptability, showing that certain strains are more suited to survive in biofilm communities the hostile environmental stress.

15.4.6 Metabolic Activity in Biofilm Formation

Metabolomics, however, is the qualitative estimation of metabolites of an organism produced as a result of specific metabolic pathways. Hence, to decipher the complex dynamics of PMBFs metabolomics is widely being used. Analysis of microbial proteome helps in understanding host–bacterial interactions and the underlying mechanisms of drug-resistance. Proteomic studies provide significant insights in biofilm behavior resulting in novel strategies to fight against biofilm-related infections leading to better survival outcomes.

Metabolomics offers a unique window merged with the intricate interplay of biochemistry, disease, and cell physiology. Metabolite profiles provide direct information on the observable phenotype of a biological system, rather than genetic or protein data, which is neither direct nor definitive about what is actually occurring in the cell. As study of biofilms expands, proteomics is expected to give significant insights concerning biofilm function. This comprehension could pave the way for innovative strategies to avert and manage infections associated with biofilms, which pose significant challenges in clinical settings. However, the complexities involved are not to be underestimated, because they often require multifaceted approaches to address. Although promising, these strategies must be carefully evaluated for effectiveness and safety (Jolivet-Gougeon & Bonnaure-Mallet, 2014; Chen & Kolodkin-Gal, 2022).

15.4.6.1 Metabolomics: Understanding Cellular Processes

Metabolomics encompasses a comprehensive analysis of metabolites (the small molecules generated as intermediates or end products of enzyme-mediated reactions in living cells). These metabolites can be classified into two main categories: primary and secondary metabolites. Primary metabolites are vital substances, which include amino acids and ethanol; they are indispensable for facilitating cell growth, development, and maintenance. Typically, these substances are produced during an organism's active growth phase, and they are fundamental for essential cellular functions. However, the overall health of the organism is heavily reliant on these metabolites because they play a significant role in sustaining life. Although secondary metabolites also contribute to various biological processes, primary metabolites are particularly crucial for the organism's survival (Sedghi et al., 2021; Zhang et al., 2013).

In contrast, secondary metabolites are more specialized compounds, such as antibiotics and signaling molecules. Unlike primary metabolites, however, they do not play a direct role in the basic metabolic processes necessary for life. Instead, secondary metabolites have unique functions that enhance an organism's ability to survive and adapt to its surroundings (this is crucial). They can provide protection against predators, facilitate communication between cells, and assist organisms in competing for resources, making them vital for ecological interactions and evolutionary success. Although their roles may not be as direct, they are essential because they contribute to the overall fitness of an organism within its environment (Zhang et al., 2013).

15.4.6.2 Metabolomics: A Comprehensive End-Point Monitoring Tool

Metabolomics constitutes a sophisticated domain that provides profound insights into the complex mechanisms of cellular physiology, pathology, and biochemistry. Unlike the data obtained from genes and proteins (which may not always accurately represent the cell's present condition), metabolite profiles establish a direct link to the observable phenotype of a biological system. These distinctive attributes render metabolomics a remarkably efficient and effective approach for end-point monitoring in numerous research and clinical applications. However, this field is still evolving and researchers must navigate its intricacies carefully. Although challenges exist, the potential benefits are substantial (Johnson et al., 2016).

15.4.6.3 Exploring Biofilm Metabolomics

Within the realm of metabolomics, the exploration of intricate communities of microorganisms that adhere to surfaces has emerged as a significant area of focus, particularly concerning antimicrobial resistance. Metabolomics analyses enable researchers to identify vital primary and secondary metabolites (which are critical) for the formation and stability of biofilms. By profiling these metabolites and utilizing biological system modeling, scientists can gain a more profound understanding of the mechanisms that underlie biofilm development and their inherent resistance to antimicrobial agents. However, this field is still evolving and there is much to uncover (Tang, 2011).

15.4.6.4 Diverse Metabolomics Approaches to Investigate Biofilms

While the diversity of metabolites differs among organisms, microbes offer a lot of potential for metabolomics studies. This difference shows how important it is to use the right analytical methods to get a full picture of the metabolome (Zhang et al., 2013). Key techniques include NMR, which allows for a deep look into the structure of metabolic compounds, and mass spectrometry, which helps detect and measure metabolites with high sensitivity. Together, these tools boost our ability to study, describe, and grasp the complex nature of biofilms and how they process metabolites (Cho, 2007).

15.4.6.5 Untargeted Metabolomics

Untargeted metabolomics constitutes a complex logical strategy designed to capture the full range of metabolites present in a sample, free from any preexisting impulses. This system is, especially salutary for medicine discovery; it aids experimenters in relating implicit biomarkers and metabolic pathways associated with a variety of conditions. Also, it's useful for thesis-generating studies that aim to probe unknown natural processes. Still, untargeted metabolomics is fraught with challenges. Specialized difficulties related to metabolite birth, maintaining perceptivity during the analysis, and directly interpreting complex data sets can significantly complicate the exploration process. Although these obstacles live, the implicit perceptivity gained from this approach are inestimable as they offer a deeper understanding of natural systems. This complexity can be dispiriting, but it also presents openings for significant advancements in the field (Johnson et al., 2016).

15.4.6.6 Targeted Metabolomics

In discrepancy, targeted metabolomics focuses on the accurate quantification of specific metabolites chosen grounded on a defined thesis or natural question. This system is guidedby a clear exploration ideal and requires previous knowledge of the metabolites of interest. To effectively conduct targeted metabolomics, experimenters must use well-characterized metabolite norms to ensure precise dimensions and confirmation of results. This approach facilitates in-depth examinations into the places of particular metabolites in natural systems, offering perceptivity that can inform remedial strategies and clinical operations (Tang, 2011).

15.4.6.7 Metabolic Flux Analysis (MFA)

Metabolic flux analysis (MFA) is a strong method that lets scientists watch and measure changes in all the metabolites in biological samples as time passes. Scientists do this using stable isotopes or radioactive isotopes as tracers. MFA helps researchers overcome the limits of regular metabolomics. This new approach makes it possible to trace pathways in detail, which allows scientists to follow specific metabolic routes and learn about how fast metabolites turn over, showing how (and well) biological systems make and use metabolites. But it's important to remember that these analyses can be tricky, as metabolites often interact in complex ways. While MFA is a useful tool, it does have its drawbacks (de Falco et al., 2022).

15.4.6.8 Applications of Metabolomics in Biofilm Research

Metabolomics is integral to advancing our appreciation of biofilm exploration, which is essential for addressing the complications presented by microbial communities. Through metabolic profiling, experimenters can interpret preliminarily uncelebrated pathways associated with biofilm conformation, thereby illuminating the intricate biochemical processes that bolster these structures (Cho, 2007). Also metabolomics enables the monitoring of antimicrobial resistance emergence within biofilms, furnishing vital perceptivity into bacterial adaption and survival in hostile surroundings. Given that, biofilms represent sophisticated and continuously evolving systems, the relinquishment of multi-omics approaches—integrating metabolomics with genomics, proteomics, and transcriptomics becomes pivotal for a comprehensive understanding of the metabolic transitions in bacteria within biofilms (Chen & Kolodkin-Gal, 2022). This holistic perspective fosters a deeper sapience into biofilm development, continuity, and its counteraccusations for health and complaint. Nonetheless, the complexity of these relations necessitates ongoing exploration to completely comprehend the ramifications of these findings. Metabolomics is set to remain a crucial element in decoding the complications of biofilm biology, antimicrobial resistance, and complaint opinion (Zhang et al.,

2013). This is largely due to its integration with colorful omics technologies, which will give significant perceptivity into cellular biology. Still, experimenters must remain active, as the challenges in this sphere are considerable. Despite the progress achieved, the trip is ongoing, and the counteraccusations of these findings could be substantial.

Metabolomics functions as a powerful instrument for investigating cellular processes, biofilm dynamics, and antimicrobial resistance. Its applications in biofilm research, disease diagnosis, and drug discovery are poised to enhance our comprehension of biological systems. However, the complexity of these interactions can be overwhelming. This complexity arises because of the myriad factors involved; thus, researchers must navigate through various challenges. Although it is a promising field, there are still limitations to consider. But with continued advancements, the potential for breakthroughs remains significant (Johnson et al., 2016). Table 15.1 represents various detection methods used for identifying polymicrobial infections (Table 15.2).

15.4.7 Applications of Multi-omics Technology and Their Prospects

The investigation of biofilms (complex communities of microorganisms that exist within a self-generated extracellular matrix) is experiencing rapid growth, fueled by the escalating demand for extensive and data-driven research. However, this burgeoning field is confronted with a significant obstacle: the lack of standardized methods for data exchange and the absence of centralized databases dedicated to

Table 15.2 Various branches in OMICS technology that are involved in diagnosing polymicrobial infections

Detection methods	Clinical samples	Clinical information
TRANSCRIPTOMICS		
The express sequence tag (EST)	Blood	Sepsis
SAGE	Peripheral blood mononuclear cell	Urinary tract infection
DNA microarrays	Several body fluids and tissues	Multiplexed pathogen detection
RNA-seq (4sU-Seq, RIP-seq, (Riboseq)	Cerebrospinal fluid	Meningitis
PROTEOMICS		
2D gels	Blood	Sepsis
Gel-free (iTRAQ, TMT) spiked-in peptides	Secretions	Urinary tract infections
Label-free quantitation	Cell-free body fluids	Respiratory infections
METABOLOMICS		
Capillary electrophoresis	Blood fluids	Sepsis
TOF-MS	Stool sample	Infections by clostridium difficile
CE-TOFMS	Cerebrospinal fluid	Meningitis

storing and sharing biofilm-related data. As a result, researchers navigate a fragmented ecosystem in which most valuable data remains locked away in private archives (with only a handful of statistics making their way into the scientific literature).

This fragmented state hampers collaboration (and complicates) the drawing of meaningful conclusions from existing research. One of the primary challenges is the inconsistency in how data are represented; for instance, experiments may report antibiotic concentrations in various units, such as milligrams per liter (mg/L) or moles per liter (mol/L). Additionally, the diversity in data processing methodologies and analytical techniques (further complicates) study comparisons. It is crucial to address these challenges to facilitate progress; however, collective understanding within the field of biofilm research suffers. To address challenges in biofilm studies, it is crucial to create bioinformatics tools that utilize publicly accessible data and standardized vocabularies. The BiofOmics platform represents a significant advancement because it organizes biofilm data for storage (searching) and analysis, regardless of research methods. This standardization aids in data comparison and the formulation of new hypotheses, although some variations may remain.

In the detection of antibiotic resistance within biofilms, metabolomics assumes a pivotal role. This involves the examination of small molecules generated during metabolism to pinpoint metabolites associated with antibiotic resistance. The exometabolome—comprising metabolites expelled from microbial cells—offers insights into microbial behavior; however, intracellular metabolic profiles deepen our understanding. The emergence of antibiotic resistance, especially concerning methicillin-resistant Staphylococcus aureus (MRSA), underscores the necessity for novel antimicrobial compounds and effective detection methodologies. Tools, such as MALDI-TOF mass spectrometry, when coupled with multi-omics strategies, are effective in identifying microbes and their resistance mechanisms.

The integration of multi-omics technologies encompasses fields (such as genomics: the study of an organism's complete set of DNA), proteomics (the large-scale study of proteins and their functions), and metabolomics (the analysis of metabolic responses and metabolites)—offers expansive and detailed understanding of biofilms, complex communities of microorganisms. However, despite the remarkable potential these approaches have for illuminating intricate dynamics of biofilms, there are considerable challenges associated with merging data from these diverse omics layers. The primary obstacles stem from discrepancies in data formats (varied computational requirements) and the complexity of interpreting large datasets. Although to derive meaningful insights from this wealth of data, it is essential to employ robust data normalization techniques and sophisticated integration methods. Techniques such as machine learning algorithms and advanced statistical methods are critical for reducing dimensionality and aiding in identification of significant biomarkers and biological pathways that may be pivotal for understanding biofilm behavior, because this can enhance our comprehension of microbial ecosystems.

Although the application of multi-omics technologies in biofilm research has been somewhat limited (to date), there have been promising advancements. For instance, certain studies have successfully combined metagenomics—which examines the collective genomic data from microbial communities—and metabolomics to gain insights into the functional output of biofilm systems. Additionally, other research initiatives have explored how environmental stressors, including heavy metals, such as copper, impact the behavior and composition of biofilm-forming bacteria. However, while some researchers have effectively utilized two omics platforms together—such as transcriptomics (the study of RNA transcripts) combined with metabolomics—very few have achieved a comprehensive integration of data across three or more omics layers: transcriptomics, proteomics, and metabolomics. This complexity presents challenges, but it also opens doors for future exploration.

Gazing into the future, the prospects of biofilm research hinge on further advancements in data integration technologies: development of innovative AI-driven tools is designed to model the complex behaviors exhibited by biofilms. A holistic integration of omics data, spanning multiple levels, will enable researchers to delve deeper into processes of biofilm formation, mechanisms of antibiotic resistance, and other critical microbial characteristics. Although software and computational power continue to improve, integration of multi-omics data is poised to dramatically transform biofilm research—paving the way for new diagnostic and therapeutic strategies that can effectively address pressing challenges, such as antibiotic resistance. However, evolution remains contingent upon overcoming existing barriers.

In conclusion, multi-omics approaches signify transformative leap forward in biofilm research; offering deeper insights into development, architecture, and resistance mechanisms of microbial communities. The establishment of shared databases, however, and the creation of advanced data analysis tools will be crucial for fostering breakthroughs in biofilm-related discoveries—ultimately contributing to enhanced medical diagnostics and more effective treatment options. Although this progress is promising, challenges remain but the potential benefits are significant.

Acknowledgments SLP gratefully acknowledges DAE-BRNS, DBT (BT/PR30629/BIC/101/1093/2018), New Delhi; UGC (Ref No: No.F.30-456/2018 (BSR) and SERB (Ref No.: PDF/2015/000867) for the financial support. PC gratefully acknowledges DBT (BT/PR30629/BIC/101/1093/2018), New Delhi for the junior research fellowship.

Conflict of Interest None declared.

References

Adil, A., Kumar, V., Jan, A. T., & Asger, M. (2021). Single-cell transcriptomics: Current methods and challenges in data acquisition and analysis. *Frontiers in Neuroscience, 15*, 591122.

Ali, A., Zahra, A., Kamthan, M., Husain, F. M., Albalawi, T., Zubair, M., et al. (2023). Microbial biofilms: Applications, clinical consequences, and alternative therapies. *Microorganisms, 11*(8), 1934.

Calderaro, A., & Chezzi, C. (2024). MALDI-TOF MS: A reliable tool in the real life of the clinical microbiology laboratory. *Microorganisms, 12*(2), 322.

Chen, Y., & Kolodkin-Gal, I. (2022). Host–Biofilm Interactions. *Microorganisms, 10*(8), 1641.

Cheng, M. P., Domingo, M. C., Lévesque, S., & Yansouni, C. P. (2016). A case report of a deep surgical site infection with Terrisporobacter glycolicus/T. Mayombei and review of the literature. *BMC Infectious Diseases, 16*, 1–4.

Cho, W. C. (2007). Proteomics technologies and challenges. *Genomics, Proteomics & Bioinformatics, 5*(2), 77–85.

de Falco, B., Giannino, F., Carteni, F., Mazzoleni, S., & Kim, D. H. (2022). Metabolic flux analysis: A comprehensive review on sample preparation, analytical techniques, data analysis, computational modelling, and main application areas. *RSC Advances, 12*(39), 25528–25548.

Dong, Z., & Chen, Y. (2013). Transcriptomics: Advances and approaches. *Science China Life Sciences, 56*, 960–967.

Duhan, L., Kumari, D., Naime, M., Parmar, V. S., Chhillar, A. K., Dangi, M., & Pasrija, R. (2024). Single-cell transcriptomics: Background, technologies, applications, and challenges. *Molecular Biology Reports, 51*(1), 600.

Gabrilska, R. A., & Rumbaugh, K. P. (2015). Biofilm models of polymicrobial infection. *Future Microbiology, 10*(12), 1997–2015.

Guo, X. K., Wang, J., van Hensbergen, V. P., Liu, J., Xu, H., & Hu, X. (2023). Interactions between host and intestinal crypt-resided biofilms are controlled by epithelial fucosylation. *Cell Reports, 42*(7), 112754.

Guzmán-Soto, I., McTiernan, C., Gonzalez-Gomez, M., Ross, A., Gupta, K., Suuronen, E. J., et al. (2021). Mimicking biofilm formation and development: Recent progress in in vitro and in vivo biofilm models. *Iscience, 24*(5), 102443.

Johnson, C. H., Ivanisevic, J., & Siuzdak, G. (2016). Metabolomics: Beyond biomarkers and towards mechanisms. *Nature Reviews Molecular Cell Biology, 17*(7), 451–459.

Jolivet-Gougeon, A., & Bonnaure-Mallet, M. (2014). Biofilms as a mechanism of bacterial resistance. *Drug Discovery Today: Technologies, 11*, 49–56.

Jovic, D., Liang, X., Zeng, H., Lin, L., Xu, F., & Luo, Y. (2022). Single-cell RNA sequencing technologies and applications: A brief overview. *Clinical and Translational Medicine, 12*(3), e694.

Kamimura, R., Kanematsu, H., Ogawa, A., Kogo, T., Miura, H., Kawai, R., et al. (2022). Quantitative analyses of biofilm by using crystal violet staining and optical reflection. *Materials, 15*(19), 6727.

Lohse, M. B., Gulati, M., Johnson, A. D., & Nobile, C. J. (2018). Development and regulation of single-and multi-species Candida albicans biofilms. *Nature Reviews Microbiology, 16*(1), 19–31.

Maeda, T., Fukushima, Y., Yoshida, H., Goto, M., Fujita, T., Tsuyuki, Y., & Takahashi, T. (2021). Biofilm production ability and associated characteristics of Streptococcus agalactiae isolates from companion animals and humans. *Journal of Infection and Chemotherapy, 27*(11), 1571–1577.

Mahnic, A., Breznik, V., Bombek Ihan, M., & Rupnik, M. (2021). Comparison between cultivation and sequencing based approaches for microbiota analysis in swabs and biopsies of chronic wounds. *Frontiers in Medicine, 8*, 607255.

Marco-Puche, G., Lois, S., Benítez, J., & Trivino, J. C. (2019). RNA-Seq perspectives to improve clinical diagnosis. *Frontiers in Genetics, 10*, 1152.

Martin-Loeches, I., Torres, A., Nagavci, B., Aliberti, S., Antonelli, M., Bassetti, M., et al. (2023). ERS/ESICM/ESCMID/ALAT guidelines for the management of severe community-acquired pneumonia. *Intensive Care Medicine, 49*(6), 615–632.

Micheel, C. M., Nass, S. J., & Omenn, G. S. (2012). Committee on the review of omics-based tests for predicting patient outcomes in clinical trials; Board on Health Care Services; Board on Health Sciences Policy; Institute of Medicine. In *Evolution of translational omics: Lessons learned and the path forward*. National Academies Press.

Peters, B. M., Jabra-Rizk, M. A., O'May, G. A., Costerton, J. W., & Shirtliff, M. E. (2012). Polymicrobial interactions: Impact on pathogenesis and human disease. *Clinical Microbiology Reviews, 25*(1), 193–213.

Rather, M. A., Gupta, K., & Mandal, M. (2021). Microbial biofilm: Formation, architecture, antibiotic resistance, and control strategies. *Brazilian Journal of Microbiology, 52*, 1–18.

Rupp, M., Kern, S., Weber, T., Menges, T. D., Schnettler, R., Heiß, C., & Alt, V. (2020). Polymicrobial infections and microbial patterns in infected nonunions–a descriptive analysis of 42 cases. *BMC Infectious Diseases, 20*, 1–8.

Sedghi, L., DiMassa, V., Harrington, A., Lynch, S. V., & Kapila, Y. L. (2021). The oral microbiome: Role of key organisms and complex networks in oral health and disease. *Periodontology 2000, 87*(1), 107–131.

Selway, C. A., Eisenhofer, R., & Weyrich, L. S. (2020). Microbiome applications for pathology: Challenges of low microbial biomass samples during diagnostic testing. *The Journal of Pathology: Clinical Research, 6*(2), 97–106.

Subramanian, I., Verma, S., Kumar, S., Jere, A., & Anamika, K. (2020). Multi-omics data integration, interpretation, and its application. *Bioinformatics and Biology Insights, 14*, 1177932219899051.

Tang, J. (2011). Microbial metabolomics. *Current Genomics, 12*(6), 391–403.

Tebani, A., Afonso, C., Marret, S., & Bekri, S. (2016). Omics-based strategies in precision medicine: Toward a paradigm shift in inborn errors of metabolism investigations. *International Journal of Molecular Sciences, 17*(9), 1555.

Zachary, J. F. (2017). Mechanisms of microbial infections. In *Pathologic basis of veterinary disease* (p. 132). Elsevier.

Zhang, A., Sun, H., Xu, H., Qiu, S., & Wang, X. (2013). Cell metabolomics. *OMICS: A Journal of Integrative Biology, 17*(10), 495–501.

Zhao, A., Sun, J., & Liu, Y. (2023). Understanding bacterial biofilms: From definition to treatment strategies. *Frontiers in Cellular and Infection Microbiology, 13*, 1137947.

Computational and Systems Biology Approaches in Biofilm Research 16

Suseela Lanka, K. Mounika, K. Anitha, and Santhi Latha Pandrangi

16.1 Introduction

Biofilms are structured communities of microbial cells encased within a self-produced extracellular matrix, which attach to biotic or abiotic surfaces (Costerton et al., 1995). These communities are highly dynamic, with properties that differ significantly from their planktonic counterparts. Biofilms are involved in a range of critical processes across various domains, such as healthcare, environmental management, and industrial operations. For instance, they play a key role in chronic infections, such as those caused by *Pseudomonas aeruginosa* in cystic fibrosis patients (Singh et al., 2000), and in the contamination of medical devices like catheters and implants (Donlan, 2001). Conversely, biofilms are instrumental in bioremediation, wastewater treatment, and in the degradation of environmental pollutants (Flemming & Wingender, 2010).

Understanding the biofilms—their formation, structural organization, metabolic processes, and resistance mechanisms—requires a multidisciplinary approach. Traditional microbiological methods, while essential, provide limited insights into the complex, multi-scale interactions within biofilm communities. Systems biology, a holistic discipline that integrates data from genomics, transcriptomics, proteomics, and metabolomics, allows researchers to decipher the intricate molecular networks underlying biofilm behavior (Kitano, 2002). By combining experimental data with computational modeling, systems biology provides a comprehensive understanding of biofilm systems, spanning from gene regulation to population dynamics.

S. Lanka · K. Mounika · K. Anitha
Department of Biosciences and Biotechnology, Krishna University, Machilipatnam, Andhra Pradesh, India

S. L. Pandrangi (✉)
Department of Biochemistry and Bioinformatics, GITAM School of Science, GITAM (Deemed to be) University, Visakhapatnam, Andhra Pradesh, India

S. Busi et al. (eds.), *Omics Approaches in Biofilm Research*,
https://doi.org/10.1007/978-3-031-91863-6_16

Biofilm formation is a highly regulated, multi-step process involving attachment, maturation, and dispersion phases (Stoodley et al., 2002). This process is governed by quorum sensing—a cell–cell communication mechanism mediated by signaling molecules—which coordinates collective behaviors, such as biofilm development and antimicrobial resistance (Parsek & Greenberg, 2005). Omics technologies have been instrumental in unraveling these molecular mechanisms. For example, transcriptomic analyses have revealed gene expression changes associated with biofilm formation, while proteomic studies have identified biofilm-specific proteins involved in adhesion and matrix production (Vilain et al., 2004).

Despite significant advancements, studying biofilms remains challenging due to their inherent heterogeneity, both spatial and temporal. Computational systems biology addresses these challenges by enabling the simulation and prediction of biofilm behavior under diverse conditions. Agent-based models (ABMs), for instance, provide insights into individual cell behaviors and their interactions within biofilms, while continuum models describe biofilm growth and nutrient diffusion at the macroscopic level (Klapper & Dockery, 2010). These models complement experimental data, facilitating a deeper understanding of biofilm structure–function relationships.

In this chapter, we explore how systems biology and computational approaches have transformed biofilm research. By integrating experimental and computational methodologies, researchers can gain unprecedented insights into biofilm dynamics, paving the way for innovative strategies to manage biofilm-associated problems and harness their beneficial applications.

16.2 Biofilm Formation

Biofilm formation is a highly coordinated and dynamic process that allows microbial communities to establish themselves on surfaces and thrive in diverse environments (Thiele et al., 2013). This process can be broadly divided into five key stages: initial attachment, irreversible attachment, EPS production and maturation, biofilm growth and development, and dispersal. Understanding these stages is crucial for unraveling the mechanisms that contribute to biofilm resilience and functionality (Selvarajoo, 2020).

16.2.1 Initial Attachment

In the first step, planktonic (free-floating) microbial cells approach a surface through physical forces, such as Brownian motion, van der Waals interactions, and electrostatic attraction. The properties of the surface, including its roughness, hydrophobicity, and charge, significantly influence microbial adhesion. Microbial surface structures, such as pili, flagella, and fimbriae, play critical roles in this initial attachment phase.

16.2.2 Irreversible Attachment

Once initial contact is established, microbial cells begin to adhere more strongly to the surface through specific interactions mediated by adhesins and extracellular appendages. At this stage, the attachment becomes irreversible due to the secretion of extracellular polymeric substances (EPS), a matrix primarily composed of polysaccharides, proteins, lipids, and extracellular DNA (eDNA). The EPS acts as a scaffold, anchoring the cells to the surface and forming the foundational structure for biofilm development (Picioreanu et al., 2007).

16.2.3 EPS Production and Maturation

As biofilm formation progresses, the microbial cells produce and secrete large quantities of EPS. This matrix not only provides structural integrity but also protects the biofilm from environmental stressors, such as desiccation, antibiotics, and host immune responses. The EPS matrix facilitates cell–cell interactions and communication through signaling molecules like quorum-sensing autoinducers, which regulate collective behaviors essential for biofilm development (Rudge et al., 2012).

16.2.4 Biofilm Growth and Development

During this phase, the biofilm matures into a three-dimensional structure characterized by microenvironments with gradients of nutrients, oxygen, and waste products. These gradients create metabolic heterogeneity, enabling different cells within the biofilm to specialize in various functions. This differentiation enhances the biofilm's resilience and adaptability, allowing it to survive under harsh conditions (Vital-Lopez et al., 2015).

16.2.5 Dispersal

The final step in biofilm formation is dispersal, where individual cells or small clusters detach from the biofilm and return to a planktonic state. Dispersal can occur in response to nutrient depletion, mechanical forces, or specific environmental signals (Dzianach et al., 2019). This phase is critical for biofilm expansion and the colonization of new surfaces.

16.3 Systems Biology in Biofilm Research

Systems biology employs a multi-omics approach to dissect biofilm systems, providing an integrative perspective on their complex and dynamic nature (Kiran et al., 2018). The application of various omics technologies enables the identification and characterization of key genetic, molecular, and metabolic processes involved in biofilm formation and maintenance (Fig. 16.1).

16.3.1 Genomics

Genomics plays a pivotal role in identifying genes associated with biofilm formation, structure, and resistance mechanisms. High-throughput sequencing has revealed genes encoding components of the extracellular matrix, signaling

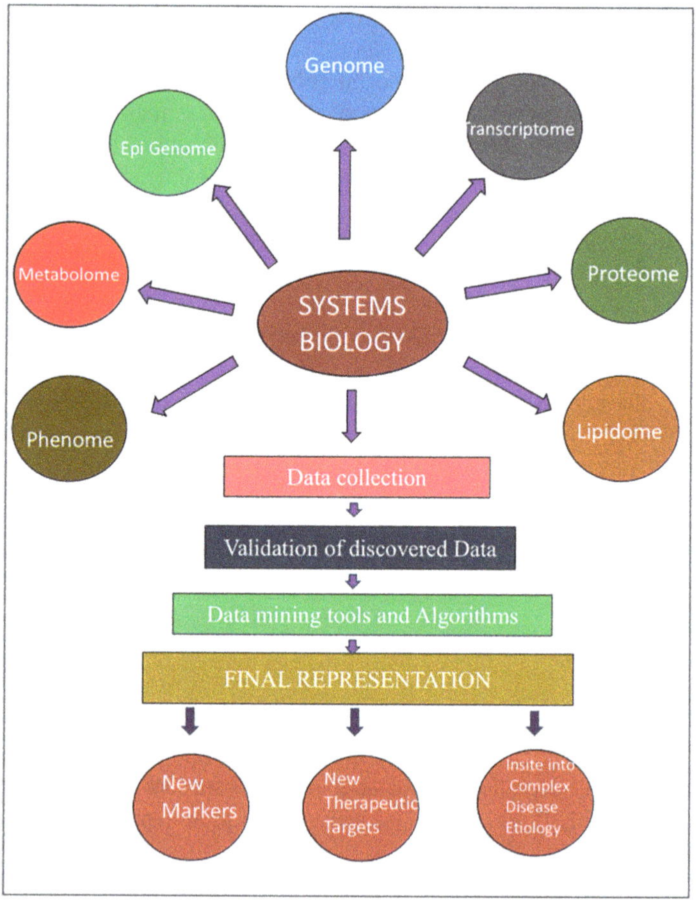

Fig. 16.1 Diagrammatic view of systems biology. Flow chart illustrating the process of systems biology. Central node labelled "Systems Biology" connects to various omics fields: Genome, Transcriptome, Proteome, Lipidome, Metabolome, Epi Genome, and Phenome. Arrows lead to "Data Collection," followed by "Validation of Discovered Data," "Data Mining Tools and Algorithms," and "Final Representation." The final outputs are "New Markers," "New Therapeutic Targets," and "Insight into Complex Disease Etiology"

pathways, such as quorum sensing, and antimicrobial resistance determinants (O'Toole et al., 2000). For instance, whole-genome sequencing of *Pseudomonas aeruginosa* has uncovered regulatory genes involved in biofilm development, such as those encoding the Las and Rhl quorum-sensing systems (Parsek & Greenberg, 2005).

16.3.2 Transcriptomics

Transcriptomic analyses provide insights into gene expression profiles during different stages of biofilm development. RNA sequencing (RNA-seq) has been employed to identify upregulated and downregulated genes during the attachment, maturation, and dispersion phases of biofilms (Stoodley et al., 2002). For example, studies on *Escherichia coli* biofilms have revealed the differential expression of stress response genes and genes encoding adhesion factors during biofilm formation (Beloin et al., 2008).

16.3.3 Proteomics

Proteomics complements transcriptomics by identifying proteins specifically associated with biofilm growth. Mass spectrometry-based approaches have revealed biofilm-specific proteins involved in adhesion, matrix production, and antimicrobial resistance. Vilain et al. (2004) used proteomic profiling to identify adhesins and enzymes that contribute to biofilm stability in *Bacillus cereus*.

16.3.4 Metabolomics

Metabolomics examines the small-molecule metabolites produced within biofilms, offering insights into metabolic pathways and chemical exchanges among microbial communities. Studies using nuclear magnetic resonance (NMR) and mass spectrometry have highlighted the metabolic adaptations of biofilms, such as shifts in central carbon metabolism and the production of signaling molecules (Flemming & Wingender, 2010). For instance, the accumulation of cyclic di-GMP, a second messenger regulating biofilm formation, has been quantified using metabolomics approaches (Jenal et al., 2017).

16.3.5 Integration of Multi-omics Data

The integration of genomics, transcriptomics, proteomics, and metabolomics provides a holistic view of biofilm systems. Systems biology platforms such as Cytoscape and STRING enable the construction and analysis of interaction networks, linking molecular components to biofilm phenotypes (Szklarczyk et al.,

2019). These integrative approaches facilitate the identification of potential targets for biofilm control strategies.

Key Studies

- O'Toole et al. (2000) demonstrated the role of type IV pili in initial attachment during biofilm formation in *Pseudomonas fluorescens*.
- Parsek and Greenberg (2005) elucidated quorum-sensing networks and their impact on biofilm dynamics.
- Vilain et al. (2004) identified biofilm-specific proteins involved in adhesion and structural stability.
- Flemming and Wingender (2010) highlighted the metabolic adaptations of biofilms under nutrient-limited conditions.

16.4 Computational Systems Biology Approaches

Computational approaches provide powerful frameworks for modeling and understanding biofilm behavior. These methods complement experimental studies by simulating complex interactions and predicting outcomes under varying conditions. Here, we delve into three major approaches with specific examples:

16.4.1 Agent-Based Models (ABMs)

Agent-based models (ABMs) are used to simulate the behavior of individual microbial cells within biofilms, accounting for their interactions and spatial heterogeneity. For example, Kreft et al. (2001) developed iDynoMiCS, an ABM framework, to investigate nutrient gradients and spatial competition in *Pseudomonas aeruginosa* biofilms. Another application can be seen in the work of Xavier et al. (2005), who used ABMs to study quorum sensing-mediated biofilm development in *Vibrio cholerae*, revealing insights into the role of signaling in biofilm robustness. Additionally, ABMs have been applied to *Staphylococcus aureus* biofilms to predict the impact of antibiotic gradients on microbial survival and resistance mechanisms. These studies highlight the ability of ABMs to replicate complex biofilm dynamics and predict outcomes under various conditions (Lardon et al., 2011). Agent-Based Models (ABMs) are used to simulate the behavior of individual microbial cells within biofilms, accounting for their interactions and spatial heterogeneity. For instance, Kreft et al. (2001) developed iDynoMiCS, an ABM framework, to study nutrient distribution and microbial competition in biofilms. This approach has been instrumental in uncovering emergent behaviors such as quorum sensing-driven collective movement and the impact of antimicrobial treatments on biofilm resilience.

16.4.2 Network-Based Models

Network-based approaches, such as gene regulatory and protein–protein interaction networks, elucidate molecular pathways driving biofilm formation and maintenance. Alm and Arkin (2003) utilized gene regulatory networks to predict how environmental changes affect *Pseudomonas aeruginosa* biofilm dynamics. Additionally, integrating transcriptional regulatory networks with proteomic data has revealed critical nodes and pathways that can be targeted to disrupt biofilm integrity (Sigurdsson et al., 2012). A notable real-world application involves the study of quorum sensing networks in *Vibrio cholerae*, where key regulators, such as HapR, were identified as critical nodes affecting biofilm formation and dispersion (Zhu & Mekalanos, 2003). Additionally, the integration of transcriptional networks with metabolomic data has uncovered metabolic bottlenecks in *Escherichia coli* biofilms, enabling targeted disruption strategies. These examples highlight the potential of network-based models in unravelling complex biofilm behaviors and identifying actionable targets for intervention. Gene regulatory networks and protein interaction networks elucidate molecular interactions underlying biofilm dynamics (Alm & Arkin, 2003).

16.4.3 Continuum Models

Continuum models focus on macroscopic biofilm properties, describing growth, nutrient diffusion, and mechanical properties. Klapper and Dockery (2010) employed these models to simulate the formation of biofilms in flow environments, providing insights into how shear forces and nutrient gradients shape biofilm architecture. Continuum models have also been used to predict the penetration and efficacy of antibiotics in biofilm treatment scenarios.

Each of these computational methods offers unique advantages and, when integrated, can provide a comprehensive understanding of biofilm systems. Computational tools model and simulate biofilm behavior, complementing experimental data. Major approaches include in (Table 16.1).

16.5 Integration of Experimental and Computational Approaches

The integration of experimental and computational approaches has revolutionized our understanding of biofilm systems (An et al., 2021). This synergy combines high-resolution experimental data with sophisticated computational tools to generate predictive models and actionable insights into biofilm behavior. Key aspects include:

Multi-omics Data Integration Comprehensive integration of genomic, transcriptomic, proteomic, and metabolomic data into computational frameworks provides a

Table 16.1 Different modeling techniques used to study biofilms

Computational approach	Description	References
Agent-based models (ABMs)	Simulate individual cells' interactions within biofilms, accounting for spatial heterogeneity	Kreft et al. (2001), Xavier et al. (2005), Lardon et al. (2011), Sweeney et al. (2019), Koshy-Chenthittayil et al. (2021), Nagarajan et al. (2022)
Network-based models	Explore gene regulatory and protein-protein interaction networks driving biofilm dynamics	Alm and Arkin (2003), Zhu and Mekalanos (2003), Sigurdsson et al. (2012), Zhang et al. (2024), Ratwatte et al. (2025)
Continuum models	Describe macroscopic biofilm properties, including growth and nutrient diffusion	Picioreanu et al. (1998), Alpkvist and Klapper (2007), Picioreanu et al. (2007), Boltz et al. (2010), Klapper and Dockery (2010): Mattei et al. (2018), D'Acunto et al. (2019), Guibert et al. (2025)
CFD (Computational Fluid Dynamics) model	Gives information on transport phenomena taking place within packed-bed reactors	Haddadi et al. (2017), Aminian-Dehkordi et al. (2018)

holistic understanding of biofilm dynamics. Genome-scale metabolic models (GSMMs) can be used to predict metabolic flux distributions in biofilms under varying environmental conditions. By incorporating transcriptomic data, GSMMs can identify bottlenecks and optimize metabolic pathways for targeted biofilm control (Henry et al., 2010).

Hybrid Modeling Approaches Combining agent-based models (ABMs) with continuum models allows researchers to capture both micro- and macro-scale biofilm phenomena. Nadell et al. (2010) employed such hybrid approaches to study biofilm structure–function relationships in *Vibrio cholerae*, elucidating the interplay between quorum sensing and mechanical stability. This dual approach enables a more nuanced understanding of biofilm dynamics.

Real-Time Data Integration Advanced imaging techniques, such as confocal laser scanning microscopy (CLSM), have been coupled with computational models to validate and refine predictions. For example, CLSM data of *Pseudomonas aeruginosa* biofilms were used to parameterize ABMs, enhancing their accuracy in simulating biofilm responses to antimicrobial treatments (Haagensen et al., 2015).

Predictive Frameworks Predictive models integrating experimental datasets enable scenario testing for biofilm control strategies. For instance, predictive simulations of quorum sensing inhibitor efficacy in *Staphylococcus aureus* biofilms have guided the development of novel therapeutic agents (Hentzer et al., 2003). These models help translate laboratory findings into practical applications.

16.6 Applications of Systems Biology in Biofilm Control

Systems biology offers a comprehensive framework to understand and manipulate biofilms, which are structured communities of microorganisms encased in a self-produced extracellular matrix. Biofilm-associated infections are notoriously resistant to antibiotics, making biofilm control a critical area of research, particularly in medical and industrial contexts.

16.6.1 Understanding Biofilm Formation and Development

Systems biology approaches integrate genomics, transcriptomics, proteomics, and metabolomics to map the regulatory networks involved in biofilm formation. This comprehensive approach helps identify key molecular players and their interactions in the development of biofilms.

- *Gene expression analysis* of biofilm formation pathways can reveal critical genes involved in adhesion, quorum sensing, and matrix production. For example, in *Pseudomonas aeruginosa*, a major pathogen associated with biofilm infections, systems biology has highlighted the roles of the las and rhl quorum-sensing systems in regulating biofilm formation (Davies et al., 1998).
- *Regulatory networks* are often explored using computational modeling, which can simulate biofilm development and predict the effects of perturbations. A systems biology model of *Escherichia coli* biofilm formation demonstrated the role of stress responses and nutrient availability in controlling biofilm growth (Klumpp et al., 2009)

16.6.2 Quorum Sensing and Signal Transduction

Quorum sensing (QS) is a key regulatory mechanism in biofilm formation, where bacteria communicate through chemical signals to coordinate group behavior. Systems biology approaches have elucidated the complex networks of QS signals and their impact on biofilm architecture.

- Systems biology has enabled the identification of novel QS inhibitors (QSIs) that could be used to control biofilm formation. For instance, the role of the LasR-LasI system in *P. aeruginosa* biofilms was highlighted in systems-based studies, where disrupting this system was found to significantly reduce biofilm formation (Rutherford & Bassler, 2012).

- The integration of *omic technologies* with computational models has advanced our understanding of the interconnectedness of QS with other microbial processes, such as metabolism and stress response. These studies have led to the discovery of potential therapeutic targets, like the inhibition of the QS receptor.

16.6.3 Biofilm Resistance Mechanisms

Biofilms are highly resistant to antibiotics and other antimicrobial agents due to the physical barrier provided by the extracellular matrix and altered gene expression within the biofilm. Systems biology offers a deeper understanding of these resistance mechanisms.

- *Antibiotic tolerance* in biofilms is partly due to nutrient deprivation and slow growth in deeper layers of the biofilm. Omics approaches have been used to identify genes involved in antibiotic resistance, such as those regulating efflux pumps and stress response pathways (Mah & O'Toole, 2001).
- *Evolutionary dynamics* of biofilm communities, studied through systems biology, have revealed how biofilms adapt to antimicrobial treatments by acquiring resistance. For example, a systems biology approach was used to investigate the adaptation of *Staphylococcus aureus* biofilms to vancomycin, highlighting the role of the agr system in modulating resistance (Peleg et al., 2012).

16.7 Inhibition of Biofilm Formation

Systems biology facilitates the development of targeted therapies to inhibit biofilm formation or disrupt existing biofilms. These approaches often involve the identification of biofilm-associated pathways and targeting these pathways with small molecules or nanomaterials.

- *Nanoparticles*, such as gold or silver nanoparticles, are being studied as potential biofilm disruptors. A systems biology approach can be used to understand the interaction of nanoparticles with biofilm-producing microorganisms at the molecular level, guiding the design of more effective biofilm-targeting treatments (Zhang et al., 2014).
- *Antibiotic combination therapies* are another focus of systems-based studies. By integrating data from multiple omics layers, researchers can identify combinations of antibiotics that work synergistically to disrupt biofilms. For instance, a systems biology model of *Streptococcus mutans* biofilms identified several antibiotic combinations that could effectively prevent biofilm formation (Suwannakham et al., 2018).

16.8 Applications in Medical and Industrial Biofilm Control

- *Medical biofilm control*: Systems biology is used to develop therapeutic strategies for chronic infections, such as those associated with *P. aeruginosa* in cystic fibrosis patients. Modeling the interactions between host immune responses and biofilm-forming bacteria has led to more personalized and effective treatment strategies (Boles et al., 2005).
- *Industrial biofilm control*: Biofilm formation in industrial systems, such as water treatment plants, can lead to fouling and corrosion. Systems biology approaches help in designing biocides that target the metabolic networks of biofilm-forming microbes. Understanding the metabolic shifts within biofilms under different environmental conditions can assist in optimizing cleaning procedures and reducing microbial fouling (Sauer et al., 2002).

16.9 Conclusion

Biofilms, complex microbial communities encased within self-produced extracellular matrices, present a dual-edged sword in various biological, medical, and industrial domains. Their ubiquitous nature and significant impact on human health, environmental systems, and industrial processes underscore the importance of understanding their formation, structure, and behavior. This thesis has explored the roles of systems biology and computational systems biology as transformative approaches for biofilm research, shedding light on their intricate dynamics and paving the way for innovative solutions to biofilm-associated challenges.

The integration of multi-omics technologies—genomics, transcriptomics, proteomics, and metabolomics—has revolutionized the study of biofilms. Genomic studies have identified key genes associated with biofilm formation, while transcriptomics has unraveled stage-specific gene expression profiles during biofilm development. Proteomics and metabolomics have provided insights into biofilm-specific proteins and metabolic pathways, revealing adaptive strategies such as quorum sensing and antimicrobial resistance mechanisms. The ability to integrate these diverse datasets into a cohesive framework has been instrumental in understanding the molecular intricacies of biofilm systems.

Computational systems biology approaches have complemented experimental methods by providing predictive and mechanistic insights into biofilm behavior. Agent-based models (ABMs) have been particularly valuable in simulating individual cell interactions within biofilms, allowing researchers to study spatial heterogeneity, nutrient gradients, and quorum sensing mechanisms. Network-based models have elucidated gene regulatory and protein-protein interaction networks, offering a systems-level perspective on biofilm dynamics. Continuum models, on the other hand, have been effective in describing macroscopic biofilm properties such as growth, nutrient diffusion, and detachment processes. These computational frameworks bridge the gap between experimental observations and theoretical predictions, enabling a deeper understanding of biofilm structure-function relationships.

The integration of experimental and computational methodologies has proven to be a powerful strategy for biofilm research. Real-time data from advanced imaging techniques, such as confocal laser scanning microscopy, have been used to refine computational models, enhancing their accuracy and applicability. Hybrid modeling approaches, which combine micro-scale agent-based models with macro-scale continuum models, have captured the complexity of biofilm systems more comprehensively. These integrative methods have provided actionable insights for managing biofilms in diverse contexts, from medical treatment to industrial applications.

The practical applications of systems biology in biofilm control are diverse and impactful. In healthcare, systems-level analyses of quorum sensing networks and biofilm matrix components have identified novel targets for anti-biofilm therapies. For instance, quorum sensing inhibitors have emerged as promising agents for disrupting communication pathways essential for biofilm formation and maintenance. In industrial settings, biofilm models have been utilized to optimize processes such as wastewater treatment and bioremediation, enhancing the efficiency and sustainability of these operations. Synthetic biology has further enabled the engineering of biofilms with customized functionalities, such as pollutant degradation and biosensing capabilities, showcasing the potential of biofilm-inspired technologies.

Despite these advancements, challenges remain. The inherent heterogeneity of biofilms, both spatial and temporal, complicates their study and modeling. Additionally, the translation of computational predictions into real-world applications often requires interdisciplinary collaboration and iterative refinement. Future research should focus on advancing data integration techniques, improving the scalability of hybrid models, and incorporating machine learning and artificial intelligence to enhance predictive capabilities. Addressing these challenges will further deepen our understanding of biofilms and their dynamics.

In conclusion, the integration of systems biology and computational approaches has transformed biofilm research, offering unprecedented insights into their complex nature. These methodologies not only provide a comprehensive understanding of biofilm systems but also translate fundamental findings into practical applications, addressing global challenges in healthcare, industry, and environmental management. The continued development of integrative frameworks and computational tools will undoubtedly enhance our ability to study, control, and harness biofilms, contributing to scientific progress and societal well-being.

References

Alm, E. J., & Arkin, A. P. (2003). Biological networks. *Current Opinion in Structural Biology, 13*(2), 193–202.

Alpkvist, E., & Klapper, I. (2007). A multidimensional multispecies continuum model for heterogeneous biofilm development. *Bulletin of Mathematical Biology, 69*(2), 765–789. https://doi.org/10.1007/s11538-006-9168-7

An, A. Y., Choi, K. Y. G., Baghela, A. S., & Hancock, R. E. (2021). An overview of biological and computational methods for designing mechanism-informed anti-biofilm agents. *Frontiers in Microbiology, 12*, 640787.

Beloin, C., Roux, A., & Ghigo, J.-M. (2008). Bacterial biofilms. In *Escherichia coli Biofilms* (Current topics in microbiology and immunology, volume 322) (pp. 249–289). Springer.

Boles, B. R., et al. (2005). Modeling the effects of biofilm structure and growth on host-pathogen interactions. *Infection and Immunity, 73*(5), 3085–3094. https://doi.org/10.1128/IAI.73.5.3085-3094.2005

Boltz, J. P., Morgenroth, E., & Sen, D. (2010). Mathematical modelling of biofilms and biofilm reactors for engineering design. *Water Science and Technology, 62*(8), 1821–1836. https://doi.org/10.2166/wst.2010.076. PMID: 20962398.

Costerton, J. W., et al. (1995). Microbial biofilms. *Annual Review of Microbiology, 49*, 711–745.

D'Acunto, B., Frunzo, L., Klapper, I., Mattei, M. R., & Stoodley, P. (2019). Mathematical modeling of dispersal phenomenon in biofilms. *Mathematical Biosciences, 307*, 70–87. https://doi.org/10.1016/j.mbs.2018.07.009

Davies, D. G., et al. (1998). The involvement of cell-to-cell signals in the development of a bacterial biofilm. *Science, 280*(5361), 295–298. https://doi.org/10.1126/science.280.5361.295

Dehkordi, J. A., et al. (2018). Kinetic studies on extra heavy crude oil upgrading using nanocatalysts by applying CFD techniques. *Chinese Journal of Chemical Engineering, 26.2*, 343–355.

Donlan, R. M. (2001). Biofilm formation: A clinically relevant microbiological process. *Clinical Infectious Diseases, 33*(8), 1387–1392.

Dzianach, P. A., Dykes, G. A., Strachan, N. J., Forbes, K. J., & Pérez-Reche, F. J. (2019). Challenges of biofilm control and utilization: Lessons from mathematical modelling. *Journal of the Royal Society Interface, 16*(155), 20190042.

Flemming, H.-C., & Wingender, J. (2010). The biofilm matrix. *Nature Reviews Microbiology, 8*(9), 623–633.

Guibert, R., Horgue, P., & Debenest, G. (2025). Micro-continuum modeling of biofilm growth coupled with hydrodynamics in OpenFOAM. *SoftwareX, 29*, 102011.

Haagensen, J. A., et al. (2015). Confocal laser scanning microscopy and biofilm imaging. *Journal of Visualized Experiments, 103*, e53404.

Haddadi, B., Jordan, C., & Harasek, M. (2017). Cost efficient CFD simulations: Proper selection of domain partitioning strategies. *Computer Physics Communications, 219*, 121–134. https://doi.org/10.1016/j.cpc.2017.05.014

Henry, C., DeJongh, M., Best, A., et al. (2010). High-throughput generation, optimization and analysis of genome-scale metabolic models. *Nature Biotechnology, 28*, 977–982. https://doi.org/10.1038/nbt.1672

Hentzer, M., et al. (2003). Attenuation of Pseudomonas aeruginosa virulence by quorum sensing inhibitors. *EMBO Journal, 22*(15), 3803–3815.

Jenal, U., et al. (2017). The second messenger cyclic di-GMP: A signaling molecule in bacterial biofilms. *Nature Reviews Microbiology, 15*(10), 576–588.

Kiran, G. S., Ramasamy, P., Sekar, S., Ramu, M., Hassan, S., Ninawe, A. S., & Selvin, J. (2018). Synthetic biology approaches: Towards sustainable exploitation of marine bioactive molecules. *International Journal of Biological Macromolecules, 112*, 1278–1288.

Kitano, H. (2002). Systems biology: A brief overview. *Science, 295*(5560), 1662–1664.

Klapper, I., & Dockery, J. (2010). Mathematical description of microbial biofilms. *SIAM Review, 52*(2), 221–265.

Klumpp, S., et al. (2009). Systems biology of microbial biofilm formation: Insights from modeling biofilm dynamics and control. *Trends in Microbiology, 17*(5), 205–211. https://doi.org/10.1016/j.tim.2009.02.002

Koshy-Chenthittayil, S., Archambault, L., Senthilkumar, D., Laubenbacher, R., Mendes, P., & Dongari-Bagtzoglou, A. (2021). Agent based models of polymicrobial biofilms and the microbiome—A review. *Microorganisms, 9*(2), 417.

Kreft, J.-U., et al. (2001). Individual-based modelling of biofilms. *Microbiology, 147*(11), 2897–2912.

Lardon, L. A., Merkey, B. V., Martins, S., Dötsch, A., Picioreanu, C., Kreft, J. U., & Smets, B. F. (2011). iDynoMiCS: next-generation individual-based modelling of biofilms. *Environmental Microbiology, 13*(9), 2416–2434.

Mah, T. F., & O'Toole, G. A. (2001). Mechanisms of biofilm resistance to antimicrobial agents. *Trends in Microbiology, 9*(1), 34–39. https://doi.org/10.1016/S0966-842X(00)01913-3

Mattei, M. R., Frunzo, L., D'acunto, B., Pechaud, Y., Pirozzi, F., & Esposito, G. (2018). Continuum and discrete approach in modeling biofilm development and structure: A review. *Journal of Mathematical Biology, 76*, 945–1003.

Nadell, C. D., et al. (2010). Emergent biofilm properties revealed by computational modeling. *Trends in Microbiology, 18*(11), 531–539.

Nagarajan, K., Ni, C., & Lu, T. (2022). Agent-based modeling of microbial communities. *ACS Synthetic Biology, 11*(11), 3564–3574.

O'Toole, G. A., et al. (2000). Biofilm formation as microbial development. *Annual Review of Microbiology, 54*, 49–79.

Parsek, M. R., & Greenberg, E. P. (2005). Sociomicrobiology: The connections between quorum sensing and biofilms. *Trends in Microbiology, 13*(1), 27–33.

Peleg, A. Y., et al. (2012). The agr system and its role in the pathogenesis of biofilm-associated infections in Staphylococcus aureus. *Journal of Clinical Investigation, 122*(8), 3028–3038. https://doi.org/10.1172/JCI62993

Picioreanu, C., Van Loosdrecht, M. C., & Heijnen, J. J. (1998). Mathematical modeling of biofilm structure with a hybrid differential-discrete cellular automaton approach. *Biotechnology and Bioengineering, 58*(1), 101–116.

Picioreanu, C., Head, I. M., Katuri, K. P., van Loosdrecht, M. C., & Scott, K. (2007). A computational model for biofilm-based microbial fuel cells. *Water Research, 41*(13), 2921–2940.

Ratwatte, A., Somathilaka, S., Balasubramaniam, S., Taggart, M., Nair, K. M., O'Riordan, A., & Dooley, J. (2025). Reconfiguring gene regulatory neural network computing for regulating biofilm formation. *IEEE Transactions on Network Science and Engineering*. https://doi.org/10.1109/TNSE.2025.3555962

Rudge, T. J., Steiner, P. J., Phillips, A., & Haseloff, J. (2012). Computational modeling of synthetic microbial biofilms. *ACS Synthetic Biology, 1*(8), 345–352.

Rutherford, S. T., & Bassler, B. L. (2012). Bacterial quorum sensing: Its role in virulence and opportunities for its control. *Nature Reviews Microbiology, 10*(4), 249–262. https://doi.org/10.1038/nrmicro2711

Sauer, K., et al. (2002). Biofilm formation by Pseudomonas aeruginosa in a flow cell: Temporal control of gene expression and phenotypic variation. *Journal of Bacteriology, 184*(15), 3559–3569. https://doi.org/10.1128/JB.184.15.3559-3569.2002

Selvarajoo, K. (2020). Systems biology approaches for understanding biofilm response. In *Quorum sensing: Microbial rules of life* (pp. 9–29). American Chemical Society.

Sigurdsson, G., Fleming, R. M., Heinken, A., & Thiele, I. (2012). A systems biology approach to drug targets in Pseudomonas aeruginosa biofilm. *PLoS One, 7*(4), e34337.

Singh, P. K., et al. (2000). Quorum-sensing signals indicate that cystic fibrosis lungs are infected with bacterial biofilms. *Nature, 407*(6805), 762–764.

Stoodley, P., et al. (2002). Biofilms as complex differentiated communities. *Annual Review of Microbiology, 56*, 187–209.

Suwannakham, S., et al. (2018). Use of systems biology for novel therapeutic strategies in biofilm control. *FEMS Microbiology Letters, 365*(7), fny097. https://doi.org/10.1093/femsle/fny097

Sweeney, E. G., Nishida, A., Weston, A., Bañuelos, M. S., Potter, K., Conery, J., & Guillemin, K. (2019). Agent-based modeling demonstrates how local chemotactic behavior can shape biofilm architecture. *Msphere, 4*(3), 10–1128.

Szklarczyk, D., et al. (2019). STRING v11: Protein–protein association networks with increased coverage, supporting functional discovery in genome-wide experimental datasets. *Nucleic Acids Research, 47*(D1), D607–D613.

Thiele, I., Heinken, A., & Fleming, R. M. (2013). A systems biology approach to studying the role of microbes in human health. *Current Opinion in Biotechnology, 24*(1), 4–12.

Vilain, S., et al. (2004). The fate of a Bacillus subtilis population in the presence of surfactin-producing strains. *Environmental Microbiology, 6*(8), 756–764.

Vital-Lopez, F. G., Reifman, J., & Wallqvist, A. (2015). Biofilm formation mechanisms of Pseudomonas aeruginosa predicted via genome-scale kinetic models of bacterial metabolism. *PLoS Computational Biology, 11*(10), e1004452.

Xavier, J. B., et al. (2005). Biofilm-control strategies based on mathematical modeling. *Communications in Computational Physics, 4*, 459–471.

Zhang, Z., et al. (2014). Nanoparticles for biofilm control: Principles and applications. *Journal of Nanobiotechnology, 12*(1), 1–10. https://doi.org/10.1186/s12951-014-0072-9

Zhang, Z., Chen, G., Hussain, W., Pan, Y., Yang, Z., Liu, Y., & Li, E. (2024). Machine learning and network analysis with focus on the biofilm in Staphylococcus aureus. *Computational and Structural Biotechnology Journal, 23*, 4148–4160.

Zhu, J., & Mekalanos, J. J. (2003). Quorum sensing-dependent biofilms enhance colonization in Vibrio cholerae. *Developmental Cell, 5*(4), 647–656.

Synthetic Biology and Biofilms 17

Paramita Pakhira, Siddhardha Busi,
and Rajaneesh Anupam

17.1 Introduction

Microorganisms are ubiquitous in the environment and can attach to various surfaces, both biotic and abiotic. In many cases, the formation of biofilms plays a beneficial role for bacteria, allowing them a transition from a planktonic lifestyle to a community-oriented mode of living. Biofilm is a collection of bacteria, and it significantly influences the organisms' physiological and phenotypic characteristics. Biofilms consist mainly of two components: bacterial cells and an extracellular polymeric substance known as the EPS matrix (Arnaouteli et al., 2021).

Bacterial biofilms can be visible on a variety of surfaces, including rocks and reefs along the coastline, plant roots, the epidermis of animals, water pipelines, equipment used in food production, and medical facilities (Fujishige et al., 2006). The formation of biofilms leads to several issues, including commercial losses and health-related concerns. Microorganisms within the biofilm are protected from harsh environments, shear stress from fluid flow, antibiotic treatments, and other chemical interventions (Cheng et al., 2019).

According to a report, approximately US$56 million is spent annually on cleaning and repairing damage caused by mature biofilms on navy ships (Schultz et al., 2011). In the food processing sector, bacteria that form biofilms are a primary source of food contamination and are linked to various foodborne illnesses.

P. Pakhira · S. Busi (✉)
Department of Microbiology, School of Life Sciences, Pondicherry University, Kalapet, Puducherry, India

R. Anupam
Department of Biochemistry and Molecular Biology, School of Life Sciences, Pondicherry University, Puducherry, India
e-mail: ranupam@pondiuni.ac.in

S. Busi et al. (eds.), *Omics Approaches in Biofilm Research*,
https://doi.org/10.1007/978-3-031-91863-6_17

Furthermore, biofilm formation on medical devices poses significant risks, as one of the main challenges in healthcare is the antibiotic resistance of multidrug-resistant biofilm-forming bacteria, which is 1000 times more than that of their planktonic counterparts (Abebe, 2020).

The need to inhibit or eliminate biofilms has driven significant research advancements. Techniques, such as drug modification, and the development of new antimicrobial agents have been thoroughly documented and applied in this field (Kyzioł et al., 2020). However, biofilms also serve beneficial roles in certain industries, such as wastewater treatment and agriculture, where they promote efficient growth, self-repair, and responsive adaptations to environmental changes (Pandit et al., 2020).

The biotransformation capabilities of biofilms can be effectively utilized for wastewater treatment. Within biofilms, the density of microbial cells is high, and the cells are immobilized, resulting in increased stability (Kyzioł et al., 2020). Furthermore, biofilms exhibit reduced sensitivity to conventional antibiotics and toxins, making them valuable biocatalysts in the production of various biological products, like antibiotics, enzymes, and bioenergy, and in biorefinery applications (Todhanakasem, 2017).

To effectively control biofilm development, it is essential to understand several major factors that can influence its formation, which are crucial for both prevention and application. Biofilm formation can be impacted by various conventional elements, including flow rate, atmospheric pH, temperature, and gravitational force (Karimi et al., 2015). In this chapter, rather than categorizing biofilms purely as environmental threats or beneficial entities, we will focus on strategies for controlling biofilm formation through synthetic biology. By gaining insight into the precise constituents of biofilms, we can explore how manipulation of these entities can optimize the use of biofilms as productive resources, facilitated by advances in synthetic biology.

17.2 Definition of Biofilms

Microbial biofilms are three-dimensional structures formed by individual bacteria or various bacterial species that attach to both biotic and abiotic surfaces. Typically, these biofilms consist of single cells or small clusters of cells that can proliferate and evolve into more complex communities, aided by the presence of EPS (Branda et al., 2005), extracellular proteins (Flemming and Wingender, 2010), lipids (Gehrke et al., 1998), and surrounded extracellular nucleic acids (Zhang et al., 1998; Whitchurch et al., 2002). Additionally, some biofilms may also contain humic substances (Nielsen et al., 1996; Mojica et al., 2007) and uronic acids (Mojica et al., 2007). Biofilms enhance bacterial fitness by increasing surface adherence, providing protection against predation (Matz, 2009), and offering defense against desiccation (Wong & O'Toole, 2011), as well as antibiotics and starvation, through carbon storage. This ability to form biofilms can also elevate the pathogenic potential of bacteria (Van Hoek, 2013).

Biofilms formed by different bacteria exhibit a variety of structural features. For example, biofilms of *Pseudomonas aeruginosa* typically present mushroom-like elevated structures and distinct networks facilitating water circulation. In contrast, other bacteria, such as *Francisella tularensis*, develop smoother and less distinguishable biofilms (Champion et al., 2019). Recent studies have identified several proteins, carbohydrates, and nucleic acids present in the extracellular matrix as the primary components of biofilm structure. Notably, it was observed that degrading these central extracellular components significantly reduced the biofilm size (Chung et al., 2014; Kaplan et al., 2004; Yu et al., 2015).

Extracellular polymeric substances (EPS) are found at the outermost layer of biofilms. EPS is composed of extracellular lipids, a variety of carbohydrates, proteins, and several acids (Yu et al., 2015). Recent research on EPS has primarily focused on carbohydrates, as they are the dominant component of biofilm biomass. The mechanisms by which sugar components are linked vary significantly between species and depend on environmental changes. These carbohydrate chains play a crucial role in the diversity of biofilm formation as observed in nature (Leme et al., 2006).

The complex structures of biofilms are associated with several diseases and are responsible for biocorrosion (Beech & Sunner, 2004) and biofouling (Baker & Dudley, 1998; Geesey, 1994). While biofilms can contribute to these negative outcomes, they also play a beneficial role in food manufacturing and the maintenance of human health (Olanbiwoninu and Popoola, 2023). Compared to planktonic cells, biofilms exhibit distinct properties, such as increased resistance to antibiotics (Nickel et al., 1985), antiseptics (Cargill et al., 1992), phages (Costerton et al., 1995), and disinfectants (Cargill et al., 1992). Additionally, biofilms provide protection against shear forces (Rittman, 1982), heat (Almatroudi et al., 2018), and ultraviolet radiation (Pozos et al., 2004). While not all biofilms possess all these characteristics, their formation enhances bacterial resilience. Recent transcriptomic studies have highlighted the differences between biofilm and planktonic cells (Lo et al., 2009).

17.3 Biofilm Macromolecules

17.3.1 Protein

Proteins embedded within bacterial cells represent a significant portion of biofilm biomass (Jahn et al., 1999). Current evidence indicates that most proteins found in biofilms are not secreted from lysed cells but are predominantly located in the outer membrane. They may be transported to the biofilm via vesicles that originate from this membrane (Toyofuku et al., 2012). The secreted proteins specifically cling to certain biotic surfaces, such as fibronectin, chitin (Brandl et al., 2011), or mannose (Pratt & Kolter, 1998), to enhance attachment.

Alternative strategies for attachment include the expression of pili (Tomaras et al., 2003), flagella (Kim et al., 2008), and fimbriae. These unique and versatile methods of attachment can be engineered using synthetic biology tools. For instance, pili, when induced by rhamnose, can lead to an increased production of biofilm in the presence of this sugar. This serves two distinct purposes: first, it provides a cost-effective means of forming multispecies biofilms; second, it acts as a leak-type structure, allowing biofilms topreferentially form in areas with higher concentrations of the inducer (Hantus et al., 2024).

Various biosurfactants can reduce formation of biofilm. The production of these biosurfactants may differ significantly among species. One specific fimbria subtype known as curli (Cookson et al., 2002) is produced by certain bacteria and can form amyloid fibers (Xavier et al., 2011). Notably, certain bacterial strains can elongate each other's amyloid fibers. Research indicates that these fibers contribute structural strength to the biofilm matrix, making the removal of biofilms challenging through physical methods (Flemming et al., 2023).

These fibers can retain specific chemical compounds and form bonding with them. Engineered binding sites within amyloid fibers can capture a larger number of substrates. Proteins that promote cellular aggregation or repel grazing protists can exhibit a range of beneficial functions, particularly within biofilms. An engineered bacterial biofilm may serve as a reservoir for the controlled release of certain chemicals (Li et al., 2014). This method offers advantages over conventional biofilm dispersion, as the reconstruction of a biofilm is dependent on resources; often, a minimal biomass loss linked to product release is necessary. The use of engineered bacterial biofilm could be particularly advantageous in systems where the collection process incurs high costs (Nguyen et al., 2014).

Cells within a biofilm commonly produce an S-layer, which is located closer to the cell membrane than the amyloid fibers. This protein layer is formed by most archaea and few bacteria, serving to protect the outer side of the cell wall (Sleytr & Beveridge, 1999). Engineered S-layers have significantly contributed to various production processes. Modifying S-layer proteins can serve as a platform for precise interactions involving several enzymes. Additionally, S-layers provide protection to bacteria in adverse environmental conditions and facilitate bacterial attachment to different surfaces (Gerbino et al., 2015). They can also function as a mesh that prevents bigger molecules from penetrating the layers of peptidoglycan. A major concern with both S-layers and amyloid fibers is their potential to become clogged with various compounds (Sleytr & Beveridge, 1999).

17.3.2 Lipids

The primary lipid constituents in bacterial cells are outer membrane vesicles (OMVs) and lipid bilayers. OMVs are recognized for their significant role in overall biofilm biomass; however, their specific impact on biofilm lipids remains unclear (Wang et al., 2015). Research has shown that biofilm-forming cells can modify their lipid membranes to a greater extent than planktonic cells (Benamara et al., 2011).

Additionally, some studies have demonstrated that the removal or substantial reduction of fatty acids can effectively hinder biofilm formation (Lattif et al., 2011). Since the field of lipidomics is still in its infancy, these early findings pose challenges for synthetic biologists seeking to efficiently alter the lipid composition in biofilms by enhancing lipid production (Zara et al., 2012).

17.3.3 Nucleic Acids

Extracellular DNA (eDNA) is present within the matrix of bacterial biofilms, referred to as extracellular polymeric substances (EPS), which exist outside the cellular membrane. The methods and techniques for isolating, manipulating, interpreting, and constructing nucleic acids are significantly more advanced than those for their lipid components (Flemming and Wingender, 2010). Initially, it was believed that nucleic acids linked with the biofilm matrix were merely remnants of lysed cells (Kadurugamuwa & Beveridge, 1995). However, it has been observed that during the lysis process, some molecules are also released within outer membrane vesicles (OMVs) (Mann et al., 2009). The regulatory mechanisms behind the release of DNA through OMVs and quorum sensing remain unclear. Additionally, OMVs are known to release various intracellular proteins (Allesen-Holm et al., 2006; Thomas et al., 2009; Song et al., 2008).

Current studies have highlighted the significance of extracellular DNA (eDNA) in the formation of biofilm (Harmsen et al., 2010). eDNA not only enhances biofilm attachment but also contributes to structural stability (Gödeke et al., 2011). Typically, eDNA is a complex mixture of genomic DNA (Steinberger & Holden, 2005); however, in depth research is required to explore structural forms of DNA other than the B-form in relation to biofilm stability and attachment. Future research could open new avenues for utilizing eDNA in biofilm development, especially considering the potential role of fragmented DNA length in modulating attachment.

In one study involving *Listeria monocytogenes*, it was demonstrated that small fragments of DNA can prevent the attachment of the bacteria. However, when larger DNA fragments were present, these smaller pieces were able to inhibit the successive attachment of the biofilm, aided by the longer DNA (Harmsen et al., 2010). The release of extracellular DNA (eDNA) in mixed-species biofilms is significantly more complex. This complexity arises from the presence of diverse organisms, each exhibiting varying tendencies to release eDNA, which can differ based on the specific species within the biofilm (Steinberger & Holden, 2005).

In one study, it was observed that the amounts of extracellular DNA (eDNA) and peptidoglycan are mutually dependent in the attachment process, highlighting the potential role of eDNA in biofilm formation (Harmsen et al., 2010). Modern research has demonstrated that the cross-linking machinery of eDNA contributes to both the structural strength and flexibility of newly formed biofilms (Devaraj et al., 2019). eDNA presents remarkable opportunities for controlling attachment of biofilm and providing structural stability.

The release of eDNA can vary depending on several factors. Specific protein mechanisms, lysis processes influenced by phages, or quorum sensing mediated by outer membrane vesicles (OMVs), along with antibiotic-induced OMV release, can modify mechanical properties of biofilms. Beyond establishment and maintenance, eDNA may also play significant roles in biofilm resistance (Wang et al., 2015).

The DNA of lysed cells can serve as a carbon source to enhance survival, providing a beneficial mechanism for cells at the stationary phase. Competence enables a pathway for gene and plasmid transfer among bacteria (Finkel & Kolter, 2001) and is also a means of developing antibiotic resistance. One notable mechanism of antibiotic resistance involves extracellular DNA (eDNA), where surrounding cells detect eDNA via two-component systems, leading to the upregulation of various genes in response (Mulcahy et al., 2008).

Moreover, structural mechanisms related to eDNA can contribute to antibiotic resistance. The negative charge of DNA can shield biofilm cells from cationic antimicrobial peptides and aminoglycosides, both of which possess a positive charge. Sequential removal of eDNA can facilitate the formation of biofilms that exhibit heightened resistance against attacks from positively charged molecules (Arciola et al., 2005).

eDNA presents an exciting opportunity for synthetic biologists. Even though the biofilm machinery may be compromised, there remains a practical possibility of integrating portions of eDNA into a new system (Pier et al., 2001). This intriguing potential is yet to be fully explored. eDNA can facilitate various physical processes, such as biofilm stabilization, and exopolysaccharides can effectively counteract both antibiotic-based disruptions and attacks from the immune system (Campoccia et al., 2021).

A study in 2021 reported that eDNA-based architecture depended on the Z-form. During the biofilms maturations Z-form of DNA accumulate and stabilized by the DNABII proteins. Z-forms of eDNA provide structural integrity to the biofilm matrix. Significantly, they also demonstrated that the DNABII family of proteins can stabilizes both bacterial eDNA and host-eDNA in the Z-form in situ (Buzzo et al., 2021).

17.3.4 Carbohydrates

Various exopolysaccharides play a crucial role in trapping micronutrients essential for biofilm formation in specific environments. The carbohydrates located in the external portion of exopolysaccharides are identified for their ability to enhance attachment to surfaces, thereby promoting biofilm development (Frederix et al., 2014). A variety of complex polysaccharides are utilized within the EPS matrix of the biofilm (Chu et al., 2008).

There are several methods to manipulate exopolysaccharide production within the biofilm matrix. This can be achieved at the transcriptional level through the use of repressors, as well as at the post-transcriptional level via the modification of precursors or through the degradation of already secreted sugar moieties. These strategies may facilitate the production of desired substances, such as glucosamine through the use of chitin (Bao et al., 2018).

17.4 Definition of Synthetic Biology

Synthetic biology has been articulated by various researchers and engineers across multiple disciplines (Benner & Sismour, 2005). In essence, synthetic biology can be called "the design of life for useful purposes." The functions of synthetic biology can be categorized into two main types: the first focuses on studying life itself, while the second involves designing life to achieve beneficial outcomes. Significant advancements have been made in the synthetic biology field through the standardization of processes (Anderson et al., 2010) and the optimization (Canton et al., 2008) for creating modules or logic gates (Slusarczyk et al., 2012). These components can be engineered sequentially to develop comprehensive systems (Hasunuma et al., 2011). In this area of biology, entire systems are being established to achieve precise outputs (Rudge et al., 2012). Furthermore, biofilms possess several properties compared to planktonic bacterial cells, and synthetic biologists are leveraging these properties to boost existing biofilm production systems (Brenner et al., 2008).

17.5 Genetic Manipulation of Biofilm Architecture

Recent advancements in biofilm genetics focus on modifying the genomes of planktonic bacterial cells to enable them to produce biofilm. By manipulating quorum sensing, the transcription of specific genes accountable for biofilm formation can be directly or indirectly influenced. Genetic manipulation offers a promising approach to either eliminate biofilm or prevent its formation entirely (Flemming et al., 2016).

While there are novel conjugation-based methodologies against biofilms produced by multi-species, these techniques have yet to be developed into modular tools. Consequently, the most effective strategy is to edit the genomes of distinct planktonic bacteria and encourage their biofilm formation. To enhance biofilm architecture, further improvements in assembly are needed; this includes examining various aspects of biofilm design, such as the distribution and roles of simple macromolecules, the dispersal of cells throughout the biofilm, their interactions with different bacterial species, and the overall three-dimensional structure of the biofilm (Flemming et al., 2016).

17.6 Synthetic Biology Approaches for Controlling Biofilm

With a better understanding of biofilms and an expanding toolkit in synthetic biology, our ability to control biofilms is continually improving (Wood et al., 2011). (EPS). This approach may lead to various biofilm-related challenges across several domains, including environmental, industrial, and medical fields (Flemming et al., 2016). The eradication of biofilms through protein engineering and synthetic biology is discussed in detail below.

17.7 Bacterial Biofilm and Protein Engineering

Hha, a protein recognized for its high hemolysin activity, which was engineered to enhance the dispersion of biofilm, that can facilitate biofilm dispersal. Proteins possess the ability to bind with signaling molecules that are engineered for both the inhibition and dispersion of biofilms (Hong et al., 2010). While in this study *E. coli* does not synthesize AHLs due to the absence of an enzyme called AHL synthase, it features a receptor known as SdiA, which is analogous to LuxR (Dyszel et al., 2010). The SdiA receptor was engineered using random and site-directed mutagenesis techniques to regulate biofilm development in the presence of molecules such as AHLs or indole (Lee et al., 2009).

In one more research, major food-contaminating bacteria *Salmonella enterica*, which is unable to synthesize AHL signaling molecules like *E. coli* was studied. However, it does possess the SdiA receptor for AHL signaling, which can regulate *S. enterica* adhesion and confer resistance against host immunity (JB & RR, 2016). Additionally, another c-di-GMP-binding protein known as BdcA has been engineered to enhance dispersal of biofilm through a specific amino acid change, E50Q.

BdcA from *E. coli* can regulate biofilm dispersal in various microorganisms, including *P. aeruginosa* (Ma et al., 2011). Thus, the protein engineering of universal regulators and proteins that interact with signaling molecules can significantly suppress biofilm formation, potentially providing a valuable approach for modulating various microbial activities related to biofilm development.

17.8 Quorum Sensing Mechanisms in Bacteria and Synthetic Biology

Quorum sensing (QS) systems in bacteria have emerged as pivotal mechanisms for a diverse range of engineered biological strategies. Autoinducers are advantageous because they can respond to signals by easily diffusing in liquid media and readily entering the cell (Choudhary & Schmidt-Dannert, 2010).

Engineered cells can generate their individual QS systems, enabling them to regulate cell density and control various mechanisms autonomously. Synthetic QS systems hold potential for the development of artificial genetic frameworks, whether by establishing predator-prey dynamics or by creating symbiotic relationships (Wood et al., 2011).

The primary quorum sensing systems in *P. aeruginosa* are LasI/R and RhlI/R; both have been widely utilized in synthetic genetic research. The LasI system produces the autoinducer C_{12}-HSL, which is recognized by the quorum-sensing regulator LasR. Similarly, the RhlI system generates an autoinducer known as C_4-HSL, which is detected by RhlR (Pesci et al., 1997). In the biofilm development, the RhlI/R quorum sensing system has been employed to identify essential components for bacterial interaction and accumulation within an artificial biofilm environment (Hong et al., 2012).

The quorum sensing system of LasI/R, in conjunction with Hha and BdcA, facilitates biofilm dispersal and demonstrates remarkable biofilm dislocation upon the detection of QS signals (Hong et al., 2012). Secondary biofilm can proliferate alongside preexisting biofilms, with QS signaling molecules produced by the LasI system accumulating within these dual-species biofilm structures. These QS molecules form a complex with LasR that can trigger biofilm dispersion by increasing levels of c-di-GMP, a process aided by BdcA. Subsequently, the dispersed biofilm is disrupted through elevated levels of Hha, induced by a chemical shift, leading to cell loss within the preformed biofilm (Hong et al., 2012).

The QS system, modified through synthetic biology, was employed to prevent membrane biofouling and effectively degrade components that contribute to environmental pollution. This beneficial biofilm demonstrated the ability to reduce the individual thickness of membranes during wastewater treatment by detecting signaling molecules that regulate c-di-GMP production, a process facilitated by BdcA (Wood et al., 2016).

Additionally, engineered biofilms can inhibit the development of harmful bacteria by producing nitric oxide (NO) and are capable of diminishing environmental pollutants such as epichlorohydrin through the action of epoxide hydrolase (Wood et al., 2016).

17.9 Importance of Enzymes Related to Quorum Quenching (QQ)

Uncovering a technique to disrupt microorganisms by interfering with their communication systems is crucial in the battle against antibiotic resistance and various biofilm-associated conditions (Marx, 2014). Quorum sensing (QS) can be inhibited through the process of quorum quenching (QQ). Strategies in this area involve the development, modification, or degradation of multiple signaling molecules essential for bacterial cellular communication, thereby preventing biofilm accumulation. The primary QQ mechanisms focus on the hydrolysis of N-acyl homoserine lactones through the action of lactonase enzymes, which can cleave the lactone rings (Oh & Lee, 2018).

The QQ mechanisms utilized by bacterial systems have proven to be effective for antifouling strategies in membrane bioreactors (MBRs) used for wastewater treatment. For instance, the production of AHL by bacteria was diminished with the assistance of the lactonase enzyme, derived from recombinant *E. coli* strains, AiiA from *Bacillus thuringiensis*, and AiiO from *Agrobacterium tumefaciens*. This enzymatic QQ approach led to a decrease in EPS production and gene expression related to microbial adhesion and accumulation (Fang et al., 2020).

The W2 strain of *Rhodococcus erythropolis* has demonstrated the ability to degrade AHLs through the action of oxidoreductase enzymes and AHL-acylase. Additionally, LsrK, a luxS-regulated kinase, can phosphorylate AI-2, functioning as a QQ enzyme. The application of pure LsrK, in combination with ATP, notably decreased AI-2 signaling in *S. typhimurium* and *V. harveyi* (Roy et al., 2010).

Candida albicans secretes a compound known as farnesol, which can repress AI-2 synthesis and modify biofouling in MBRs. In an innovative approach, metagenomics has been employed to identify strategies for altering AI-2 levels (Oh & Lee, 2018). Furthermore, a native strain of *Acinetobacter* sp. DKY-1 was identified to deactivate AI-2 by producing a hydrophilic compound that serves as a QQ substrate with a lower molecular weight. The integration of QQ enzyme production systems with synthetic genetic pathways could enable significant control over biofilm disruption and elimination (Lee et al., 2018).

17.10 Effect of Bacteriophages

Bacteriophages can able to enter in the inner layers of biofilms due to their phage depolymerase enzymes, which are capable of degrading extracellular polymeric substances (EPS). Both single-type phages and phage cocktails have been shown to inhibit biofilm formation (Fu et al., 2010).

They present significant potential for the development of antimicrobial agents and can also serve as vehicles for drug delivery. Engineered bacteriophages may be utilized for the design of new materials (Pires et al., 2016).

Synthetic biologists are focusing on creating enhanced bacteriophages that can effectively eliminate harmful biofilm cells by producing biofilm-inhibiting enzymes or improving the penetration of antibiotics. For instance, dispersin B, an enzyme capable of degrading biofilms, was engineered from the T7 phage during phage infection. *Actinobacillus actinomycetemcomitans* produced a variant of dspB that was incorporated into the phage genome under the control of a T7 φ10 (Lu & Collins, 2007).

This setup enhances dspB transcription through the activity of T7 RNA polymerase during *E. coli* TG1 biofilm infection by the phage. The phages not only have the capability to destroy cells but also allow DspB to target the biofilm matrix through hydrolyzing the biofilm-stimulating adhesin known as β-1,6-N-acetyl-D-glucosamine, which is produced by *E. coli* (Lu & Collins, 2007).

This engineered phage can reduce biofilm levels of *E. coli* by two orders of magnitude compared to the previously utilized wild-type phage, which lacks enzymatic action. Infection by the engineered phage increased the bactericidal effects of ofloxacin antibiotics, making it more effective in eliminating both biofilm and planktonic cells than non-modified phage treatments (Lu & Collins, 2007).

To address the substrate specificity limitations of enzymes used for biofilm degradation, DspB was combined with bacteriophages, resulting in a more effective strategy for preventing multi-species biofilms (Lu & Collins, 2007).

Quorum quenching (QQ) phage therapy is crucial for inhibiting biofilms of *P. aeruginosa* and *E. coli*, either through cell lysis or by degrading AHL molecules (Pei & Lamas-Samanamud, 2014).

17.11 Probiotics

Probiotics are known to enhance the host's immunity and inhibit pathogens, as well as possess the capability to prevent formation of biofilm (Hill et al., 2014). Probiotics have been considered for biotechnological applications and therapeutic purposes in humans. The *E. coli* strain Nissle 1917 is regarded as one of the most effective probiotics, utilized in various clinical treatments related to intestinal disorders. Engineered EcN also enhances live biotherapeutics, making it applicable for tumor detection, treatment of hyperammonemia, and serving as a drug delivery vehicle (Henker et al., 2007).

Wild-type EcN has been shown to prevent *E. coli* biofilm. EcN inhibits the formation of multi-species biofilms involving Gram-positive bacteria such as *Staphylococcus aureus* and *S. epidermidis* (Fang et al., 2018). Engineered EcN has also been developed for sensing, targeting, and inhibiting biofilms to protect *Caenorhabditis elegans* from *P. aeruginosa* (Hwang et al., 2017).

The *alr* and *dadx* genes have been deleted from all mutant strains of EcN, creating a D-alanine auxotroph that stabilizes the maintenance of the plasmid expressing *alr*. This modified EcN is regulated through a synthetic genetic system. In the presence of C_{12}-HSL from *P. aeruginosa*, engineered EcN produces E7 lysis protein to disrupt host cells. Additionally, S5 pyocin is generated to kill *P. aeruginosa*, while DspB is produced to degrade the biofilm matrix. This engineered EcN can disrupt preformed biofilms of *P. aeruginosa* and prevent the new biofilm's formation (Hwang et al., 2017).

Overall, this innovative research underscores the potential of genetic studies in synthetic biology to improve the prophylactic and therapeutic efficacy of probiotics against biofilm-forming microorganisms.

17.12 Biofilm Engineering by Synthetic Biology

While the removal of harmful cells in biofilms is crucial, it is significant that they can also possess beneficial properties if produced under controlled conditions, with careful consideration of pattern, thickness, composition, and metabolism (Mangwani et al., 2016). Engineered biofilms created through synthetic biology can be applied in various areas such as bioremediation, wastewater treatment, biocorrosion control, biofuel production (Heimann, 2016), specialty and bulk chemical biorefineries, biomedical microelectromechanical systems (bio-MEMS devices), and pharmaceutical testing (Fernandes et al., 2010). The design and development of systems that manage biofilm behavior can be facilitated by synthetic genetic pathways.

Artificial biofilms demonstrate favorable electrochemical properties due to the integration of a redox-active cross-linking network with the adjacent polymeric matrix surrounding the bacterial cells (Hubenova et al., 2019). It is acknowledged that artificially produced biofilms may influence the phylogenetic composition of emerging biofilms, potentially leading to a reduction in phylogenetic diversity and in the elimination of unwanted bacterial species (Rijavec et al., 2019).

17.13 Monitoring of Biofilm Formation with the Help of Synthetic Biology

The bacterial genome has a crucial role in regulating biofilm formation. Synthetic biology represents a promising approach for precisely manipulating biological systems. Through genetic modification, both bacteria and their associated biofilms can be tailored to obtain specific functions (El Karoui et al., 2019).

17.13.1 Amyloids

The extracellular polymeric substances (EPS) within biofilms consist of polysaccharides, eDNA, and proteins. The primary component of the polysaccharides includes glucan, fructan homopolymers, and heteropolymers of glucose, mannose, or rhamnose, as well as alginate and N-acetylglucosamine, all of which are integral to the biofilm biomass (Nguyen, 2017). Amyloids, essential proteins found in the biofilm matrix, play a crucial role as structural elements that facilitate interactions among bacterial cells (Nguyen, 2017).

Furthermore, amyloids contribute to morphological variation, affect surface properties, and provide protection against viral infections. A notable example of these amyloids is curli proteins, which are highly stable and robust, making them suitable for applications in nanotechnology. *E. coli* can produce curli proteins either autonomously or with the aid of an external control system. In the biofilm of *Bacillus subtilis*, fibers are composed of proteins known as TasA and TapA (Romero et al., 2014).

17.13.2 Biomineralization

Hierarchical architectural mineralized constituents play a crucial role in both nature and human activities and can form through the regulation of the mineralization process. This process involves an organic matrix that helps to create a framework, following the establishment of nucleation sites, culminating in growth and termination. Mineralization is primarily categorized into two types: biologically induced and biologically controlled processes (Connett et al., 1983).

Various bacteria can mediate different types of mineralization depending on the availability of elements such as iron and manganese. The biomineralization process has applications across numerous fields, including environmental pollutant removal, wastewater treatment, material construction, cancer therapy, and tissue engineering. Notably, biofilm-associated coatings have demonstrated superior mineralization and greater durability compared to hydroxyapatite spray samples (Qin et al., 2020a, b).

17.13.3 Cell-Based Medicine

Species capable of producing biofilms and functioning as probiotics play an essential role as gut-associated microbiota (Duraj-Thatte et al., 2018). The engineering of biofilms can lead to the development of novel bacteria that may have diagnostic and therapeutic applications (Fig. 17.1). A biofilm-producing strain of *E. coli* known as EcN has been engineered to compete against various pathogenic species, including

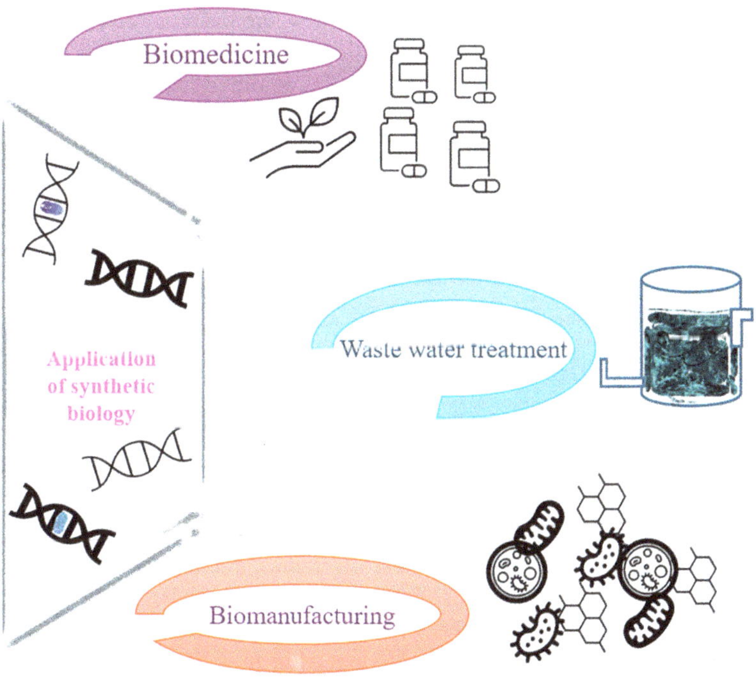

Fig. 17.1 Application of synthetic biology in biofilm research

enterohemorrhagic *E. coli* and *Staphylococcus epidermidis*, by expressing and secreting the protease DegP (Fang et al., 2018).

This engineered EcN has established a synthetic biosensing module that enables these strains to colonize the mouse gut and naturally monitor inflammation (Riglar et al., 2017). Additionally, the extracellular matrix (ECM) of biofilms has been utilized to coat probiotic *B. subtilis* cells, enhancing gut adhesion and bioavailability (Wang et al., 2020). Many biofilm-forming species also secrete neuroactive molecules. In the *C. elegans* gut bacteria can modulate host sensory responses through the production of tyramine (O'Donnell et al., 2020).

17.13.4 Biomanufacturing Process

Synthetic biology has the potential to advance more effective systems for synthesizing biochemical compounds (Fig. 17.1). One of the key challenges in expressing multiple enzymes along a single pathway is minimizing toxicity and the metabolic problem on the host cell. However, in the context of biofilms, this issue can be mitigated through labor division among the population. This cooperative behavior is primarily observed in the production of various extracellular matrix (ECM) constituents within the biofilm of *Bacillus subtilis*, where bacterial cells collaborate to generate specific products (Dragoš et al., 2018).

Additionally, the ECM of the biofilm can be modified to exhibit precise attractions and crosslinking, transforming it into a renewable biomaterial. The amyloid fibers of both *E. coli* and *B. subtilis* have been genetically engineered to express domains of peptides that enhance the matrix of the biofilm, making it a robust and renewable biomaterial (Nguyen et al., 2014; Zhang et al., 2019; Huang et al., 2019). Furthermore, these strategies can be integrated with three-dimensional printing to quickly and accurately shape biofilms into preferred structures.

17.13.5 Remediation of Environment

The ability of biofilms to thrive in challenging environments makes them a perfect platform for organizing engineered bacteria capable of mitigating pollutants (Tay et al., 2018). Currently, bacterial biofilms are widely employed in mercury contaminated water treatment, where they effectively break down organic pollutants (Tay et al., 2017). Additionally, in the context of on-site remediation and the upstream capture of pollutants, the functionality of biofilms can be enhanced through the application of synthetic biology (Benedetti et al., 2016). Research has also demonstrated that biofilms can sequester pollutants from their surrounding environment (Mukherjee et al., 2018).

17.14 Conclusion and Future Perspectives

Synthetic biology in biofilm research has significantly advanced our understanding of biofilm formation, structure, and behavior. By leveraging synthetic biology tools, such as gene editing, synthetic gene circuits, and microbial engineering, researchers have been able to manipulate biofilm-forming organisms to study their underlying molecular mechanisms and to design new strategies for controlling or enhancing biofilm-related processes. These approaches have implications across a variety of fields, including healthcare (e.g., treatment of biofilm-associated infections), environmental management (e.g., wastewater treatment), and industrial applications (e.g., biofouling control). The ability to engineer microbes and their biofilm-producing capabilities opens new avenues for targeted intervention, optimization of biofilm-related processes, and the development of novel materials with biofilm-like properties.

References

Abebe, G. M. (2020). The role of bacterial biofilm in antibiotic resistance and food contamination. *International Journal of Microbiology, 2020*(1), 1705814.

Allesen-Holm, M., Barken, K. B., Yang, L., Klausen, M., Webb, J. S., Kjelleberg, S., et al. (2006). A characterization of DNA release in Pseudomonas aeruginosa cultures and biofilms. *Molecular Microbiology, 59*(4), 1114–1128.

Almatroudi, A., Tahir, S., Hu, H., Chowdhury, D., Gosbell, I. B., Jensen, S. O., ... & Vickery, K. (2018). Staphylococcus aureus dry-surface biofilms are more resistant to heat treatment than traditional hydrated biofilms. *Journal of Hospital Infection, 98*(2), 161–167.

Anderson, J., Dueber, J. E., Leguia, M., Wu, G. C., Goler, J. A., Arkin, A. P., & Keasling, J. D. (2010). BglBricks: A flexible standard for biological part assembly. *Journal of Biological Engineering, 4*, 1–12.

Arciola, C. R., Campoccia, D., Gamberini, S., Donati, M. E., Pirini, V., Visai, L., et al. (2005). Antibiotic resistance in exopolysaccharide-forming Staphylococcus epidermidis clinical isolates from orthopaedic implant infections. *Biomaterials, 26*(33), 6530–6535.

Arnaouteli, S., Bamford, N. C., Stanley-Wall, N. R., & Kovács, Á. T. (2021). Bacillus subtilis biofilm formation and social interactions. *Nature Reviews Microbiology, 19*(9), 600–614.

Baker, J. S., & Dudley, L. Y. (1998). Biofouling in membrane systems—a review. *Desalination, 118*(1–3), 81–89.

Bao, J., Liu, N., Zhu, L., Xu, Q., Huang, H., & Jiang, L. (2018). Programming a biofilm-mediated multienzyme-assembly-cascade system for the biocatalytic production of glucosamine from chitin. *Journal of Agricultural and Food Chemistry, 66*(30), 8061–8068.

Beech, I. B., & Sunner, J. (2004). Biocorrosion: Towards understanding interactions between biofilms and metals. *Current Opinion in Biotechnology, 15*(3), 181–186.

Benamara, H., Rihouey, C., Jouenne, T., & Alexandre, S. (2011). Impact of the biofilm mode of growth on the inner membrane phospholipid composition and lipid domains in Pseudomonas aeruginosa. *Biochimica et Biophysica Acta (BBA)-Biomembranes, 1808*(1), 98–105.

Benedetti, I., de Lorenzo, V., & Nikel, P. I. (2016). Genetic programming of catalytic Pseudomonas putida biofilms for boosting biodegradation of haloalkanes. *Metabolic Engineering, 33*, 109–118.

Benner, S. A., & Sismour, A. M. (2005). Synthetic biology. *Nature Reviews Genetics, 6*(7), 533–543.

Branda, S. S., Vik, Å., Friedman, L., & Kolter, R. (2005). Biofilms: The matrix revisited. *Trends in Microbiology, 13*(1), 20–26.

Brandl, M. T., Carter, M. Q., Parker, C. T., Chapman, M. R., Huynh, S., & Zhou, Y. (2011). Salmonella biofilm formation on Aspergillus Niger involves cellulose–chitin interactions. *PLoS One, 6*(10), e25553.

Brenner, K., You, L., & Arnold, F. H. (2008). Engineering microbial consortia: A new frontier in synthetic biology. *Trends in Biotechnology, 26*(9), 483–489.

Buzzo, J. R., Devaraj, A., Gloag, E. S., Jurcisek, J. A., Robledo-Avila, F., Kesler, T., et al. (2021). Z-form extracellular DNA is a structural component of the bacterial biofilm matrix. *Cell, 184*(23), 5740–5758.

Canton, B., Labno, A., & Endy, D. (2008). Refinement and standardization of synthetic biological parts and devices. *Nature Biotechnology, 26*(7), 787–793.

Cargill, K. L., Pyle, B. H., Sauer, R. L., & McFeters, G. A. (1992). Effects of culture conditions and biofilm formation on the iodine susceptibility of Legionella pneumophila. *Canadian Journal of Microbiology, 38*(5), 423–429.

Campoccia, D., Montanaro, L., & Arciola, C. R. (2021). Extracellular DNA (eDNA). A major ubiquitous element of the bacterial biofilm architecture. *International Journal of Molecular Sciences, 22*(16), 9100.

Champion, A. E., Catanzaro, K. C. F., Bandara, A. B., & Inzana, T. J. (2019). Formation of the Francisella tularensis biofilm is affected by cell surface glycosylation, growth medium, and a glucan exopolysaccharide. *Scientific Reports, 9*(1), 12252.

Cheng, Y., Feng, G., & Moraru, C. I. (2019). Micro-and nanotopography sensitive bacterial attachment mechanisms: A review. *Frontiers in Microbiology, 10*, 191.

Choudhary, S., & Schmidt-Dannert, C. (2010). Applications of quorum sensing in biotechnology. *Applied Microbiology and Biotechnology, 86*, 1267–1279.

Chu, F., Kearns, D. B., McLoon, A., Chai, Y., Kolter, R., & Losick, R. (2008). A novel regulatory protein governing biofilm formation in Bacillus subtilis. *Molecular Microbiology, 68*(5), 1117–1127.

Chung, M. C., Dean, S., Marakasova, E. S., Nwabueze, A. O., & van Hoek, M. L. (2014). Chitinases are negative regulators of Francisella novicida biofilms. *PLoS One, 9*(3), e93119.

Cookson, A. L., Cooley, W. A., & Woodward, M. J. (2002). The role of type 1 and curli fimbriae of Shiga toxin-producing Escherichia coli in adherence to abiotic surfaces. *International Journal of Medical Microbiology, 292*(3–4), 195–205.

Connett, P. H., Follmann, H., Lammers, M., Mann, S., Odom, J. D., Wetterhahn, K. E., & Mann, S. (1983). Mineralization in biological systems. *Inorganic Elements in Biochemistry*, 125–174.

Costerton, J. W., Lewandowski, Z., Caldwell, D. E., Korber, D. R., & Lappin-Scott, H. M. (1995). Microbial biofilms.

Devaraj, A., Buzzo, J. R., Mashburn-Warren, L., Gloag, E. S., Novotny, L. A., Stoodley, P., et al. (2019). The extracellular DNA lattice of bacterial biofilms is structurally related to Holliday junction recombination intermediates. *Proceedings of the National Academy of Sciences, 116*(50), 25068–25077.

Dragoš, A., Kiesewalter, H., Martin, M., Hsu, C. Y., Hartmann, R., Wechsler, T., et al. (2018). Division of labor during biofilm matrix production. *Current Biology, 28*(12), 1903–1913.

Duraj-Thatte, A. M., Praveschotinunt, P., Nash, T. R., Ward, F. R., Nguyen, P. Q., & Joshi, N. S. (2018). Modulating bacterial and gut mucosal interactions with engineered biofilm matrix proteins. *Scientific Reports, 8*(1), 3475.

Dyszel, J. L., Soares, J. A., Swearingen, M. C., Lindsay, A., Smith, J. N., & Ahmer, B. M. (2010). E. coli K-12 and EHEC genes regulated by SdiA. *PLoS One, 5*(1), e8946.

El Karoui, M., Hoyos-Flight, M., & Fletcher, L. (2019). Future trends in synthetic biology—a report. *Frontiers in Bioengineering and Biotechnology, 7*, 471916.

Fang, K., Jin, X., & Hong, S. H. (2018). Probiotic Escherichia coli inhibits biofilm formation of pathogenic E. coli via extracellular activity of DegP. *Scientific Reports, 8*(1), 4939.

Fang, K., Park, O. J., & Hong, S. H. (2020). Controlling biofilms using synthetic biology approaches. *Biotechnology Advances, 40*, 107518.

Fernandes, R., Luo, X., Tsao, C. Y., Payne, G. F., Ghodssi, R., Rubloff, G. W., & Bentley, W. E. (2010). Biological nanofactories facilitate spatially selective capture and manipulation of quorum sensing bacteria in a bioMEMS device. *Lab on a Chip, 10*(9), 1128–1134.

Flemming, H. C., & Wingender, J. (2010). The biofilm matrix. *Nature Reviews Microbiology, 8*(9), 623–633.

Flemming, H. C., Wingender, J., Szewzyk, U., Steinberg, P., Rice, S. A., & Kjelleberg, S. (2016). Biofilms: An emergent form of bacterial life. *Nature Reviews Microbiology, 14*(9), 563–575.

Flemming, H. C., van Hullebusch, E. D., Neu, T. R., Nielsen, P. H., Seviour, T., Stoodley, P., ... & Wuertz, S. (2023). The biofilm matrix: Multitasking in a shared space. *Nature Reviews Microbiology, 21*(2), 70–86.

Finkel, S. E., & Kolter, R. (2001). DNA as a nutrient: Novel role for bacterial competence gene homologs. *Journal of Bacteriology, 183*(21), 6288–6293.

Frederix, M., Edwards, A., Swiderska, A., Stanger, A., Karunakaran, R., Williams, A., et al. (2014). Mutation of praR in R hizobium leguminosarum enhances root biofilms, improving nodulation competitiveness by increased expression of attachment proteins. *Molecular Microbiology, 93*(3), 464–478.

Fujishige, N. A., Kapadia, N. N., & Hirsch, A. M. (2006). A feeling for the micro-organism: Structure on a small scale. Biofilms on plant roots. *Botanical Journal of the Linnean Society, 150*(1), 79–88.

Fu, W., Forster, T., Mayer, O., Curtin, J. J., Lehman, S. M., & Donlan, R. M. (2010). Bacteriophage cocktail for the prevention of biofilm formation by Pseudomonas aeruginosa on catheters in an in vitro model system. *Antimicrobial Agents and Chemotherapy, 54*(1), 397–404.

Geesey, G. G. (1994). *Biofouling and biocorrosion in industrial water systems*. CRC Press.

Gehrke, T., Telegdi, J., Thierry, D., & Sand, W. (1998). Importance of extracellular polymeric substances from Thiobacillus ferrooxidans for bioleaching. *Applied and Environmental Microbiology, 64*(7), 2743–2747.

Gerbino, E., Carasi, P., Mobili, P., Serradell, M. A., & Gómez-Zavaglia, A. (2015). Role of S-layer proteins in bacteria. *World Journal of Microbiology and Biotechnology, 31*, 1877–1887.

Gödeke, J., Paul, K., Lassak, J., & Thormann, K. M. (2011). Phage-induced lysis enhances biofilm formation in Shewanella oneidensis MR-1. *The ISME Journal, 5*(4), 613–626.

Harmsen, M., Lappann, M., Knøchel, S., & Molin, S. (2010). Role of extracellular DNA during biofilm formation by Listeria monocytogenes. *Applied and Environmental Microbiology, 76*(7), 2271–2279.

Hantus, C. E., Moppel, I. J., Frizzell, J. K., Francis, A. E., Nagashima, K., & Ryno, L. M. (2024). L-rhamnose globally changes the transcriptome of planktonic and biofilm Escherichia coli cells and modulates biofilm growth. *Microorganisms, 12*(9), 1911.

Hasunuma, T., Sanda, T., Yamada, R., Yoshimura, K., Ishii, J., & Kondo, A. (2011). Metabolic pathway engineering based on metabolomics confers acetic and formic acid tolerance to a recombinant xylose-fermenting strain of Saccharomyces cerevisiae. *Microbial Cell Factories, 10*, 1–13.

Heimann, K. (2016). Novel approaches to microalgal and cyanobacterial cultivation for bioenergy and biofuel production. *Current Opinion in Biotechnology, 38*, 183–189.

Henker, J., Laass, M., Blokhin, B. M., Bolbot, Y. K., Maydannik, V. G., Elze, M., ... & Schulze, J. (2007). The probiotic Escherichia coli strain Nissle 1917 (EcN) stops acute diarrhoea in infants and toddlers. *European journal of pediatrics, 166*, 311–318.

Hill, C., Guarner, F., Reid, G., Gibson, G. R., Merenstein, D. J., Pot, B., et al. (2014). Expert consensus document: The International Scientific Association for Probiotics and Prebiotics consensus statement on the scope and appropriate use of the term probiotic. *Nature Reviews Gastroenterology & Hepatology, 11, 506*.

Hong, S. H., Hegde, M., Kim, J., Wang, X., Jayaraman, A., & Wood, T. K. (2012). Synthetic quorum-sensing circuit to control consortial biofilm formation and dispersal in a microfluidic device. *Nature Communications, 3*(1), 613.

Huang, J., Liu, S., Zhang, C., Wang, X., Pu, J., Ba, F., et al. (2019). Programmable and printable Bacillus subtilis biofilms as engineered living materials. *Nature Chemical Biology, 15*(1), 34–41.

Hubenova, Y., Hubenova, E., Burdin, B., Vladikova, D., & Mitov, M. (2019). Development of coupled redox active network in Ca-alginate polymer for immobilization of Pseudomonas putida 1046 on electrode surface. *Electrochimica Acta, 312*, 432–440.

Hwang, I. Y., Koh, E., Wong, A., March, J. C., Bentley, W. E., Lee, Y. S., & Chang, M. W. (2017). Engineered probiotic Escherichia coli can eliminate and prevent Pseudomonas aeruginosa gut infection in animal models. *Nature Communications, 8*(1), 15028.

JB, A., & RR, V. (2016). Effect of small chain N acyl homoserine lactone quorum sensing signals on biofilms of food-borne pathogens. *Journal of Food Science and Technology, 53*(9), 3609–3614.

Jahn, A., Griebe, T., & Nielsen, P. H. (1999). Composition of Pseudomonas putida biofilms: Accumulation of protein in the biofilm matrix. *Biofouling, 14*(1), 49–57.

Kadurugamuwa, J. L., & Beveridge, T. J. (1995). Virulence factors are released from Pseudomonas aeruginosa in association with membrane vesicles during normal growth and exposure to gentamicin: A novel mechanism of enzyme secretion. *Journal of Bacteriology, 177*(14), 3998–4008.

Kaplan, J. B., Ragunath, C., Velliyagounder, K., Fine, D. H., & Ramasubbu, N. (2004). Enzymatic detachment of Staphylococcus epidermidis biofilms. *Antimicrobial Agents and Chemotherapy, 48*(7), 2633–2636.

Karimi, A., Karig, D., Kumar, A., & Ardekani, A. M. (2015). Interplay of physical mechanisms and biofilm processes: review of microfluidic methods. *Lab on a Chip, 15*(1), 23–42.

Kim, T. J., Young, B. M., & Young, G. M. (2008). Effect of flagellar mutations on Yersinia enterocolitica biofilm formation. *Applied and Environmental Microbiology, 74*(17), 5466–5474.

Kyzioł, A., Khan, W., Sebastian, V., & Kyzioł, K. (2020). Tackling microbial infections and increasing resistance involving formulations based on antimicrobial polymers. *Chemical Engineering Journal, 385*, 123888.

Lattif, A. A., Mukherjee, P. K., Chandra, J., Roth, M. R., Welti, R., Rouabhia, M., & Ghannoum, M. A. (2011). Lipidomics of Candida albicans biofilms reveals phase-dependent production of phospholipid molecular classes and role for lipid rafts in biofilm formation. *Microbiology, 157*(11), 3232–3242.

Lee, J., Maeda, T., Hong, S. H., & Wood, T. K. (2009). Reconfiguring the quorum-sensing regulator SdiA of Escherichia coli to control biofilm formation via indole and N-acylhomoserine lactones. *Applied and Environmental Microbiology, 75*(6), 1703–1716.

Lee, K., Kim, Y. W., Lee, S., Lee, S. H., Nahm, C. H., Kwon, H., et al. (2018). Stopping autoinducer-2 chatter by means of an indigenous bacterium (Acinetobacter sp. DKY-1): A new antibiofouling strategy in a membrane bioreactor for wastewater treatment. *Environmental Science & Technology, 52*(11), 6237–6245.

Leme, A. P., Koo, H., Bellato, C. M., Bedi, G., & Cury, J. A. (2006). The role of sucrose in cariogenic dental biofilm formation—new insight. *Journal of dental research, 85*(10), 878–887.

Li, D., Furukawa, H., Deng, H., Liu, C., Yaghi, O. M., & Eisenberg, D. S. (2014). Designed amyloid fibers as materials for selective carbon dioxide capture. *Proceedings of the National Academy of Sciences, 111*(1), 191–196.

Lo, A. W., Seers, C. A., Boyce, J. D., Dashper, S. G., Slakeski, N., Lissel, J. P., & Reynolds, E. C. (2009). Comparative transcriptomic analysis of Porphyromonas gingivalis biofilm and planktonic cells. *BMC Microbiology, 9*, 1–11.

Lu, T. K., & Collins, J. J. (2007). Dispersing biofilms with engineered enzymatic bacteriophage. *Proceedings of the National Academy of Sciences, 104*(27), 11197–11202.

Ma, Q., Zhang, G., & Wood, T. K. (2011). Escherichia coli BdcA controls biofilm dispersal in Pseudomonas aeruginosa and Rhizobium meliloti. *BMC Research Notes, 4*, 1–10.

Mangwani, N., Kumari, S., & Das, S. (2016). Bacterial biofilms and quorum sensing: Fidelity in bioremediation technology. *Biotechnology and Genetic Engineering Reviews, 32*(1–2), 43–73.

Mann, E. E., Rice, K. C., Boles, B. R., Endres, J. L., Ranjit, D., Chandramohan, L., et al. (2009). Modulation of eDNA release and degradation affects Staphylococcus aureus biofilm maturation. *PLoS One, 4*(6), e5822.

Marx, V. (2014). Stop the microbial chatter. *Nature, 511*(7510), 493–497.

Matz, C. (2009). Biofilms as refuge against predation. *Lab. Chip, 10*, 238.

Mojica, K., Elsey, D., & Cooney, M. J. (2007). Quantitative analysis of biofilm EPS uronic acid content. *Journal of Microbiological Methods, 71*(1), 61–65.

Mukherjee, M., Hu, Y., Tan, C. H., Rice, S. A., & Cao, B. (2018). Engineering a light-responsive, quorum quenching biofilm to mitigate biofouling on water purification membranes. *Science Advances, 4*(12), eaau1459.

Mulcahy, H., Charron-Mazenod, L., & Lewenza, S. (2008). Extracellular DNA chelates cations and induces antibiotic resistance in Pseudomonas aeruginosa biofilms. *PLoS Pathogens, 4*(11), e1000213.

Nguyen, P. Q. (2017). Synthetic biology engineering of biofilms as nanomaterials factories. *Biochemical Society Transactions, 45*(3), 585–597.

Nguyen, P. Q., Botyanszki, Z., Tay, P. K. R., & Joshi, N. S. (2014). Programmable biofilm-based materials from engineered curli nanofibres. *Nature Communications, 5*(1), 4945.

Nickel, J. C., Ruseska, I., Wright, J. B., & Costerton, J. W. (1985). Tobramycin resistance of Pseudomonas aeruginosa cells growing as a biofilm on urinary catheter material. *Antimicrobial Agents and Chemotherapy, 27*(4), 619–624.

Nielsen, P. H., Frølund, B., & Keiding, K. (1996). Changes in the composition of extracellular polymeric substances in activated sludge during anaerobic storage. *Applied Microbiology and Biotechnology, 44*, 823–830.

O'Donnell, M. P., Fox, B. W., Chao, P. H., Schroeder, F. C., & Sengupta, P. (2020). A neurotransmitter produced by gut bacteria modulates host sensory behaviour. *Nature, 583*(7816), 415–420.

Oh, H. S., & Lee, C. H. (2018). Origin and evolution of quorum quenching technology for biofouling control in MBRs for wastewater treatment. *Journal of Membrane Science, 554*, 331–345.

Olanbiwoninu, A., & Popoola, B. M. (2023). Biofilms and their impact on the food industry. *Saudi Journal of Biological Sciences, 30*(2), 103523.

Pandit, A., Adholeya, A., Cahill, D., Brau, L., & Kochar, M. (2020). Microbial biofilms in nature: Unlocking their potential for agricultural applications. *Journal of Applied Microbiology, 129*(2), 199–211.

Pei, R., & Lamas-Samanamud, G. R. (2014). Inhibition of biofilm formation by T7 bacteriophages producing quorum-quenching enzymes. *Applied and Environmental Microbiology, 80*(17), 5340–5348.

Pesci, E. C., Pearson, J. P., Seed, P. C., & Iglewski, B. H. (1997). Regulation of las and rhl quorum sensing in Pseudomonas aeruginosa. *Journal of Bacteriology, 179*(10), 3127–3132.

Pier, G. B., Coleman, F., Grout, M., Franklin, M., & Ohman, D. E. (2001). Role of alginate O acetylation in resistance of mucoid Pseudomonas aeruginosa to opsonic phagocytosis. *Infection and Immunity, 69*(3), 1895–1901.

Pires, D. P., Cleto, S., Sillankorva, S., Azeredo, J., & Lu, T. K. (2016). Genetically engineered phages: A review of advances over the last decade. *Microbiology and Molecular Biology Reviews, 80*(3), 523–543.

Pozos, N., Scow, K., Wuertz, S., & Darby, J. (2004). UV disinfection in a model distribution system:: Biofilm growth and microbial community. *Water Research, 38*(13), 3083–3091.

Pratt, L. A., & Kolter, R. (1998). Genetic analysis of Escherichia coli biofilm formation: Roles of flagella, motility, chemotaxis and type I pili. *Molecular Microbiology, 30*(2), 285–293.

Qin, B., Fei, C., Bridges, A. A., Mashruwala, A. A., Stone, H. A., Wingreen, N. S., & Bassler, B. L. (2020a). Cell position fates and collective fountain flow in bacterial biofilms revealed by light-sheet microscopy. *Science, 369*(6499), 71–77.

Qin, W., Wang, C. Y., Ma, Y. X., Shen, M. J., Li, J., Jiao, K., et al. (2020b). Microbe-mediated extracellular and intracellular mineralization: Environmental, industrial, and biotechnological applications. *Advanced Materials, 32*(22), 1907833.

Riglar, D. T., Giessen, T. W., Baym, M., Kerns, S. J., Niederhuber, M. J., Bronson, R. T., et al. (2017). Engineered bacteria can function in the mammalian gut long-term as live diagnostics of inflammation. *Nature Biotechnology, 35*(7), 653–658.

Rijavec, T., Zrimec, J., van Spanning, R., & Lapanje, A. (2019). Natural microbial communities can be manipulated by artificially constructed biofilms. *Advanced Science, 6*(22), 1901408.

Rittman, B. E. (1982). The effect of shear stress on biofilm loss rate. *Biotechnology and Bioengineering, 24*(2), 501–506.

Romero, D., Vlamakis, H., Losick, R., & Kolter, R. (2014). Functional analysis of the accessory protein TapA in Bacillus subtilis amyloid fiber assembly. *Journal of Bacteriology, 196*(8), 1505–1513.

Roy, V., Fernandes, R., Tsao, C. Y., & Bentley, W. E. (2010). Cross species quorum quenching using a native AI-2 processing enzyme. *ACS Chemical Biology, 5*(2), 223–232.

Rudge, T. J., Steiner, P. J., Phillips, A., & Haseloff, J. (2012). Computational modeling of synthetic microbial biofilms. *ACS Synthetic Biology, 1*(8), 345–352.

Schultz, M. P., Bendick, J. A., Holm, E. R., & Hertel, W. M. (2011). Economic impact of biofouling on a naval surface ship. *Biofouling, 27*(1), 87–98. https://doi.org/10.1080/08927014.2010.542809

Sleytr, U. B., & Beveridge, T. J. (1999). Bacterial S-layers. *Trends in Microbiology, 7*(6), 253–260.

Slusarczyk, A. L., Lin, A., & Weiss, R. (2012). Foundations for the design and implementation of synthetic genetic circuits. *Nature Reviews Genetics, 13*(6), 406–420.

Song, T., Mika, F., Lindmark, B., Liu, Z., Schild, S., Bishop, A., et al. (2008). A new Vibrio cholerae sRNA modulates colonization and affects release of outer membrane vesicles. *Molecular Microbiology, 70*(1), 100–111.

Steinberger, R. E., & Holden, P. (2005). Extracellular DNA in single-and multiple-species unsaturated biofilms. *Applied and Environmental Microbiology, 71*(9), 5404–5410.

Tay, P. K. R., Nguyen, P. Q., & Joshi, N. S. (2017). A synthetic circuit for mercury bioremediation using self-assembling functional amyloids. *ACS Synthetic Biology, 6*(10), 1841–1850.

Tay, P. K. R., Manjula-Basavanna, A., & Joshi, N. S. (2018). Repurposing bacterial extracellular matrix for selective and differential abstraction of rare earth elements. *Green Chemistry, 20*(15), 3512–3520.

Thomas, V. C., Hiromasa, Y., Harms, N., Thurlow, L., Tomich, J., & Hancock, L. E. (2009). A fratricidal mechanism is responsible for eDNA release and contributes to biofilm development of Enterococcus faecalis. *Molecular Microbiology, 72*(4), 1022–1036.

Todhanakasem, T. (2017). Developing microbial biofilm as a robust biocatalyst and its challenges. *Biocatalysis and Biotransformation, 35*(2), 86–95. https://doi.org/10.1080/10242422.2017.1295230

Tomaras, A. P., Dorsey, C. W., Edelmann, R. E., & Actis, L. A. (2003). Attachment to and biofilm formation on abiotic surfaces by Acinetobacter baumannii: Involvement of a novel chaperone-usher pili assembly system. *Microbiology, 149*(12), 3473–3484.

Toyofuku, M., Roschitzki, B., Riedel, K., & Eberl, L. (2012). Identification of proteins associated with the Pseudomonas aeruginosa biofilm extracellular matrix. *Journal of Proteome Research, 11*(10), 4906–4915.

Van Hoek, M. L. (2013). Biofilms: An advancement in our understanding of Francisella species. *Virulence, 4*(8), 833–846.

Wang, W., Chanda, W., & Zhong, M. (2015). The relationship between biofilm and outer membrane vesicles: a novel therapy overview. *FEMS Microbiology Letters, 362*(15), fnv117.

Wang, X., Cao, Z., Zhang, M., Meng, L., Ming, Z., & Liu, J. (2020). Bioinspired oral delivery of gut microbiota by self-coating with biofilms. *Science Advances, 6*(26), eabb1952.

Whitchurch, C. B., Tolker-Nielsen, T., Ragas, P. C., & Mattick, J. S. (2002). Extracellular DNA required for bacterial biofilm formation. *Science, 295*(5559), 1487–1487.

Wong, G. C., & O'Toole, G. A. (2011). All together now: Integrating biofilm research across disciplines. *MRS Bulletin, 36*(5), 339–342.

Wood, T. K., Hong, S. H., & Ma, Q. (2011). Engineering biofilm formation and dispersal. *Trends in Biotechnology, 29*(2), 87–94.

Wood, T. L., Guha, R., Tang, L., Geitner, M., Kumar, M., & Wood, T. K. (2016). Living biofouling-resistant membranes as a model for the beneficial use of engineered biofilms. *Proceedings of the National Academy of Sciences, 113*(20), E2802–E2811.

Xavier, J. B., Kim, W., & Foster, K. R. (2011). A molecular mechanism that stabilizes cooperative secretions in Pseudomonas aeruginosa. *Molecular Microbiology, 79*(1), 166–179.

Yu, S., Su, T., Wu, H., Liu, S., Wang, D., Zhao, T., et al. (2015). PslG, a self-produced glycosyl hydrolase, triggers biofilm disassembly by disrupting exopolysaccharide matrix. *Cell Research, 25*(12), 1352–1367.

Zara, G., Goffrini, P., Lodi, T., Zara, S., Mannazzu, I., & Budroni, M. (2012). FLO11 expression and lipid biosynthesis are required for air–liquid biofilm formation in a Saccharomyces cerevisiae flor strain. *FEMS Yeast Research, 12*(7), 864–866.

Zhang, X., Bishop, P. L., & Kupferle, M. J. (1998). Measurement of polysaccharides and proteins in biofilm extracellular polymers. *Water Science and Technology, 37*(4–5), 345–348.

Zhang, C., Huang, J., Zhang, J., Liu, S., Cui, M., An, B., et al. (2019). Engineered Bacillus subtilis biofilms as living glues. *Materials Today, 28*, 40–48.

Omics-Based Therapeutics Against Biofilm Mechanics

18

Hemavathy Nagarajan, Sampathkumar Ranganathan, Srujana Chitipothu, Sangavi Pandi, and Ramya Chandar Charles Mariasoosai

18.1 Introduction

Biofilms are complex assemblies of microorganisms, possibly bacteria and fungi, thriving together as a well-arranged colony (Hall-Stoodley et al., 2004). Existing in an extracellular polymeric substance (EPS) matrix, these microbial aggregates have the potency to adhere to surfaces of inert or living origins. Unlike the freely floating planktonic microorganisms, biofilms have structured social structures and cooperative behaviors contributing to variable deterministic characteristics. A solid-like and liquid-like mechanical behavior deriving from viscoelastic properties furnishes

H. Nagarajan
Centre for Bioinformatics, Kamalnayan Bajaj Institute for Research in Vision and Ophthalmology, Vision Research Foundation, Chennai, Tamil Nadu, India

S. Ranganathan
Centre for Bioinformatics, Kamalnayan Bajaj Institute for Research in Vision and Ophthalmology, Vision Research Foundation, Chennai, Tamil Nadu, India

Department of Chemical Engineering, Konkuk University, Seoul, Republic of Korea

S. Chitipothu (✉)
Centre for Bioinformatics, Kamalnayan Bajaj Institute for Research in Vision and Ophthalmology, Vision Research Foundation, Chennai, Tamil Nadu, India

Central Research Instrumentation Facility, Kamalnayan Bajaj Institute for Research in Vision and Ophthalmology, Vision Research Foundation, Chennai, Tamil Nadu, India
e-mail: drsrujana@snmail.org

S. Pandi
Department of Microbiology, Karpagam Academy of Higher Education, Coimbatore, Tamil Nadu, India

Center for Bioinformatics, Karpagam Academy of Higher Education, Coimbatore, Tamil Nadu, India

R. C. C. Mariasoosai (✉)
Department of Chemistry and Biochemistry, University of Texas at Dallas, Austin, TX, USA

S. Busi et al. (eds.), *Omics Approaches in Biofilm Research*,
https://doi.org/10.1007/978-3-031-91863-6_18

biofilms with their ability to respond to external forces (Dunsmore et al., 2002). Interestingly, the EPS matrix engenders a high level of dynamicity, with changes in its composition reportable for differing growth stages and conditions. The matrix consists of various structural components constituting microbial cell membranes, biogenic particles of several sizes, and cytoplasmic entities (Souza et al., 2020). Fungal biofilms are poorly studied compared to bacterial biofilms and share similar structures and mechanical features operative via the extracellular matrix. For example, viscoelastic biofilms of *Candida krusei* or *Rhodotorula mucilaginosa* have solid-like properties, indicating that rheological studies are essential if more mechanistic analyses of the biofilm are to be undertaken (Costa-Orlandi et al., 2017; Nazir et al., 2020). Concerns arise about the formidable environmental presence of fungal-derived biofilms, particularly in healthcare settings. These biofilms exhibit an unusual resilience to common forms of medical treatment and contribute to establishing resistant infections; hence, advanced research is urgently needed to unveil their nature, mechanics, and possible weaknesses. Due to the antibiotic resistance of biofilm-causing pathogens, it remains a potential cause of high mortality and worse pathological circumstances. There are many different approaches to preventing, treating, and eradicating biofilms in clinical practice. There is still a lack of uniform understanding regarding biofilm-induced inflammation's physiological and pathological processes, resulting in limited effective treatments for serious bacterial infections and chronic inflammation. Omics approaches have emerged as powerful tools for monitoring and understanding cellular processes, addressing the needs in the life sciences, particularly in the context of regulatory mechanisms at various cellular levels. Combining omics approaches, microbial population phylogeny, and conceptual biology is ultimately used to detect infections caused by multispecies biofilms. Recent genomic techniques, such as DNA microarray analysis, have uncovered several additional *Escherichia coli* genes associated with biofilm formation. Multi-omics approaches enable detailed characterization of the molecular pathways that underlie interactions between genes, proteins, and metabolites. Bacterial omics techniques present a promising strategy for mitigating the emergence of bacterial multidrug resistance by integrating genomic mapping of various drug-resistance gene clusters, transcriptome pathways, and alternative molecular targets (Niba et al., 2008).

18.2 Challenges in the Proliferation of Biofilms

Biofilms are responsible for many microbial contagions as they are perceived to elude the body's immune system and resist antimicrobials (Cerca et al., 2005; Donelli & Vuotto, 2014; Glage et al., 2017; Hall-Stoodley et al., 2004; Koseki et al., 2014; Li et al., 2021; Oliveira et al., 2018; Paharik & Horswill, 2016; Rahim et al., 2016; Schmoldt et al., 1975; Zaborowska et al., 2017). They are structured, and therefore, biofilm infections are hard to detect, as there are currently no universally accepted clinical procedures for visualizing the presence of a biofilm-associated pathogen. Such diagnostic ambiguity leads to inappropriate treatment and even lethal consequences (Jamal et al., 2018; Mooney et al., 2018). In addition, under stress conditions, biofilms enhance microbial survival and, in this way, bacterial biofilms achieve persistence by switching their metabolism, gene expression, and protein synthesis (Lebeaux et al., 2014; Zarnowski et al., 2021). The ability to characterize and understand biofilms and their interactions with the environment or host organisms has significantly advanced in recent years. This progress has been facilitated by technological breakthroughs such as high-throughput, data-driven "omics" techniques, high-resolution imaging, and improved spectroscopy. Omics analysis may become a promising strategy for diagnosing biofilms in clinical laboratories. It enables the identification of pathogens in a shorter timeframe than conventional techniques and allows for the comprehensive analysis of the pathogen's entire microbiome (Coenye et al., 2024).

18.3 Quorum Sensing in Biofilm Development

Quorum sensing is another essential mechanism that participates in biofilm development in that it enables the bacteria to ascertain cell density and their coordinated actions through AIs (autoinducers) (Ng & Bassler, 2009; Solano et al., 2014). In Gram-positive bacteria, a captivating form of communication unfolds through autoinducing peptides; Gram-negative bacteria use acyl-homoserine lactones (AHL) synthesized by Lux-type proteins (Bharatula et al., 2020). The AI-2 system is a universal way to communicate with both types of bacteria, thus complicating relationships among biofilms some more (Tselepi et al., 2024; Waters & Bassler, 2005).

18.4 Formation and Characteristics of Biofilms

Five steps largely govern the development of biofilms: (1) initial attachment, (2) aggregation, (3) formation of microcolony, (4) maturing of the biofilm, and (5) detachment of planktonic cells (Balasubramanian et al., 2018). The EPS-based matrix encasing the biofilm protects and supports the whole biofilm colony. Consequently, the biofilm forms a protective barrier against the entry of antimicrobial agents. The extensive phenotypic expression of the free-floating bacteria distinguishes biofilm bacteria from other members of the environment because adhesion to the surface induces rapid gene expression and EPS production (Maskarinec et al., 2018). The quantitative proteomics approach employing iTRAQ is efficacious for detecting bacterial biofilms and determining potential targets for biofilm elimination. Additionally, metabolomics studies provide a systematic technique for illustrating complex bacterial communities, exhibiting bacterial cell behavior within biofilms, and enhancing the understanding of biofilms (Zhao et al., 2023).

18.5 Key Mechanisms and Factors

- In Quorum Sensing (QS), populations of quorum sensing systems induce the regulation of biofilms that mature and develop. QS inhibition (quorum quenching) can result in the potential therapy of biofilms-related antibiotic resistance (Mourer et al., 2023).
- Some environmental forces, mainly temperature, pH, hydrodynamics, and surface properties, are critical for biofilm demographics (Roscini et al., 2018).
- EPS is structural support for biofilms while protecting them from stress, mediating cell–cell interaction (Slezak et al., 2022).
- Syntrophic interactions, such as cooperation between species in biofilms, lead to optimizing metabolism in adaptation to environments with less competition, leading to a sure degree of resilience (Ezzat et al., 2021).

18.6 Significance and Impact of Biofilms

Biofilms present formidable challenges for medical and industrial settings. They shield the pathogen from antimicrobial agents and host immune responses. This, in turn, causes chronic infections where medical devices are contaminated (Ali et al., 2024). About 70 percent of all microbial contagions are related to biofilms, veritably trapping device-oriented infections such as prosthetic heart valves, central venous catheters, and urinary catheters (Siddam et al., 2020). Biofilms participate in disease development by biofilm detachment, antimicrobial resistance, endotoxin production, and immune evasion (Tuck et al., 2022).

18.7 Applications and Challenges

Biofilms are an emerging area for continuous and device-related infections and antibiotic resistance. Some research on biofilm decomposition and quorum sensing mechanisms can lead to the growth of new therapies that can prevent the formation of biofilm or improve existing ones (Thompson et al., 2020). Regarding industrial impact, biofilms negatively influence pipelines, water treatment, and food production in different branches of industry. Developing methods aimed at biofilm prevention will reduce biofouling and contamination, thus increasing the effectiveness and safety of the end products (Dutra et al., 2018). Moreover, recent studies have concentrated on using quorum-sensing inhibitors and quorum quenchers to fight biofilm proliferation. Predictive mathematical models describing biofilm dynamics and intervention strategies' efficacy are being implemented. Additionally, enzymes that cleave extracellular polymeric substances (EPS) are being developed to improve biofilm removal and antimicrobial efficacy (Madani et al., 2022).

18.8 Strategies for Addressing Biofilm-Embedded Infections

Traditional antibiotics frequently fail to cure biofilm-associated infections, necessitating the development of novel strategies. Potential approaches include quorum-sensing inhibitors, antibody-based therapies, natural product-based treatments, and gene-editing approaches (Gupta et al., 2016). Understanding biofilm dynamics is crucial for addressing their role in infections and developing effective mitigation strategies (Table 18.1).

18.9 Multi-species Biofilm Formation

Biofilms of more than one species have specific relationships and spatial arrangements due to eco- and biocommunity factors. The creation of multi-species biofilms happens in phases. Initially, primary microbial colonization on a surface creates a

Table 18.1 Strategies and applications for disrupting and controlling biofilms

Methods	Applications
1. *Physical Methods*	*Ionizing and ultraviolet radiation* can damage the materials that biofilm forms and also disrupt the biofilm's structure (Lahiri et al., 2021). Treatment with *ultrasound*, which relies on pressure, vibrational energy, and shock waves, is used to disrupt bacterial cells and the substances that makeup biofilm matrices (Sharma & Pagedar Singh, 2018). In *plasma technology,* the use of charged particles, synthetic ions, and ultraviolet rays is employed to destroy bacteria along with the biofilm matrices (Falcón García et al., 2020).
2. *Chemical Methods*	Due to the slight stability, conditions, and toxic byproducts, chemical methods are less preferred (Maheswari & Kalaiselvi, 2024).
3. *Biological Methods*	*AMPs*, known as *antimicrobial peptides*, can prevent adhesion, disrupt quorum sensing, and down-regulate biofilm genes. This category includes leucine-leucine 37 and lactoferrin peptides (Blackman et al., 2021). *Nanoparticles* increase antimicrobial delivery and take apart biofilms through oxidative stress and damage. This category includes silver, gold, and chitosan nanoparticles (Voinescu et al., 2024). Bacteria modified to infect other bacteria, penetrate biofilms, and produce enzymes to destroy biofilm matrices are referred to as *bacteriophages* (Iñiguez-Moreno et al., 2018). Blockers of biofilms, known as *quorum-sensing inhibitors* (QSI), possess the capability to block any biofilm regulating systems. These include NSAIDs and certain antibiotics (Salvatore et al., 2021). *Enzymes* that use oxidases, polysaccharides, and proteolytic enzymes are known to break down exopolymeric substances, which are organized within the biofilm matrix (Steinhäußer et al., 2024). *Aptamers* bind to target molecules, exhibiting high specificity and interfering with biofilm formation (Ding, 2024). *Peptide Nucleic Acids (PNAs)* are synthetic molecules that attach to bacteria DNA/RNA, preventing biofilm development (Hall et al., 2021). *Vaccines* are implemented for protective immunes using biofilm-generated antigens (Simcox et al., 2019).

microcolony that EPS maintains. This is called primary colonization. In secondary colonization, other microbial species integrate into the biofilm because of the stability and nutrients in the EPS matrix. In the maturation phase, dispersion can occur due to environmental factors stimulating a signal that triggers the lysis of biofilm or cells. This allows for further colonization by releasing biofilm fragments to other surfaces (Li et al., 2023). Multi-species biofilms are also organized spatially into multi-systems. A certain kind of biofilm formation called side by side is when microorganisms are put beside each other to make different separated biofilms, like *Burkholderia* and *Pseudomonas*, which exist together without commensal relationships. In some other cases, biofilm species may be layered such that *Acinetobacter* is found in the uppermost layers and *Pseudomonas putida* in the lower layers, facilitating efficient metabolic interactions. The metabolic layering further optimizes biofilm productivity by arranging microbial species according to their metabolic roles, such as oxygen utilization, nutrient cycling, and byproduct consumption (Costerton et al., 1994; Foster & Kolenbrander, 2004; Otto et al., 2012; Ryan et al., 2009).

18.10 Implications and Future Directions

Therefore, understanding multi-species biofilm dynamics is crucial for developing strategies to control their impacts. In clinical settings, understanding the formation of biofilms and the interactions between species can inform targeted antimicrobial therapies and biofilm detection methods. In industrial applications, harnessing beneficial biofilms for processes such as wastewater treatment can provide considerable advantages while mitigating risks associated with forming biofilms on infrastructure. Future research should be directed toward integrating advanced technologies, such as multi-omics approaches, to unravel the complexities of multi-species biofilms and their interactions (Rouillard et al., 2022).

18.11 Decoding Biofilms with Omics

Omics-based approaches revealed an outline of the molecular pathways that drive biofilm development. These approaches identify genes, proteins, and metabolites that play a role in the formation of biofilm, offering acumens into the regulatory networks supporting this process (Vohra et al., 2024). Recent advancements in omics technology have facilitated the capture of systems-level information on biofilms. The integrated omics, or multi-omics, approach provides a comprehensive outlook on variations within biofilm systems. Nevertheless, there are inherent challenges associated with the integration of multi-omics data (Peyrusson et al., 2020). Biofilm research has yet to utilize multi-omics techniques to explore system-level relationships extensively. Currently, most studies in this area amalgamate metagenomics and metabolomics analyses. A scarcity of literature encompasses multi-omics investigations of biofilms, particularly those that integrate transcriptomics, proteomics, and metabolomics. Currently, the majority of multi-omics research integrates data from two techniques: transcriptomics and metabolomics or proteomics. This trend might be attributed to the limitations of existing data analysis tools, which primarily support two-level data integration. However, advancements in software development and computing power now facilitate data integration across multiple layers. These significant developments provide a comprehensive understanding of biofilm biology.

18.12 Multi-omics Technology in Biofilm Detection and Characterization

It has revolutionized the study of microbial systems, thus affording unprecedented insights into complexities of microbial behavior, such as biofilm formation. The approach integrates layers of omics data: genomics, transcriptomics, proteomics, and metabolomics. As a result, researchers can now analyze microbial physiology, behavior, and host–pathogen interactions at a whole-systems level (Otto et al., 2012; Ryan et al., 2009; Thakur et al., 2008). Genomics provides insights into the genetic arrangement of organisms and thus helps identify druggable genomes and genetic elements associated with pathogenicity and biofilm formation (Desai et al., 2010). Transcriptomics involves the study of gene expression through the quantification of mRNA levels; researchers can identify genes that are related with biofilm formation and mechanisms of resistance to antimicrobials (Chiu & Miller, 2019). Combinations of techniques such as ESTs (Express Sequence Tags) and RNA sequencing (RNA-Seq) help identify biofilm-specific genes that are linked with a variety of infections, including sepsis, urinary tract infections, meningitis, and others (Hrdlickova et al., 2017). Proteomics involves the protein content and its functions within a cell, emphasizing the molecular mechanisms that introduce biofilm formation, stress responses, and antibiotic resistance (Otto et al., 2012). 2D gels (DIGE) and gel-free techniques like iTRAQ and TMT aid in diagnosing infections like sepsis and urinary tract infections by analyzing protein expression profiles (Gil et al., 2014). Metabolomics provides information about cellular processes, such as carbohydrate and lipid metabolism that are essential for the development and persistence of biofilm (Chiu & Miller, 2019; Desai et al., 2010; Gil et al., 2014; Hrdlickova et al., 2017; Seneviratne et al., 2010). Metabolic changes in infections like sepsis and meningitis can be identified using CE and TOF-MS. These techniques also play a vital role in investigating the metabolic dynamics of biofilm-forming pathogens (Date et al., 2010).

18.13 Applications of Multi-omics in Biofilm Research

1. *Biofilm Formation Mechanisms:*

   One can use multi-omics technologies to analyze the planktonic cell phenotypic switch into biofilm-forming states in detail. The biofilm extracellular matrix confers mechanical strength and allows microorganisms to resist antibiotic therapy and host immune responses.

2. *Mechanism of Biofilm Resistance:*
   - *Persister Cells:* One of the most essential characteristics of biofilm resistance is the existence of persister cells, quiescent subpopulations that are metabolically dormant and resistant to antibiotic therapy. These cells have been reported in pathogens like *Staphylococcus* spp., *Candida* spp., *Escherichia coli, Lactobacillus acidophilus*, and *Pseudomonas aeruginosa* (Ellepola et al., 2019).
   - *Drug Efflux Pumps:* The activation of drug efflux pumps is another distinctive feature of multidrug-resistant biofilm-forming microorganisms. Nosocomial pathogens such *as S. aureus, Bacillus subtilis*, and *Acinetobacter baumannii* have overexpressed pumps linked with resistance (Miyaue et al., 2018).
3. *Biofilm Treatment Complexity:* Physical and genetic heterogeneity in biofilms makes treatment challenging. A complex system of genetic alterations and enzymatic cell wall modifications governs the resistance of the biofilm to antibiotics. Single-target traditional therapeutic approaches are impractical because of biofilms' complex internal microenvironments and cellular interactions.
4. *Multi-Omics in the Battle Against Biofilm Resistance:* Multi-omics methods have been developed because of their capacity to decipher the complex regulation of biofilm development and drug resistance. The combination of metabolomics, proteomics, and transcriptomics assists in investigating different genetic and phenotypic regulation mechanisms at systemic levels. These assist in identifying possible targets for therapeutic interventions and strategies that may be employed to combat infections caused by the biofilm (Fig. 18.1).

18.14 Transcriptomics in Biofilm Monitoring and Drug Resistance Identification

Transcriptomics aids in explaining gene expression alterations happening with biofilm maturation and establishment and the development of antimicrobial resistance. It examines the entirety of all mRNA transcripts produced from the genome with the

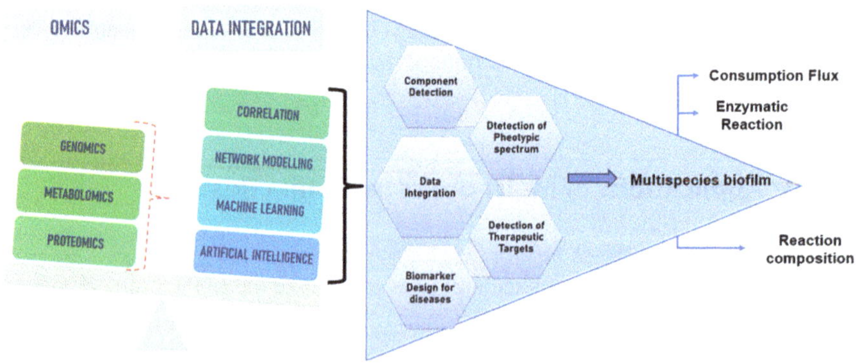

Fig. 18.1 Omics-driven data integration for Targeting Multispecies Biofilms

capacity to unravel the process of gene control during biofilm formation, encompassing the conversion between planktonic and biofilm life phases of microbes and alterations leading it to be drug resistant.

18.14.1 Transcriptomics Approaches for Biofilm Analysis

The most sophisticated method for transcriptomics is RNA-Seq, which has greater sensitivity than the older methods. It can identify rare transcripts and quantify gene expression at single-base resolution. Consequently, RNA-Seq became the new darling, even overtaking DNA microarrays because of its higher applications and capacity to identify novel or rare transcripts (Singh et al., 2017). Quantitative PCR (qPCR) was employed to confirm gene expression profiles from RNA-Seq and microarray experiments. For example, qPCR can identify the expression of precise genes in biofilms, such as the esp gene in *Enterococcus faecalis* biofilms exposed to zinc oxide nanoparticles (Lowe et al., 2017). Although less popular today than RNA-Seq, DNA microarrays remain proficient at comparing gene expression between biofilm and planktonic conditions in various microbial species (Partoazar et al., 2019).

18.14.2 Case Studies on Biofilm Inhibition Using Transcriptomic Techniques

RNA-Seq represents a high-throughput technique that effectively assesses gene regulation and expression levels. RNA-Seq was used to investigate the influence of antibiotics and potent anti-biofilm drugs on biofilm development. Utilizing RNA-Seq technology, the anti-biofilm properties of exopolysaccharide EPS273 on *Pseudomonas aeruginosa* were investigated. According to the findings, EPS273 may function by downregulating the expression of genes linked with the formation of biofilm in the PhoP-PhoQ two-component system, as well as in the quorum sensing systems LasI/LasR and RhlI/RhlR. A metagenomic and metatranscriptomic investigation utilizing RNA-seq of subgingival biofilm samples obtained from patients with periodontitis has notorious dissimilar molecular profiles associated with the development of the disease. A comprehensive RNA-seq analysis revealed the presence of 27 metabolically active taxa within the dental plaque community. Among these, *Streptococcus* species were identified as the most prevalent, followed by *Veillonella* species and *Capnocytophaga* species. Using mutagenesis and RNA-seq, studies examined the role of the internalin-family protein InlJ in facilitating adhesion between *Candida albicans* and *Porphyromonas gingivalis*. Transcriptional profiling revealed that 57 genes were specifically upregulated in an InlJ-dependent manner, including components of the type IX secretion system. These outcomes suggest that *C. albicans* may enhance the pathogenicity of *P. gingivalis* by activating the type IX secretion system.

qPCR is an effective method for monitoring changes in gene expression within biofilms over time, even following treatment interventions, such as examining the esp gene in Enterococcus faecalis biofilms exposed to zinc oxide nanoparticles. A DNA microarray-based transcriptome analysis was conducted to investigate differential gene expression in *Streptococcus mutans* biofilms as influenced by biofilm thickness. The findings revealed that WapA, a significant surface protein, exhibited higher expression levels in 200-micron and 400-micron biofilms than in 100-micron biofilms. This indicates a thickness-dependent regulation of gene expression during the growth of biofilms. DNA microarrays can be employed to evaluate the immune responses of oral epithelial cells infected with various species of oral pathogenic bacteria in both biofilm and planktonic states. High-throughput sequencing and de novo assembly methods were used to analyze the transcriptome of the biofilm produced by methicillin-sensitive *Staphylococcus aureus* (Tan et al., 2015). This study generated 394 million sequencing reads, leading to the assembly of 3054 transcripts and the identification of 3819 coding sequences (CDSs). Remarkably, 99.69% of these sequences matched known proteins in the Nr database. The findings suggest that the transcriptome of *S. aureus* is both highly conserved and functionally diverse, making it a valuable genetic resource for further investigations into its biology and highlighting potential targets for future research and therapeutic development.

A recent study investigated the effects of benzyl isothiocyanate (BITC) on movement and biofilm formation in *Salmonella typhimurium* (Niu et al., 2020). The researchers treated the bacteria with a low dose of BITC and sequenced the RNA from both the control and BITC-treated groups. This process resulted in six high-quality cDNA libraries containing over 109 million accurate reads. The study intended to recognize variations in gene expression related to the effects of BITC. The findings indicate that BITC significantly alters gene activity in *S. typhimurium*, impacting its movement and biofilm growth. This research provides valuable insights into the mechanisms by which BITC functions as an antimicrobial agent.

18.14.3 Detection of Drug Resistance Via Biofilms Using Transcriptomics

Transcriptomics has been highly useful for shedding light on drug resistance mechanisms in microbial biofilms. RNA-Seq is the gold standard for transcriptional analysis, with high sensitivity and resolution, enabling the detection of rare transcripts and measuring gene expression to single-base resolution. The method has been modified for screening biofilm- and resistance-related genes, including efflux pump genes and sporulation regulators. For instance, it has been shown that RNA-Seq was able to detect some of the significant biofilm- and drug resistance-related genes of *E. faecalis*, including paiA (sporulation-regulating gene), arsD and arsR (arsenic resistance genes), and ATP-binding cassette (ABC) drug transporters (Ebersole et al., 2019). In Staphylococcus aureus, research indicated that low ampicillin concentrations stimulated biofilm development, enhanced biofilm cell viability, and upregulated genes involved in penicillin-binding proteins and drug resistance efflux pumps (Seneviratne et al., 2017). An aqueous *G. chinensis* extract inhibited biofilm-related genes and carbohydrate metabolism-related genes in MRSA, demonstrating the promise of transcriptomics in investigating the impact of plant-derived treatments on biofilm and drug resistance (Liu et al., 2018).

18.15 Single-Cell Transcriptomics

Conventional transcriptomic analyses tend to assume that biofilms are uniform, but biofilms are heterogeneous, with the cells having differing metabolic activities and resistance to antibiotics. Single-cell transcriptomics allows for examining gene expression on a cell-by-cell basis and can reveal how individual cells drive biofilm function. For instance, in *S. aureus* biofilms, long-term exposure to antibiotics led to the downregulation of metabolic activity in persister cells. These cells were, however, viable and metabolically active because of transcriptome reprogramming (Peterson et al., 2014). Recent advancements in machine learning have examined transcriptomic data from *P. aeruginosa* infections in vitro, human and mouse models. These analyses pinpointed genes for iron acquisition and central carbon metabolism, essential for biofilm formation and antibiotic resistance (Peyrusson et al., 2020).

While transcriptomics offers precise information on gene expression and biofilm dynamics, it is not without its drawbacks, such as (a) *Model Selection:* The selection of in vivo or in vitro models highly impacts transcriptomic profiles since environmental conditions directly affect gene expression. (b) *Data Integration:* A comprehensive understanding of the complex mechanisms behind biofilm formation and drug resistance requires integrating transcriptomic data with proteomics and metabolomics. This holistic approach can reveal the complete biofilm phenotype.

18.16 Proteomics

Proteomics is the study of the whole protein profile of an organism under given conditions, providing more detailed information than transcriptomics, which is limited to the measurement of the gene expression level. Using proteomics, scientists can track how microbes alter their proteome in reaction to environmental stresses, essential for understanding biofilm formation, maturation, and drug resistance development.

18.16.1 Proteomics Methods for Biofilm Analysis

- Mass spectrometry (MS) -based methods are now favored because they can detect and quantify many proteins, including intricate proteins, such as membrane-bound proteins, within a limited time frame. Gel-free approaches, for example, tandem mass tag (TMT), isobaric tags for relative and absolute quantitation (iTRAQ), and stable-isotope labeling by aminoacids in cell culture (SILAC), are employed for high-throughput proteomics analysis (Cornforth et al., 2018).

- Differential isotopic labeling methods, such as TMT and iTRAQ, enable the detection of differential protein expression in biofilms, for example, of *Aeromonas hydrophila* and *Candida*, including antibiotic-sensitive and resistant strains (Date et al., 2010).
- Label-free methods utilize spectral intensities to quantify proteins and have been used in research of bacterial and fungal biofilms, for example, subgingival biofilm models, and have shown protein upregulation and downregulation of different microbial species (Soufi & Soufi, 2016).

18.16.2 Proteomics in Biofilm-Host Interaction

Few studies have used proteomics to examine the interaction between biofilms and host tissues, but recent work has provided valuable insights. For example, label-free proteomics showed reduced host protein release upon interaction with biofilms in a multi-layered gingival epithelium co-culture model with a 10-species subgingival biofilm. Inflammation and apoptosis pathways were upregulated, whereas breakdown of epithelial integrity and tissue turnover were downregulated, indicating how biofilms modulate host immune responses (Bao et al., 2015).

18.16.3 Proteomics and Drug Resistance in Biofilms

Biofilms exhibit a remarkable resistance to antimicrobial agents, a phenomenon that proteomics has helped to elucidate. One significant mechanism contributing to this resistance is the reduced metabolic activity within biofilms. Proteomics research has identified down-regulation of energy generation pathways, for example, glycolysis and pentose-phosphate pathways, in *E. coli* and *S. mutans* biofilms relative to their planktonic forms (Bostanci et al., 2015; Fulaz et al., 2019). In *Candida glabrata* biofilms, there was reduced expression of carbohydrate metabolism pathways compared to planktonic cells, emphasizing metabolic changes in biofilm modes (Gil et al., 2014). Several biofilm-forming species demonstrate high antioxidant protein levels, leading to their high resistance to oxidative stress and antimicrobial agents. For instance, antioxidant proteins, such as Ahp1p and Trx1p, are more expressed in *Candida* biofilms (Ebersole et al., 2019; Gil et al., 2014). The same was observed in the case of *E. coli* O157:H7 and *P. aeruginosa* biofilms, which had higher concentrations of periplasmic antioxidants (Kim et al., 2006; Rathsam et al., 2005). The discriminating antioxidative potential of biofilms may be associated with augmented resistance to antimicrobial compounds, as observed in *Listeria monocytogenes* and *Salmonella enterica* biofilms, where superoxide dismutase activity was upregulated (Patrauchan et al., 2007).

18.16.4 Proteomics of Biofilm EPS (Extracellular Polymeric Substances) and Drug Response

Proteomics has been used to study the constituents of the EPS matrix, which is involved in the drug resistance of biofilms. In *S. enterica* biofilms, even though the EPS composition and biofilm structure change under different flow regimes, there were no observable changes in the overall protein expression system. The EPS matrix has been associated with resistance to antimicrobials, with specific studies targeting the polysaccharide content, while others target the protein components of the matrix. Proteomics research on *A. hydrophila* biofilms has associated elevated fatty acid production with antibiotic resistance, implying that biofilm cells' metabolic changes are responsible for drug resistance (Zhang et al., 2015). Exoproteomic analysis in *P. aeruginosa* biofilms detected the presence of type III secretion system effectors, which can contribute to pathogenesis, stress resistance, and nutrient uptake, further suggesting the intricate interaction between biofilm matrix elements and drug resistance (Mangalappalli-Illathu et al., 2008).

18.16.5 Case Study

Cell-shaving proteomics was conducted via proteases to cleave peptides visible on the cell surface to illustrate the surface proteins expressed by *Staphylococcus aureus*. These peptides were subsequently identified through liquid chromatography-mass spectrometry (LC-MS). Additionally, matrix proteins were isolated from biofilms by employing centrifugation and filtration techniques to separate the cells from the matrix, followed by thorough proteomic analysis (An et al., 2021). Proteomic analysis of *Enterococcus faecalis* biofilm and planktonic growth has demonstrated the biofilm-specific upregulation of proteins associated with glycolysis, quorum sensing, rhamnopolysaccharide synthesis, and arginine metabolism. This research has identified potential targets for anti-biofilm strategies. Similar methodologies have been employed to explore targets for *Haemophilus influenzae* and *Mycobacterium tuberculosis* (Suryaletha et al., 2019; Wang et al., 2019). Metaproteomics has emerged as a potent methodological approach for investigating multi-species biofilms. This is exemplified by analyzing a four-species soil bacterial community comprising *Xanthomonas retroflexus*, *Stenotrophomonas rhizophila*, *Microbacterium oxydans*, and *Paenibacillus amylolyticus*. The study revealed enhanced biofilm formation and identified critical proteins facilitating synergistic interactions during co-cultivation (Herschend et al., 2017).

18.17 Metabolomics in Biofilm Research

Metabolomics is the investigation of small molecules (metabolites) formed during cellular metabolism, and it plays a crucial role in unraveling the biochemical complexity of biofilms. By examining both intracellular and extracellular metabolites, metabolomics provides a detailed understanding of the metabolic adaptations that biofilm-forming microorganisms employ to survive in hostile environments. One specialized branch, exometabolomics, focuses on metabolites released into the extracellular environment, revealing how biofilms interact with their surroundings and adapt to antibiotic stress. This approach has been instrumental in identifying biofilm-specific metabolic pathways and potential biomarkers of resistance. For instance, the biofilms of *Pseudomonas aeruginosa* increase phenazine production, which is a secondary metabolite that allows for electron transfer in low-oxygen environments and, therefore, maintains metabolic activity while contributing to resistance to antibiotics (Mielko et al., 2019). Metabolic signatures determined by metabolomics help a researcher focus on specific resistance mechanisms and discover new targets for intervention. Metabolomics further shows how biofilms react to antibiotic use. Under stress, the bacteria forming the biofilm typically reprogram their metabolic pathways for minimum damage. Examples include downregulating energy-intensive processes, such as glycolysis, and instead using fermentation or alternative respiratory pathways. These shifts not only support the persistence of the biofilm but also complicate treatment since many antibiotics target actively growing cells. Metabolomic profiling has shown that biofilms upregulate antioxidant production, such as glutathione and superoxide dismutase, to neutralize reactive oxygen species generated during antibiotic treatment (Shen et al., 2020). Understanding these metabolic adaptations is crucial for designing drugs that target biofilm-specific vulnerabilities.

18.17.1 Antibiotic Resistance and the Role of Metabolites

Overuse and misuse of antibiotics have driven the evolution of resistant strains, including multi-drug-*resistant Pseudomonas aeruginosa* and methicillin-resistant *Staphylococcus aureus* (MRSA). Biofilms contribute to this further, creating a protected environment where resistance genes can be exchanged and expressed more readily. Therefore, metabolomics has been critical in identifying distinctive metabolic signatures associated with resistance. For instance, *E. coli* and *Klebsiella pneumoniae* produce enhanced amounts of polyamines that stabilize the matrix and protect cells against antibiotics (Garcia Mendez et al., 2022). Alteration in lipid metabolism has also been associated with the development of resistance, where the biofilms create specialized fatty acids that make the membranes more stable and difficult for drugs to pass through (Shen et al., 2020). Beyond understanding resistance mechanisms, metabolomics has helped scientists discover biomarkers for the early detection of resistant biofilms. For example, specific metabolites, such as pyocyanin in *Pseudomonas aeruginosa* biofilm or rhamnolipids in *Staphylococcus aureus* biofilms are used to indicate biofilm formation and related resistance profiles. These biomarkers can be exploited to produce diagnostic tools, where clinicians could rapidly identify resistant biofilms for better treatment strategies (Sharifian et al., 2020).

18.17.2 Case Study

Metabolomic analysis of biofilms provides essential insights into their metabolic activities. Research indicates that *Staphylococcus aureus* biofilms demonstrate a downregulation of arginine metabolites, which implies that these metabolites are utilized within the urea pathway to maintain pH balance (Harrison et al., 2019). This finding is reliable, as transcriptomic data shows proteins' upregulation in the urea cycle. Consequently, targeting the urea cycle may represent a novel strategy for disrupting the development of *S. aureus* biofilms. Furthermore, analogous metabolomic studies associating biofilm-associated and planktonic bacteria underscore significant metabolic alterations integral to biofilm-specific functions (Stipetic et al., 2016).

18.18 BiofOmics: High-Throughput Biofilm Data Analysis

BiofOmics is a next-generation web-based platform for streamlined and optimized high-throughput collection and data analysis pertinent to biofilm-related resources (Lourenço et al., 2012). Given that the area of research in biofilms has been expanding rapidly, the complexity and variability of the data make it difficult. In most cases, fragmented data storage results from traditional approaches, with isolated datasets held by researchers in non-consistent formats. Therefore, a lack of standardization and limited data sharing cause the integration and comparison of studies across different findings to be relatively slow. BiofOmics fosters efficient, collaborative research in developing therapeutic strategies and a deeper understanding of biofilm formation, behavior, and resistance mechanisms by incorporating features, such as standardized data practices, improved data documentation sharing, and integration of multidimensional information. This platform is a step forward in resolving complexities in biofilm research and overcoming biofilm-associated challenges.

18.19 Combining Omics Technologies for Antibiotic Resistance Detection

While metabolomics yields valuable information in biofilm metabolism, integrating other omics technologies adds further resolution and depth to the biofilm studies. For instance, correlating metabolic changes with specific genetic resistance determinants could be possible by combining metabolomics with genomics. Genomic sequencing of resistance genes often identifies upregulation of genes encoding efflux pumps or beta-lactamases. However, proteomics informs on the functional proteins expressed during biofilm formation and resistance. Through linking genomic, proteomic, and metabolomic data, scientists can construct a comprehensive model of biofilm behavior, revealing interactions between genes, proteins, and metabolites that drive resistance (Inchingolo et al., 2022). Mass spectrometry, or MALDI-TOF MS, has become a critical tool for integrating multi-omics approaches.

The MALDI-TOF MS, which first identified the species, has been modified for specific proteins, peptides, and metabolites associated with resistance mechanisms. It enables rapid, precise profiling of biofilms, thereby defining virulence factors and resistance pathways relevant to combating biofilm-associated infections. Some of its examples in this regard include detecting efflux pump proteins, antibiotic-degrading enzymes, and the metabolic byproducts of resistant biofilms. Such integration of omics technologies provides a holistic view of biofilm resistance mechanisms, potentially allowing multiple targets to be designed in a therapeutic strategy (Sapi et al., 2019).

18.20 Advantages and Disadvantages of Multi-omics Technology

Multi-omics technologies bring many advantages to the study of biofilm. They provide a superior understanding of the biofilm system from molecular layers to data integration, providing insights into the complex genetic interplay between genetic, proteomic, and metabolic factors. This system-level insight is critical to understanding biofilm behavior since it captures heterogeneity in microbial populations and their responses to environmental stimuli or antibiotics. Integration of omics data improves biomarker identification, allowing the development of diagnostic tools and targeted therapies. Multi-omics approaches may also reveal sophisticated pathways that single-omics methods may be unable to trace, such as the interaction of metabolic pathways and protein expression, which are crucial for maintaining biofilms (Cruickshank et al., 2024).

In contrast, some challenges are imposed on multi-omics methods. The integration of large datasets becomes computationally very complex, and so does the analysis, which requires advanced tools as well as sophisticated expertise and is thus not available, everywhere. Data normalization and integration across platforms are still significant challenges because each omics technology produces data in different formats. Additionally, multi-omics studies are resource-intensive, involving high-end equipment, specialized software, and skilled personnel. The demand for sufficient biomass is another limitation, particularly in biofilm studies in which material is often limited in supply. There is no standardized protocol for data sharing and processing to facilitate the comparison of results across studies (Middelveen et al., 2019).

18.21 Applications and Future Prospects of Multi-omics in Biofilm Research

The immense benefits of integrating multi-omics technologies in understanding biofilm-associated infections and resistance have already been gained. For example, combining metagenomics with metabolomics has uncovered metabolic interactions in biofilm communities and revealed how species interact to enhance resistance.

Multi-omics has also been applied to identify therapeutic targets, including enzymes involved in EPS synthesis or proteins critical for biofilm integrity. Future Integration of AI into multi-omics platforms would revolutionize data analysis by automatically identifying patterns and biomarkers. Biofilm behavior would be predicted with AI-driven models under different conditions, thus developing a personalized approach to treatment strategies (Larras et al., 2018). One promising direction would be integrating three or more omics layers- proteomics, genomics, metabolomics, and transcriptomics to map the dynamics of biofilms. This approach can provide the key node in the networks of resistance for identifying the key vulnerabilities exploited in treatment. Ultimately, it will allow customized biofilm treatment by tailoring interventions according to patient-specific biofilm profiles. The great potential of a multi-omics approach to contribute to advancements in biofilm-related research, diagnosis, and therapy is enormous (Nishanth & Prasanna, 2022).

18.22 Conclusion

Irrespective of geographical locations, biofilms seem to represent a highly organized community of microorganisms enveloped in an EPS that possess extreme resistance against antimicrobial agents and drive biofilm-associated infections; thus, developing an understanding of biofilm formation, mean values, and resistance mechanisms, especially in the clinical isolates. Some measures one might undertake to curb biofilms mean targeting components of the extracellular matrix, some combination of mechanical and chemical treatments, and targeting biofilm resilience. New multi-omics technologies like genomics, proteomics, transcriptomics, and metabolomics provide a comprehensive framework for studying biofilms, giving insight into their development and resistance mechanisms. Emerging technologies such as RNA-Seq, single-cell transcriptomics, and MALDI-TOF MS help identify resistant bacteria quickly and type-targeted antimicrobial therapies. Integrating multi-omics data remains an obstacle due to computational complexity, problems related to data normalization, high cost, technical issues, massive amounts of biomass, and a real need. Yet, they severely impede the broader acceptance and application in biofilm research. However, the progress in innovation could prove pivotal to managing biofilm-related infections.

References

Ali, S., Stavropoulos, A., Jenkins, B., Graves, S., Che, G., Cheng, J., Tan, H., Wei, X., Egan, S., Neufeld, J. D., Eckhard, U., Charles, T. C., & Doxey, A. C. (2024). *Comparative proteomics of biofilm development in Pseudoalteromonas tunicata discovers a distinct family of Ca 2+ -dependent adhesins*. https://doi.org/10.1101/2024.10.22.619756

An, A. Y., Choi, K.-Y. G., Baghela, A. S., & Hancock, R. E. W. (2021). An overview of biological and computational methods for designing mechanism-informed anti-biofilm agents. *Frontiers in Microbiology, 12*, 640787. https://doi.org/10.3389/fmicb.2021.640787

Balasubramanian, S., Skaf, J., Holzgrabe, U., Bharti, R., Förstner, K. U., Ziebuhr, W., Humeida, U. H., Abdelmohsen, U. R., & Oelschlaeger, T. A. (2018). A new bioactive compound from the marine sponge-derived Streptomyces sp. SBT348 inhibits Staphylococcal growth and biofilm formation. *Frontiers in Microbiology, 9*, 1473. https://doi.org/10.3389/fmicb.2018.01473

Bao, K., Bostanci, N., Selevsek, N., Thurnheer, T., & Belibasakis, G. N. (2015). Quantitative proteomics reveal distinct protein regulations caused by Aggregatibacter actinomycetemcomitans within subgingival biofilms. *PLoS One, 10*(3), e0119222. https://doi.org/10.1371/journal.pone.0119222

Bharatula, L. D., Marsili, E., Rice, S. A., & Kwan, J. J. (2020). Influence of high intensity focused ultrasound on the microstructure and c-di-GMP signaling of Pseudomonas aeruginosa biofilms. *Frontiers in Microbiology, 11*, 599407. https://doi.org/10.3389/fmicb.2020.599407

Blackman, L. D., Qu, Y., Cass, P., & Locock, K. E. S. (2021). Approaches for the inhibition and elimination of microbial biofilms using macromolecular agents. *Chemical Society Reviews, 50*(3), 1587–1616. https://doi.org/10.1039/d0cs00986e

Bostanci, N., Bao, K., Wahlander, A., Grossmann, J., Thurnheer, T., & Belibasakis, G. N. (2015). Secretome of gingival epithelium in response to subgingival biofilms. *Molecular Oral Microbiology, 30*(4), 323–335. https://doi.org/10.1111/omi.12096

Cerca, N., Pier, G. B., Vilanova, M., Oliveira, R., & Azeredo, J. (2005). Quantitative analysis of adhesion and biofilm formation on hydrophilic and hydrophobic surfaces of clinical isolates of Staphylococcus epidermidis. *Research in Microbiology, 156*(4), 506–514. https://doi.org/10.1016/j.resmic.2005.01.007

Chiu, C. Y., & Miller, S. A. (2019). Clinical metagenomics. *Nature Reviews Genetics, 20*(6), 341–355. https://doi.org/10.1038/s41576-019-0113-7

Coenye, T., Ahonen, M., Anderson, S., Cámara, M., Chundi, P., Fields, M., Foidl, I., Gnimpieba, E. Z., Griffin, K., Hinks, J., Loka, A. R., Lushbough, C., MacPhee, C., Nater, N., Raval, R., Slater-Jefferies, J., Teo, P., Wilks, S., Yung, M., & Webb, J. S. (2024). Global challenges and microbial biofilms: Identification of priority questions in biofilm research, innovation and policy. *Biofilms, 8*, 100210. https://doi.org/10.1016/j.bioflm.2024.100210

Cornforth, D. M., Dees, J. L., Ibberson, C. B., Huse, H. K., Mathiesen, I. H., Kirketerp-Møller, K., Wolcott, R. D., Rumbaugh, K. P., Bjarnsholt, T., & Whiteley, M. (2018). Pseudomonas aeruginosa transcriptome during human infection. *Proceedings of the National Academy of Sciences of the United States of America, 115*(22), E5125. https://doi.org/10.1073/pnas.1717525115

Costa-Orlandi, C. B., Sardi, J. C. O., Pitangui, N. S., de Oliveira, H. C., Scorzoni, L., Galeane, M. C., Medina-Alarcón, K. P., Melo, W. C. M. A., Marcelino, M. Y., Braz, J. D., Fusco-Almeida, A. M., & Mendes-Giannini, M. J. S. (2017). Fungal biofilms and polymicrobial diseases. *Journal of Fungi (Basel, Switzerland), 3*(2). https://doi.org/10.3390/jof3020022

Costerton, J. W., Lewandowski, Z., DeBeer, D., Caldwell, D., Korber, D., & James, G. (1994). Biofilms, the customized microniche. *Journal of Bacteriology, 176*(8), 2137–2142. https://doi.org/10.1128/jb.176.8.2137-2142.1994

Cruickshank, D., Hamilton, D. E., Iloba, I., & Jensen, G. S. (2024). Secreted metabolites from pseudomonas, staphylococcus, and Borrelia biofilm: Modulation of immunogenicity by a nutraceutical enzyme and botanical blend. *Microorganisms, 12*(5). https://doi.org/10.3390/microorganisms12050991

Date, Y., Nakanishi, Y., Fukuda, S., Kato, T., Tsuneda, S., Ohno, H., & Kikuchi, J. (2010). New monitoring approach for metabolic dynamics in microbial ecosystems using stable-isotope-labeling technologies. *Journal of Bioscience and Bioengineering, 110*(1), 87–93. https://doi.org/10.1016/j.jbiosc.2010.01.004

Desai, C., Pathak, H., & Madamwar, D. (2010). Advances in molecular and \textquotedbl-omics\textquotedbl technologies to gauge microbial communities and bioremediation at xenobiotic/anthropogen contaminated sites. *Bioresource Technology, 101*(6), 1558–1569. https://doi.org/10.1016/j.biortech.2009.10.080

Ding, Y. (2024). *A bibliography study of biofilm life cycle*. https://doi.org/10.20944/preprints202401.1451.v1.

Donelli, G., & Vuotto, C. (2014). Biofilm-based infections in long-term care facilities. *Future Microbiology, 9*(2), 175–188. https://doi.org/10.2217/fmb.13.149

Dunsmore, B. C., Jacobsen, A., Hall-Stoodley, L., Bass, C. J., Lappin-Scott, H. M., & Stoodley, P. (2002). The influence of fluid shear on the structure and material properties of sulphate-reducing bacterial biofilms. *Journal of Industrial Microbiology & Biotechnology, 29*(6), 347–353. https://doi.org/10.1038/sj.jim.7000302

Dutra, T. V., Da Fernandes, M. S., Perdoncini, M. R. F. G., Anjos, M. M. dos, & de Abreu Filho, B. A. (2018). Capacity of Escherichia coli and Staphylococcus aureus to produce biofilm on stainless steel surfaces in the presence of food residues. *Journal of Food Processing and Preservation, 42*(4), e13574. https://doi.org/10.1111/jfpp.13574

Ebersole, J. L., Peyyala, R., & Gonzalez, O. A. (2019). Biofilm–induced profiles of immune response gene expression by oral epithelial cells. *Molecular Oral Microbiology, 34*(1). https://doi.org/10.1111/omi.12251

Ellepola, K., Truong, T., Liu, Y., Lin, Q., Lim, T. K., Lee, Y. M., Cao, T., Koo, H., & Seneviratne, C. J. (2019). Multi-omics analyses reveal synergistic carbohydrate metabolism in Streptococcus mutans-Candida albicans mixed-species biofilms. *Infection and Immunity, 87*(10). https://doi.org/10.1128/iai.00339-19

Ezzat, W. F., Askoura, A., Falah, H., & Negm, A. (2021). Pathological role of fungal biofilms in fungal Rhinosinusitis: A case-control study. *International Journal of Otolaryngology and Head & Neck Surgery, 10*(02), 75–84. https://doi.org/10.4236/ijohns.2021.102008

Falcón García, C., Kretschmer, M., Lozano-Andrade, C. N., Schönleitner, M., Dragoŝ, A., Kovács, Á. T., & Lieleg, O. (2020). Metal ions weaken the hydrophobicity and antibiotic resistance of Bacillus subtilis NCIB 3610 biofilms. *npj Biofilms and Microbiomes, 6*, 1. https://doi.org/10.1038/s41522-019-0111-8

Foster, J. S., & Kolenbrander, P. E. (2004). Development of a multispecies oral bacterial community in a saliva-conditioned flow cell. *Applied and Environmental Microbiology, 70*(7), 4340–4348. https://doi.org/10.1128/aem.70.7.4340-4348.2004

Fulaz, S., Vitale, S., Quinn, L., & Casey, E. (2019). Nanoparticle–biofilm interactions: The role of the EPS matrix. *Trends in Microbiology, 27*(11), 915–926. https://doi.org/10.1016/j.tim.2019.07.004

Garcia Mendez, D. F., Rengifo Herrera, J. A., Sanabria, J., & Wist, J. (2022). Analysis of the metabolic response of planktonic cells and biofilms of Klebsiella pneumoniae to sublethal disinfection with sodium hypochlorite measured by NMR. *Microorganisms, 10*(7). https://doi.org/10.3390/microorganisms10071323

Gil, C., Solano, C., Burgui, S., Latasa, C., García, B., Toledo-Arana, A., Lasa, I., & Valle, J. (2014). Biofilm matrix exoproteins induce a protective immune response against Staphylococcus aureus biofilm infection. *Infection and Immunity, 82*(3), 1017–1029. https://doi.org/10.1128/iai.01419-13

Glage, S., Paret, S., Winkel, A., Stiesch, M., Bleich, A., Krauss, J. K., & Schwabe, K. (2017). A new model for biofilm formation and inflammatory tissue reaction: Intraoperative infection of a cranial implant with Staphylococcus aureus in rats. *Acta Neurochirurgica, 159*(9), 1747–1756. https://doi.org/10.1007/s00701-017-3244-7

Gupta, P., Sarkar, S., Das, B., Bhattacharjee, S., & Tribedi, P. (2016). Biofilm, pathogenesis and prevention—A journey to break the wall: A review. *Archives of Microbiology, 198*(1), 1–15. https://doi.org/10.1007/s00203-015-1148-6

Hall, D. C., Palmer, P., Ji, H.-F., Ehrlich, G. D., & Król, J. E. (2021). Bacterial biofilm growth on 3D-printed materials. *Frontiers in Microbiology, 12*, 646303. https://doi.org/10.3389/fmicb.2021.646303

Hall-Stoodley, L., Costerton, J. W., & Stoodley, P. (2004). Bacterial biofilms: From the natural environment to infectious diseases. *Nature Reviews. Microbiology, 2*(2), 95–108. https://doi.org/10.1038/nrmicro821

Harrison, A., Hardison, R. L., Wallace, R. M., Fitch, J., Heimlich, D. R., Bryan, M. O., Dubois, L., John-Williams, L. S., Sebra, R. P., White, P., Moseley, M. A., Thompson, J. W., Justice, S. S., & Mason, K. M. (2019). Reprioritization of biofilm metabolism is associated with nutrient adaptation and long-term survival of Haemophilus influenzae. *npj Biofilms and Microbiomes, 5*(1), 33. https://doi.org/10.1038/s41522-019-0105-6

Herschend, J., Damholt, Z. B. V., Marquard, A. M., Svensson, B., Sørensen, S. J., Hägglund, P., & Burmølle, M. (2017). A meta-proteomics approach to study the interspecies interactions affecting microbial biofilm development in a model community. *Scientific Reports, 7*(1), 16483. https://doi.org/10.1038/s41598-017-16633-6

Hrdlickova, R., Toloue, M., & Tian, B. (2017). RNA –Seq methods for transcriptome analysis. *WIREs RNA, 8*(1). https://doi.org/10.1002/wrna.1364

Inchingolo, A. D., Malcangi, G., Semjonova, A., Inchingolo, A. M., Patano, A., Coloccia, G., Ceci, S., Marinelli, G., Di Pede, C., Ciocia, A. M., Mancini, A., Palmieri, G., Barile, G., Settanni, V., de Leonardis, N., Rapone, B., Piras, F., Viapiano, F., Cardarelli, F., et al. (2022). Oralbiotica/Oralbiotics: The impact of oral microbiota on dental health and demineralization: A systematic review of the literature. *Children (Basel, Switzerland), 9*(7). https://doi.org/10.3390/children9071014

Iñiguez-Moreno, M., Gutiérrez-Lomelí, M., Guerrero-Medina, P. J., & Avila-Novoa, M. G. (2018). Biofilm formation by Staphylococcus aureus and Salmonella spp. Under mono and dual-species conditions and their sensitivity to cetrimonium bromide, peracetic acid and sodium hypochlorite. *Brazilian Journal of Microbiology: [Publication of the Brazilian Society for Microbiology], 49*(2), 310–319. https://doi.org/10.1016/j.bjm.2017.08.002

Jamal, M., Ahmad, W., Andleeb, S., Jalil, F., Imran, M., Nawaz, M. A., Hussain, T., Ali, M., Rafiq, M., & Kamil, M. A. (2018). Bacterial biofilm and associated infections. *Journal of the Chinese Medical Association, 81*(1), 7–11. https://doi.org/10.1016/j.jcma.2017.07.012

Kim, Y. H., Lee, Y., Kim, S., Yeom, J., Yeom, S., Seok Kim, B., Oh, S., Park, S., Jeon, C. O., & Park, W. (2006). The role of periplasmic antioxidant enzymes (superoxide dismutase and thiol peroxidase) of the Shiga toxin-producing Escherichia coli O157:H7 in the formation of biofilms. *Proteomics, 6*(23), 6181–6193. https://doi.org/10.1002/pmic.200600320

Koseki, H., Yonekura, A., Shida, T., Yoda, I., Horiuchi, H., Morinaga, Y., Yanagihara, K., Sakoda, H., Osaki, M., & Tomita, M. (2014). Early staphylococcal biofilm formation on solid orthopaedic implant materials: In vitro study. *PLoS One, 9*(10), e107588. https://doi.org/10.1371/journal.pone.0107588

Lahiri, D., Nag, M., Banerjee, R., Mukherjee, D., Garai, S., Sarkar, T., Dey, A., Sheikh, H. I., Pathak, S. K., Edinur, H. A., Pati, S., & Ray, R. R. (2021). Amylases: Biofilm inducer or biofilm inhibitor? *Frontiers in Cellular and Infection Microbiology, 11*, 660048. https://doi.org/10.3389/fcimb.2021.660048

Larras, F., Billoir, E., Baillard, V., Siberchicot, A., Scholz, S., Wubet, T., Tarkka, M., Schmitt-Jansen, M., & Delignette-Muller, M.-L. (2018). DRomics: A Turnkey tool to support the use of the dose-response framework for Omics Data in ecological risk assessment. *Environmental Science & Technology, 52*(24), 14461–14468. https://doi.org/10.1021/acs.est.8b04752

Lebeaux, D., Ghigo, J.-M., & Beloin, C. (2014). Biofilm-related infections: Bridging the gap between clinical management and fundamental aspects of recalcitrance toward antibiotics. *Microbiology and Molecular Biology Reviews, 78*(3), 510–543. https://doi.org/10.1128/MMBR.00013-14

Li, Z., Tu, J., Han, G., Liu, N., & Sheng, C. (2021). Novel Carboline Fungal Histone Deacetylase (HDAC) inhibitors for combinational treatment of Azole-resistant candidiasis. *Journal of Medicinal Chemistry, 64*(2), 1116–1126. https://doi.org/10.1021/acs.jmedchem.0c01763

Li, M., Nahum, Y., Matouš, K., Stoodley, P., & Nerenberg, R. (2023). Effects of biofilm heterogeneity on the apparent mechanical properties obtained by shear rheometry. *Biotechnology and Bioengineering, 120*(2), 553–561. https://doi.org/10.1002/bit.28276

Liu, J., Yang, L., Hou, Y., Soteyome, T., Zeng, B., Su, J., Li, L., Li, B., Chen, D., Li, Y., Wu, A., Shirtliff, M. E., Harro, J. M., Xu, Z., & Peters, B. M. (2018). Transcriptomics study on Staphylococcus aureus biofilm under low concentration of Ampicillin. *Frontiers in Microbiology, 9*. https://doi.org/10.3389/fmicb.2018.02413

Lourenço, A., Ferreira, A., Veiga, N., Machado, I., Pereira, M. O., & Azevedo, N. F. (2012). BiofOmics: A web platform for the systematic and standardized collection of high-throughput biofilm data. *PLoS One, 7*(6), e39960. https://doi.org/10.1371/journal.pone.0039960

Lowe, R., Shirley, N., Bleackley, M., Dolan, S., & Shafee, T. (2017). Transcriptomics technologies. *PLoS Computational Biology, 13*(5), e1005457. https://doi.org/10.1371/journal.pcbi.1005457

Madani, A., Esfandiari, Z., Shoaei, P., & Ataei, B. (2022). Evaluation of virulence factors, antibiotic resistance, and biofilm formation of Escherichia coli isolated from milk and dairy products in Isfahan, Iran. *Foods (Basel, Switzerland), 11*(7). https://doi.org/10.3390/foods11070960

Maheswari, B. U., & Kalaiselvi, G. (2024). Exploring Neophytadiene from Ampelocissus araneosa: A molecular docking approach to inhibit biofilm formation in Staphylococcus aureus. *Uttar Pradesh Journal of Zoology, 45*(3), 59–70. https://doi.org/10.56557/upjoz/2024/v45i33875

Mangalappalli-Illathu, A. K., Vidović, S., & Korber, D. R. (2008). Differential adaptive response and survival of Salmonella enterica serovar enteritidis planktonic and biofilm cells exposed to benzalkonium chloride. *Antimicrobial Agents and Chemotherapy, 52*(10), 3669–3680. https://doi.org/10.1128/aac.00073-08

Maskarinec, S. A., Parlak, Z., Tu, Q., Levering, V., Zauscher, S., López, G. P., Fowler, V. G., & Perfect, J. R. (2018). On-demand release of Candida albicans biofilms from urinary catheters by mechanical surface deformation. *Biofouling, 34*(6), 595–604. https://doi.org/10.1080/08927014.2018.1474461

Middelveen, M. J., Filush, K. R., Bandoski, C., Kasliwala, R. S., Melillo, A., Stricker, R. B., & Sapi, E. (2019). Mixed Borrelia burgdorferi and Helicobacter pylori biofilms in Morgellons disease dermatological specimens. *Healthcare (Basel, Switzerland), 7*(2). https://doi.org/10.3390/healthcare7020070

Mielko, K. A., Jabłoński, S. J., Milczewska, J., Sands, D., Łukaszewicz, M., & Młynarz, P. (2019). Metabolomic studies of Pseudomonas aeruginosa. *World Journal of Microbiology & Biotechnology, 35*(11), 178. https://doi.org/10.1007/s11274-019-2739-1

Miyaue, S., Suzuki, E., Komiyama, Y., Kondo, Y., Morikawa, M., & Maeda, S. (2018). Bacterial memory of Persisters: Bacterial Persister cells can retain their phenotype for days or weeks after withdrawal from Colony–biofilm culture. *Frontiers in Microbiology, 9*. https://doi.org/10.3389/fmicb.2018.01396

Mooney, J. A., Pridgen, E. M., Manasherob, R., Suh, G., Blackwell, H. E., Barron, A. E., Bollyky, P. L., Goodman, S. B., & Amanatullah, D. F. (2018). Periprosthetic bacterial biofilm and quorum sensing. *Journal of Orthopaedic Research, 36*(9), 2331–2339. https://doi.org/10.1002/jor.24019

Mourer, T., El Ghalid, M., Pehau-Arnaudet, G., Kauffmann, B., Loquet, A., Brûlé, S., Cabral, V., d'Enfert, C., & Bachellier-Bassi, S. (2023). The Pga59 cell wall protein is an amyloid forming protein involved in adhesion and biofilm establishment in the pathogenic yeast Candida albicans. *npj Biofilms and Microbiomes, 9*(1), 6. https://doi.org/10.1038/s41522-023-00371-x

Nazir, N., et al. (2020). Inhibitory effects of antifungal drugs on biofilm producing Aspergillus spp. recovered from drinking water system. *Journal of Bacteriology and Mycology, 7*(2). https://doi.org/10.26420/jbacteriolmycol.2020.1128

Ng, W.-L., & Bassler, B. L. (2009). Bacterial quorum-sensing network architectures. *Annual Review of Genetics, 43*, 197–222. https://doi.org/10.1146/annurev-genet-102108-134304

Niba, E. T. E., Naka, Y., Nagase, M., Mori, H., & Kitakawa, M. (2008). A genome-wide approach to identify the genes involved in biofilm formation in E. coli. *DNA Research, 14*(6), 237–246. https://doi.org/10.1093/dnares/dsm024

Nishanth, S., & Prasanna, R. (2022). Untargeted GC-MS reveals differential regulation of metabolic pathways in cyanobacterium Anabaena and its biofilms with Trichoderma viride and Providencia sp. *Current Research in Microbial Sciences, 3*, 100174. https://doi.org/10.1016/j.crmicr.2022.100174

Niu, T.-X., Wang, X.-N., Wu, H.-Y., Bi, J.-R., Hao, H.-S., Hou, H.-M., & Zhang, G.-L. (2020). Transcriptomic analysis, motility and biofilm formation characteristics of Salmonella typhimurium exposed to benzyl Isothiocyanate treatment. *International Journal of Molecular Sciences, 21*(3), 1025. https://doi.org/10.3390/ijms21031025

Oliveira, W. F., Silva, P. M. S., Silva, R. C. S., Silva, G. M. M., Machado, G., Coelho, L. C. B. B., & Correia, M. T. S. (2018). Staphylococcus aureus and Staphylococcus epidermidis infections on implants. *The Journal of Hospital Infection, 98*(2), 111–117. https://doi.org/10.1016/j.jhin.2017.11.008

Otto, A., Bernhardt, J., Hecker, M., & Becher, D. (2012). Global relative and absolute quantitation in microbial proteomics. *Current Opinion in Microbiology, 15*(3), 364–372. https://doi.org/10.1016/j.mib.2012.02.005

Paharik, A. E., & Horswill, A. R. (2016). The Staphylococcal biofilm: Adhesins, regulation, and host response. *Microbiology Spectrum, 4*(2). https://doi.org/10.1128/microbiolspec.vmbf-0022-2015

Partoazar, A., Talaei, N., Bahador, A., Pourhajibagher, M., Dehpour, S., Sadati, M., & Bakhtiarian, A. (2019). Antibiofilm activity of natural zeolite supported Nanozno: Inhibition of Esp gene expression of enterococcus Faecalis. *Nanomedicine, 14*(6), 675–687. https://doi.org/10.2217/nnm-2018-0173

Patrauchan, M. A., Sarkisova, S. A., & Franklin, M. J. (2007). Strain-specific proteome responses of Pseudomonas aeruginosa to biofilm-associated growth and to calcium. *Microbiology, 153*(Pt 11), 3838–3851. https://doi.org/10.1099/mic.0.2007/010371-0

Peterson, S. N., Meissner, T., Su, A. I., Snesrud, E., Ong, A. C., Schork, N. J., & Bretz, W. A. (2014). Functional expression of dental plaque microbiota. *Frontiers in Cellular and Infection Microbiology, 4*. https://doi.org/10.3389/fcimb.2014.00108

Peyrusson, F., Varet, H., Nguyen, T. K., Legendre, R., Sismeiro, O., Coppée, J.-Y., Wolz, C., Tenson, T., & van Bambeke, F. (2020). Intracellular Staphylococcus aureus persisters upon antibiotic exposure. *Nature Communications, 11*(1), 2200. https://doi.org/10.1038/s41467-020-15966-7

Rahim, M. I., Rohde, M., Rais, B., Seitz, J.-M., & Mueller, P. P. (2016). Susceptibility of metallic magnesium implants to bacterial biofilm infections. *Journal of Biomedical Materials Research. Part A, 104*(6), 1489–1499. https://doi.org/10.1002/jbm.a.35680

Rathsam, C., Eaton, R. E., Simpson, C. L., Browne, G. V., Berg, T., Harty, D. W. S., & Jacques, N. A. (2005). Up-regulation of competence- but not stress-responsive proteins accompanies an altered metabolic phenotype in Streptococcus mutans biofilms. *Microbiology, 151*(6), 1823–1837. https://doi.org/10.1099/mic.0.27830-0

Roscini, L., Vassiliou, A., Corte, L., Casagrande Pierantoni, D., Robert, V., Tascini, C., Mattana, S., Alunni Cardinali, M., Orfanos, S. E., Fioretto, D., & Cardinali, G. (2018). Yeast biofilm as a bridge between medical and environmental microbiology across different detection techniques. *Infectious Diseases and Therapy, 7*(Suppl 1), 27–34. https://doi.org/10.1007/s40121-018-0191-4

Rouillard, K. R., Kissner, W. J., Markovetz, M. R., & Hill, D. B. (2022). *Effects of mucin and DNA concentrations in airway mucus on Pseudomonas aeruginosa biofilm recalcitrance*. https://doi.org/10.1101/2022.03.24.485733.

Ryan, R. P., McCarthy, Y., Watt, S. A., Niehaus, K., & Dow, J. M. (2009). Intraspecies signaling involving the diffusible signal factor BDSF (cis-2-dodecenoic acid) influences virulence in Burkholderia cenocepacia. *Journal of Bacteriology, 191*(15), 5013–5019. https://doi.org/10.1128/jb.00473-09

Salvatore, M. M., Maione, A., Albarano, L., de Alteriis, E., Carraturo, F., Andolfi, A., Salvatore, F., Galdiero, E., & Guida, M. (2021). An integrated analysis of intracellular metabolites and virulence gene expression during biofilm development of a clinical isolate of Candida tropicalis on distinct surfaces. *International Journal of Molecular Sciences, 22*(16). https://doi.org/10.3390/ijms22169038

Sapi, E., Gupta, K., Wawrzeniak, K., Gaur, G., Torres, J., Filush, K., Melillo, A., & Zelger, B. (2019). Borrelia and Chlamydia can form mixed biofilms in infected human skin tissues. *European Journal of Microbiology & Immunology, 9*(2), 46–55. https://doi.org/10.1556/1886.2019.00003

Schmoldt, A., Benthe, H. F., & Haberland, G. (1975). Digitoxin metabolism by rat liver microsomes. *Biochemical Pharmacology, 24*(17), 1639–1641.

Seneviratne, C. J., Wang, Y., Jin, L., Abiko, Y., & Samaranayake, L. P. (2010). Proteomics of drug resistance in Candida glabrata biofilms. *Proteomics, 10*(7), 1444–1454. https://doi.org/10.1002/pmic.200900611

Seneviratne, C. J., Suriyanarayanan, T., Swarup, S., Chia, K. H. B., Nagarajan, N., & Zhang, C. (2017). Transcriptomics analysis reveals putative genes involved in biofilm formation and biofilm-associated drug resistance of enterococcus faecalis. *Journal of Endodontics, 43*(6), 949–955. https://doi.org/10.1016/j.joen.2017.01.020

Sharifian, P., Yaslianifard, S., Fallah, P., Aynesazi, S., Bakhtiyari, M., & Mohammadzadeh, M. (2020). Investigating the effect of nano-curcumin on the expression of biofilm regulatory genes of Pseudomonas aeruginosa. *Infection and Drug Resistance, 13*, 2477–2484. https://doi.org/10.2147/idr.s263387

Sharma, K., & Pagedar Singh, A. (2018). *Antibiofilm effect of DNase against single and mixed species biofilm* , *7*(3) (Vol. 7). https://doi.org/10.3390/foods7030042

Shen, F., Ge, C., & Yuan, P. (2020). Metabolomics study reveals inhibition and metabolic dysregulation in Staphylococcus aureus Planktonic cells and biofilms induced by Carnosol. *Frontiers in Microbiology, 11*, 538572. https://doi.org/10.3389/fmicb.2020.538572

Siddam, A. D., Zaslow, S. J., Wang, Y., Phillips, K. S., Silverman, M. D., Regan, P. M., & Amarasinghe, J. J. (2020). Characterization of biofilm formation by Mycobacterium chimaera on medical device materials. *Frontiers in Microbiology, 11*, 586657. https://doi.org/10.3389/fmicb.2020.586657

Simcox, L. J., Pereira, R. P. A., Wellington, E. M. H., & Macpherson, J. V. (2019). Boron doped diamond as a low biofouling material in aquatic environments: Assessment of Pseudomonas aeruginosa biofilm formation. *ACS Applied Materials & Interfaces, 11*(28), 25024–25033. https://doi.org/10.1021/acsami.9b07245

Singh, S., Singh, S. K., Chowdhury, I., & Singh, R. (2017). Understanding the mechanism of bacterial biofilms resistance to antimicrobial agents. *The Open Microbiology Journal, 11*(1), 53–62. https://doi.org/10.2174/1874285801711010053

Slezak, C., Anderson, K., Hillock, T., Miller, M., Dungel, P., Kopp, O., Sterflinger, K., & Slezak, P. (2022). Shockwaves increase in vitro resilience of Rhizopus oryzae biofilm under Amphotericin B treatment. *International Journal of Molecular Sciences, 23*(16). https://doi.org/10.3390/ijms23169226

Solano, C., Echeverz, M., & Lasa, I. (2014). Biofilm dispersion and quorum sensing. *Current Opinion in Microbiology, 18*, 96–104. https://doi.org/10.1016/j.mib.2014.02.008

Soufi, Y., & Soufi, B. (2016). Mass spectrometry-based bacterial proteomics: Focus on dermatologic microbial pathogens. *Frontiers in Microbiology, 7*. https://doi.org/10.3389/fmicb.2016.00181

Souza, J. G. S., Bertolini, M., Thompson, A., Barão, V. A. R., & Dongari-Bagtzoglou, A. (2020). Biofilm interactions of Candida albicans and Mitis Group Streptococci in a Titanium-Mucosal Interface model. *Applied and Environmental Microbiology, 86*(9). https://doi.org/10.1128/aem.02950-19

Steinhäußer, L., König, U., Fietzke, F., & Gotzmann, G. (2024). Influence of a UVA-activated TiO2 coating on bacterial surface colonization in water-bearing systems. *Coatings, 14*(4), 454. https://doi.org/10.3390/coatings14040454

Stipetic, L. H., Dalby, M. J., Davies, R. L., Morton, F. R., Ramage, G., & Burgess, K. E. V. (2016). A novel metabolomic approach used for the comparison of Staphylococcus aureus planktonic cells and biofilm samples. *Metabolomics, 12*(4), 75. https://doi.org/10.1007/s11306-016-1002-0

Suryaletha, K., Narendrakumar, L., John, J., Radhakrishnan, M. P., George, S., & Thomas, S. (2019). Decoding the proteomic changes involved in the biofilm formation of Enterococcus faecalis SK460 to elucidate potential biofilm determinants. *BMC Microbiology, 19*(1), 146. https://doi.org/10.1186/s12866-019-1527-2

Tan, X., Qin, N., Wu, C., Sheng, J., Yang, R., Zheng, B., Ma, Z., Liu, L., Peng, X., & Jia, A. (2015). Transcriptome analysis of the biofilm formed by methicillin-susceptible Staphylococcus aureus. *Scientific Reports, 5*(1), 11997. https://doi.org/10.1038/srep11997

Thakur, N. L., Jain, R., Natalio, F., Hamer, B., Thakur, A. N., & Müller, W. E. G. (2008). Marine molecular biology: An emerging field of biological sciences. *Biotechnology Advances, 26*(3), 233–245. https://doi.org/10.1016/j.biotechadv.2008.01.001

Thompson, A. F., English, E. L., Nock, A. M., Willsey, G. G., Eckstrom, K., Cairns, B., Bavelock, M., Tighe, S. W., Foote, A., Shulman, H., Pericleous, A., Gupta, S., Kadouri, D. E., & Wargo, M. J. (2020). Characterizing species interactions that contribute to biofilm formation in a multispecies model of a potable water bacterial community. *Microbiology, 166*(1), 34–43. https://doi.org/10.1099/mic.0.000849

Tselepi, V., Sarkiris, P., Nioras, D., Tsouko, E., Sarris, D., Gogolides, E., & Ellinas, K. (2024). Functional surfaces for passive fungal proliferation control: Effect of surface micro- and nanotopography, material, and wetting properties. *ACS Applied Bio Materials, 7*(7), 4510–4518. https://doi.org/10.1021/acsabm.4c00387

Tuck, B., Watkin, E., Somers, A., Forsyth, M., & Machuca, L. L. (2022). Enhancing biocide efficacy: Targeting extracellular DNA for marine biofilm disruption. *Microorganisms, 10*(6). https://doi.org/10.3390/microorganisms10061227

Vohra, M., Kour, A., Kalia, N. P., Kumar, M., Sharma, S., Jaglan, S., Kamath, N., & Sharma, S. (2024). A comprehensive review of genomics, transcriptomics, proteomics, and metabolomic insights into the differentiation of Pseudomonas aeruginosa from the planktonic to biofilm state: A multi-omics approach. *International Journal of Biological Macromolecules, 257*, 128563. https://doi.org/10.1016/j.ijbiomac.2023.128563

Voinescu, A., Licker, M., Muntean, D., Musuroi, C., Musuroi, S. I., Izmendi, O., Vulpie, S., Jumanca, R., Munteanu, M., & Cosnita, A. (2024). A comprehensive review of microbial biofilms on contact lenses: Challenges and solutions. *Infection and Drug Resistance, 17*, 2659–2671. https://doi.org/10.2147/idr.s463779

Wang, C., Zhang, Q., Wang, Y., Tang, X., An, Y., Li, S., Xu, H., Li, Y., Luan, W., Wang, X., Liu, M., & Yu, L. (2019). Comparative proteomics analysis between biofilm and planktonic cells of *Mycobacterium tuberculosis*. *Electrophoresis, 40*(20), 2736–2746. https://doi.org/10.1002/elps.201900030

Waters, C. M., & Bassler, B. L. (2005). Quorum sensing: Cell-to-cell communication in bacteria. *Annual Review of Cell and Developmental Biology, 21*, 319–346. https://doi.org/10.1146/annurev.cellbio.21.012704.131001

Zaborowska, M., Tillander, J., Brånemark, R., Hagberg, L., Thomsen, P., & Trobos, M. (2017). Biofilm formation and antimicrobial susceptibility of staphylococci and enterococci from osteomyelitis associated with percutaneous orthopaedic implants. *Journal of Biomedical Materials Research. Part B, Applied Biomaterials, 105*(8), 2630–2640. https://doi.org/10.1002/jbm.b.33803

Zarnowski, R., Noll, A., Chevrette, M. G., Sanchez, H., Jones, R., Anhalt, H., Fossen, J., Jaromin, A., Currie, C., Nett, J. E., Mitchell, A., & Andes, D. R. (2021). Coordination of fungal biofilm development by extracellular vesicle cargo. *Nature Communications, 12*(1), 6235. https://doi.org/10.1038/s41467-021-26525-z

Zhang, W., Jing, H., Xu, L., Zhao, L., Han, Y., & Li, C. (2015). Numerical investigation of creep crack initiation in P92 steel pipes with embedded spherical defects under internal pressure at 650°C. *Engineering Fracture Mechanics, 139*, 40–55. https://doi.org/10.1016/j.engfracmech.2015.03.043

Zhao, A., Sun, J., & Liu, Y. (2023). Understanding bacterial biofilms: From definition to treatment strategies. *Frontiers in Cellular and Infection Microbiology, 13*, 1137947. https://doi.org/10.3389/fcimb.2023.1137947

Index

S. Busi et al. (eds.), *Omics Approaches in Biofilm Research*,
https://doi.org/10.1007/978-3-031-91863-6

www.ingramcontent.com/pod-product-compliance
Lightning Source LLC
Chambersburg PA
CBHW081552110725
29474CB00003B/20
9783031918629